DNA METHYLATION AND COMPLEX HUMAN DISEASE

DNA METHYLATION AND COMPLEX HUMAN DISEASE

MICHEL NEIDHART
University Hospital Zurich, Switzerland

AMSTERDAM • BOSTON • HEIDELBERG • LONDON
NEW YORK • OXFORD • PARIS • SAN DIEGO
SAN FRANCISCO • SINGAPORE • SYDNEY • TOKYO
Academic Press is an imprint of Elsevier

Academic Press is an imprint of Elsevier
525 B Street, Suite 1800, San Diego, CA 92101-4495, USA
225 Wyman Street, Waltham, MA 02451, USA
The Boulevard, Langford Lane, Kidlington, Oxford OX5 1GB, UK

Library of Congress Cataloging-in-Publication Data
A catalog record for this book is available from the Library of Congress

British Library Cataloguing-in-Publication Data
A catalogue record for this book is available from the British Library

ISBN: 978-0-12-420194-1

For information on all Academic Press publications
visit our website at http://store.elsevier.com

Publisher: Mica Haley
Acquisition Editor: Catherine Van Der Laan
Editorial Project Manager: Lisa Eppich
Production Project Manager: Melissa Read
Designer: Mark Rogers

Printed and bound in the United States of America

Contents

Preface

There is a considerable interest in whether environmental factors modulate the establishment and maintenance of epigenetic modifications, and could thereby influence gene expression and phenotype. Chemical pollutants, dietary components, temperature changes, and other external stresses can indeed have long-lasting effects on development, metabolism, and health, sometimes even in subsequent generations. The goal of this book is to provide the first comprehensive analysis of DNA methylation in human diseases, including cancer and non-neoplastic diseases. We review the possibilities of methyl group-based epigenetic biomarkers of major diseases, tailored epigenetic therapies, and the future uses of high-throughput methylome technologies. In cancer, the distinction between mutations and epimutations becomes important in the context of possible therapeutic strategies. Thus, DNA methylation also became relevant in blood tests for non-invasive screening, and diagnostic and prognostic tests, as compared to biopsy-driven gene expression analysis. In many diseases, not only cancer, a global DNA hypomethylation is accompanied by hypermethylation of specific genes. The possibility of using methyl donors in therapeutic strategies is discussed. This book is intended for those with interests ranging from the fundamental basis of DNA methylation to therapeutic interventions. It should be a motivation for basic and applied researchers to enter this exciting and growing field.

Michel Neidhart

About the Author

Professor **Michel Neidhart** graduated in experimental biology in 1982, as well as nutrition and endocrinology in Montreal in 1986 and, finally, zoology at the University of Basel in 1988. After research training in Zurich and Bern, he became a researcher at the Clinic of Rheumatology at the University Hospital in Zurich (1989), senior researcher at the Center of Experimental Rheumatology (1995), and associate professor for experimental rheumatology at the Faculty of Medicine in Zurich (2009). As a researcher, Michel Neidhart has published more than 70 peer-reviewed scientific papers in the fields of endocrinology, immunology, cardiology, and rheumatic diseases.

The Center of Experimental Rheumatology has been part of the EC-supported EURO-RA Marie Curie Actions Research Training Networks (RTNs), such as EURO-RA, and was part of the EC-FP6 supported AUTOCURE project (2006–2011) and the EC FP7 MASTERSWITCH project (2008–2013). Currently the Center is supported by the Institute of Arthritis Research (2010–2016), the EC IMI BTCure (2011–2016), EC Marie Curie Osteoimmune (2012–2016), and the EC EuroTEAM (2012–2016).

CHAPTER

1

DNA Methylation – Introduction

OUTLINE

1.1 EPIGENETICS

In this first chapter we introduce the concept of epigenetics, especially the members of the DNA methylation machinery. Epigenetics is the study of cellular and physiological traits that are not caused by changes in the DNA sequence. Epigenetics describes the study of stable, long-term alterations in the transcriptional potential of a cell, but also

M. Neidhart: *DNA Methylation and Complex Human Disease*.
DOI: http://dx.doi.org/10.1016/B978-0-12-420194-1.00001-4

can induce transient changes. Some of those alterations are heritable. For example, during embryogenesis, totipotent stem cells become the various pluripotent cell lines of the embryo, which in turn become fully differentiated cells. This process is regulated by epigenetics [1].

1.2 HISTONE MODIFICATIONS

Histones are the core protein components of chromatin complexes and they provide the structural backbone around which DNA wraps at regular intervals, generating chromatin. The nucleosome represents the first level of chromatin organization and is composed of two of each of histones H2A, H2B, H3, and H4, assembled in an octameric core with DNA tightly wrapped around the octamer [2]. The first epigenetic mechanism is the post-translational modification of the amino acids that make up histone proteins. If the amino acids in the chain are changed, the shape of the histone might be modified. DNA is not completely unwound during replication, and thus it is possible that the modified histones may be carried into each new copy of the DNA. Once there, these histones may act as templates, initiating shaping of the surrounding new histones in the new manner. By altering the shape of the histones around them, these modified histones ensure that a lineage-specific transcription program is maintained after cell division. Although histone modifications occur throughout the entire sequence, the histone tails are particularly highly modified. These modifications include acetylation, methylation, ubiquitylation, phosphorylation, sumoylation, ribosylation, and citrullination, of which acetylation and methylation are the most highly studied.

1.2.1 Histone Acetylation and Deacetylation

Histone modifications are linked to essentially every cellular process requiring DNA access, including transcription, replication, and repair. Histone acetylation is carried out by enzymes called histone acetyltransferases (HATs) that are responsible for adding acetyl groups to lysine residues on histone tails, while histone deacetylases (HDACs) are those that remove acetyl groups from acetylated lysines [3,4]. For example, acetylation of the K14 and K9 lysines of the tail of histone H3 by HATs is generally related to transcriptional competence. The presence of acetylated lysine on histone tails leads to a relaxed chromatin state that promotes transcriptional activation of selected genes; in contrast, deacetylation of lysine residues leads to chromatin compaction and transcriptional inactivation [5].

1.2.2 Histone Methylation

Histone methylation is a process by which methyl groups are transferred to amino acids of histones. Depending on the target site, methylation can modify histones so that different portions of chromatin are activated or inactivated. In most cases methylation and demethylation of histones turn the genes in DNA "off" and "on," respectively, either by loosening or encompassing their tails, thereby allowing transcription factors to access or blocking them from accessing the DNA. Trimethylation of histone H3 at lysine 4 (H3K4me3) is an active mark for transcription [6]. However, dimethylation of histone H3 at lysine 9 (H3K9me2) is a signal for transcriptional silencing [7]. Histone methyltransferases (e.g., SUV4-20H1/KMT5B, SUV4-20H2/KMT5C, ATXR5) or demethylases (UTX/KDM6A, JMJD3/KDM6B, JMJD2D/KDM4D) actively add or remove various methylation marks in a cell type-specific and context-dependent way [8,9]. SET domain lysine methyltransferases (SUVAs/KMTs) catalyze the site- and state-specific methylation of lysine residues in histone and non-histone substrates [10]. These modifications play fundamental roles in transcriptional regulation, heterochromatin formation, X chromosome inactivation, and DNA damage response, and have been implicated in the epigenetic regulation of cell identity and fate.

1.3 DNA METHYLATION

The second epigenetic mechanism is the addition of methyl groups to the DNA, mostly at CpG sites, to convert cytosine to 5-methylcytosine (5-mC). 5-mC performs much like a regular cytosine, pairing with a guanine in double-stranded DNA. Transcription of most protein coding genes in mammals is initiated at promoters rich in CG sequences, where cytosine is positioned next to a guanine nucleotide linked by a phosphate called a CpG site. Such short stretches of CpG-dense DNA are known as CpG islands. In the human genome, 60–80% of 28 million CpG dinucleotides are methylated [11]. The chromatin structure adjacent to CpG island promoters facilitates transcription, while methylated CpG islands impart a tight compaction to chromatin that prevents onset of transcription and, therefore, gene expression. Some areas of the genome are methylated more heavily than others, and highly methylated areas tend to be less transcriptionally active. Methylation of cytosines can also persist from the germline of one of the parents into the zygote, marking the chromosome as being inherited from one parent or the other; this is called genetic imprinting [12]. DNA methylation frequently occurs in repeated sequences, and helps to suppress the expression and mobility of transposable elements such as LINE-1 [13]. DNA methylation is associated with histone modifications, particularly the absence of histone H3 lysine 4 methylation (H3K4me0) and the presence of H3 lysine 9 dimethylation (H3K9me2) [14].

1.3.1 DNA Methyltransferases

DNA methylation patterns are known to be established and modified in response to environmental factors by a complex interplay of at least three independent DNA methyltransferases (DNMTs): DNMT1, DNMT3A, and DNMT3B [15]. These catalyze the methyl group transfer from S-adenosyl-L-methionine to cytosine bases on the DNA [16].

By preferentially modifying hemimethylated DNA, DNA methyltransferase 1 (DNMT1) transfers patterns of methylation to a newly synthesized strand after DNA replication; it is therefore often referred to as the "maintenance" methyltransferase [17]. DNMT1 is essential for proper embryonic development, imprinting, and X-inactivation [18].

DNMT3 is a family of DNA methyltransferases that can methylate hemimethylated and unmethylated CpG at the same rate. The architecture of DNMT3 enzymes is similar to that of DNMT1, with a regulatory region attached to a catalytic domain [19]. There are three known members of the DNMT3 family: DNMT3A, 3B, and 3L. DNMT3A and 3B can mediate methylation-independent gene repression, while DNMT3A can co-localize with heterochromatin protein (HP1) [20] and methyl-CpG-domain binding proteins (MBDs). They can also interact with DNMT1, which might be a co-operative event during DNA methylation. DNMT3L contains DNA methyltransferase motifs and is required for establishing maternal genomic imprints, despite being catalytically inactive. DNMT3L is expressed during gametogenesis when genomic imprinting takes place, but also plays a role in stem cell biology [21].

1.4 METHYL-BINDING DOMAIN PROTEINS

DNA methylation may affect the transcription of genes in two ways. First, the methylation of DNA itself may physically impede the binding of transcriptional proteins to the gene; second, and likely more important, methylated DNA may be bound by proteins known as MBDs [22]. MBDs then recruit additional proteins to the locus, such as histone deacetylases and other chromatin remodeling proteins that can modify histones, thereby forming compact, inactive chromatin, termed heterochromatin. This link between DNA methylation and chromatin structure is very important. In particular, loss of methyl-CpG-binding protein 2 (MeCP2) has been implicated in

Rett syndrome, and methyl-CpG-binding domain protein 2 (MBD2) mediates the transcriptional silencing of hypermethylated genes in cancer.

1.5 DNA DEMETHYLATION

DNA demethylation is the process of removal of a methyl group from cytosines. DNA demethylation can be passive or active. The passive process takes place in the absence of methylation of newly synthesized DNA strands by DNMT1 during several replication rounds – for example, upon 5-azacytidine treatment [23]. These pharmaceutical demethylating agents act via inhibition of DNMT1 and favoring its degradation by proteosomes [24]. This approach is currently used in cancer to demethylate tumor suppressor genes [25]. Active DNA demethylation occurs via direct removal of a methyl group independently of DNA replication [26].

1.6 DNA HYDROXYMETHYLATION

Oxidation of the methyl group generates 5-hydroxymethylcytosine. Several mechanisms have been proposed to mediate demethylation of 5-hydroxymethylcytosines. This base can be deaminated by AID/Apobec enzymes to give 5-hydroxymethyluracil [27]; alternatively, ten–eleven translocation family enzymes (TETs) can further oxidize 5-hydroxymethylcytosine to 5-formylcytosine and 5-carboxylcytosine [26,28]. DNA hydroxymethylation has been proposed to act as a specific epigenetic mark opposing DNA methylation, rather than a passive intermediate in the demethylation pathway. DNA hydroxymethylation *in vivo* is sometimes associated with labile nucleosomes, which are more easy to disassemble and to out-compete by transcription factors during cell development [29].

1.7 DIFFERENTIALLY METHYLATED REGIONS

Differentially methylated regions (DMRs) are genomic regions with different methylation statuses among multiple samples (tissues, cells, individuals, etc.), and are regarded as possible functional regions involved in gene transcriptional regulation [30]. The identification of DMRs among multiple tissues provides a comprehensive survey of epigenetic differences among human tissues. DMRs between cancer and normal samples demonstrate the aberrant methylation in cancers [31].

It is well known that DNA methylation is associated with cell differentiation and proliferation. Many DMRs have been found in the development stages and in the reprogramming progress [32]. In addition, there are intra-individual DMRs with longitudinal changes in global DNA methylation along with the increase of age in a given individual [33]. The following chapters will focus mainly on DMRs in various diseases. Some of them are used as biomarkers, other are targets of epigenetic therapy.

1.8 NUTRIEPIGENOMICS

It has been identified that several lifestyle factors, such as diet, obesity, physical activity, tobacco smoking, alcohol consumption, environmental pollutants, and psychological stress, might modify epigenetic patterns [34]. Nutriepigenomics has emerged as a new and promising field in current epigenetics research in the past few years [35]. In the viable yellow agouti (A^{vy}) mouse, maternal diet affects the coat-color distribution of offspring by perturbing the establishment of methylation at the A^{vy} metastable epiallele [36]. Hence, the A^{vy} mouse can be employed as a sensitive epigenetic biosensor to assess the effects of dietary methionine supplementation on locus-specific DNA methylation. As a consequence of such observations, maternal diets supplemented with L-methionine are under investigation [37]. Furthermore, methyl donors have shown to be beneficial in several complex human diseases [38]. In some chapters, this promising approach is presented in more detail.

References

[1] Kobayashi H, Sakurai T, Sato S, Nakabayashi K, Hata K, Kono T. Imprinted DNA methylation reprogramming during early mouse embryogenesis at the Gpr1-Zdbf2 locus is linked to long cis-intergenic transcription. FEBS Lett 2012;586:827–33.
[2] Luger K, Mader AW, Richmond RK, Sargent DF, Richmond TJ. Crystal structure of the nucleosome core particle at 2.8 A resolution. Nature 1997;389:251–60.
[3] Davie JR, Spencer VA. Control of histone modifications. J Cell Biochem 1999;Suppl 32–33:141–8.
[4] Roth SY, Denu JM, Allis CD. Histone acetyltransferases. Annu Rev Biochem 2001;70:81–120.
[5] Zentner GE, Henikoff S. Regulation of nucleosome dynamics by histone modifications. Nat Struct Mol Biol 2013;20:259–66.
[6] Vermeulen M, Timmers HT. Grasping trimethylation of histone H3 at lysine 4. Epigenomics 2010;2:395–406.
[7] Pinskaya M, Morillon A. Histone H3 lysine 4 di-methylation: a novel mark for transcriptional fidelity? Epigenetics 2009;4:302–6.
[8] Roidl D, Hacker C. Histone methylation during neural development. Cell Tissue Res 2014;356:539–52.

[9] Del Rizzo PA, Trievel RC. Molecular basis for substrate recognition by lysine methyltransferases and demethylases. Biochimica et biophysica acta 2014;1839(12):1404–15.
[10] Del Rizzo PA, Trievel RC. Substrate and product specificities of SET domain methyltransferases. Epigenetics 2011;6:1059–67.
[11] Ziller MJ, Gu H, Muller F, Donaghey J, Tsai LT-Y, Kohlbacher O, et al. Charting a dynamic DNA methylation landscape of the human genome. Nature 2013;500:477–81.
[12] Plasschaert RN, Bartolomei MS. Genomic imprinting in development, growth, behavior and stem cells. Development 2014;141:1805–13.
[13] Schulz WA, Steinhoff C, Florl AR. Methylation of endogenous human retroelements in health and disease. Curr Top Microbiol Immunol 2006;310:211–50.
[14] Hashimoto H, Vertino PM, Cheng X. Molecular coupling of DNA methylation and histone methylation. Epigenomics 2010;2:657–69.
[15] Jeltsch A. Molecular enzymology of mammalian DNA methyltransferases. Curr Top Microbiol Immunol 2006;301:203–25.
[16] Pradhan S, Esteve PO. Mammalian DNA (cytosine-5) methyltransferases and their expression. Clin Immunol 2003;109:6–16.
[17] Jeltsch A. On the enzymatic properties of Dnmt1: specificity, processivity, mechanism of linear diffusion and allosteric regulation of the enzyme. Epigenetics 2006;1:63–6.
[18] Bestor TH. The DNA methyltransferases of mammals. Hum Mol Genet 2000;9:2395–402.
[19] Chedin F. The DNMT3 family of mammalian de novo DNA methyltransferases. Prog Mol Biol Transl Sci 2011;101:255–85.
[20] Yang Y, Liu R, Qiu R, Zheng Y, Huang W, Hu H, et al. CRL4B promotes tumorigenesis by coordinating with SUV39H1/HP1/DNMT3A in DNA methylation-based epigenetic silencing. Oncogene 2015;34(1):104–18.
[21] Liao HF, Tai KY, Chen WS-C, Cheng LCW, Ho H-N, et al. Functions of DNA methyltransferase 3-like in germ cells and beyond. Biol Cell 2012;104:571–87.
[22] Clouaire T, Stancheva I. Methyl-CpG binding proteins: specialized transcriptional repressors or structural components of chromatin? Cell Mol Life Sci: CMLS 2008;65: 1509–22.
[23] Gnyszka A, Jastrzebski Z, Flis S. DNA methyltransferase inhibitors and their emerging role in epigenetic therapy of cancer. Anticancer Res 2013;33:2989–96.
[24] Ghoshal K, Datta J, Majumder S, Bai S, Kutay H, Motiwala T, et al. ST. 5-Aza-deoxycytidine induces selective degradation of DNA methyltransferase 1 by a proteasomal pathway that requires the KEN box, bromo-adjacent homology domain, and nuclear localization signal. Mol Cell Biol 2005;25:4727–41.
[25] Di Costanzo A, Del Gaudio N, Migliaccio A, Altucci L. Epigenetic drugs against cancer: an evolving landscape. Arch Toxicol 2014;88:1651–68.
[26] Sadakierska-Chudy A, Kostrzewa RM, Filip M. A Comprehensive View of the Epigenetic Landscape Part I: DNA Methylation, Passive and Active DNA Demethylation Pathways and Histone Variants. Neurotox Res 2015;27(1):84–97.
[27] Gavin DP, Chase KA, Sharma RP. Active DNA demethylation in post-mitotic neurons: a reason for optimism. Neuropharmacology 2013;75:233–45.
[28] Guibert S, Weber M. Functions of DNA methylation and hydroxymethylation in mammalian development. Curr Top Dev Biol 2013;104:47–83.
[29] Teif VB, Beshnova DA, Vainshtein Y, Marth C, Mallm JP, Hofer T, et al. Nucleosome repositioning links DNA (de)methylation and differential CTCF binding during stem cell development. Genome Res 2014;24:1285–95.
[30] Docherty LE, Rezwan FI, Poole RL, Jagoe H, Lake H, Lockett GA, et al. Genome-wide DNA methylation analysis of patients with imprinting disorders identifies

differentially methylated regions associated with novel candidate imprinted genes. J Med Genet 2014;51:229–38.

[31] Tang MH, Varadan V, Kamalakaran S, Zhang MQ, Dimitrova N, Hicks J. Major chromosomal breakpoint intervals in breast cancer co-localize with differentially methylated regions. Front Oncol 2012;2:197.

[32] He W, Kang XJ, Du HZ, Song B, Lu ZY, Huang Y, et al. Defining differentially methylated regions specific for the acquisition of pluripotency and maintenance in human pluripotent stem cells via microarray. PloS One 2014;9:e108350.

[33] Bell JT, Tsai PC, Yang TP, Pidsley R, Nisbet J, Glass D, et al. Epigenome-wide scans identify differentially methylated regions for age and age-related phenotypes in a healthy ageing population. PLoS Genet 2012;8:e1002629.

[34] Alegria-Torres JA, Baccarelli A, Bollati V. Epigenetics and lifestyle. Epigenomics 2011;3:267–77.

[35] Remely M, Lovrecic L, de la Garza AL, Migliore L, Peterlin B, Milagro FI, et al. Therapeutic perspectives of epigenetically active nutrients. Br J Pharmacol 2015;172 (11):2756–68.

[36] Waterland RA. Assessing the effects of high methionine intake on DNA methylation. J Nutr 2006;136:1706S–10S.

[37] O'Neill RJ, Vrana PB, Rosenfeld CS. Maternal methyl supplemented diets and effects on offspring health. Front Genet 2014;5:289.

[38] Sharma S, Litonjua A. Asthma, allergy, and responses to methyl donor supplements and nutrients. J Allergy Clin Immunol 2014;133:1246–54.

CHAPTER

2

DNA Methylation and Epigenetic Biomarkers in Cancer

OUTLINE

M. Neidhart: *DNA Methylation and Complex Human Disease.*
DOI: http://dx.doi.org/10.1016/B978-0-12-420194-1.00002-6

2.1 INTRODUCTION

Biological markers need objectively to measure and evaluate normal biological processes in comparison to pathogenic processes, or pharmacologic responses to a therapeutic intervention. More specifically, a biomarker indicates a change in expression or state of a molecular entity (i.e., the target, which can be DNA, RNA, or protein) that correlates with the risk (i.e., detection) or progression of a disease, or with the susceptibility of the disease to a given treatment. This target can be analyzed in a qualitative or quantitative manner, depending on its nature and the available technologies, but the measurement must be objective to rule out inter-observer errors. Biomarkers may be physiological measurements (e.g., blood pressure), but nowadays are usually molecular measurements applied to tissue or body fluids. While protein markers are the most commonly used and studied, to overcome limitations for certain applications, many biomarkers based on RNA or DNA (or DNA modifications, such as epigenetic alterations) have been discovered in the last decade and currently are being tested on patient cohorts [1]. This chapter discusses genetic versus epigenetic origin of cancer, including cancer susceptibility, and how epigenetic biomarkers can be used in clinical practice.

2.1.1 Diagnosis Biomarkers

For diagnosis or screening, a biomarker should be a tool to identify those individuals with a disease or premalignant conditions (e.g., gastrointestinal cancer or precursor lesions). In gastrointestinal tumors, diagnosis is mostly based on endoscopic procedures; nevertheless, biomarkers such as blood- or stool-based methods which depend on molecular markers could help to improve this diagnosis or replace the endoscopic procedure for a first population-based screening approach for esophageal, gastric, or colon cancer. While these tests can include DNA (e.g., stool tests for colorectal cancer detection) or protein-based markers (e.g., glycosylated hemoglobin to identify the average plasma glucose concentration in blood), most biomarkers have failed to achieve clinical usefulness so far. No blood-based tests for esophageal or gastric cancer have been developed to date, and though blood-based tests for colon cancer exist, they are not yet standard in clinical practice [1].

2.1.2 Progression Biomarkers

Progression or surveillance markers could be used in addition to endoscopic surveillance and biopsies in order to predict a risk for a patient with a premalignant condition, such as metaplasia or adenoma, to identify certain types of precancerous lesions that are deemed more aggressive (e.g., high-grade dysplasia in the esophagus, or villous polyps in the colon). These biomarkers could include mutations and epimutations, as well as protein expression in tissue biopsies and blood- or stool-based tests [1].

2.1.3 Predictive Biomarkers

Response prediction markers analyze the potential effect of certain treatments on a specific patient cohort and include mutations or specific DNA methylation in genes which affect the metabolism of a chemotherapeutic agent, such as *MGMT* methylation for temozolomide or *KRAS* mutations for the use of cetuximab, or *HER2* overamplification for the indication of trastuzumab. Another prediction utility for a biomarker can also be recurrence or relapse after treatment, and thus will influence stricter surveillance strategies. Of note, while most applications allude to disease-related diagnostic and prognostic biomarkers, predictive biomarkers are drug-related biomarkers that indicate whether a drug will be effective in a specific patient. These markers are probably the only ones that have been implemented in the clinical routine so far [1].

2.1.4 Discovery of Novel Biomarkers

Two approaches are usually utilized for the discovery of novel biomarkers: hypothesis-driven identification is mostly by using biological key players that are important molecular alterations during carcinogenesis [2], and a systematic approach usually employs genome-wide screening technologies for target identification. Examples for this course of action are *IFFO1* in ovarian cancer [3], *SEPT9* in colorectal cancer [4], and *TFF3* in Barrett's esophagus [5]. This approach is less biased but might lead to high false-positive rates with low reproducibility rates, and will likely result in generating marker panels (or gene signatures). A validation phase is critical to guarantee reproducible results in independent studies; this seems to have been an important issue, as few biomarkers have progressed to a clinical phase yet [1].

2.2 CANCEROGENESIS

Cancer is traditionally viewed as a disease of abnormal cell proliferation controlled by a series of mutations. Mutations typically affect oncogenes or tumor suppressor genes, thereby conferring growth advantage. Genomic instability facilitates mutation accumulation [6]. An early hypothesis of cancer pathogenesis claimed that inactivation of tumor suppressor genes always requires two hits, the first occurring either in somatic cells or in the germline (i.e., hereditary) while the second is always somatic [7].

A vast majority of common cancers occur sporadically. In patients with such diseases, multiple common alleles may increase cancer risk, each having only a weak effect [8]. In familial cancers, which account for 10–30% of cancers, the frequency of risk alleles in the population varies from low to high and the effect on cancer risk from weak to moderate. Less than 5% of cancers represent monogenic disorders with highly pathogenic mutations in single genes; the same changes are practically absent in the normal population.

More on specific mechanisms leading to cancerogenesis is presented in Chapter 7.

2.3 EPIMUTATIONS IN CANCER

Epigenetics has emerged as a new and promising field in recent years. Lifestyle, stress, drugs, physiopathological situations, and pharmacological interventions have a great impact on the epigenetic code of the cells by altering the methylome, microRNA expression, and the covalent histone modifications [9]. The Human Genome Project [10] has dramatically facilitated the understanding of genetic factors behind human disease, including cancer. At present, some 100 genes (corresponding to 0.5% of all genes in the human genome) display highly or moderately penetrant mutations that underlie hereditary cancer syndromes [11]. These include the main DNA mismatch repair (MMR) genes *MLH1* and *MSH2* underlying Lynch syndrome, the *APC* (adenomatous polyposis coli) gene underlying familial adenomatous polyposis, and the *BRCA1* and *BRCA2* genes underlying hereditary breast and ovarian cancer syndrome. Moreover, systematic screening of the entire cancer genomes utilizing new technologies such as next-generation sequencing has identified some 400 genes that are somatically mutated in cancer and likely to contribute to the neoplastic process [12].

Activation of oncogenes and inactivation of tumor suppressor genes, as well as genomic instability, can be achieved by epigenetic mechanisms as well. Unlike genetic mutations, epimutations do not change the base sequence of DNA and are potentially reversible. Similar to genetic mutations, epimutations are associated with specific patterns of gene expression that are heritable through cell divisions [6].

2.3.1 Esophageal Adenocarcinoma

Esophageal adenocarcinoma arises from a premalignant condition called Barrett's esophagus (BE), which in turn is thought to be an adaptation to gastroesophageal reflux disease. Like many aggressive cancers, unless diagnosed at very early stages the survival rates are very low. Thus, biomarkers are urgently needed as prognostic markers to identify those patients with endoscopically diagnosed and histologically confirmed Barrett's esophagus who have increased risk of developing esophageal adenocarcinoma [1]. It has been suggested that hypermethylation of *AKAP12* [13] may be useful as a screening marker, since the methylation occurs early in the metaplasia–dysplasia–carcinoma sequence, and early detection increases the survival chances of patients. Other promising markers for early detection of esophageal cancer are hypermethylation of *Reprimo* (*RPRM*) [14] and *TAC1* [15]. Regarding the progression of Barrett's esophagus to esophagal cancer, several DNA methylation biomarkers have been described. For example, a retrospective study demonstrated that hypermethylation of CDKN2A, RUNX3, and HPP1 in patients with non-dysplastic Barrett's esophagus or low-grade dysplasia is independently associated with an increased risk of malignant transformation to high-grade dysplasia and esophageal cancer [16]. Another study showed that APC, TIMP3, and TERT were significantly more methylated in progressors versus non-progressors, making hypermethylation of these genes promising predictive markers [17]. Hypermethylation of both APC and CDKN2A is also associated with an increased risk of progression to high-grade dysplasia or esophageal cancer [18], and a methylation biomarker panel comprising eight genes (*p16, RUNX3, HPP1, NELL1, TAC1, SST, AKAP12,* and *CDH13*) could predict 50% of progressors [19].

As well as progression markers, prognostic biomarkers are useful to determine which patients can benefit from aggressive therapies, as measurement of survival is usually the clinical endpoint. Patients with EAC whose tumors had more than 50% of a seven-gene profile (*APC, E-cadherin, MGMT, ER, CDKN2A, DAPK, TIMP3*) methylated had significantly poorer survival and earlier tumor recurrence [20]. Other methylated genes linked with reduced survival include APC [21], NELL1 [22], and methylation of DAPK and APC promoter in cell-free serum [23]. Predicting therapy response is important in order to give each patient the best therapy and avoid ineffective treatments. Therefore, response prediction markers are very important in clinical practice when it comes to chemotherapy. A nine-gene methylation panel (*CDKN2A, RPRM, p57, p73, RUNX3, CHFR, MGMT, TIMP3,* and *HPP1*) showed a significantly higher number of methylated genes in non-responders than in responders [24]. DAPK promoter methylation has also been linked with response to neoadjuvant radiochemotherapy [25].

2.3.2 Gastric Cancer

Gastric cancer is a major health burden, comprising around 8% of all cancer cases worldwide and over 10% of all cancer deaths. While *Helicobacter pylori* infection is believed to be the cause of most stomach cancer cases, only 1–2% of infected individuals develop this malignancy during their lifetime. Some epigenetic changes, such as aberrant methylation of CpG islands in promoter regions, are commonly detected in gastric adenocarcinoma, and may be considered a reasonable molecular biomarker for the diagnosis of gastric cancer. Most studies analyze biomarkers in endoscopic biopsy tissues. *RPRM* methylation has been detected in plasma and stool samples of gastric cancer patients, and therefore could be such a potential diagnostic marker [26]. *CDKN2A* [27] and *RASSF1A* [28] have also been discussed as diagnostic markers, some of them even as serum or plasma markers, but larger validation studies are missing. Several combinations of altered genes, such as a panel of *CDH1, CDKN2A, RARB* [29,30], or a panel of *APC, CDH1, GSTP 1,* hMLH1, *MGMT, CDKN2A, CDKN2B, SOCS1, TIMP3, TGFBR2* [31], or *RUNX3, CDKN2A, CDH1,* and *RASSF1A* [32], have also been tested and achieved sensitivity rates up to 55% while keeping high specificity. Commonly detected epigenetic changes, such as aberrant methylation of CpG islands in promoter regions, may be considered another molecular phenotype of gastric cancer [33]. Methylation of the CpG island in the *hMLH1* promoter region is a frequent event in gastric cancer, and is associated with the loss of MLH1 expression in the majority of gastric cancers exhibiting microsatellite instability, thus linking genomic instability and epigenetic alterations [34]. Tumors with concurrent hypermethylation in multiple loci have been defined as a CpG island methylation phenotype (CIMP), and this is associated with a microsatellite instability-high phenotype in gastric cancer [35] and DNMT1 overexpression [36]. Methylation of *RUNX3* has been mentioned as a potential diagnostic marker in gastric cancer [37–39], as well as a marker for poor prognosis [40], and has been associated with genomic instability [41] as well as *H. pylori* infection [42] and even tested in serum [43]. At the moment, no diagnostic biomarker test for gastric cancer is available. Nevertheless it remains to be seen (when analyzed in larger cohorts) whether any of the above markers could function as a prognostic marker or help with clinical treatment decisions for certain subtypes of gastric cancer that have an expression of distinct marker panels. Some genes, such as *CDKN2A, RARB,* and *CDH1,* can provide not only diagnostic but also prognostic information in gastric cancer [44]. Hypermethylation of *DAPK, PYACRD,* and *BNIP3* has also been indicated as not only a prognostic but also a predictive marker for chemotherapy [45], and *DAPK* has been linked to interleukin

6 serum levels – an important cytokine in cancer initiation, progression, and metastasis [46]. Numerous studies has been performed regarding biomarkers for prognosis and survival in gastric cancer patients. Methylated *DKK3* [47], *TFPI2* [48], and *PAX6* [49] are the most promising candidates as prognostic biomarkers.

2.3.3 Colorectal Carcinoma

Colorectal carcinoma provides a useful model to study genetic and epigenetic events in cancer because it develops through multiple steps from normal mucosa via benign precursor lesions and polyps; additionally, well-known hereditary forms exist where tumorigenesis is in many ways similar but accelerated compared to sporadic disease. Knudson's two-hit theory [7] postulates a mechanistic link between dominantly inherited cancers and their sporadic counterparts. In the hereditary form, one mutation (the first hit) in a tumor suppressor gene is inherited and the second hit occurs in (and is restricted to) the somatic cancer progenitor cell in target tissue. In the sporadic form, two inactivating hits, one in each allele, occur somatically prior to tumor initiation. The accumulating evidence suggests that either hit, or both, may be genetic or epigenetic in both hereditary and sporadic settings. Furthermore, although tumor suppressor genes are typically recessive on the cellular level, exceptions to the rule exist, including dominant negative effects (where the mutant gene product antagonizes the wild-type product), haploinsufficiency (where inactivation of a single allele is sufficient to initiate tumorigenesis), or residual function of the mutant allele (where more than two hits may be necessary to initiate tumorigenesis). In such situations in which a subtle shift in gene dosage may trigger the cell to a cancer path, epigenetic mechanisms may have a special role to play [6].

SEPT9, a gene involved in cytokinesis and cell cycle control, was identified as a potential biomarker for blood-based colorectal cancer screening [4]: in contrast to healthy controls, colorectal cancer-affected patients showed *SEPT9* promoter methylation in DNA derived from blood plasma [50]. Although such tests are on the market, their specificity has been questioned [51].

2.3.4 Lynch Syndrome

Lynch syndrome, a multi-organ cancer syndrome, is the prototype example of constitutional epimutation. This notion refers to constitutional hypermethylation at the promoter of one allele of a given (non-imprinted) gene leading to silencing of expression from that allele in all main somatic tissues. If a separate inductor (e.g., an adjacent genetic alteration) for the epigenetic change exists, epimutation can be classified as secondary or primary [52]. Lynch syndrome is associated with germline mutations in DNA mismatch repair genes [53], namely primary epimutation of *MLH1* [54] or *MSH2* [55,56]. In secondary epimutation, transcription of an altered gene lacking a stop codon (for example, *PTPRJ* among the genes listed in Table 2.1) reads into an adjacent, structurally normal gene (*MSH2* and *PTPRJ*, respectively) and induces promoter hypermethylation of the latter gene [56,57]. In another example, a heat shock factor binding to a single nucleotide variant in the 5′UTR of *MLH1* was possibly implicated in the induction of the epigenetic change in *MLH1* [58].

Constitutional epimutations may explain around 10% of Lynch syndrome-suspected families in which MLH1 or MSH2 protein is absent in tumor tissue and no genetic mutations in the respective genes are detectable in the germline [59,60]. Preliminary observations from some other genes (e.g., *KLLN* [61,62]) suggest that constitutional epimutation may even be the predominant mechanism of cancer susceptibility.

Examples of epimutations conferring susceptibility to cancer are given in Table 2.1 [13–15,26–28,38,45,47–49,56–83].

2.3.5 Endometrial and Ovarian Cancers

Endometrial cancer is the most common gynecological malignancy among women worldwide. Hypermethylation of the *APC* promoter region is an early event in endometrial tumorigenesis [65]. Methylation of the promoter region of *CDKNA2* inactivation in endometrial carcinoma defines a subgroup of patients with high proliferation and aggressiveness [74]. Together with inactivation of *MLH1* and *MGMT*, it is an early event in the development of synchronous primary endometrial and ovarian cancers [75]. Detection of *MGMT* hypermethylation may be useful to define a set of gynecologic malignancies with a specific sensitivity to alkylating chemotherapy. *MLH1* mutation or epimutation is also strongly associated with aberrant methylations of *SESN3* and *TITF1* [81]. Hypermethylation of the CDH13 promoter is associated with aneuploidy [66]. Molecular changes in endometrial tissue are detectable several years before endometrial carcinoma in genetically predisposed individuals, including methylation changes in *CDH13*, *GSTP1*, and *RASSF1* [67].

Furthermore, epigenetic silencing by hypermethylation of the connective tissue growth factor (*CTGF*) promoter leads to a loss of CTGF function, which can be a factor in the development of ovarian cancer [76]. Ovarian carcinoma is also associated to hypermethylation cell differentiation genes, in particular *HOXA10* and *HOXA1* [79].

TABLE 2.1 Epimutations (i.e., Hypermethylation of the Promoter) that Can Be Used as Biomarkers in Cancer

Genes	Associated cancer susceptibility phenotype	Frequency of epimutations	Type of biomarker
AHR	Mantle cell lymphoma	32% [63]	Diagnosis
ARF4	Follicular lymphoma	87% [64]	Diagnosis
	Mantle cell lymphoma	58% [64]	Diagnosis
AKAP12	Esophageal carcinoma	53% [13]	Diagnosis
APC	Endometrial cancer	38–77% [65–67]	Diagnosis
	B-cell leukemia	46% [68]	Diagnosis
APP	Chronic lymphocytic leukemia	27% [69]	Prognosis
BTG4	Chronic lymphocytic leukemia	54% [70]	Diagnosis
BNIP3	Gastric cancer	39% [71]	Prediction
CDH1	B-cell leukemia	59% [68]	Diagnosis
	Gastric cancer	8% [72]	Prediction
CDH13	Endometrial cancer	68–71% [66,67]	Prediction
CDKN2A	B-cell leukemia	52–54% [68,73]	Diagnosis
	Gastric cancer	27% [27]	Diagnosis
	Endometrial cancer	4–41% [66,74,75]	Prognosis
CD38	Chronic lymphocytic leukemia	77% [70]	Diagnosis
CTCF	Ovarian cancer	59% [76]	Diagnosis
DAPK	B-cell leukemia	63% [68]	Diagnosis
	Gastric cancer	22–41% [45,71]	Prediction
DLC1	Non-Hodgkin's lymphoma	87% [77]	Diagnosis
DKK3	Gastric cancer	68% [47]	Prognosis
	Esophageal cancer	52% [47]	Prognosis
	Chronic lymphocytic leukemia	44% [78]	Diagnosis
GSTP1	B-cell leukemia	65% [68]	Diagnosis
	Endometrial cancer	17–40% [67]	Diagnosis
HOXA4	Chronic lymphocytic leukemia	30% [70]	Diagnosis
HOXA9	Mantle cell lymphoma	41% [73]	Diagnosis
HOXC10	Ovarian cancer	44% [79]	Diagnosis
	Mantle cell lymphoma	42% [64]	Diagnosis
	Follicular lymphoma	27% [64]	Diagnosis

(Continued)

TABLE 2.1 (Continued)

Genes	Associated cancer susceptibility phenotype	Frequency of epimutations	Type of biomarker
HOXC11	Ovarian cancer	38% [79]	Diagnosis
KLLN	Cowden syndrome	37% [61]	Prediction
	Renal carcinoma	48% [62]	Prediction
LHX2	Follicular lymphoma	73% [64]	Diagnosis
	B-cell leukemia	47% [64]	Diagnosis
	Mantle cell lymphoma	42% [64]	Diagnosis
LRP1b	Follicular lymphoma	87% [64]	Diagnosis
	Mantle cell lymphoma	33% [64]	Diagnosis
	B-cell leukemia	13% [64]	Diagnosis
MAD2	B-cell leukemia	50% [73]	Diagnosis
MGMT	Endometrial cancer	9–48% [66,75]	Prediction
	B-cell leukemia	46% [73]	Prognosis
	Glioblastoma	40% [80]	Prediction
MLH1	Endometrial cancer	32–39% [66,75]	Diagnosis
	Lynch syndrome	3–11% [54,58–60]	Prediction
MLH2	Lynch syndrome	9–40% [56,60]	Prediction
NKX6-1	Follicular lymphoma	66% [64]	Diagnosis
	Mantle cell lymphoma	42% [64]	Diagnosis
NRP2	Follicular lymphoma	87% [64]	Diagnosis
	B-cell leukemia	15% [64]	Diagnosis
PAX6	Gastric cancer	37% [49]	Prognosis
PCDHGB7	Non-Hodgkin's lymphoma	78% [77]	Diagnosis
POU3F3	Follicular lymphoma	87% [64]	Diagnosis
	Mantle cell lymphoma	42% [64]	Diagnosis
	B-cell leukemia	20% [64]	Diagnosis
PRKCE	Follicular lymphoma	33% [64]	Diagnosis
	B-cell leukemia	27% [64]	Diagnosis
	Mantle cell lymphoma	25% [64]	Diagnosis
PTPRJ	Colorectal cancer	>1% [57]	Prediction
PYCARD	Gastric cancer	32% [45]	Prediction

(*Continued*)

TABLE 2.1 (Continued)

Genes	Associated cancer susceptibility phenotype	Frequency of epimutations	Type of biomarker
P57KIP2	B-cell leukemia	48% [73]	Prognosis
RASSF1A	Endometrial cancer	58–82% [66,67]	Diagnosis
	Gastric cancer	34% [28]	Diagnosis
	B-cell leukemia	15% [68]	Diagnosis
RB1	B-cell leukemia	20% [68]	Diagnosis
RPRM	Gastric cancer	96% [26]	Diagnosis
	Esophageal cancer	63% [14]	Diagnosis
	Barrett's esophagus	36% [14]	Diagnosis
RUNX3	Gastric cancer	65% [38]	Diagnosis
SESN3	Endometrial cancer	20% [81]	Diagnosis
SFRP1	Endometrial cancer	16% [66]	Diagnosis
	Chronic lymphocytic leukemia	11% [78]	Diagnosis
SFRP2	Endometrial cancer	51% [66]	Diagnosis
	Chronic lymphocytic leukemia	14% [78]	Diagnosis
SFRP4	Chronic lymphocytic leukemia	5–20% [78,82]	Diagnosis
	Endometrial cancer	5% [66]	Diagnosis
SFRP5	Endometrial cancer	48% [66]	Diagnosis
	Chronic lymphocytic leukemia	7% [78]	Diagnosis
SHOX2	Lung cancer	78% [83]	Diagnosis
SHP1	B-cell leukemia	78% [68]	Diagnosis
SOX9	Mantle cell lymphoma	35% [63]	Diagnosis
SOX11	Chronic lymphocytic leukemia	29% [69]	Diagnosis
TAC1	Esophageal cancer	30–61% [15]	Diagnosis
TFPI2	Gastric cancer	81% [48]	Prognosis
TIMP3	B-cell leukemia	18% [68]	Diagnosis
TITF1	Endometrial cancer	70% [81]	Diagnosis
VHL	B-cell leukemia	44% [68]	Diagnosis
WIF1	Endometrial cancer	43% [66]	Diagnosis
	Chronic lymphocytic leukemia	11% [78]	Diagnosis

2.3.6 Lung Cancer

Lung cancer is one of the deadliest cancers worldwide, with the highest incidence and mortality. While the prognosis of lung cancer is generally grim, with 5-year survival rates of only 15%, there is hope, and evidence, that early detection of lung cancer can reduce mortality. A number of approaches have resulted in a diverse set of molecular biomarkers for lung carcinoma, particularly in recent years with the advent of next-generation sequencing technologies combined with advances in bronchoscopic and imaging techniques [84]. These have classically been genetic or epigenetic (e.g., abnormal methylation of *APC, TMS1, RASSF1, CDKN2A, DAPK*) [85,86]. Recently, short stature homeobox 2 (*SHOX2*) was found to be hypermethylated in lung cancer when lung tumor tissues were compared with normal lung tissue in genome-wide methylation profiling [83,87].

2.3.7 Glioblastoma

The DNA repair protein O^6-methylguanine-DNA methyltransferase (MGMT) removes O^6-alkyl-guanine from DNA, a lesion induced by alkylating mutagens. In 2000, it was reported that *MGMT* methylation predicts the outcome of glioma patients treated with the alkylating agent carmustine. Methylation of the *MGMT* promoter was associated with regression of the tumor, and with prolonged overall and disease-free survival, which can be attributed to the increased carmustine-sensitivity of cancer cells due to silencing of *MGMT* upon methylation [80].

2.3.8 Hematopoietic Cancers

The introduction of high-throughput technologies that measure DNA methylation has facilitated multiple studies with a focus on the molecular classification of lymphoid malignancies. One early study clearly segregated small B-cell lymphoma subtypes based on patterns of methylation [88]. In addition, candidate genes were also identified and further studied to gain insight into the pathogenesis of follicular lymphoma, mantle cell lymphoma, and chronic lymphocytic leukemia (*LRP1B, ARF4, POUF3, HOXC10, LHX2, MLLT2*) [64]. It was also determined that subtypes of lymphoma have variable amounts of methylated DNA. For example, it was observed that mantle cell lymphoma and chronic lymphocytic leukemia (both non-germinal center tumors) have many fewer methylated genes than follicular lymphoma (a germinal center lymphoma). An additional study was designed to determine if indolent versus rapidly progressing chronic lymphocytic leukemia could be segregated according to patterns of DNA methylation [64].

Using next-generation sequencing technologies, promoter methylation profiles were generated for 25 candidate genes that may play a role in the development of lymphoid malignancy [89], especially methylations of the *DLC-1* and *LRP1B* promoters [77,89]. These studies recognized that in addition to being able to segregate lymphoid malignancies based upon differential patterns of DNA methylation, there are also quantitative differences in methylation at particular CpG loci among subtypes of lymphomas. Different methylation patterns exist among molecular subtypes of large diffuse B-cell lymphoma [90]. These results raise a question regarding the germinal center reaction and its role in DNA methylation.

Aberrantly methylated loci can also serve as biomarkers for early diagnosis of lymphoma, for evaluating the progression of lymphoma, and as prognostic indicators in lymphoma patients. Irving and colleagues used a panel of three genes (*CD38*, *HOXA4*, and *BTG4*) and created a combined methylation score that incorporated the methylation status of each of the individual genes to predict outcome in chronic lymphocytic leukemia patients [70]. Each of the individual genes displayed a correlation with patient outcome, but when all three genes were used together the correlation became stronger. An additional study identified 280 targets of aberrant methylation in chronic lymphocytic leukemia using methylated CpG island amplification coupled with hybridization to a microarray [69]. Of these 280 aberrantly methylated genes, the methylation status of two genes (*LINE1*, *APP*) was found to be correlated with an overall shorter survival. A genome-wide screen after pharmacologic reversal of DNA methylation identified genes that were putatively silenced by DNA methylation [63]. This screen identified 252 potentially methylated genes. The investigators then identified a set of five genes whose degree of methylation correlates with aggressive clinical features of mantle cell lymphoma (*SOX9*, *HOXA9*, *AHR*, *NR2F2*, and *ROBO1*) by examining 25 candidate genes in 38 primary mantle cell lymphoma samples. Associations between methylation biomarkers and aggressive disease have also been shown in diffuse large B-cell lymphoma. Hypermethylation of *CDKN2A*, *VHL*, *DAPK*, and *SHP1* was associated with aggressive phenotype in diffuse large B-cell lymphoma [68]. An additional study found that methylation in the promoter of *MGMT* and *P57KIP2* could be used to predict survival in diffuse large B-cell lymphoma patients treated with cyclophosphamide, hydroxydaunorubicin, vincristin, and prednisone based chemotherapy [73]. It is important to point out that the utility of methylation biomarkers is independent of the functional relevance of the observed methylation. Therefore, panels can be developed that will aid in predicting the course of a patient's disease even if the genes/loci involved have no currently known functional relevance.

Constitutive Wnt signaling has been reported in lymphoid malignancies, and multiple studies have shown that genes within this pathway are aberrantly methylated [64,82]. The canonical Wnt/B-catenin pathway is well understood, and mainly controls proliferation and differentiation. For this pathway to be active, Wnt genes must bind to their receptors, which leads to inactivation of the B-catenin destruction complex and allows B-catenin to stabilize and translocate to the nucleus, where it regulates the transcription of Wnt target genes. Many genes comprise the Wnt pathway, and they can generally be divided into either positive or negative regulators of Wnt signaling. The negative regulators are often downregulated in lymphoid malignancies while the positive regulators are often upregulated. To date, many Wnt antagonists have been identified that are aberrantly methylated in lymphoma (e.g., *WIF1, DKK3, APC, SFRPs, SOX9, SOX11*) [64,78,82]. The identification of these genes is important because it may lead to a better understanding of the mechanisms that are responsible for aberrant Wnt signaling in lymphoma. For example, it is quite likely that inactivation of these "brakes" in the pathway may contribute to a permissively activated pathway. Since many genes in the pathway may have the same ultimate role of activating the pathway, it is important that we consider the functional consequences of aberrations in multiple genes for the activation of downstream Wnt targets. For example, it has been shown that more than 50% of chronic lymphocytic leukemia patients have aberrant methylation in at least one of the seven Wnt antagonist genes [78].

2.4 CONCLUSION

To evaluate the prognosis in general, we need markers that can indicate better or worse outcomes (e.g., likelihood of survival over a certain time span). Prognostic markers besides tumor size (or shrinkage after therapy) include molecular and histological markers, as the more aggressive cancer types with a worse prognosis have distinct genetic and epigenetic alterations. Biomarkers can help to decide between therapeutic options, and to evaluate the prediction of response in the future [1]. We need more prospective studies that treat with new inhibitors or antibodies or small molecules in correlation to a complete genetic characterization prior to the start of any treatment in order to evaluate the impact of genetic and epigenetic changes on treatment outcome and survival. In general, molecular biomarkers will help to improve personalized medicine and in the near future could be able to characterize patients for specific therapeutic options.

References

[1] Tanzer M, Liebl M, Quante M. Molecular biomarkers in esophageal, gastric, and colorectal adenocarcinoma. Pharmacol Ther 2013;S0163–72.
[2] Galipeau PC, Li X, Blount PL, Maley CC, Sanchez CA, Odze RD, et al. NSAIDs modulate CDKN2A, TP53, and DNA content risk for progression to esophageal adenocarcinoma. PLoS Med 2007;4:e67.
[3] Campan M, Moffitt M, Houshdaran S, Shen H, Widschwendter M, Daxenbichler G, et al. Genome-scale screen for DNA methylation-based detection markers for ovarian cancer. PLoS One 2011;6:e28141.
[4] Lofton-Day C, Model F, Devos T, Tetzner R, Distler J, Schuster M, et al. DNA methylation biomarkers for blood-based colorectal cancer screening. Clin Chem 2008;54:414–23.
[5] Lao-Sirieix P, Boussioutas A, Kadri SR, O'Donovan M, Debiram I, Das M, et al. Non-endoscopic screening biomarkers for Barrett's oesophagus: from microarray analysis to the clinic. Gut 2009;58:1451–9.
[6] Peltomaki P. Mutations and epimutations in the origin of cancer. Exp Cell Res 2012;318:299–310.
[7] Knudson Jr. AG. Mutation and cancer: statistical study of retinoblastoma. Proc Natl Acad Sci USA 1971;68:820–3.
[8] Foulkes WD. Inherited susceptibility to common cancers. N Engl J Med 2008;359: 2143–53.
[9] Sandoval J, Peiro-Chova L, Pallardo FV, Garcia-Gimenez JL. Epigenetic biomarkers in laboratory diagnostics: emerging approaches and opportunities. Expert Rev Mol Diagn 2013;13:457–71.
[10] International Human Genome Sequencing Consortium. Finishing the euchromatic sequence of the human genome. Nature 2004;431:931–45.
[11] Garber JE, Offit K. Hereditary cancer predisposition syndromes. J Clin Oncol 2005;23: 276–92.
[12] Stratton MR. Exploring the genomes of cancer cells: progress and promise. Science 2011;331:1553–8.
[13] Jin Z, Hamilton JP, Yang J, Mori Y, Olaru A, Sato F, et al. Hypermethylation of the AKAP12 promoter is a biomarker of Barrett's-associated esophageal neoplastic progression. Cancer Epidemiol Biomarkers Prev 2008;17:111–17.
[14] Hamilton JP, Sato F, Jin Z, Greenwald BD, Ito T, Mori T, et al. Reprimo methylation is a potential biomarker of Barrett's-associated esophageal neoplastic progression. Clin Cancer Res 2006;12:6637–42.
[15] Jin Z, Olaru A, Yang J, Sato F, Cheng Y, Kan T, et al. Hypermethylation of tachykinin-1 is a potential biomarker in human esophageal cancer. Clin Cancer Res 2007;13:6293–300.
[16] Schulmann K, Sterian A, Berki A, Yin J, Sato F, Xu Y, et al. Inactivation of p16, RUNX3, and HPP1 occurs early in Barrett's-associated neoplastic progression and predicts progression risk. Oncogene 2005;24:4138–48.
[17] Clement G, Braunschweig R, Pasquier N, Bosman FT, Benhattar J. Methylation of APC, TIMP3, and TERT: a new predictive marker to distinguish Barrett's oesophagus patients at risk for malignant transformation. J Pathol 2006;208:100–7.
[18] Wang JS, Guo M, Montgomery EA, Thompson RE, Cosby H, Hicks L, et al. DNA promoter hypermethylation of p16 and APC predicts neoplastic progression in Barrett's esophagus. Am J Gastroenterol 2009;104:2153–60.
[19] Jin Z, Cheng Y, Gu W, Zheng Y, Sato F, Mori Y, et al. A multicenter, double-blinded validation study of methylation biomarkers for progression prediction in Barrett's esophagus. Cancer Res 2009;69:4112–15.

[20] Brock MV, Guo M, Akiyama Y, Muller A, Wu TT, Montgomery E, et al. Prognostic importance of promoter hypermethylation of multiple genes in esophageal adenocarcinoma. Clin Cancer Res 2003;9:2912–19.
[21] Kawakami K, Brabender J, Lord RV, Groshen S, Greenwald BD, Krasna MJ, et al. Hypermethylated APC DNA in plasma and prognosis of patients with esophageal adenocarcinoma. J Natl Cancer Inst 2000;92:1805–11.
[22] Jin Z, Mori Y, Yang J, Sato F, Ito T, Cheng Y, et al. Hypermethylation of the nel-like 1 gene is a common and early event and is associated with poor prognosis in early-stage esophageal adenocarcinoma. Oncogene 2007;26:6332–40.
[23] Hoffmann AC, Vallbohmer D, Prenzel K, Metzger R, Heitmann M, Neiss S, et al. Methylated DAPK and APC promoter DNA detection in peripheral blood is significantly associated with apparent residual tumor and outcome. J Cancer Res Clin Oncol 2009;135:1231–7.
[24] Hamilton JP, Sato F, Greenwald BD, Suntharalingam M, Krasna MJ, Edelman MJ, et al. Promoter methylation and response to chemotherapy and radiation in esophageal cancer. Clin Gastroenterol Hepatol 2006;4:701–8.
[25] Brabender J, Arbab D, Huan X, Vallbohmer D, Grimminger P, Ling F, et al. Death-associated protein kinase (DAPK) promoter methylation and response to neoadjuvant radiochemotherapy in esophageal cancer. Ann Surg Oncol 2009;16: 1378–83.
[26] Bernal C, Aquayo F, Villarroel C, Vargas M, Diaz I, Ossandon FJ, et al. Reprimo as a potential biomarker for early detection in gastric cancer. Clin Cancer Res 2008;14:6264–9.
[27] Abbaszadegan MR, Moaven O, Sima HR, Ghafarzadegan K, A'rabi A, Forghani MN, et al. p16 promoter hypermethylation: a useful serum marker for early detection of gastric cancer. World J Gastroenterol 2008;14:2055–60.
[28] Wang YC, Yu ZH, Liu C, Xu LZ, Yu W, Lu J, et al. Detection of RASSF1A promoter hypermethylation in serum from gastric and colorectal adenocarcinoma patients. World J Gastroenterol 2008;14:3074–80.
[29] Ichikawa D, Koike H, Ikoma H, Ikoma D, Tani N, Otsuji E, et al. Detection of aberrant methylation as a tumor marker in serum of patients with gastric cancer. Anticancer Res 2004;24:2477–81.
[30] Koike H, Ichikawa D, Ikoma H, Tani N, Ikoma D, Otsuji E, et al. Comparison of serum aberrant methylation and conventional tumor markers in gastric cancer patients. Hepatogastroenterology 2005;52:1293–6.
[31] Leung WK, To KF, Chu ESH, Chan MWY, Bai AHC, Ng EKW, et al. Potential diagnostic and prognostic values of detecting promoter hypermethylation in the serum of patients with gastric cancer. Br J Cancer 2005;92:2190–4.
[32] Tan SH, Ida H, Lau QC, Goh BC, Chieng WS, Loh M, et al. Detection of promoter hypermethylation in serum samples of cancer patients by methylation-specific polymerase chain reaction for tumour suppressor genes including RUNX3. Oncol Rep 2007;18:1225–30.
[33] Ushijima T, Sasako M. Focus on gastric cancer. Cancer Cell 2004;5:121–5.
[34] Herman JG, Baylin SB. Gene silencing in cancer in association with promoter hypermethylation. N Engl J Med 2003;349:2042–54.
[35] An C, Choi IS, Yao JC, Worah S, Xie K, Mansfield PF, et al. Prognostic significance of CpG island methylator phenotype and microsatellite instability in gastric carcinoma. Clin Cancer Res 2005;11.
[36] Etoh T, Kanai Y, Ushijima S, Nakagawa T, Nakanishi Y, Sasako M, et al. Increased DNA methyltransferase 1 (DNMT1) protein expression correlates significantly with poorer tumor differentiation and frequent DNA hypermethylation of multiple CpG islands in gastric cancers. Am J Pathol 2004;164:689–99.

[37] Waki T, Tamura G, Sato M, Terashima M, Nishizuka S, Motoyama T. Promoter methylation status of DAP-kinase and RUNX3 genes in neoplastic and non-neoplastic gastric epithelia. Cancer Sci 2003;94:360–4.
[38] Kim TY, Lee HJ, Hwang KS, Lee M, Kim JW, Bang YJ, et al. Methylation of RUNX3 in various types of human cancers and premalignant stages of gastric carcinoma. Lab Invest 2004;84:479–84.
[39] Oshimo Y, Oue N, Mitani Y, Nakayama H, Kitadai Y, Yoshida K, et al. Frequent loss of RUNX3 expression by promoter hypermethylation in gastric carcinoma. Pathobiology 2004;71:137–43.
[40] Wei D, Gong W, Oh SC, Li Q, Kim WD, Wang L, et al. Loss of RUNX3 expression significantly affects the clinical outcome of gastric cancer patients and its restoration causes drastic suppression of tumor growth and metastasis. Cancer Res 2005;65:4809–16.
[41] Gargano G, Calcara D, Corsale S, Agnese V, Intrivici C, Fulfaro F, et al. Aberrant methylation within RUNX3 CpG island associated with the nuclear and mitochondrial microsatellite instability in sporadic gastric cancers. Results of a GOIM (Gruppo Oncologico dell'Italia Meridionale) prospective study. Ann Oncol 2007;18(Suppl. 6):vi103–109.
[42] Kitajima Y, Ohtaka K, Mitsuno M, Tanaka M, Sato S, Nakafusa Y, et al. Helicobacter pylori infection is an independent risk factor for Runx3 methylation in gastric cancer. Oncol Rep 2008;19:197–202.
[43] Sakakura C, Hamada T, Miyagawa K, Nishio M, Miyashita A, Nagata H, et al. Quantitative analysis of tumor-derived methylated RUNX3 sequences in the serum of gastric cancer patients. Anticancer Res 2009;29:2619–25.
[44] Ikoma H, Ichikawa D, Koike H, Ikoma D, Tani N, Okamoto K, et al. Correlation between serum DNA methylation and prognosis in gastric cancer patients. Anticancer Res 2006;26:2313–16.
[45] Kato K, Iida S, Uetake H, Takagi Y, Yamashita T, Inokuchi M, et al. Methylated TMS1 and DAPK genes predict prognosis and response to chemotherapy in gastric cancer. Int J Cancer 2008;122:603–8.
[46] Tang LP, Cho CH, Hui WM, Huang C, Chu KM, Xia HH, et al. An inverse correlation between Interleukin-6 and select gene promoter methylation in patients with gastric cancer. Digestion 2006;74:85–90.
[47] Yu J, Tao Q, Cheng YY, Lee KY, Ng SS, Cheung KF, et al. Promoter methylation of the Wnt/beta-catenin signaling antagonist Dkk-3 is associated with poor survival in gastric cancer. Cancer 2009;115:49–60.
[48] Jee CD, Kim MA, Jung EJ, Kim J, Kim WH. Identification of genes epigenetically silenced by CpG methylation in human gastric carcinoma. Eur J Cancer 2009;45:1282–93.
[49] Yao D, Shi J, Shi B, Wang N, Liu W, Zhang G, et al. Quantitative assessment of gene methylation and their impact on clinical outcome in gastric cancer. Clin Chim Acta 2012;413:787–94.
[50] Warren JD, Xiong W, Bunker AM, Vaughn CP, Furtado LV, Roberts WL, et al. Septin 9 methylated DNA is a sensitive and specific blood test for colorectal cancer. BMC Med 2011;9:133.
[51] von Kanel T, Huber AR. DNA methylation analysis. Swiss Med Wkly 2013;143:w13799.
[52] Whitelaw NC, Whitelaw E. Transgenerational epigenetic inheritance in health and disease. Curr Opin Genet Dev 2008;18:273–9.
[53] Peltomaki P. Lynch syndrome genes. Fam Cancer 2005;4:227–32.
[54] Gazzoli I, Loda M, Garber J, Syngal S, Kolodner RD. A hereditary nonpolyposis colorectal carcinoma case associated with hypermethylation of the MLH1 gene in normal tissue and loss of heterozygosity of the unmethylated allele in the resulting microsatellite instability-high tumor. Cancer Res 2002;62:3925–8.

[55] Chan TL, Yuen ST, Kong CK, Chan YW, Chan AS, Ng WF, et al. Heritable germline epimutation of MSH2 in a family with hereditary nonpolyposis colorectal cancer. Nat Genet 2006;38:1178–83.

[56] Ligtenberg MJ, Kuiper RP, Chan TL, Goossens M, Hebeda KM, Voorendt M, et al. Heritable somatic methylation and inactivation of MSH2 in families with Lynch syndrome due to deletion of the 3′ exons of TACSTD1. Nat Genet 2009;41:112–17.

[57] Venkatachalam R, Ligtenberg MJ, Hoogerbrugge N, Schackert HK, Gorgens H, Hahn MM, et al. Germline epigenetic silencing of the tumor suppressor gene PTPRJ in early-onset familial colorectal cancer. Gastroenterology 2010;139:2221–4.

[58] Hitchins MP, Rapkins RW, Kwok CT, Srivastava S, Wong JJ, Khachigian LM, et al. Dominantly inherited constitutional epigenetic silencing of MLH1 in a cancer-affected family is linked to a single nucleotide variant within the 5′UTR. Cancer Cell 2011;20:200–13.

[59] Gylling A, Ridanpaa M, Vierimaa O, Aittomaki K, Avela K, Kaariainen H, et al. Large genomic rearrangements and germline epimutations in Lynch syndrome. Int J Cancer 2009;124:2333–40.

[60] Niessen RC, Hofstra RM, Westers H, Lightenberg MJ, Kooi K, Jager PO, et al. Germline hypermethylation of MLH1 and EPCAM deletions are a frequent cause of Lynch syndrome. Genes Chromosomes Cancer 2009;48:737–44.

[61] Bennett KL, Mester J, Eng C. Germline epigenetic regulation of KILLIN in Cowden and Cowden-like syndrome. JAMA 2010;304:2724–31.

[62] Bennett KL, Campbell R, Ganapathi S, Zhou M, Rini B, Ganapathi R, et al. Germline and somatic DNA methylation and epigenetic regulation of KILLIN in renal cell carcinoma. Genes Chromosomes Cancer 2011;50:654–61.

[63] Enjuanes A, Fernandez V, Hernandez L, Navarro A, Bea S, Pinyol M, et al. Identification of methylated genes associated with aggressive clinicopathological features in mantle cell lymphoma. PLoS One 2011;6:e19736.

[64] Rahmatpanah FB, Carstens S, Hooshmand SI, Welsh EC, Sjahputera O, Taylor KH, et al. Large-scale analysis of DNA methylation in chronic lymphocytic leukemia. Epigenomics 2009;1:39–61.

[65] Ignatov A, Bischoff J, Ignatov T, Schwarzenau C, Krebs T, Kuester D, et al. APC promoter hypermethylation is an early event in endometrial tumorigenesis. Cancer Sci 2010;101:321–7.

[66] Suehiro Y, Okada T, Okada T, Anno K, Okayama N, Ueno K, et al. Aneuploidy predicts outcome in patients with endometrial carcinoma and is related to lack of CDH13 hypermethylation. Clin Cancer Res 2008;14:3354–61.

[67] Nieminen TT, Gylling A, Abdel-Rahman WM, Nuorva K, Aarnio M, Renkonen-Sinisalo L, et al. Molecular analysis of endometrial tumorigenesis: importance of complex hyperplasia regardless of atypia. Clin Cancer Res 2009;15:5772–83.

[68] Amara K, Trimeche M, Ziadi S, Laatiri A, Hachana M, Korbi S. Prognostic significance of aberrant promoter hypermethylation of CpG islands in patients with diffuse large B-cell lymphomas. Ann Oncol 2008;19:1774–86.

[69] Tong WG, Wierda WG, Lin E, Kuang SQ, Bekele BN, Estrov Z, et al. Genome-wide DNA methylation profiling of chronic lymphocytic leukemia allows identification of epigenetically repressed molecular pathways with clinical impact. Epigenetics 2010;5:499–508.

[70] Irving L, Mainou-Fowler T, Parker A, Ibbotson RE, Oscier DG, Strathdee G. Methylation markers identify high risk patients in IGHV mutated chronic lymphocytic leukemia. Epigenetics 2011;6:300–6.

[71] Sugita H, Iida S, Inokuchi M, Kato K, Ishiguro M, Ishikawa T, et al. Methylation of BNIP3 and DAPK indicates lower response to chemotherapy and poor prognosis in gastric cancer. Oncol Rep 2011;25:513–18.

[72] Pinheiro H, Bordeira-Carrico R, Seixas S, Carvalho J, Senz J, Oliveira P, et al. Allele-specific CDH1 downregulation and hereditary diffuse gastric cancer. Hum Mol Genet 2010;19:943–52.

[73] Lee SM, Lee EJ, Ko YH, Lee SH, Maeng L, Kim KM. Prognostic significance of O6-methylguanine DNA methyltransferase and p57 methylation in patients with diffuse large B-cell lymphomas. APMIS 2009;117:87–94.
[74] Salvesen HB, Das S, Akslen LA. Loss of nuclear p16 protein expression is not associated with promoter methylation but defines a subgroup of aggressive endometrial carcinomas with poor prognosis. Clin Cancer Res 2000;6:153–9.
[75] Furlan D, Carnevali I, Marcomini B, Cerutti R, Dainese E, Capella C, et al. The high frequency of de novo promoter methylation in synchronous primary endometrial and ovarian carcinomas. Clin Cancer Res 2006;12:3329–36.
[76] Kikuchi R, Tsuda H, Kanai Y, Kasamatsu T, Sengoku K, Hirohashi S, et al. Promoter hypermethylation contributes to frequent inactivation of a putative conditional tumor suppressor gene connective tissue growth factor in ovarian cancer. Cancer Res 2007;67:7095–105.
[77] Shi H, Guo J, Duff DJ, Rahmatpanah F, Chitima-Matsiga R, Al-Kuhlani M, et al. Discovery of novel epigenetic markers in non-Hodgkin's lymphoma. Carcinogenesis 2007;28:60–70.
[78] Chim CS, Pang R, Liang R. Epigenetic dysregulation of the Wnt signalling pathway in chronic lymphocytic leukaemia. J Clin Pathol 2008;61:1214–19.
[79] Fiegl H, Windbichler G, Mueller-Holzner E, Goebel G, Lechner M, Jacobs IJ, et al. HOXA11 DNA methylation – a novel prognostic biomarker in ovarian cancer. Int J Cancer 2008;123:725–9.
[80] Esteller M, Garcia-Foncillas J, Andion E, Goodman SN, Hidalgo OF, Vanaclocha V, et al. Inactivation of the DNA-repair gene MGMT and the clinical response of gliomas to alkylating agents. N Engl J Med 2000;343:1350–4.
[81] Zighelboim I, Goodfellow PJ, Schmidt AP, Walls KC, Mallon MA, Mutch DG, et al. Differential methylation hybridization array of endometrial cancers reveals two novel cancer-specific methylation markers. Clin Cancer Res 2007;13:2882–9.
[82] Liu TH, Raval A, Chen SS, Matkovic JJ, Byrd JC, Plass C. CpG island methylation and expression of the secreted frizzled-related protein gene family in chronic lymphocytic leukemia. Cancer Res 2006;66:653–8.
[83] Dietrich D, Kneip C, Raji O, Liloglou T, Seegebarth A, Schlegel T, et al. Performance evaluation of the DNA methylation biomarker SHOX2 for the aid in diagnosis of lung cancer based on the analysis of bronchial aspirates. Int J Oncol 2012;40:825–32.
[84] Subramaniam S, Thakur RK, Yadav VK, Nanda R, Chowdhury S, Agrawal A. Lung cancer biomarkers: State of the art. J Carcinog 2013;12:3.
[85] Belinsky SA. Gene-promoter hypermethylation as a biomarker in lung cancer. Nat Rev Cancer 2004;4:707–17.
[86] Esteller M, Sanchez-Cespedes M, Rosell R, Sidransky D, Baylin SB, Herman JG. Detection of aberrant promoter hypermethylation of tumor suppressor genes in serum DNA from non-small cell lung cancer patients. Cancer Res 1999;59:67–70.
[87] Schmidt B, Liebenberg V, Dietrich D, Schlegel T, Kneip C, Seegebarth A, et al. SHOX2 DNA methylation is a biomarker for the diagnosis of lung cancer based on bronchial aspirates. BMC Cancer 2010;10:600.
[88] Rahmatpanah FB, Carstens S, Guo J, Sjahputera O, Taylor KH, Duff D, et al. Differential DNA methylation patterns of small B-cell lymphoma subclasses with different clinical behavior. Leukemia 2006;20:1855–62.
[89] Taylor KH, Kramer RS, Davis JW, Guo J, Duff DJ, Xu D, et al. Ultradeep bisulfite sequencing analysis of DNA methylation patterns in multiple gene promoters by 454 sequencing. Cancer Res 2007;67:8511–18.
[90] Shaknovich R, Geng H, Johnson NA, Tsikitas L, Cerchietti L, Greally JM, et al. DNA methylation signatures define molecular subtypes of diffuse large B-cell lymphoma. Blood 2010;116:e81–89.

CHAPTER

3

DNA Methylation and Epigenetic Biomarkers in Non-Neoplastic Diseases

OUTLINE

M. Neidhart: DNA Methylation and Complex Human Disease.
DOI: http://dx.doi.org/10.1016/B978-0-12-420194-1.00003-8

3.1 INTRODUCTION

Epigenetics has emerged as a new and promising field in recent years. Lifestyle, stress, drugs, physiopathological situations, and pharmacological interventions have a great impact on the epigenetic code of the cells by altering the methylome, miRNA expression and the covalent histone modifications. A biomarker is a molecular target analyzed in a qualitative or quantitative manner to detect and diagnose the presence of a disease, to predict the outcome and the response to a specific treatment, allowing personalized tailoring of patient management. Since there exists a need to find new biomarkers and improve diagnosis for several diseases, the research on epigenetic biomarkers for molecular diagnostics encourages the translation of this field from the bench to clinical practice, not only in cancer (Chapter 2) but also in non-neoplastic diseases such as autoimmunity and developmental diseases. In this context, deciphering intricate epigenetic modifications involved in several molecular processes is a challenge that will be solved in the near future. Traditionally, proteins, measured either in serum, plasma, or the tissue of interest, have been used as biomarkers to identify individuals with increased predisposition to develop a disease, screen for early detection of disease, assess its gravity, or predict the response to a therapy. The increasing interest in the role of DNA methylation in the pathogenesis of various diseases has coincided with advances in platform-based DNA methylation array technologies, which have superseded candidate gene methylation profiling techniques (Chapters 28 and 29).

3.2 AUTOIMMUNITY

The use of DNA methylation as a biomarker is promising not only for cancer patients but also for individuals with autoimmune, inflammatory, or other complex diseases. Autoimmune and inflammatory diseases are characterized by the immune system's failure to recognize self-antigens. Genome-wide hypomethylation has been found to be associated with complex diseases such as type 1 diabetes [1], rheumatoid arthritis [2–4], systemic lupus erythematosus, systemic sclerosis (SSc), dermatomyositis, and ulcerative colitis [5], where the aberrant DNA methylation influences expression of pro-inflammatory genes and/or is combined with autoreaction of T helper cells.

3.2.1 Type 1 Diabetes

Epigenetic regulation has been proposed as a key mechanism by which environmental influences interact with genetic factors to trigger type 1 diabetes (T1D) [1,6]. Using whole blood samples, a genome-wide DNA methylation study [7] identified 19 potential CpG sites associated with risk for for T1D-associated nephropathy, many of them involved in transcription regulation: *NRBF2, RUNX3, COBRA1, ZBTB5,* and *ZNF639.* More recently, a genome-wide DNA methylation analysis in lymphocyte DNA from monozygotic twins discordant and concordant for T1D has been performed [8]. This identified six genes, known to be associated with T1D from previous studies, that showed differential CpG methylation between T1D affected and non-affected twins. Of these, *HLA-E* was the only T1D-associated gene showing hypomethylated CpG sites in affected twins; the differential methylation was in the promoter CpG island. Three genes, *HLA-DOB*, insulin (*INS*), and *IL-2RB*, showed hypermethylated variable positions in the majority of affected discordant and concordant twins (at least two out of three discordant twin pairs and at least four out of six concordant twin pairs), and two T1D genes (*HLA-DQA2* and *CD226*) showed hypermethylated CpG sites in affected discordant twins. Another study [9] also reported methylation-specific patterns at CpG sites upstream of the INS gene transcription start site in T1D patients. Interestingly, polymorphisms in the *IL-2RA* promoter are associated with the methylation of a specific CpG site (at position −373) [10].

3.2.2 Rheumatoid Arthritis

In rheumatoid arthritis (RA), *LINE-1* expression and promoter hypomethylation were the first reported observation that epigenetics play a crucial role in this disease [2]. Despite the fact that global hypomethylation is now a well-established event in autoimmune and inflammatory diseases, there is increased interest in the search for disease-specific hypermethylation of promoter regions. Hypermethylation of death receptor 3 (*DR3*) in a Japanese RA cohort [11] has linked epigenetics and disease phenotype, since gene silencing would prohibit DR3s normal function as an initiator of the intrinsic apoptotic network. One of the current hypotheses is that RA synovial fibroblasts are resistant to apoptosis, at least under certain circumstances. Other possible useful methylation-sensitive biomarkers in RA include CXCL12 [12], IL-6 [13], and microRNA-203 [14]. More on DNA methylation in RA is presented in Chapters 22 and 23.

3.2.3 Systemic Lupus Erythematosis

Many studies have indicated correlation of DNA methylation with the pathogenesis of systemic lupus erythematosis (SLE). A hypomethylated state of genes in T and B lymphocytes has been generally observed in SLE patients [15,16] – for instance, the lymphocytes from SLE patients display gene hypomethylation when compared to normal controls. Studies using demethylating agents indicate that DNA hypomethylation plays a pathophysiological role in SLE. The hypomethylated state and demethylated DNA fragments in the serum of SLE patients can induce the production of anti-DNA antibodies, which are involved in the pathophysiology of SLE [15,17]. The degree of hypomethylation is thought to be correlated with lupus disease activity. Numerous methylation-sensitive genes are over-expressed in lupus $CD4^+$ T cells. Also, $CD8^+$ T lymphocyte and NK cell-specific perforin (*PRF1*) was overexpressed due to hypomethylation modification of DNA in SLE patients [18], and overexpression of the serine/threonine protein phosphatase gene *PP2A* was reported in SLE patients [19]. Reduced CD5 expression caused by hypomethylation of an intracellularly expressed truncated CD5 variant (*CD5-E1B*) on the surface of B lymphocytes promotes autoreactivity in SLE [20]. The involvement of DNA methylation in the X chromosome provides a potential explanation for the female predominance in SLE. In $CD4^+$ T cells of female SLE patients, demethylation with 5-azacytidine results in overexpression of the CD40 ligand, which is a B lymphocyte co-stimulatory molecule encoded on the X chromosome. However, hypomethylation is not associated with SLE for male SLE patients due to the fact that the male X chromosome is demethylated under physiological conditions [21]. More observations regarding DNA methylation in SLE are presented in Chapters 19, 20, and 21.

3.2.4 Inflammatory Bowel Disease, Crohn's Disease, and Ulcerative Colitis

In inflammatory bowel disease (IBD), several key developments in molecular studies have led us from genetics to explore epigenetics. DNA methyltransferase (DNMT) 3A and 3B were identified as susceptibility genes [22]. DNA methylation changes in colonic epithelial cells that normally occur with aging are accelerated in IBD because of higher cell turnover in inflammation [23]. Increased age-related DNA methylation, observed in colon cells of patients with colitis, could lead to genetic instability and development of cancer. Increased DNA methylation has been shown in dysplastic and surrounding non-dysplastic colon tissues from patients with ulcerative colitis (UC) compared with control subjects or patients with UC who do not have dysplasia. Four of 15 loci associated with development of cancer (*CDH1*, *GDNF*, *HPP1*, and *MYOD1*) were differentially methylated in surgical resection samples from patients with active UC compared with those with quiescent mucosa [24]. *CDH1* encodes the cell adhesion molecule E-cadherin, which is associated with IBD-associated cancer. Hypermethylation of the CDH1 promoter region has been shown in dysplastic and cancerous gut samples of patients with UC compared with non-dysplastic tissues [25–27]. In addition, hypermethylation of *PAR-2* [28] and *MDR1* [29] in mucosal and ileum biopsies has been associated with UC. DNA methylation has also been reported in colonic tissue. A study [30] of intestinal biopsy samples from 20 monozygotic twins discordant for UC identified 61 differentially methylated loci, with several containing genes that regulate inflammation (for example, *CFI, SPINKK4, THY1/CD90*). This study had an interesting design in that after the loci were identified in the analysis of discordant monozygotic twins (to exclude differences in genetic factors), they were validated in an independent cohort.

To overcome the heterogeneity of cell types in tissues, a methylation-wide profiling study of whole rectal biopsy specimens from patients with active and quiescent UC and Crohn's disease (CD) was validated using isolated epithelial cells from rectal biopsy specimens [31]. Many differentially methylated genes were identified in whole tissue, encoding proteins including *DOK2* (involved in IL-4 mediated cell proliferation), *TAP1* (a major histocompatibility complex class I transport molecule), and members of the TNF family (*TNFSF4* and *TNFSF12*). ULK1 was methylated only in patients with CD; its product has a role in autophagy. Some genes identified as being differentially methylated in this study were identified previously as susceptibility genes; for example, *CDH1, ICAM3, IL8RA*, and *CARD9*.

Whole-blood DNA analysis revealed 50 genes showing significantly different levels of methylation between patients with IBD and controls [32], including some involved in immune system activation (*MAPK, RPIK3*, and *IL21R*). Ontology analysis highlighted several pathways associated with IBD, including immune system processes, immune response, and host response to bacteria, whereas canonical pathway analysis indicated the involvement of Th17 cell pathways. In addition, peripheral blood mononuclear cells (PBMCs) from patients with IBD showed hypermethylation at the *TEPP* locus, which encodes testes-, prostate-, and placenta-expressed [33]. A study that analyzed DNA methylation in Epstein-Barr virus-transformed B cells from 18 patients with IBD versus non-affected siblings identified 49 differentially methylated CpG sites. More than half the differentially methylated loci contained genes that regulate immune functions, including several (*BCL3, STAT3, OSM, STAT5*) involved in the IL-12 and IL-23 pathways [34].

3.3 METABOLIC DISEASES

3.3.1 Type 2 Diabetes

Multiple signaling systems and transcription factor cascades control pancreas development and endocrine cell fate determination. Epigenetic processes contribute to the control of this transcriptional hierarchy, involving both histone modifications and DNA methylation. Changes in DNA methylation in insulin resistance, type 2 diabetes (T2D), and obesity are presented in more detail in Chapters 10 and 11.

A pool-based study has established and identified differentially methylated sites in type 2 diabetes mellitus [35]. An in-depth analysis of a CpG site in the first intron of *FTO* showed significant hypomethylation in disease cases compared to controls. This effect was independent of the sequence polymorphisms in the region. Associations of the *FTO* risk allele with hypermethylation have now been demonstrated and replicated in several studies [35–37], but it remains unclear whether methylation is the causal link between the *FTO* risk allele and T2D [35].

Prospective studies have shown that low levels of circulating insulin-like growth factor binding protein-1 (IGFBP-1) are associated with the risk of type 2 diabetes. A recent study [38] using peripheral blood cells showed that increased *IGFBP1* promoter methylation and decreased IGFBP-1 serum levels are features of type 2 diabetes with a short duration. Furthermore, insulin-like growth factor-binding protein 7 (IGFBP-7) is able to interact with insulin-like growth factor 1 (IGF-1) as well as insulin. Thus, serum IGFBP-7 levels could be associated with insulin resistance in T2D. *GFBP7* DNA methylation levels are increased in men with newly diagnosed T2D [39]. The correlation between IGFBP-7 and IGFBP-1 suggests that low IGFBP-7 may be associated with insulin resistance. In peripheral blood leukocytes, other gene promoters that are hypermethylated in T2D include, for example, protein kinase C epsilon zeta (*PRKCZ*) [40].

Conversely, other gene promoters can be hypomethylated. Gastric inhibitory peptide (GIP) action is reduced in T2D; this is associated with striking hypomethylation in the GIP receptor (*GIPR*) promoter [41]. Decreased methylation in this promoter is associated with higher insulin resistance and higher fasting glucose.

Insulin secretion is enhanced upon the binding of glucagon-like peptide-1 (GLP-1) to its receptor (GLP1R) in pancreatic beta cells. In T2D, expression of GLP1R in pancreatic islets is reduced, caused by an increased methylation of its promoter [42].

An interesting example of the challenges involved in designing, executing, and interpreting candidate gene methylation studies comes from analyses targeting the leptin-responsive gene *POMC* (pro-opiomelanocortin). It encodes several key neuropeptides and hormones known to play a central role in body-weight regulation within the hypothalamus. Mutations in *POMC* are one cause of severe early-onset obesity, and the *POMC* region also harbors common genetic variants that are associated with obesity. A hypermethylated CpG site at the intron 2/exon 3 boundary of *POMC* identified in human peripheral blood leukocytes has shown replicated association with childhood obesity [43]. This hypermethylated position overlaps with a putative binding site for the histone acetyltransferase P300 complex, which is known to be involved in chromatin acetylation and gene activation, raising the idea that the methylation state influences P300 binding and thereby POMC expression. Although peripheral blood leukocytes are not likely to be directly involved in weight regulation, the suggestion is that similar methylation effects are present within the POMC-expressing neurons of the hypothalamic arcuate nucleus. The inaccessibility of this tissue means this hypothesis cannot be directly tested in humans.

3.3.2 Heart Failure

In a genome-wide methylation profile study of left ventricle tissue from heart failure (HF) patients [44], three differentially methylated angiogenesis-related loci have been identified and correlated to differential expression levels of the corresponding gene. Thus, hypermethylation of the 5′ regulatory region of platelet endothelial cell adhesion molecule 1 (*PECAM1*) and hypomethylation of the angiomotin-like protein 2 (*AMOTL2*) in failing hearts correlated with reduced expression of those genes, while hypermethylation within the Rho GTPase activating protein 24 gene (*ARHGAP24*) is correlated with increased expression of ARHGAP24 in failing hearts. Moreover, a follow-up study [45] generated a genome-wide DNA methylation map of human hearts and revealed a significant decrease in global promoter methylation of genes with increased expression in failing hearts. The genome-wide methylation profile of patients with idiopathic dilated cardiomyopathy was recently generated [46]. Such approach allowed the identification of two novel genes with differential methylation profiles between patient and control subjects: lymphocyte antigen 75 (*LY75*) and adenosine A2a receptor (*ADORA2A*). Furthermore, DNA methylation was found to be responsible for the hypermutability of distinct cardiac genes [47]. This is the case for the cardiac isoform of the myosin binding protein C gene (*MYBPC3*), which has a significantly higher level of exonic methylation of CpG sites than the skeletal isoform (*MYBPBC2*). This suggests that there are unique aspects of the *MXBPC3* gene or its epigenetic environment that are prone to generate genetic mutations. More about DNA methylation and cardiology is presented in Chapter 15.

3.4 DEVELOPMENTAL DISEASES

3.4.1 Imprinting Disorders

Imprinted genes are a small subset of genes (up to 200) in mammalian genomes that are, at least in one tissue, at specific time points of development, expressed in a parent-of-origin specific manner. Most imprinted genes are regulated by differentially methylated imprinting control regions. Loss of methylation of the imprinted region located at chromosome 11p15, which harbors *IGF2, CDKN1C,* and *H19,* is a major cause of Silver-Russell syndrome, while hypermethylation of the same region is found in a proportion of patients with Beckwith-Wiedemann syndrome (BWS) [48] (see also Chapters 4, 10, 13, and 14). Analysis of DNA methylation aberrations in the imprinting disorders Prader-Willi or Angelman syndrome (PW/AS) on 15q11−q13 is already today part of the clinical routine, and guidelines for the workflow of molecular genetic tests, including DNA methylation analysis, have been developed [49]. Accurate methylation of the *IGF2/H19* alleles is essential for correct development (Chapter 4).

DNA methylation analysis using quantitative technologies such as pyrosequencing, high-resolution melting analysis, or MLPA have been proposed as diagnostic tools for the screening of imprinting disorders and placental abnormalities [50]. However, it should be noted that only some of the cases will show aberrant DNA methylation profiles while the same phenotype can also be caused by a variety of genetic changes, such as microdeletions and/or uniparental disomies. For example, DNA methylation defects at the two imprinting centers at 11p15 are observed in only 60% of Beckwith-Wiedemann and Silver-Russell syndrome patients, requiring integrated genetic and epigenetic analyses to identify comprehensively the causative abnormalities [48]. More on DNA methylation and developmental diseases is presented in Chapter 13.

3.4.2 Arrhythmia

The *KCNQ1* gene is located on chromosome 11 in a region that contains a cluster of six genes that are expressed from either only the maternal or only the paternal allele. In mice, the KCNQ1 overlapping transcript (*KCNQ1ot1*) is transcribed from a promoter located in intron 10 of the *KCNQ1* gene. This promoter region is a CpG island and undergoes methylation on the maternal chromosome, preventing transcription and therefore allowing expression of the gene cluster. However, this promoter region is not methylated on the paternal chromosome, allowing expression of the regulatory transcript and suppressing the expression of the gene cluster [51]. The maternal allele is transcribed in early embryogenesis, with the paternal allele being progressively methylated and therefore only activated during late embryogenesis.

In mice, variable imprinting of the *KCNQ1* gene provides a possible explanation for the existence of long QT syndrome (LQTS) in the absence of a coding sequence mutation in *KCNQ1*. Paternal imprinting is probably relieved in cardiac tissue, meaning that, during differentiation, methylation of the paternal chromosome must occur to block production of the suppressive KCNQot1 transcript. Mutations that disrupt the CpG island could prevent methylation and silence the paternal allele in the heart [51,52]. A more recent study [53] associates epigenetic modifications with regulation of the ATP-sensitive potassium channel (*KATP*). In murine cardiac myocytes, different isoform combinations of the SURx (*SUR1*, *SUR2*) and Kir6.2 (*KCNJ11*) are responsible for distinct physiological and pharmacological properties, depending on the isoforms expressed. Promotor CpG methylation appears to be one of the regulators of SURx isoform expression, and therefore regulated or aberrant CpG methylation might play a role in controlling channel structure and function under different conditions. This has yet to be confirmed in humans.

3.5 PSYCHIATRIC DISEASES

3.5.1 Major Depressive Disorder

Epigenetic modification of gene expression provides a mechanism for understanding the link between long-term effects of adverse life events and the changes in gene expression that are associated with depression [54]. Although still a developing field, in the future epigenetic modifications of gene expression may provide novel biomarkers to predict future susceptibility and/or onset of major depressive disorder (MDD), improve diagnosis, and aid in the development of epigenetics-based therapies for depression [54].

Alterations in DNA methylation of promoter regions of *BDNF* have been reported in the blood of patients with major depressive disorder (MDD) [55]. In the Wernicke's area in brains of suicide victims, some of whom had a diagnosis of depression, an increase in DNA methylation at brain-derived neurotrophic factor (BDNF) promoter occurred that is associated with decreased *BDNF* mRNA [56]. In addition, decreased trombomyosin-related kinase B (Trk B-T1) mRNA (a receptor for BDNF) in the frontal cortex is associated with increased methylation of CpG sites in the promoter region in suicide completers [57], some of whom had depression.

3.5.2 Schizophrenia

Post-mortem DNA methylation analysis of human brains of individuals diagnosed with schizophrenia has suggested the involvement of the genes *RELN* [58], *SOX10* [59], *HTR2A* [60], and *FOXP2* [61] in the disease phenotype, as these were found to be hypermethylated compared to normal healthy individuals. In contrast, the promoter region of MB-COMT has been found hypomethylated [62]. However, it still has to be established whether these differential methylated regions are the drivers of the disease, or a second event due to disease state, drug effects, or an abnormal lifestyle of the patient. Furthermore, whether non-invasive epigenetic biomarkers can be used in this disease has yet to be explored. More about DNA methylation and psychiatric diseases is presented in Chapter 17.

3.6 CONCLUSION

Although there is a vast potential for DNA methylation-based biomarkers and some promising candidates have been identified in research laboratories (Table 3.1), their use in clinical settings is still very limited. This is in great part due to the material that is needed for analysis requiring an invasive procedure (i.e., biopsy). Biomarkers from peripheral blood are useful in a limited number of diseases, particularly autoimmune diseases, such as T1D and SLE. Unfortunately, this is rarely clinical relevant for infiltrative diseases, such as RA and IBD/UC, where a biopsy is essential. Often, post-mortem samples are analyzed in cardiology and psychiatry, although, interesting, there is a long way to go until a DNA methylation-based biomarker can be used in clinical practice. In addition, there is often an insufficient sensitivity and specificity required for a diagnostic test, which might be caused by our so-far limited knowledge on the complexity of DNA methylation patterns, their heterogeneity, and the most predictive CpGs in a gene of interest. Indeed, research on the use of DNA methylation-based biomarkers in complex diseases is still at a very early stage, requiring far more well-designed investigations. However, the recent advances in technology permitting the precise assessment of more subtle changes in DNA methylation patterns, the rise of interest in epigenetic epidemiology [63] (Chapter 5), and statistical approaches to analyze DNA methylation based quantitative traits will lead to rapid catch-up in the field of epigenomics for non-neoplastic diseases. Nonetheless, as changes will probably frequently be found outside CpG islands, the large interindividual variation in

TABLE 3.1 Potential DNA Methylation Biomarkers in Non-Neoplastic Diseases

T1D	CD4$^+$ T cells	Global hypomethylation [1]
T1D	Blood cells	*NRBF2, PIGU, RUNX3, COBRA1, MAP3K9, MFSD3, HIT1H3I, DAPK3, KTI12, UNC13B, ZBTB5, TRPS1, DCUN1D4, SFXN4, PPAPR3, CCNB2, DOC2A, ZNF639, VPS26* [7]
T1D	PBL	*CD226, HLA-E, HLA-DOB, HLA-DQA2* [8], *INS* [8,9], *IL-2RA* [10], *IL-2RB* [8]
RA	Synovial fibroblasts	Global hypomethylation [2–4], *LINE-1* [2,3], *DR3* [11], *CXCL12* [12], *IL-6* [13], *microRNA-203* [14]
SLE	CD4$^+$ T cells	Global hypomethylation [5], *CD40L* [21]
SLE	CD8$^+$ T cells	*PRF1* [18], *PP2A* [19]
SLE	B cells	*CD5-E1B* [20]
SSc	CD4$^+$ T cells	Global hypomethylation [5]
IBD/UC	Biopsies	Global hypomethylation [5], *CDH1, GDNF, HPP1, MYOD1* [23], *PAR-2* [28], *MDR1* [29], *CFI, SPINKK4, THY1/CD90* [30], *THRAP2, FANCC, TNFSF4, TNFSF12, FUT7, CARD9, ICAM3, IL8RB* [31]
IBD/UC	Blood cells	*MAPK13, FASLG, PRF1, S100A13, RIPK3, IL-21R* [32], *TEPP* [33], *BGN, SERPINA, TNFSF1A, AATK, GABRA5, MAPK10, STAT5A* [34]
T2D	PBL	*FTO* [35–37], *IGFBP-1* [38], *IGFBP-7* [39], *GIPR* [41], *PRKCZ* [40], *POMC* [43]
T2D	Pancreatic islets	*GLP1R* [42]
HF	Heart	*PECAM1, AMOTL2, ARHGAP24* [44], *LY75, ADORA2A* [46], *MYBPC3* [47]
BWS	Biopsies	*IGF2, H19, CDKN1C* [48], *KCNQ1OT1* [50]
PW/AS	Biopsies	*SNRPN* [49]
MDD	Brain	*BDNF* [55,56], *Trk B-T1* [57]
Schiz	Brain	*RELN* [58], *SOX10* [59], HTR2A [60], *FOXP2* [61], *MB-COMT* [62]

Arryth, arrythmia; AS, Angelman syndrome; BWS, Beckwith-Wiedemann syndrome; HF, heart failure; IBD, inflammatory bowel disease; MDD, major depressive disorders; PBL, peripheral blood leukocytes; PW, Prader-Willi syndrome; RA, rheumatoid arthritis; Schiz, schizophrenia; SLE, systemic lupus erythematosus; SSc, systemic sclerosis; T1D, type 1 diabetes; T2D, type 2 diabetes; UC, ulcerative colitis.

DNA methylation in these regions needs to be taken into account [64] and requires adequately powered epigenetic association studies as well as high-resolution DNA methylation technologies [65] (Chapters 28–30).

References

[1] Richardson B. Effect of an inhibitor of DNA methylation on T cells. II. 5-Azacytidine induces self-reactivity in antigen-specific $T4^+$ cells. Hum Immunol 1986;17:456–70.
[2] Neidhart M, Rethage J, Kuchen S, Kunzler P, Crowl RM, Billingham ME, et al. Retrotransposable L1 elements expressed in rheumatoid arthritis synovial tissue: association with genomic DNA hypomethylation and influence on gene expression. Arthritis Rheum 2000;43:2634–47.
[3] Karouzakis E, Gay RE, Michel BA, Gay S, Neidhart M. DNA hypomethylation in rheumatoid arthritis synovial fibroblasts. Arthritis Rheum 2009;60:3613–22.
[4] Liu CC, Fang TJ, Ou TT, Wu CC, Li RN, Lin YC, et al. Global DNA methylation, DNMT1, and MBD2 in patients with rheumatoid arthritis. Immunol Lett 2011;135:96–9.
[5] Lei W, Luo Y, Lei W, Luo Y, Yan K, Zhao S, et al. Abnormal DNA methylation in $CD4^+$ T cells from patients with systemic lupus erythematosus, systemic sclerosis, and dermatomyositis. Scand J Rheumatol 2009;38:369–74.
[6] MacFarlane AJ, Strom A, Scott FW. Epigenetics: deciphering how environmental factors may modify autoimmune type 1 diabetes. Mamm Genome 2009;20:624–32.
[7] Bell CG, Teschendorff AE, Rakyan VK, Maxwell AP, Beck S, Savage DA. Genome-wide DNA methylation analysis for diabetic nephropathy in type 1 diabetes mellitus. BMC Med Genomics 2010;3.
[8] Stefan M, Zhang W, Concepcion E, Yi Z, Tomer Y. DNA methylation profiles in type 1 diabetes twins point to strong epigenetic effects on etiology. J Autoimmun. 2013;50: 33–7.
[9] Fradin D, Le Fur S, Mille C, Naoui N, Groves C, Zelenika D, et al. Association of the CpG methylation pattern of the proximal insulin gene promoter with type 1 diabetes. PLoS One 2012;7:e36278.
[10] Belot MP, Fradin D, Mai N, Le Fur S, Zelenika D, Kerr-Conte J, et al. CpG methylation changes within the IL2RA promoter in type 1 diabetes of childhood onset. PLoS One 2013;8:e68093.
[11] Takami N, Osawa K, Miura Y, Komai K, Taniguchi M, Shiraishi M, et al. Hypermethylated promoter region of DR3, the death receptor 3 gene, in rheumatoid arthritis synovial cells. Arthritis Rheum 2006;54:779–87.
[12] Karouzakis E, Rengel Y, Jungel A, Kolling C, Gay RE, Michel BA, et al. DNA methylation regulates the expression of CXCL12 in rheumatoid arthritis synovial fibroblasts. Genes Immun 2011;12.
[13] Nile CJ, Read RC, Akil M, Duff GW, Wilson AG. Methylation status of a single CpG site in the IL6 promoter is related to IL6 messenger RNA levels and rheumatoid arthritis. Arthritis Rheum 2008;58:2686–93.
[14] Stanczyk J, Ospelt C, Karouzakis E, Filer A, Raza K, Kolling C, et al. Altered expression of microRNA-203 in rheumatoid arthritis synovial fibroblasts and its role in fibroblast activation. Arthritis Rheum 2011;63:373–81.
[15] Renaudineau Y, Youinou P. Epigenetics and autoimmunity, with special emphasis on methylation. Keio J Med 2011;60:10–16.
[16] Zouali M. Epigenetics in lupus. Ann NY Acad Sci 2011;1217:154–65.
[17] Hedrich CM, Tsokos GC. Epigenetic mechanisms in systemic lupus erythematosus and other autoimmune diseases. Trends Mol Med 2011;17:714–24.
[18] Kaplan MJ, Lu Q, Wu A, Attwood J, Richardson B. Demethylation of promoter regulatory elements contributes to perforin overexpression in $CD4^+$ lupus T cells. J Immunol 2004;172:3652–61.
[19] Sunahori K, Juang YT, Kyttaris VC, Tsokos GC. Promoter hypomethylation results in increased expression of protein phosphatase 2A in T cells from patients with systemic lupus erythematosus. J Immunol 2011;186:4508–17.

[20] Garaud S, Le Dantec C, Jousse-Joulin S, Hanrote-Saliou C, Saraux A, Mageed RA, et al. IL-6 modulates CD5 expression in B cells from patients with lupus by regulating DNA methylation. J Immunol 2009;182:5623–32.
[21] Lu Q, Wu A, Tesmer L, Ray D, Yousif N, Richardson B. Demethylation of CD40LG on the inactive X in T cells from women with lupus. J Immunol 2007;179:6352–8.
[22] Franke A, McGovern DP, Barrett JC, Wang K, Radford-Smith GL, Ahmad T, et al. Genome-wide meta-analysis increases to 71 the number of confirmed Crohn's disease susceptibility loci. Nat Genet 2010;42:1118–25.
[23] Issa JP, Ahuja N, Toyota M, Bronner MP, Brentnall TA. Accelerated age-related CpG island methylation in ulcerative colitis. Cancer Res 2001;61:3573–7.
[24] Saito S, Kato J, Hiraoka S, Horii J, Suzuki H, Higashi R, et al. DNA methylation of colon mucosa in ulcerative colitis patients: correlation with inflammatory status. Inflamm Bowel Dis 2011;17:1955–65.
[25] Barrett JC, Lee JC, Lees CW, Prescott NJ, Anderson CA, Phillips A, et al. Genome-wide association study of ulcerative colitis identifies three new susceptibility loci, including the HNF4A region. Nat Genet 2009;41:1330–4.
[26] Azarschab P, Porschen R, Gregor M, Blin N, Holzmann K. Epigenetic control of the E-cadherin gene (CDH1) by CpG methylation in colectomy samples of patients with ulcerative colitis. Genes Chromosomes Cancer 2002;35:121–6.
[27] Wheeler JM, Kim HC, Efstathiou JA, Ilyas M, Mortensen NJ, Bodmer WF. Hypermethylation of the promoter region of the E-cadherin gene (CDH1) in sporadic and ulcerative colitis associated colorectal cancer. Gut 2001;48:367–71.
[28] Tahara T, Shibata T, Nakamura M, Yamashita H, Yoshioka D, Okubo M, et al. Promoter methylation of protease-activated receptor (PAR2) is associated with severe clinical phenotypes of ulcerative colitis (UC). Clin Exp Med 2009;9:125–30.
[29] Tahara T, Shibata T, Nakamura M, Yamashita H, Yoshioka D, Okubo M, et al. Effect of MDR1 gene promoter methylation in patients with ulcerative colitis. Int J Mol Med 2009;23:521–7.
[30] Hasler R, Feng Z, Backdahl L, Spehlmann ME, Franke A, Teschendorff A, et al. A functional methylome map of ulcerative colitis. Genome Res 2012;22:2130–7.
[31] Cooke J, Zhang H, Greger L, Silva AL, Massey D, Dawson C, et al. Mucosal genome-wide methylation changes in inflammatory bowel disease. Inflamm Bowel Dis 2012; 18:2128–37.
[32] Nimmo ER, Prendergast JG, Aldhous MC, Kennedy NA, Henderson P, Drummond HE, et al. Genome-wide methylation profiling in Crohn's disease identifies altered epigenetic regulation of key host defense mechanisms including the Th17 pathway. Inflamm Bowel Dis 2012;18:889–99.
[33] Harris RA, Nagy-Szakal D, Pedersen N, Opekun A, Bronsky J, Munkholm P, et al. Genome-wide peripheral blood leukocyte DNA methylation microarrays identified a single association with inflammatory bowel diseases. Inflamm Bowel Dis 2012;18:2334–41.
[34] Lin Z, Hegarty JP, Yu W, Cappel JA, Chen X, Faber PW, et al. Identification of disease-associated DNA methylation in B cells from Crohn's disease and ulcerative colitis patients. Dig Dis Sci 2012;57:3145–53.
[35] Toperoff G, Aran D, Kark JD, Rosenberg M, Dubnikov T, Nissan B, et al. Genome-wide survey reveals predisposing diabetes type 2-related DNA methylation variations in human peripheral blood. Hum Mol Genet 2012;21:371–83.
[36] Bell CG, Finer S, Lindgren CM, Wilson GA, Rakyan VK, Teschendorff AE, et al. Integrated genetic and epigenetic analysis identifies haplotype-specific methylation in the FTO type 2 diabetes and obesity susceptibility locus. PLoS One 2010;5:e14040.
[37] Almen MS, Jacobsson JA, Moschonis G, Benedict C, Chrousos GP, Fredriksson R, et al. Genome wide analysis reveals association of a FTO gene variant with epigenetic changes. Genomics 2012;99:132–7.

[38] Gu T, Gu HF, Hilding A, Sjoholm LK, Ostenson C-G, Ekstrom TJ, et al. Increased DNA methylation levels of the insulin-like growth factor binding protein 1 gene are associated with type 2 diabetes in Swedish men. Clin Epigenetics 2013;5:21.
[39] Gu HF, Gu T, Hilding A, Zhu Y, Karvestedt L, Ostenson C-G, et al. Evaluation of IGFBP-7 DNA methylation changes and serum protein variation in Swedish subjects with and without type 2 diabetes. Clin Epigenetics 2013;5:20.
[40] Zou L, Yan S, Guan X, Pan Y, Qu X. Hypermethylation of the PRKCZ Gene in Type 2 Diabetes Mellitus. J Diabetes Res 2013;2013:721493.
[41] Canivell S, Ruano EG, Siso-Almirall A, Kostov B, Gonzalez-de Paz L, Fernandez-Rebollo E, et al. Gastric inhibitory polypeptide receptor methylation in newly diagnosed, drug-naive patients with type 2 diabetes: a case–control study. PLoS One 2013;8:e75474.
[42] Hall E, Dayeh T, Kirkpatrick CL, Wollheim CB, Nitert MD, Ling C. DNA methylation of the glucagon-like peptide 1 receptor (GLP1R) in human pancreatic islets. BMC Med Genet 2013;14:76.
[43] Kuehnen P, Mischke Mo, Wiegand S, Sers C, Horsthemke B, Lau S, et al. An Alu element-associated hypermethylation variant of the POMC gene is associated with childhood obesity. PLoS Genet 2012;8:e1002543.
[44] Movassagh M, Choy MK, Goddard M, Bennett MR, Down TA, Foo RS-Y. Differential DNA methylation correlates with differential expression of angiogenic factors in human heart failure. PLoS One 2010;5:e8564.
[45] Movassagh M, Choy MK, Knowles DA, Cordeddu L, Haider S, Down T, et al. Distinct epigenomic features in end-stage failing human hearts. Circulation 2011;124: 2411–22.
[46] Haas J, Frese KS, Park YJ, Keller A, Vogel B, Lindroth AM, et al. Alterations in cardiac DNA methylation in human dilated cardiomyopathy. EMBO Mol Med 2013;5: 413–29.
[47] Meurs KM, Kuan M. Differential methylation of CpG sites in two isoforms of myosin binding protein C, an important hypertrophic cardiomyopathy gene. Environ Mol Mutagen 2011;52:161–4.
[48] Demars J, Gicquel C. Epigenetic and genetic disturbance of the imprinted 11p15 region in Beckwith-Wiedemann and Silver-Russell syndromes. Clin Genet 2012;81: 350–61.
[49] Ramsden SC, Clayton-Smith J, Birch R, Buiting K. Practice guidelines for the molecular analysis of Prader-Willi and Angelman syndromes. BMC Med Genet 2010;11:70.
[50] Alders M, Bliek J, vd Lip K, vd Bogaard R, Mannens M. Determination of KCNQ1OT1 and H19 methylation levels in BWS and SRS patients using methylation-sensitive high-resolution melting analysis. Eur J Hum Genet 2009;17: 467–73.
[51] Mancini-DiNardo D, Steele SJ, Ingram RS, Tilghman SM. A differentially methylated region within the gene Kcnq1 functions as an imprinted promoter and silencer. Hum Mol Genet 2003;12:283–94.
[52] Bokil NJ, Baisden JM, Radford DJ, Summers KM. Molecular genetics of long QT syndrome. Mol Genet Metab 2010;101:1–8.
[53] Fatima N, Schooley Jr. JF, Claycomb WC, Flagg TP. Promoter DNA methylation regulates murine SUR1 (Abcc8) and SUR2 (Abcc9) expression in HL-1 cardiomyocytes. PLoS One 2012;7:e41533.
[54] Dalton VS, Kolshus E, McLoughlin DM. Epigenetics and depression: return of the repressed. J Affect Disord 2013;155:1–12.
[55] Fuchikami M, Morinobu S, Segawa M, Okamoto Y, Yamawaki S, Ozaki N, et al. DNA methylation profiles of the brain-derived neurotrophic factor (BDNF) gene as a potent diagnostic biomarker in major depression. PLoS One 2011;6:e23881.

[56] Keller S, Sarchiapone M, Zarrilli F, Videtic A, Ferraro A, Carli V, et al. Increased BDNF promoter methylation in the Wernicke area of suicide subjects. Arch Gen Psychiatry 2010;67:258–67.

[57] Ernst C, Deleva V, Deng X, Sequeira A, Pomarenski A, Klempan T, et al. Alternative splicing, methylation state, and expression profile of tropomyosin-related kinase B in the frontal cortex of suicide completers. Arch Gen Psychiatry 2009;66:22–32.

[58] Abdolmaleky HM, Cheng KH, Russo A, Smith CL, Faraone SV, Wilcox M, et al. Hypermethylation of the reelin (RELN) promoter in the brain of schizophrenic patients: a preliminary report. Am J Med Genet B Neuropsychiatr Genet 2005;134B: 60–6.

[59] Iwamoto K, Bundo M, Yamada K, Takao H, Iwayama-Shigeno T, Toshikawa T, et al. DNA methylation status of SOX10 correlates with its downregulation and oligodendrocyte dysfunction in schizophrenia. J Neurosci 2005;25:5376–81.

[60] Abdolmaleky HM, Yagubi S, Papageorgis P, Lambert AW, Ozturk S, Sivaraman V, et al. Epigenetic dysregulation of HTR2A in the brain of patients with schizophrenia and bipolar disorder. Schizophr Res 2011;129:183–90.

[61] Tolosa A, Sanjuan J, Dagnall AM, Molto MD, Herrero N, de Frutos R. FOXP2 gene and language impairment in schizophrenia: association and epigenetic studies. BMC Med Genet 2010;11:114.

[62] Abdolmaleky HM, Cheng KH, Faraone SV, Wilcox M, Glatt SJ, Gao F, et al. Hypomethylation of MB-COMT promoter is a major risk factor for schizophrenia and bipolar disorder. Hum Mol Genet 2006;15:3132–45.

[63] Christensen BC, Marsit CJ. Epigenomics in environmental health. Front Genet 2011;2:84.

[64] Bock C, Walter J, Paulsen M, Lengauer T. Inter-individual variation of DNA methylation and its implications for large-scale epigenome mapping. Nucleic Acids Res 2008;36:e55.

[65] Rakyan VK, Down TA, Balding DJ, Beck S. Epigenome-wide association studies for common human diseases. Nat Rev Genet 2011;12:529–41.

CHAPTER

4

DNA Methylation and Environmental Factors

OUTLINE

M. Neidhart: DNA Methylation and Complex Human Disease.
DOI: http://dx.doi.org/10.1016/B978-0-12-420194-1.00004-X

4.1 INTRODUCTION

A growing body of evidence has drawn the attention of the scientific community by indicating the potential vulnerability to environmental changes of epigenetic mechanisms that control gene expression. Being critical components of normal development, the importance of epigenetic mechanisms for normal biology is illustrated by the fact that abnormal epigenetic patterns have increasingly been linked to the etiology of various diseases, including cancer, pediatric syndromes, genetic disorders, autoimmune diseases, rheumatic diseases, and even the molecular process of aging. Very important to note is that the degree of vulnerability to changes in epigenetic patterns is high during early embryonic development – a period of life in which epigenetic patterns are established – and cell differentiation. Moreover, increasing amounts of relevant data and information reveal that the environment might potentially impact on epigenetic patterns at every period of life. A comparative study between monozygotic twins of different ages (3 and 50 years) showed that epigenetic differences among twins develop throughout life as a consequence of differential environment exposure [1]. Nevertheless, the questions concerning how, when, and which environmental factors can induce epigenetic modifications and modulate aging and/or predisposition to disease remain for the most part unanswered. Within this context, this chapter reviews the principles of epigenetic vulnerability to environmental changes, the impacts on development, and the association with the origin of common diseases, and also speculates on the potential of lifestyle changes to modulate epigenetic patterns and contribute to preventing common diseases. Indeed, DNA methylation can be modified by environmental factors [2], including nutritional, infectious, and sociological factors. Environments acting during early embryogenesis (e.g., when global erasure and re-establishment of DNA methylation occur) may induce extensive, soma-wide modifications of DNA methylation, whereas environments acting later during life are more likely to induce less extensive, tissue-specific modifications of DNA methylation. The former may be involved in fetal programming of adult disorders [3], whereas the latter may play a role in tissue-specific carcinogenesis [4].

4.2 IMPACT OF THE ENVIRONMENT IN EARLY LIFE

A complex phenomenon of epigenetic programming characterized by modifications of DNA methylation pattern and histone modifications that are crucial for the control of developing events, such as the establishment of totipotency of stem cells, cellular fate, and early organogenesis, takes place at the beginning of embryo development in mammals [5,6]. A few hours after fertilization, the male and female pronuclei undergo a reprogramming process marked by a wave of genomic demethylation, which is important for establishing the totipotency and the reprogramming of somatic cells of the embryo gene expression. However, some specific genomic loci are protected against this demethylation wave and maintain the parental (gamete) pattern of methylation in somatic cells. The protected loci correspond to differentially methylated regions (DMRs) (methylated only in the paternal or maternal inherited allele) involved in the control of genomic imprinting [7].

4.2.1 Genomic Imprinting

Genomic imprinting is an epigenetically controlled mechanism in which gene expression depends on the parental origin and only one of the two alleles (paternally or maternally inherited) is expressed. At a functional level, imprinted regions are effectively haploid, which makes them vulnerable to recessive mutations and epigenetic changes. Approximately 85 imprinted genes have been identified to date, with most being associated with the control of fetal growth, the transfer of nutrients from mother to fetus through the placenta, and brain functions. Generally, imprinted genes are mapped to clusters and at least one DMR is included among them, suggesting the existence of a similar control mechanism for the various imprinted regions [7,8]. Acquisition of the methylation patterns of male and female differentially methylated regions (DMRs) occurs during gametogenesis at developmentally different stages. For example, DNA methylations at the paternal loci, *H19DMR* [9], RAS protein-specific guanine nucleotide-releasing factor 1 (*Rasgrf1*), and gene trap locus 2 (*Gtl2*) [10], in male germlines are initiated during germ cell development in embryogenesis, and are completed by the pachytene phase of postnatal spermatogenesis in mice. On the other hand, methylations at the maternal loci, insulin-like growth factor 2 receptor (*IGF2R*), small nuclear ribonucleoprotein N (Snrnp1), paternally expressed 1 (*Peg1*), and paternally expressed 3 (*Peg3*), in female germlines are acquired asynchronously in a gene-specific manner only during oocyte maturation in the postnatal growth phase [7].

4.2.2 Beckwith-Wiedemann Syndrome

The Beckwith-Wiedemann syndrome (BWS) is prototypic of genomic imprinting disease, and is characterized by overgrowth, macroglossy, abdominal wall defects, and predisposition to embryonal cancer

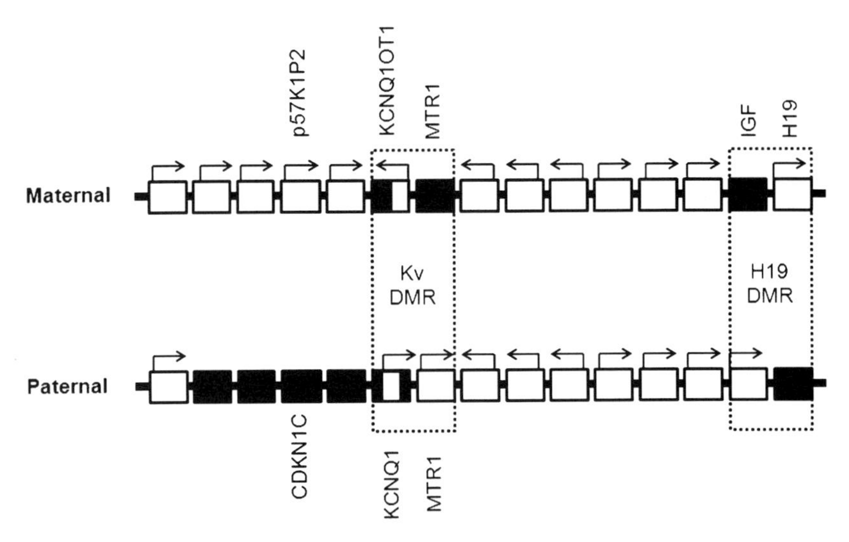

FIGURE 4.1 The imprinted region of human chromosome 11p15.1 (open boxes: expressed genes, black boxes: silenced genes by promoter methylation). This complex pattern has to be established during embryogenesis. Dyfunctions lead to diseases, such as the Beckwith-Wiedemann syndrome.

(mainly Wilms tumor in the kidney). The etiology of BWS is attributed to epigenetic alterations in a cluster of imprinted genes mapped to human chromosome 11p15.5 (Figure 4.1). This imprinted cluster is divided between centromeric and telomeric regions. The telomeric region harbors the paternally expressed *IGF*2 gene and the maternally expressed *H19* gene, whereas the centromeric region harbors the paternally potassium voltage-gated channel 1 opposite strand/antisense transcript 1 (*KCNQ1OT1*) and the maternally expressed cyclin-dependent kinase inhibitor 1C (*CDKN1C*) and potassium voltage-gated channel 1 (*KCNQ1*) genes. Imprinted gene expression patterns at the telomeric and centromeric regions are controlled by the differential methylation of the H19DMR and KvDMR, respectively. H19DMR is commonly methylated on the paternally inherited allele, whereas Beckwith-Wiedemann is methylated on the maternal allele [8,11]. Approximately 20% of all BWS cases result from paternal uniparental disomy, whereas paternally derived duplications and maternally derived translocations of 11p15.5 account for approximately 2% of the cases. Mutations in the maternally derived allele of the *CDKN1C* gene (also known as $p57^{kip2}$) are found in approximately 10% of sporadic cases. Aberrant methylation at the paternal imprinting control region H19DMR has been found in approximately 2–7% of cases, and the loss of methylation at the maternal allele of KvDMR ICR – an intronic CpG island within the *KCNQ1* gene – is the most frequent alteration, found in approximately 50% of cases. Etiology is unknown for 10–15% of BWS patients [12]. Chapter 13 further underlines the importance of imprinting and DNA methylation in developmental disorders.

4.2.3 Assisted Reproductive Technologies

Early studies in mouse models have drawn the attention of the scientific community by providing evidence of vulnerabilities of embryo reprogramming of genomic imprinting to changes in the embryo culture medium [13]. The above-mentioned association between *in vitro* reproductive technologies and epigenetic alterations was confirmed by epidemiological studies containing alarming data, reporting a four- to nine-fold increased risk for BWS observed in children conceived by assisted reproductive technologies (ART) when compared to those naturally conceived [14]. In theory, the association between epigenetic alterations and *in vitro* reproduction is based on the fact that some procedures used in ART, such as gamete handling and embryo culture, take place at two biological moments that are of vital importance in epigenetic programming: gametogenesis and postfertilization [5].

Superovulation, a common procedure in IVF protocols, has been demonstrated to induce the formation of oocytes without their correct primary imprint in both human and mouse models, thereby providing support to the proposition that hormonal induction of the ovulation with high doses of gonadotropins might generate epigenetically "immature" oocytes with incomplete methylation acquisition [15]. On the other hand, observations that abnormal methylation in BWS patients conceived by ART is not confined to 11p15 or to a specific parental inheritance serve as an argument in favor of the hypothesis that an epigenetic error might occur after fertilization, most likely due to a failure in maintaining parental-specific methylation marks [16].

4.2.4 Large Offspring Syndrome

The association between ART and abnormal epigenetics is also evidenced in bovine models conceived by IVF, with increased incidence of large offspring syndrome (LOS) – a pathology characterized by high birth weight, respiratory problems, and an increase in prenatal death – caused by loss of imprinting of the IGF2R gene being commonly observed [17]. Comparative analysis of epigenetic errors occurring in consequence of *in vitro* fertilization in both human and bovine models is important in order to exclude the influence of the parental genetic background on the origin of the epigenetic alterations, since, contrary to humans, in whom IVF is an alternative to fertility problems, in bovines the best parental genetic backgrounds are chosen and IVF is generally used as a tool for genetic improvement of offspring and genetic gain. Thus, the observation of LOS in bovine models provides clear evidence that epigenetic abnormalities consequent to *in vitro* reproduction are not dependent on an impaired parental genetic background or the cause of infertility, highlighting the influence of procedures used in assisted fertilization (such as gamete handling and embryo culture) on epigenetic acquisition or maintenance [5].

4.3 NUTRITIONAL FACTORS

4.3.1 Gestational Maternal Nutritional Factors

Studies to date have suggested that certain dietary components, such as folate, polyphenols, retinoids, fatty acids, isothiocyanates, and allyl compounds, can alter DNA methylation and histone modification levels in both systemic and target tissues [18–20].

During the DNA methylation process, the methyl group is obtained from the methyl donor SAM by DNMT enzymes, which convert SAM to S-adenosylhomocysteine. SAM levels are dependent of dietary intake, since mammals cannot synthesize major sources of one-carbon unit methionine and choline, or critical co-factors for methyl group metabolism, such as folic acid, vitamin B12, and pyridoxal phosphate. Thus, an unbalanced ratio (excess or deficit) of the major sources of one-carbon units and co-factors for methyl group metabolism may alter the supply of methyl groups and consequently establish links between nutritional aspects and epigenetic alterations. This mechanism is presented in detail in Chapter 1.

The viable yellow agouti (A^{vy}) is an interesting mouse model that has been comprehensively explored over the past few years to demonstrate the effects of gestational maternal nutritional factors and characteristics in the epigenetic programming of offspring (Figure 4.2). Basically, the agouti gene is genetically and epigenetically regulated by an intracisternal A particle retrotransposon insertion which contains a promoter that drives ectopic agouti expression and affects coat color. Using the A^{vy} model, it has been demonstrated that the excess of methyl donor supplementation before and during pregnancy can alter the methylation pattern at specific loci, indicating a direct influence of maternal nutrition on the offspring's epigenetic mechanisms [21].

In humans, the association between early life and the origin of late-onset diseases lies at the basis of the burgeoning Developmental Origin of Health and Disease (DOHaD) theory [22]. Proposed in the late 1980s, the DOHaD theory basically states that during normal early

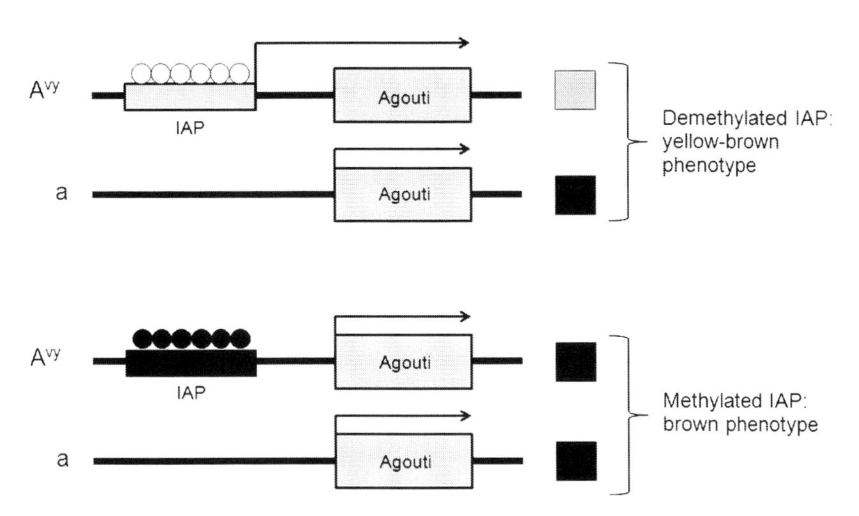

FIGURE 4.2 Epigenetic regulation of the agouti gene in A^{vy}/a mice. White-filled circles indicate unmethylated CpG sites and black-filled circles indicate methylated CpG sites. Phaeomelanin (the product of the agouti gene) is not produced from the *a* allele because the agouti gene is mutated. Two potential epigenetic states of the A^{vy} allele can occur within cells of A^{vy}/a mice. The IAP (intracisternal A particle) that lies upstream of the agouti gene can remain unmethylated, allowing ectopic expression of the gene from the IAP and resulting in a yellow-brown coat color (top). Alternatively, the IAP can be methylated, so that the gene is expressed under its normal developmental controls, leading to a brown coat color (bottom). If the IAP methylation event occurs later in development and does not affect all embryonic cells, the offspring will have a mottled appearance.

development the organs and body systems go through "critical" periods during which they are plastic and sensitive to the environment, and, depending on the degree of adaptation of the body during these periods, this phenomenon can manifest itself as diseases in adult life. Thus, undernutrition during pregnancy and rapid postnatal weight gain are associated with obesity and type 2 diabetes in the adult offspring. Moreover, increasing evidence suggests that early-life exposure to a wide range of chemicals has a significant impact on the causes of metabolic disorders. Although the underlying molecular mechanisms remain to be determined, these factors can affect epigenetic processes such as DNA methylation [23], allowing the developmental environment to modulate gene transcription. The role of DNA methylation in metabolic diseases is addressed in Chapter 11.

4.3.2 Nutritional Deficiencies

The nutritional deficiencies and starvation that affected part of the Dutch population at the end of the World War II, during a 24-week period in which people officially consumed around 900 kcal/day, provides one of the best existing human models to assess whether exposure to famine during pregnancy is related to adult diseases. According to epidemiological studies involving this group of individuals, gestational nutrition deficiency is associated with increased prevalence of overweight, hypertension, coronary heart disease, major affective disorders, antisocial personality disorders, schizophrenia, decreased intracranial volume, and congenital abnormalities of the central nervous system [17,24]. Epigenetic mechanisms are the principal candidates for the biological bridge that links early life to late phenotypes in DOHaD. Corroborating this association, individuals who had been prenatally exposed to the Dutch Famine had, six decades later, less DNA methylation of the imprinted *IGF2* gene when compared to their unexposed same-sex siblings, confirming the vulnerability of epigenetic mechanisms during early development and also indicating that epigenetic alterations that occur during early development persist over the years [25].

4.4 ADVERSE ENVIRONMENTAL FACTORS

4.4.1 Environmental Toxicants

The vulnerability window of embryo epigenetic programming and the late-onset effects have been supported by various pieces of evidence in animal models. For instance, *in utero* exposure to ethanol and bisphenol A (BPA) – a chemical product used in the manufacture of polycarbonate plastic and methylmercury – have previously been shown to affect the epigenome of the developing embryo, culminating in increased predisposition to adult disease and affecting brain development [26,27].

Recently, cadmium – an environmental toxicant that crosses the placental barrier – has been reported to modify the methylation pattern in mother and child [28]. Girls' methylome appeared to be more affected by cadmium compared to boys [29]. However, little is known about the molecular mechanism by which these environmental factors affect the methylome.

Furthermore, occupational and environmental exposures to lead are a worldwide concern. DNA methylation apparently plays an important role in the development of lead toxicity. Recently, occupational lead exposure has been reported to decrease the level of LINE-1 methylation [30]. Similarly, exposure to black carbon (a tracer of traffic particles) has been linked to decreased DNA methylation of LINE-1, but not of Alu repeat elements [31]. Furthermore, prenatal exposure to polycyclic aromatic hydrocarbons derived from backpacks worn by pregnant mothers was significantly associated with methylation of the 5′-CpG islands in the acyl-CoA synthetase long-chain family member 3 (ACSL3) gene, which encodes a key enzyme in fatty acid metabolism [32]. More recently, activation of aryl hydrocarbon receptor (AhR), a transcription factor of the basic helix–loop–helix (bHLH/PAS) family that is well known to mediate the effects of environmental chemicals, was shown to regulate differentiation of FoxP3$^+$ T regulatory cells as well as Th17 cells through epigenetic mechanisms, and to suppress dextran sodium sulfate (DSS)-induced colitis [33].

4.4.2 Ethanol Consumption

In utero exposure to ethanol alters the epigenotype and phenotype of offspring [34–36]. Exposure to this compound has been demonstrated to affect DNA methylation by interfering with the expression of DNMT-3B [37], or by an antagonistic effect on folate methyl group metabolism [38] (Figure 4.3). A first set of studies investigated whether alcohol use influences global DNA methylation in peripheral blood lymphocytes. An early study reported increased DNA methylation in men drinking alcohol versus controls [39]. This has not been confirmed in mixed gender cohorts [40,41]. Conversely, lower Alu methylation (but not LINE-1) occurred in ever-drinkers [42]. A second type of

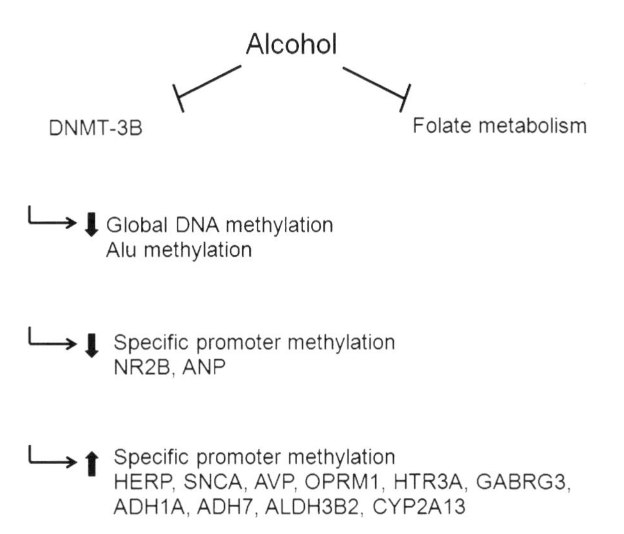

FIGURE 4.3 Effect of abusive alcohol consumption on DNA methylation and specific promoter methylation in peripheral blood lymphocytes.

studies looked at specific gene promoters in peripheral blood lymphocytes and reported decreased methylation of the N-methyl-D-apartate 2b receptor subtype (NR2B) [43] and atrial natriuretic gene (ANP) [44], as well as increased methylation of homocysteine-induced endoplasmic reticulum protein (HERP) [45], alpha synuclein (SNCA) [46], vasopressin (AVP) [44], μ-opioid receptor (OPRM1) [47], 5-hydroxytryptamine receptor 3A (HTR3A) [48], GABAα receptor γ3 (GABRG3) [49], and genes previously implicated in alcohol metabolism (ADH1A, ADH7, ALDH3B2, CYP2A13) [50]. Recently, methylome-wide analysis has been performed [48–51].

4.4.3 Smoking

Cigarette smoke is considered to be one of the most powerful environmental modifiers of DNA methylation [52]. First, cigarette smoke may modulate DNA methylation through DNA damage and subsequent recruitment of DNMTs (Figure 4.4). Carcinogens in cigarette smoke, such as arsenic, chromium, formaldehyde, polycyclic aromatic hydrocarbons, and nitrosamine, can damage DNA by causing double-stranded breaks, as shown in mouse embryonic stem cells exposed to cigarette-smoke condensate [53]. In these experiments, survivor cells display a high capacity for DNA repair and normal karyotypes. The

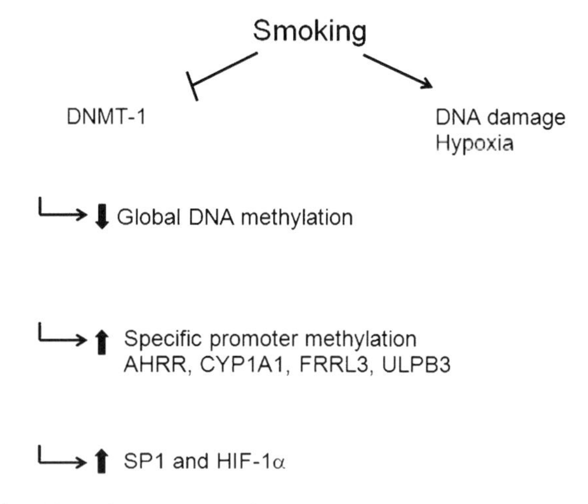

FIGURE 4.4 Effect of cigarette smoking on DNA methylation and specific promoter methylation. Involved are the SP1 transcription factor activated by the nicotinic acetylcholine receptors and the increase of S-adenosylmethionine in response to hypoxia, HIF-1α and methionine adenosyltransferase 2A.

DNA repair sites recruit DNMT-1 [54], which methylates CpGs adjacent to the repaired nucleotides [55]. Second, cigarette smoke may also modulate DNA methylation through nicotine effects on gene expression [56]. Nicotine binds to and activates the nicotinic acetylcholine receptors (present abundantly in the central and peripheral nervous systems) and thus increases intracellular calcium and leads to downstream activation of cAMP response element-binding protein, a key transcription factor for many genes. Acting possibly through this pathway, nicotine has been shown to downregulate DNMT-1 mRNA and protein expression in mouse brain neurons [57]. Third, cigarette smoke may alter DNA methylation indirectly through the modulation of expression and activity of DNA-binding factors. It has been demonstrated, for example, that cigarette-smoke condensate increases SP1 expression and binding to DNA in lung epithelial cells [58]. SP1 is a common transcription factor that binds to GC-rich motifs in gene promoters and plays a key role in early development; as such, it may prevent *de novo* methylation of CpGs within these motifs during early embryogenesis [59]. Fourth, cigarette smoke may alter DNA methylation via hypoxia – cigarette smoke contains carbon monoxide that binds to hemoglobin (competitively with oxygen) and thus decreases tissue oxygenation [60]. Hypoxia, in turn, leads to the HIF-1α-dependent upregulation of methionine adenosyltransferase 2A, which is an enzyme that synthesizes *S*-adenosylmethionine, a major biological methyl donor critical for DNA methylation processes [61]. Recent technological developments allow for the interrogation of CpG methylation status across the genome, and this has uncovered associated CpG sites within candidate genes, including those that have not been previously implicated in cigarette smoking [62]. Remarkably, in both prenatal and current cigarette smoke exposure, similar genes, such as those involved in chemical detoxification (*AHRR, CYP1A1*), are differentially methylated, suggesting that smoking effects might be targeted to specific regions of the epigenome. Cigarette smoking may also have shared global consequences, as both prenatal and current exposures are associated with DNA hypomethylation [63, 64]. In addition, a pronounced association of smoking-related methylation patterns in the coagulation factor II (thrombin) receptor-like 3 (F2RL3) gene has been associated with stable coronary heart disease [65].

Thus, methylome-wide association studies could uncover candidate genes associated with cigarette smoking that have biologically relevant functions in the etiology of smoking-related diseases [66]. Smoking even increases the susceptibility of viral infection by favoring the promoter methylation of genes involved in the antiviral response of the nasal epithelium, such as *ULPB3* [67].

4.5 IMMUNE ACTIVATION

4.5.1 Asthma and Allergy

Asthma and allergy are complex disorders influenced by both inheritance and environment, a relationship that might be further clarified by epigenetics. For instance, T cells mediate the inflammatory responses observed in asthma among genetically susceptible individuals, and they have been suspected to be prone to epigenetic regulation (Figure 4.5). Relying on a cohort of monozygotic twins discordant for asthma, a recent study [68] investigated whether epigenetic modifications in T cells are associated with current asthma and explored whether such modifications are associated with second-hand smoke exposures. Regulatory T cell and effector T cell subsets were assessed for levels of cellular function, protein expression, gene expression and CpG methylation within Forkhead box P3 (FOXP3) and interferon gamma-γ (IFN-γ) loci. Regulatory T cells from asthmatic discordant twins demonstrated decreased FOXP3 protein expression and impaired regulatory function that was associated with increased levels of CpG methylation within the FOXP3 locus when compared to their non-asthmatic twin partner. In parallel, effector T cells from discordant asthmatic twins demonstrated increased methylation of the IFN-γ locus, decreased IFN-γ expression and reduced effector function when compared to such cells from the non-asthmatic twin. Interestingly, report of current exposure to second-hand smoke is associated with modifications in both regulatory and effector T cells at the transcriptional level among asthmatics. This provided evidence for differential function of T cell subsets in monozygotic twins discordant for asthma that are regulated by changes in DNA methylation. In another pilot study [69], DNA methylation patterns change significantly in early childhood in specific asthma- and

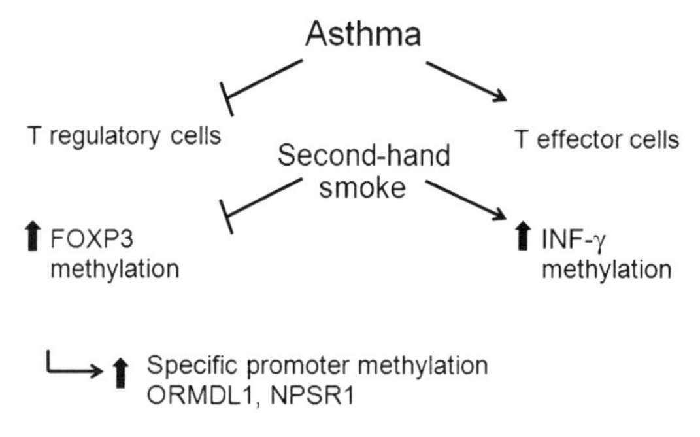

FIGURE 4.5 Decreased T regulatory and increased T effector functions in asthma, due to increased methylation of specific loci: FOXP3 and IFN-γ, respectively. Other promoters are also involved; for example, ORMDL1 and NPSR1.

allergy-related genes in peripheral blood cells, and early exposure to farm environment seems to influence methylation patterns in distinct genes, including adoplin-1 (ORMDL1) [47,69], a negative regulator of sphingolipid synthesis. Furthermore, neuropeptide S receptor 1 (NPSR1) has been associated with asthma and allergy, and the genetic risk is modulated by environmental factors. Interestingly, DNA methylation levels in the promoter of NPSR1 showed small but significant associations with asthma, both in adults and in children, as well as with allergy [70]. Both genetic variation of this gene and the methylated state of CpG sites seem to have an effect on the binding of nuclear proteins in the regulatory region of NPSR1. This raises an interesting hypothesis as to whether genetic variations in the promoter region associated with diseases modulate the binding of various transcription factors.

4.5.2 Bacterial Infections

Host–bacterial interactions not only occur upon pathogenic bacterial infection but also continuously exist between commensal bacteria and the host. These bacterial stimuli play an essential role in various biological responses involving external stimuli and in maintaining physiological homeostasis by altering epigenetic markers and machinery [71].

Epigenetic memory resulting from infection with pathogenic bacteria may be associated with diseases such as cancer, although there are few clear demonstrations of such epigenetic imprinting by pathogenic bacteria. A representative example is infection with *Helicobacter pylori*, which is known to be an acquired risk factor for gastric cancer. It has been reported that *H. pylori*-induced aberrant DNA methylation is an important mechanism in stomach carcinogenesis [72,73]. The mechanisms by which *H. pylori* induce DNA hypermethylation remain unclear, but infection-associated inflammatory responses emerged as a key explanation recently [73]. Remarkably, eradication of *H. pylori* infection in human patients leads to a decrease in but not complete disappearance of methylation of CpG islands at gene promoters closely correlated with the risk of gastric cancer development [74], suggesting that bacterial infection leaves an epigenetic imprint in a tissue to cause permanent changes in gene expression.

In addition, it has been speculated that *Escherichia coli* infection is linked with bladder carcinoma risk through methylation of the CDKN2A promoter [75]. Intestinal commensal bacteria affect DNA methylation of the Toll-like receptor 4 (TLR4) gene in the host [71]. TLR4 is a pattern-recognition receptor that senses Gram-negative bacteria by recognizing LPS, a cell-wall constituent specific to Gram-negative bacteria.

4.5.3 Viral Infections

Esophageal cancer is one of the most prevalent and deadly cancers worldwide. Along with nutrition, smoking, and alcohol consumption, human papillomavirus (HPV) infection is one of the major risk factors that is modulated by host immune response. HPV normally decreases the promoter methylation of HLA-DQB1, increasing its expression on the infected cells. This was found not to occur in cells predisposed to develop a cancerous phenotype [76].

Replication of influenza virus in the host cells results in production of immune mediators like cytokines. Excessive secretion of cytokines (hypercytokinemia) has been observed during highly pathogenic avian influenza virus (HPAI-H5N1) infections resulting in high fatality rates. Upon infection, changes in promoter methylation levels of the pro-inflammatory cytokines CXCL14, CCL25, CXCL6, and interleukins IL13, IL17, and IL4R, occurred [77]. This decrease in methylation was found to be positively correlated with the increased expression of these genes.

4.6 MATERNAL MOOD AND EXPERIENCE IN EARLY LIFE

Exposure to maternal mood disorders *in utero* may program infant neurobehavior via DNA methylation of the glucocorticoid receptor (NR3C1) and 11β-hydroxysteroid dehydrogenase type 2 (11β-HSD-2) [78], two placental genes that have been implicated in perturbations of the hypothalamic pituitary adrenocortical axis (Figure 4.6). Interestingly, infants whose mothers were depressed during pregnancy

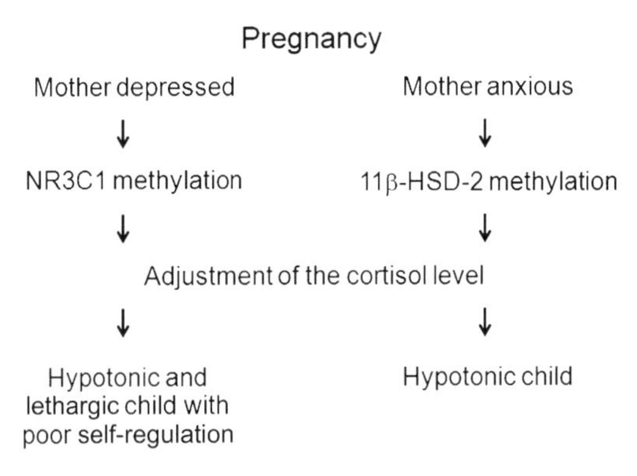

FIGURE 4.6 Effect of maternal mood disorders during pregnancy on differential DNA methylation and behavior of the child. Possibly, circulating hormones such as cortisol that are released in response to adversity act on similar targets in multiple tissues.

and showed greater methylation of placental NR3C1 promoter had poorer self-regulation, more hypotonia, and more lethargy than infants whose mothers were not depressed. On the other hand, infants whose mothers were anxious during pregnancy and showed greater methylation of placental 11β-HSD-2 promoter were more hypotonic compared with infants of mothers who were not anxious during pregnancy. These results further support the fetal programming hypothesis and suggest that fetal adjustments to cues from the intrauterine environment (in this case an environment that could be characterized by increased exposure to maternal cortisol) may lead to poor neurodevelopmental outcomes. Recent data suggest that DNA methylation is a candidate to serve as a mechanism that responds to external experiences and alters gene function and phenotypes in a stable manner [79]. The role of DNA methylation in psychiatric disorders is addressed in Chapter 16.

DNA methylation serves as a mechanism conferring specific functions to identical DNA sequences in response to different life experiences. These responses are system wide and are not limited to the brain, since social adversity has important physical implications. A possible mechanism that integrates DNA methylation responses across tissues comprises circulating hormones such as glucocorticoids that are released in response to adversity and could act on similar targets in multiple tissues. This is presented in more detail in Chapter 10.

4.7 PHYSICAL EXERCISE

The potential of simple changes in lifestyle – such as initiating a calorie restricted diet or an aerobic exercise program – to affect epigenetic patterns and act as a preventive and restorative intervention is still totally speculative [5]. However, practicing physical exercises certainly has numerous positive effects on brain functioning, including improved cognitive performance, neurogenesis, synaptogenesis, and angiogenesis [80–82]. The hippocampus, a part of the brain area that plays a key role in learning, memory, and stress response, seems to be particularly sensitive to physical exercise [83]. Recent evidence revealed that improvements in the cognitive capabilities of exercised rats, compared to sedentary counterparts, and the increased capability to face and handle psychologically stressful challenges are consequent on epigenetic changes in the dentate gyrus neurons of the hippocampus [84]. Interestingly, a significant higher level of global genomic DNA methylation has recently been observed in human individuals that engage in physical activity for 26–30 minutes per day when compared to people taking less physical activity [85]. This information warrants further investigation regarding the impact of physical activities on DNA methylation. Furthermore, a recent follow-up study reported on a very interesting interrelationship between physical exercise, epigenetic changes, and cancer prognostics. After studying the effects of 6 months' of moderate-intensity aerobic exercise on DNA methylation of peripheral blood leukocytes of breast cancer survivors, these investigators [86] showed that physical activity affects loci-specific epigenetic mechanisms that control the expression of tumor suppressor genes, such as the *lethal(3)malignant brain tumor-like protein 1* (L3MBTL1), and is linked to favorable survivor outcomes.

4.8 CONCLUSION

The links between environmental toxicants, cigarette smoking, alcohol consumption, dietary intake, and infections with genome-wide or loci-specific epigenetic changes have been investigated intensely over recent years [34]. However, the list of environmental factors capable of inducing epigenetic errors seems to be completely underestimated, and exposure to apparently inoffensive products such as hair dye should also be taken into consideration [87], as well as drug consumption and familial factors. The knowledge and understanding of molecular linkages between environment and health has been greatly improved by newly emerging data on epigenetic vulnerability. Future human studies addressing the potential of preventive interventions based on the modulation of epigenetic patterns should be encouraged, and might reveal, at the molecular level, the role of lifestyle in lifelong good health.

References

[1] Fraga MF, Ballestar E, Paz MF, Ropero S, Setien F, Ballestar ML, et al. Epigenetic differences arise during the lifetime of monozygotic twins. Proc Natl Acad Sci USA 2005;102:10604–9.

[2] Terry MB, Delgado-Cruzata L, Vin-Raviv N, Wu HC, Santella RM. DNA methylation in white blood cells: association with risk factors in epidemiologic studies. Epigenetics 2011;6:828–37.

[3] Suter MA, Anders AM, Aagaard KM. Maternal smoking as a model for environmental epigenetic changes affecting birthweight and fetal programming. Mol Hum Reprod 2013;19:1–6.

[4] Ehrlich M, Lacey M. DNA hypomethylation and hemimethylation in cancer. Adv Exp Med Biol 2013;754:31–56.

[5] Gomes MV, Pelosi GG. Epigenetic vulnerability and the environmental influence on health. Exp Biol Med (Maywood) 2013;238:859–65.

[6] Geiman TM, Muegge K. DNA methylation in early development. Mol Reprod Dev 2010;77:105–13.

[7] Lucifero D, Chaillet JR, Trasler JM. Potential significance of genomic imprinting defects for reproduction and assisted reproductive technology. Hum Reprod Update 2004;10:3–18.
[8] Bell AC, Felsenfeld G. Methylation of a CTCF-dependent boundary controls imprinted expression of the Igf2 gene. Nature 2000;405:482–5.
[9] Davis TL, Yang GJ, McCarrey JR, Bartolomei MS. The H19 methylation imprint is erased and re-established differentially on the parental alleles during male germ cell development. Hum Mol Genet 2000;9:2885–94.
[10] Li JY, Lees-Murdock DJ, Xu GL, Walsh CP. Timing of establishment of paternal methylation imprints in the mouse. Genomics 2004;84:952–60.
[11] Diaz-Meyer N, Day CD, Khatod K, Maher ER, Cooper W, Reik W, et al. Silencing of CDKN1C (p57KIP2) is associated with hypomethylation at KvDMR1 in Beckwith-Wiedemann syndrome. J Med Genet 2003;40:797–801.
[12] Weksberg R, Shuman C, Smith AC. Beckwith-Wiedemann syndrome. Am J Med Genet C Semin Med Genet 2005;137C:12–23.
[13] Khosla S, Dean W, Brown D, Reik W, Feil R. Culture of preimplantation mouse embryos affects fetal development and the expression of imprinted genes. Biol Reprod 2001;64:918–26.
[14] Maher ER, Afnan M, Barratt CL. Epigenetic risks related to assisted reproductive technologies: epigenetics, imprinting, ART and icebergs? Hum Reprod 2003; 18:2508–11.
[15] Sato A, Otsu E, Negishi H, Utsunomiya T, Arima T. Aberrant DNA methylation of imprinted loci in superovulated oocytes. Hum Reprod 2007;22:26–35.
[16] Rossignol S, Steunou V, Chalas C, Kerjean A, Rigolet M, Viegas-Pequignot E, et al. The epigenetic imprinting defect of patients with Beckwith-Wiedemann syndrome born after assisted reproductive technology is not restricted to the 11p15 region. J Med Genet 2006;43:902–7.
[17] Roseboom TJ, Painter RC, van Abeelen AF, Veenendaal MV, de Rooij SR. Hungry in the womb: what are the consequences? Lessons from the Dutch famine. Maturitas 2011;70:141–5.
[18] Anderson OS, Sant KE, Dolinoy DC. Nutrition and epigenetics: an interplay of dietary methyl donors, one-carbon metabolism and DNA methylation. J Nutr Biochem 2012;23:853–9.
[19] Ghoshal K, Li X, Datta J, Bai S, Pogribny I, Pogribny M, et al. A folate- and methyl-deficient diet alters the expression of DNA methyltransferases and methyl CpG binding proteins involved in epigenetic gene silencing in livers of F344 rats. J Nutr 2006;136.
[20] Stefanska B, Salame P, Bednarek A, Fabianowska-Majewska K. Comparative effects of retinoic acid, vitamin D and resveratrol alone and in combination with adenosine analogues on methylation and expression of phosphatase and tensin homologue tumour suppressor gene in breast cancer cells. Br J Nutr 2012; 107:781–90.
[21] Waterland RA, Jirtle RL. Transposable elements: targets for early nutritional effects on epigenetic gene regulation. Mol Cell Biol 2003;23:5293–300.
[22] Barker DJ. The developmental origins of adult disease. J Am Coll Nutr 2004; 23:588S–95S.
[23] Inadera H. Developmental origins of obesity and type 2 diabetes: molecular aspects and role of chemicals. Environ Health Prev Med 2013;18:185–97.
[24] de Rooij SR, Wouters H, Yonker JE, Painter RC, Roseboom TJ. Prenatal undernutrition and cognitive function in late adulthood. Proc Natl Acad Sci USA 2010; 107:16881–6.

[25] Heijmans BT, Tobi EW, Stein AD, Putter H, Blauw GJ, Susser ES, et al. Persistent epigenetic differences associated with prenatal exposure to famine in humans. Proc Natl Acad Sci USA 2008;105:17046–9.
[26] Palanza P, Gioiosa L, vom Saal FS, Parmigiani S. Effects of developmental exposure to bisphenol A on brain and behavior in mice. Environ Res 2008;108: 150–7.
[27] Onishchenko N, Karpova N, Sabri F, Castren E, Ceccatelli S. Long-lasting depression-like behavior and epigenetic changes of BDNF gene expression induced by perinatal exposure to methylmercury. J Neurochem 2008;106:1378–87.
[28] Sanders AP, Smeester L, Rojas D, DeBussycher T, Wu MC, Wright FA, et al. Cadmium exposure and the epigenome: Exposure-associated patterns of DNA methylation in leukocytes from mother-baby pairs. Epigenetics 2013;9:212–21.
[29] Kippler M, Engstrom K, Mlakar SJ, Bottai M, Ahmed S, Hossain MB, et al. Sex-specific effects of early life cadmium exposure on DNA methylation and implications for birth weight. Epigenetics 2013;8:494–503.
[30] Li C, Yang X, Xu M, Zhang J, Sun N. Epigenetic marker (LINE-1 promoter) methylation level was associated with occupational lead exposure. Clin Toxicol (Phila) 2013;51:225–9.
[31] Baccarelli A, Wright RO, Bollati V, Tarantini L, Litonjua AA, Suh HH, et al. Rapid DNA methylation changes after exposure to traffic particles. Am J Respir Crit Care Med 2009;179:572–8.
[32] Perera FP, Li Z, Whyatt R, Hoepner L, Wang S, Camann D, et al. Prenatal airborne polycyclic aromatic hydrocarbon exposure and child IQ at age 5 years. Pediatrics 2009;124:e195–202.
[33] Singh NP, Singh UP, Singh B, Price RL, Nagarkatti M, Nagarkatti PS. Activation of aryl hydrocarbon receptor (AhR) leads to reciprocal epigenetic regulation of FoxP3 and IL-17 expression and amelioration of experimental colitis. PLoS One 2011;6: e23522.
[34] Zhang TY, Hellstrom IC, Bagot RC, Wen X, Diorio J, Meany MJ. Maternal care and DNA methylation of a glutamic acid decarboxylase 1 promoter in rat hippocampus. J Neurosci 2010;30:13130–7.
[35] Kaminen-Ahola N, Ahola A, Maga M, Mallitt KA, Fahey P, Cox TC, et al. Maternal ethanol consumption alters the epigenotype and the phenotype of offspring in a mouse model. PLoS Genet 2010;6:e1000811.
[36] Meaney MJ. Epigenetics and the biological definition of gene x environment interactions. Child Dev 2010;81:41–79.
[37] Bonsch D, Lenz B, Fiszer R, Frieling H, Kornhuber J, Bleich S. Lowered DNA methyltransferase (DNMT-3b) mRNA expression is associated with genomic DNA hypermethylation in patients with chronic alcoholism. J Neural Transm 2006;113: 1299–304.
[38] Halsted CH, Villanueva JA, Devlin AM, Chandler CJ. Metabolic interactions of alcohol and folate. J Nutr 2002;132:2367S–72S.
[39] Bonsch D, Lenz B, Reulbach U, Kornhuber J, Bleich S. Homocysteine associated genomic DNA hypermethylation in patients with chronic alcoholism. J Neural Transm 2004;111:1611–16.
[40] Zhang FF, Cardarelli R, Carroll J, Fulda KG, Kaur M, Gonzalez K, et al. Significant differences in global genomic DNA methylation by gender and race/ethnicity in peripheral blood. Epigenetics 2011;6:623–9.
[41] Ono H, Iwasaki M, Kuchiba A, Kasuga Y, Yokoyama S, Onuma H, et al. Association of dietary and genetic factors related to one-carbon metabolism with global methylation level of leukocyte DNA. Cancer Sci 2012;103:2159–64.

[42] Zhu ZZ, Hou L, Bollati V, Tarantini L, Marinelli B, Cantone L, et al. Predictors of global methylation levels in blood DNA of healthy subjects: a combined analysis. Int J Epidemiol 2012;41:126–39.
[43] Biermann T, Reulbach U, Lenz B, Frieling H, Muschler M, Hillemacher T, et al. N-methyl-D-aspartate 2b receptor subtype (NR2B) promoter methylation in patients during alcohol withdrawal. J Neural Transm 2009;116:615–22.
[44] Hillemacher T, Frieling H, Luber K, Yazici A, Muschler MA, Lenz B, et al. Epigenetic regulation and gene expression of vasopressin and atrial natriuretic peptide in alcohol withdrawal. Psychoneuroendocrinology 2009;34:555–60.
[45] Bleich S, Lenz B, Ziegenbein M, Beutler S, Frieling H, Kornhuber J, et al. Epigenetic DNA hypermethylation of the HERP gene promoter induces down-regulation of its mRNA expression in patients with alcohol dependence. Alcohol Clin Exp Res 2006;30:587–91.
[46] Bonsch D, Lenz B, Kornhuber J, Bleich S. DNA hypermethylation of the alpha synuclein promoter in patients with alcoholism. Neuroreport 2005;16:167–70.
[47] Zhang H, Herman AI, Kranzler HR, Anton RF, Simen AA, Gelernter J. Hypermethylation of OPRM1 promoter region in European Americans with alcohol dependence. J Hum Genet 2012;57:670–5.
[48] Zhang H, Herman AI, Kranzler HR, Anton RF, Zhao H, Zheng W, et al. Array-based profiling of DNA methylation changes associated with alcohol dependence. Alcohol Clin Exp Res 2013;37(Suppl. 1):E108–115.
[49] Zhang H, Wang F, Kranzler HR, Zhao H, Gelernter J. Profiling of childhood adversity-associated DNA methylation changes in alcoholic patients and healthy controls. PLoS One 2013;8:e65648.
[50] Zhang R, Mia Q, Wang C, Zhao R, Li W, Haile CN, et al. Genome-wide DNA methylation analysis in alcohol dependence. Addict Biol 2013;18:392–403.
[51] Zhao R, Zhang R, Li W, Liao Y, Tang J, Miao Q, et al. Genome-wide DNA methylation patterns in discordant sib pairs with alcohol dependence. Asia Pac Psychiatry 2013;5:39–50.
[52] Breitling LP, Yang R, Korn B, Burwinkel B, Brenner H. Tobacco-smoking-related differential DNA methylation: 27K discovery and replication. Am J Hum Genet 2011;88:450–7.
[53] Huang J, Okuka M, Lu W, Tsibris JC, McLean MP, Keefe DL, et al. Telomere shortening and DNA damage of embryonic stem cells induced by cigarette smoke. Reprod Toxicol 2013;35:89–95.
[54] Mortusewicz O, Schermelleh L, Walter J, Cardoso MC, Leonhardt H. Recruitment of DNA methyltransferase I to DNA repair sites. Proc Natl Acad Sci USA 2005; 102:8905–9.
[55] Cuozzo C, Porcelini A, Angrisano T, Morano A, Lee B, Di Pardo A, et al. DNA damage, homology-directed repair, and DNA methylation. PLoS Genet 2007;3:e110.
[56] Lee EW, D'Alonzo GE. Cigarette smoking, nicotine addiction, and its pharmacologic treatment. Arch Intern Med 1993;153:34–48.
[57] Satta R, Maloku E, Zhubi A, Pibiri F, Hajos M, Costa E, et al. Nicotine decreases DNA methyltransferase 1 expression and glutamic acid decarboxylase 67 promoter methylation in GABAergic interneurons. Proc Natl Acad Sci USA 2008;105:16356–61.
[58] Di YP, Zhao J, Harper R. Cigarette smoke induces MUC5AC protein expression through the activation of Sp1. J Biol Chem 2012;287:27948–58.
[59] Han L, Su B, Li WH, Zhao Z. CpG island density and its correlations with genomic features in mammalian genomes. Genome Biol 2008;9:R79.
[60] Olson KR. Carbon monoxide poisoning: mechanisms, presentation, and controversies in management. J Emerg Med 1984;1:233–43.

[61] Liu Q, Liu L, Zhao Y, Zhang J, Wang D, Chen J, et al. Hypoxia induces genomic DNA demethylation through the activation of HIF-1alpha and transcriptional upregulation of MAT2A in hepatoma cells. Mol Cancer Ther 2011;10:1113–23.

[62] Joubert B, London S. Epigenomics and maternal smoking, with Bonnie Joubert and Stephanie London by Ashley Ahearn. Environ Health Perspect 2012;120:9.

[63] Guerrero-Preston R, Goldman LR, Brebi-Mieville P, Ili-Gangas C, Lebron C, Witter FR, et al. Global DNA hypomethylation is associated with *in utero* exposure to cotinine and perfluorinated alkyl compounds. Epigenetics 2010;5:539–46.

[64] Shigaki H, Baba Y, Watanabe M, Iwagami S, Miyake K, Ishimoto T, et al. LINE-1 hypomethylation in noncancerous esophageal mucosae is associated with smoking history. Ann Surg Oncol 2012;19:4238–43.

[65] Breitling LP. Current genetics and epigenetics of smoking/tobacco-related cardiovascular disease. Arterioscler Thromb Vasc Biol 2013;33:1468–72.

[66] Lee KW, Pausova Z. Cigarette smoking and DNA methylation. Front Genet 2013; 4:132.

[67] Rager JE, Bauer RN, Muller LL, Smeester L, Carson JL, Brighton LE, et al. DNA methylation in nasal epithelial cells from smokers: identification of ULBP3-related effects. Am J Physiol Lung Cell Mol Physiol 2013;305:L432–438.

[68] Runyon RS, Cachola LM, Rajeshuni N, Hunter T, Garcia M, Ahn R, et al. Asthma discordance in twins is linked to epigenetic modifications of T cells. PLoS One 2012;7:e48796.

[69] Michel S, Busato F, Genuneit J, Pekkanen J, Dalphin JC, Riedler J, et al. Farm exposure and time trends in early childhood may influence DNA methylation in genes related to asthma and allergy. Allergy 2013;68:355–64.

[70] Reinius LE, Gref A, Saaf A, Acevedo N, Joerink M, Kupczyk M, et al. DNA methylation in the Neuropeptide S Receptor 1 (NPSR1) promoter in relation to asthma and environmental factors. PLoS One 2013;8:e53877.

[71] Takahashi K. Influence of bacteria on epigenetic gene control. Cell Mol Life Sci 2013; 71:1045–54.

[72] Ding SZ, Goldberg JB, Hatakeyama M. *Helicobacter pylori* infection, oncogenic pathways and epigenetic mechanisms in gastric carcinogenesis. Future Oncol 2010; 6:851–62.

[73] Ushijima T, Hattori N. Molecular pathways: involvement of *Helicobacter pylori*-triggered inflammation in the formation of an epigenetic field defect, and its usefulness as cancer risk and exposure markers. Clin Cancer Res 2012;18:923–9.

[74] Nakajima T, Enomoto S, Yamashita S, Ando T, Nakanishi Y, Nakazawa K, et al. Persistence of a component of DNA methylation in gastric mucosae after *Helicobacter pylori* eradication. J Gastroenterol 2010;45:37–44.

[75] Tolg C, Sabha N, Cortese R, Panchal T, Ahsan A, Soliman A, et al. Uropathogenic *E. coli* infection provokes epigenetic downregulation of CDKN2A (p16INK4A) in uroepithelial cells. Lab Invest 2011;91:825–36.

[76] Feng B, Awuti I, Deng Y, Li D, Niyazi M, Aniwar J, et al. Human papillomavirus promotes esophageal squamous cell carcinoma by regulating DNA methylation and expression of HLA-DQB1. Asia Pac J Clin Oncol 2013;10:66–74.

[77] Mukherjee S, Vipat VC, Chakrabarti AK. Infection with influenza A viruses causes changes in promoter DNA methylation of inflammatory genes. Influenza Other Respir Viruses 2013;7:979–86.

[78] Conradt E, Lester BM, Appleton AA, Armstrong DA, Marsit CJ. The role of DNA methylation of NR3C1 and 11beta-HSD2 and exposure to maternal mood disorder *in utero* on newborn neurobehavior. Epigenetics 2013;8:1321–9.

[79] Szyf M. DNA methylation, behavior and early life adversity. J Genet Genomics 2013;40:331–8.
[80] Kaliman P, Parrizas M, Lalanza JF, Camins A, Escorihuela RM, Pallas M. Neurophysiological and epigenetic effects of physical exercise on the aging process. Ageing Res Rev 2011;10:475–86.
[81] Kempermann G, Fabel K, Ehninger D, Babu H, Leal-Galicia P, Garthe A, et al. Why and how physical activity promotes experience-induced brain plasticity. Front Neurosci 2010;4:189.
[82] Rolland Y, Abellan van Kan G, Vellas B. Healthy brain aging: role of exercise and physical activity. Clin Geriatr Med 2010;26:75–87.
[83] Murray PS, et al. Locus coeruleus galanin expression is enhanced after exercise in rats selectively bred for high capacity for aerobic activity. Peptides 2010;31:2264–8.
[84] Collins A, Hill LE, Chandramohan Y, Whitcomb D, Droste SK, Reul JMHM. Exercise improves cognitive responses to psychological stress through enhancement of epigenetic mechanisms and gene expression in the dentate gyrus. PLoS One 2009;4:e4330.
[85] Zhang FF, Cardarelli R, Carroll J, Zhang S, Fulda KG, Gonzalez K, et al. Physical activity and global genomic DNA methylation in a cancer-free population. Epigenetics 2011;6:293–9.
[86] Zeng H, Irwin ML, Lu L, Risch H, Mayne S, Mu L, et al. Physical activity and breast cancer survival: an epigenetic link through reduced methylation of a tumor suppressor gene L3MBTL1. Breast Cancer Res Treat 2012;133:127–35.
[87] Langevin SM, Houseman EA, Christensen BC, Wiencke JK, Nelson HH, Karagas MR, et al. The influence of aging, environmental exposures and local sequence features on the variation of DNA methylation in blood. Epigenetics 2011;6:908–19.

CHAPTER 5

DNA Methylation and Epidemiology

OUTLINE

5.1 INTRODUCTION

Epigenetic epidemiology is the integration of epigenetic analyses into population-based epidemiological research with the goal of identifying both the causes (that is, environmental, genetic or stochastic) and phenotypic consequences (that is, health and disease) of epigenomic

M. Neidhart: DNA Methylation and Complex Human Disease.
DOI: http://dx.doi.org/10.1016/B978-0-12-420194-1.00005-1

variation. Other definitions are narrower and have a more direct focus on environmental epigenetics and the specific analysis of transgenerational epigenetic effects [1] or the developmental origins of health and disease [2].

The evolution of living organisms is the expression of different and sometimes opposing forces. The main tension is between the need for stability and the need for change. The first is expressed, for example, by the great structural stability of DNA and of its coding system, conserved almost intact across species. However, without the ability to change organisms would not be able to adapt to changing environments and their continuous threats. Transposons (transposable elements), crossing-over at meiosis, and epigenetic changes are some of the mechanisms that ensure the variety of genetic configurations that allow variation, adaptation, and evolution.

As a very general rule, environmental stresses tend to increase epigenomic instability; this is both a mechanism to enhance diversity and respond to threats, and a potentially harmful mechanism leading to disease.

Over the past two decades, much research has gone into the role of epigenetic changes in the development of cancer [3–5]. The most broadly researched and studied epigenetic mechanism is DNA methylation [6]. First observed in the early 1980s in studies of carcinogenesis [3], DNA methylation mainly occurs at CpG dinucleotides, which are found grouped together in promoter regions of around 50% of all human genes; the latter are also labeled as CpG islands. DNA methylation involves the addition of a methyl group to a CpG dinucleotide. Methylation of CpG islands can cause gene silencing. If this affects tumor suppressor genes, there may be the induction of cancer. Cancer cells have been seen, in fact, to have unusual patterns of methylation, including both hypermethylation of tumor suppressor genes and hypomethylation of proto-oncogenes or transposable elements [7].

It is believed that exposure to environmental carcinogens and the associated epigenetic changes can occur *in utero*; epigenetic programming may play a major role in embryonic development and the health of an individual in the long term [6]. In rats, for example, arsenic exposure during gestation resulted in altered patterns of DNA methylation of brain cells affecting memory, thus suggesting that exposure *in utero* has long-standing effects on growth and development [8]. The epigenetic signatures of prenatal famine are apparent in peripheral blood cells many decades after exposure [9], while obesity-associated hypermethylation at a CpG island in the pro-opiomelanocortin (POMC) gene appears to become manifest early in development and then remains stable across tissues and through the life course [10].

Other convincing environmentally associated epigenetic differences are reported for tobacco smoke: consistent differentially methylated loci have been observed in studies of *in utero* [11] and adulthood exposure [12,13].

5.2 ENDOCRINE DISRUPTING CHEMICALS

Many studies have focused on the association between exposure to endocrine disrupting chemicals (EDCs) and breast cancer risk. EDCs are synthetic and natural compounds in the environment that interfere with (i.e., mimic and/or antagonize) the actions of endogenous hormones by altering hormone synthesis, secretion, transport, binding, action, or elimination, and thereby disrupt the functions of the endocrine system. Exposure to EDCs can occur in many different ways. At times it is intentional – for example, in the case of medicinal use – but mostly it is accidental. Many EDCs are found in food, which can be the main source of exposure, as in the cases of, for example, bisphenol-A (BPA), diethylstilbestrol (DES), and cadmium (Cd).

5.2.1 Bisphenol-A

Exposure to bisphenol-A has been studied in animal models, and has been associated with epigenetic changes which can pose health threats. Exposing mice perinatally to both high and low doses of BPA led to hypomethylation in the offspring [14,15]; another study showed an increased risk of prostate cancer in rats due to epigenetic alterations after BPA exposure [16,17]. More recently, studies in human breast cancer cell lines have been performed examining the association between bisphenol-A exposure and DNA methylation [18]; 170 genes were found differentially expressed after BPA exposure (57 upregulated and 113 downregulated). Moreover, hierarchical clustering indicated that the upregulated genes were mainly found in ERα-negative breast cancer cell lines while the downregulated genes were mostly detected in ERα-positive breast cancer cell lines. Furthermore, DNA methylation analysis on the candidate gene lysosomal-associated membrane protein 3 (LAMP3) showed increased DNA methylation levels in breast cancer cell lines compared to controls, primarily in tumors with a positive ERα status. Another study exposed human mammary epithelial cells (HMECs) to BPA and concluded that there was an increase in DNA methylation of six genes [19].

5.2.2 Diethylstillbestrol

Another well-known and extensively studied disrupting chemical is diethylstilbestrol (DES). The transgenerational effect after DES exposure that has been shown in epidemiological studies strongly indicates the involvement of epigenetic factors. An early study [20] showed a decrease in methylation of CpG sites in the lactoferrin gene after neonatal exposure of mice to DES, and that this hypomethylation was also detected in uterine tumors. A few years later the same investigators reported hypomethylation of the exon-4 region of the c-fos gene in mice, after neonatal exposure to DES [21]. Increase in DNA methylation of the HOXA-10 gene occurs in female offspring after exposure to DES *in utero*, but not *in vitro* or *in vivo* [22]. After postnatal exposure of mice to DES, differential DNA methylation was observed in the uterus, including nucleosomal binding protein 1 (NSBP1) [23] and paxillin [24].

5.2.3 Cadmium

Several animal studies concerning the exposure to cadmium (Cd) have been performed, as well as studies using animal cell lines. A study investigating the epigenetic effects of dietary cadmium exposure in humans reported an hypomethylation of LINE-1 in peripheral blood leukocytes [25]. The chronic myelogenous leukemia cell line K562 showed a reduction in global DNA methylation after short exposure (24 and 48 hours) to 2.0 μM of cadmium [26]. TRL1215 rat liver cells exposed to a dose of 2.5 μM of cadmium for 1 week became hypomethylated, but after 10 weeks were hypermethylated [27]; the cells probably become transformed. Similarly, human prostate epithelial cells exposed to cadmium showed global hypermethylation [28]. In those cells, the promoters of RASSF1A and p16, two tumor suppressor genes, were found hypermethylated, leading to decreased expression. Global hypermethylation was also found by a study in which human embryo lung fibroblast cells underwent long-term exposure to cadmium [29]. Another study showed that both a low (140 mg/kg $CdCl_2$) and high (210 mg/kg $CdCl_2$) dose of cadmium for 60 days increased global methylation in both the liver and kidney of hens [30]. Moreover, cadmium exposure did seem to slow down the naturally occurring hypomethylation of part of the c-fos gene in the testis [31]. An epigenome-wide study reported hypermethylation of caspase 8 (CASP8) and hypomethylation of tumor necrosis factor alpha (TNF-α), in both cases resulting in decreased gene expression [32].

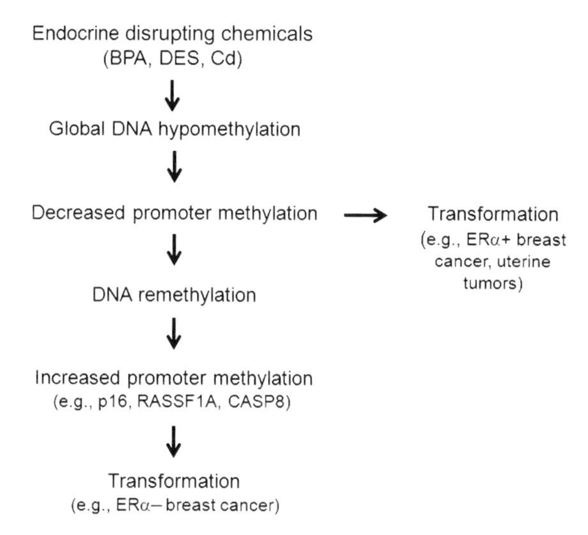

FIGURE 5.1 Endocrine-disrupting chemicals lead to global DNA hypomethylation and decreased methylation of specific genes. This can favor the development of certain forms of cancer. Errors can occur as the system tries to remethylate DNA. In the case where so-called tumor suppressor genes are affected, this leads to development of other types of tumors.

5.3 EPIGENOMICS AND BREAST CANCER

Among genes that appeared to be hypermethylated in subjects with breast cancer, two were also found to be differentially methylated after exposure to EDCs; namely, RASSF1A [33] and p16 [34] (Figure 5.1). Thus, the exposure to EDCs, hypermethylation of these tumor suppressor genes, and development of breast cancer appear to be linked. This is the best example of an epidemiological investigation in epigenetics.

5.3.1 RASSF1A

Loss of heterozygosity of a segment at 3p21.3 is frequently observed in lung cancer and several other carcinomas. Ras-association domain family 1A gene (RASSF1A), which is localized at 3p21.3, has been identified in a minimum deletion sequence [35]. *De novo* methylation of the RASSF1A promoter is one of the most frequent epigenetic inactivation events detected in human cancer, and leads to silencing of RASSF1A expression [35,36]. Hypermethylation of RASSF1A has frequently been found in most major types of human tumors, including lung, breast, prostate, pancreas, kidney, liver, cervical, thyroid, and many other cancers [36]. Methylation of many other genes (e.g., HIN-1, RIL, and CDH13) in breast cancers is associated with clinical characteristics, but only RASSF1A methylation was associated with time to first recurrence and overall survival [37].

5.3.2 P16/CDKN2A

P16 is the prototype of the gene known as cyclin-dependent kinase inhibitor 2A (CDKN2A). It has several different aliases, of which p16INK4a and CDKN2A are the most commonly used. Its location on the human genome is at 9p21, and, besides other functions, it plays an important role in cell cycle arrest and apoptosis. The gene is known as a tumor suppressor gene because it is often mutated and deleted in several types of cancer. It has been shown that the promoter region is often hypermethylated in cancer, leading to reduced expression and silencing of this gene [38]. Breast cancer-specific mortality is strongly associated with promoter methylation of p16 [39].

5.4 TOBACCO SMOKING

Tobacco smoking is responsible for substantial morbidity and mortality worldwide, in particular through cardiovascular, pulmonary, and malignant pathology. Even rheumatoid arthritis has been associated with smoking [40]. CpG methylation might plausibly play a role in a variety of smoking-related phenomena, as suggested by candidate gene promoter or global methylation studies – for example, coagulation factor II receptor-like 3 (F2RL3), Runt-related transcription factor 3 (RUNX3), aryl-hydrocarbon receptor repressor (AHRR), and cytochrome P450 isoform A1 (CYP1A1) (Figure 5.2).

5.4.1 F2RL3, Smoking, and Cardiovascular Diseases

Tobacco-smoking is one of the most important cardiovascular risk factors that is partially determined by genetic background and associated with altered epigenetic patterns [41]. Particularly, smoking-related F2RL3 promoter methylation has been associated with prognosis in patients with stable coronary heart disease [42]. In general, methylation of the F2RL3 promoter appears to be a promising biomarker for both current and long-term past tobacco exposure [43].

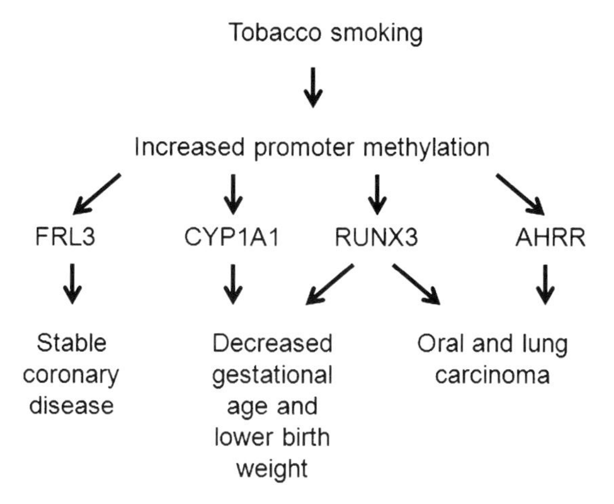

FIGURE 5.2 Pathologies associated with tobacco smoking occurring through specific promoter methylations.

5.4.2 RUNX3, Smoking, Low Birth Weight, and Lung Carcinomas

Maternal smoking during pregnancy is associated with low birth weight, preterm birth, and other complications, while exposure to cigarette smoke *in utero* has been linked to gross pathologic and molecular changes to the placenta, including differential DNA methylation in placental tissue. For example, methylation patterns of a number of loci within the RUNX3 gene are significantly associated with smoking during pregnancy, and one of these loci has been associated with decreased gestational age [44].

Epidemiological studies have demonstrated a causal link between tobacco smoking and lung cancer. Promoter hypermethylation of p16, RASSF1A, and RUNX3 are specific for non-small cell lung carcinoma [45]. The frequency of epithelial growth factor receptor (EGFR) mutation is strongly correlated with smoking status, and this in turn is associated with increased RUNX3 promoter methylation [46]. It correlates with clinical stage, lymph node metastasis, and degree of differentiation [47], and is also a potential prognostic marker in oral carcinoma [48,49].

Downregulation of RUNX3 expression in laryngeal carcinoma can also occur through increased microRNA-106b [50]. This has been found to increase the proliferation and invasion of tumor cells.

5.4.3 AHRR, Smoking, and Lung Metastasis

A recent study [11] investigated epigenome-wide methylation in cord blood of newborns in relation to maternal smoking during pregnancy. The authors found differential DNA methylation of 26 CpGs mapped to 10 genes, in particular AHRR, cytochrome P450 isoform A1 (CYP1A1), and growth factor independent 1 transcription repressor (GFI1). AHRR and CYP1A1 play a key role in the aryl hydrocarbon receptor signaling pathway, which mediates the detoxification of the components of tobacco smoke. GFI1 is involved in diverse developmental processes, but has not previously been implicated in responses to tobacco smoke.

Another recent study [13] examined the relationship of smoking to genome-wide methylation and gene expression using biomaterial from two independent samples: lymphoblast DNA and RNA, and lung alveolar macrophage DNA. In both samples, current smoking status was associated with significant changes in DNA methylation of the AHRR promoter. The aryl hydrocarbon receptor repressor (AHRR) is a bHLH/Per-ARNT-Sim transcription factor located in a region of chromosome 5 (5p15.3) that has been proposed to contain one or more tumor suppressor genes. The role of AHRR as a negative regulator of pathways involved in pleiotropic responses to environmental contaminants raises the possibility that smoking-induced hypomethylation is an adaptive response to an adverse *in utero* environmental exposure. A further study [51] analyzed AHRR methylation in three cell types – cord blood mononuclear cells, buccal epithelium, and placenta tissue – from newborn twins of mothers who smoked throughout pregnancy, and matched controls. The authors confirmed the association between maternal smoking and AHRR methylation in neonatal blood: AHRR methylation is altered in response to maternal smoking during pregnancy. Thus, AHRR promoter methylation status is a sensitive marker of smoking history and could serve as a biomarker of smoking that could even supplement self-report or existing biomarker measures in clinical or epidemiological analyses of the effects of smoking [52].

The silencing of AHRR in human malignant tissue from different anatomical origins, including colon, breast, lung, stomach, cervix, and ovary, is caused by DNA hypermethylation [53]. The downregulation of AHRR expression in the human lung cancer cell line conferred resistance to apoptotic signals and enhanced motility and invasion *in vitro* and angiogenic potential *in vivo*.

5.4.4 CYP1A1, Smoking, and Obesity

Obesity is a global epidemic, and maternal smoking has been shown to be associated with the development of childhood obesity. Overall, approximately 40% of children worldwide are exposed to tobacco smoke at home. It is well known that environmental changes within a critical window of development, such as gestation or lactation, can initiate permanent alterations in metabolism that lead to diseases in adulthood – a phenomenon called programming. It is known that programming is based on epigenetic alterations, including DNA methylation, which changes the expression pattern of several genes. Epidemiological studies of adults exposed *in utero* to calorie restriction during the Dutch Hunger in World War II were the first in which DNA hypomethylation at the imprinted inulin growth factor 2 (IGF2) region in mononuclear cells was demonstrated [54]. Those undernourished babies developed into overweight adults who had a higher cardiovascular risk, confirming the hypothesis that the early life environment can cause epigenetic changes that persist throughout life.

Several epidemiological and experimental studies confirmed the association between maternal tobacco exposure during gestation or lactation, and the development of obesity and endocrine dysfunction [55]. For example, a positive correlation was demonstrated in rodents between increases in the neonatal period and exposure of the mothers to nicotine during lactation, and the further development of leptin and insulin resistance, and thyroid and adrenal dysfunction, in adulthood in the same offspring [56]. Thus, a smoke-free environment during the lactation period is essential to improving health outcomes in adulthood and reducing the risk for future diseases.

Cytochrome P4501A1 (CYP1A1), an important drug-metabolizing enzyme, is expressed in human placenta throughout gestation as well as in fetal liver. Obesity, a chronic inflammatory condition, is known to alter cytochrome enzyme expression in non-placental tissues. *In utero*, tobacco exposure epigenetically modifies placental CYP1A1 expression through a decrease in promoter methylation, and this gene imprinting could explain the lower birth weight of babies of smoking mothers [57,58]. Thus, maternal lifestyle could have a significant impact on CYP1A1 [59]. This hints at a possible role for CYP1A1 in feto-placental growth and thereby the well-being of the fetus.

5.5 CONCLUSION

It will be essential to establish a causal role in pathology for disease-associated epigenetic changes. In that respect, epigenetic epidemiology will play an important role. We have addressed here two examples, related to endocrine disrupting chemicals and breast cancer, as well as the detrimental effect of smoking, all occurring in great part through changes in specific promoter methylation. Epigenetics provides a link between the environment, lifestyle, diet, and disease development. Each disease process results from unique profiles of genetic susceptibility, as well as epigenetic, mRNA, and protein patterns in relation to the macroenvironment and tissue microenvironment. Thus, epigenetic epidemiology may represent a logical evolution of genome-wide association studies. Although epigenome-wide association studies are attracting increasing attention, currently, there is a fundamental problem in that each cell within one individual has a unique, time-varying epigenome [60]. Having a similar conceptual framework to systems biology, the holistic epigenetic epidemiology approach enables us to link potential etiological factors to specific molecular pathology, and gain novel pathogenic insights on causality.

References

[1] Jablonka E. Epigenetic epidemiology. Int J Epidemiol 2004;33:929–35.
[2] Waterland RA, Michels KB. Epigenetic epidemiology of the developmental origins hypothesis. Annu Rev Nutr 2007;27:363–88.
[3] Feinberg AP, Tycko B. The history of cancer epigenetics. Nat Rev Cancer 2004;4:143–53.
[4] Salnikow K, Zhitkovich A. Genetic and epigenetic mechanisms in metal carcinogenesis and cocarcinogenesis: nickel, arsenic, and chromium. Chem Res Toxicol 2008;21:28–44.
[5] van Veldhoven K, Rahman S, Vineis P. Epigenetics and epidemiology: models of study and examples. Cancer Treat Res 2014;159:241–55.
[6] Christensen BC, Marsit CJ. Epigenomics in environmental health. Front Genet 2011;2:84.
[7] Tabish AM, Poels K, Hoet P, Godderis L. Epigenetic factors in cancer risk: effect of chemical carcinogens on global DNA methylation pattern in human TK6 cells. PLoS One 2012;7:e34674.
[8] Martinez L, Jimenez V, Garcia-Sepulveda C, Ceballos F, Delgado JM, Nino-Moreno P, et al. Impact of early developmental arsenic exposure on promotor CpG-island methylation of genes involved in neuronal plasticity. Neurochem Int 2011;58:574–81.
[9] Tobi EW, Lumey LH, Talens RP, Kremer D, Putter H, Stein AD, et al. DNA methylation differences after exposure to prenatal famine are common and timing- and sex-specific. Hum Mol Genet 2009;18:4046–53.
[10] Kuehnen P, Mischke M, Wiegand S, Sers C, Horsthemke B, Lau S, et al. An Alu element-associated hypermethylation variant of the POMC gene is associated with childhood obesity. PLoS Genet 2012;8:e1002543.
[11] Joubert BR, Haberg SE, Nilsen RM, Wang X, Vollset SE, Murphy SK, et al. 450 K epigenome-wide scan identifies differential DNA methylation in newborns related to maternal smoking during pregnancy. Environ Health Perspect 2012;120:1425–31.
[12] Breitling LP, Yang R, Korn B, Burwinkel B, Brenner H. Tobacco-smoking-related differential DNA methylation: 27 K discovery and replication. Am J Hum Genet 2011;88:450–7.
[13] Monick MM, Beach SR, Plume J, Sears R, Gerrard M, Brody GH, et al. Coordinated changes in AHRR methylation in lymphoblasts and pulmonary macrophages from smokers. Am J Med Genet B Neuropsychiatr Genet 2012;159B:141–51.
[14] Dolinoy DC, Huang D, Jirtle RL. Maternal nutrient supplementation counteracts bisphenol A-induced DNA hypomethylation in early development. Proc Natl Acad Sci USA 2007;104:13056–61.

[15] Yaoi T, Itoh K, Nakamura K, Ogi H, Fujiwara Y, Fushiki S. Genome-wide analysis of epigenomic alterations in fetal mouse forebrain after exposure to low doses of bisphenol A. Biochem Biophys Res Commun 2008;376:563–7.
[16] Ho SM, Tang WY, Belmonte de Frausto J, Prins GS. Developmental exposure to estradiol and bisphenol A increases susceptibility to prostate carcinogenesis and epigenetically regulates phosphodiesterase type 4 variant 4. Cancer Res 2006;66:5624–32.
[17] Prins GS, Tang WY, Belmonte J, Ho SM. Perinatal exposure to oestradiol and bisphenol A alters the prostate epigenome and increases susceptibility to carcinogenesis. Basic Clin Pharmacol Toxicol 2008;102:134–8.
[18] Weng YI, Hsu PY, Liyanarachchi S, Liu J, Deatherage DE, Huang YW, et al. Epigenetic influences of low-dose bisphenol A in primary human breast epithelial cells. Toxicol Appl Pharmacol 2010;248:111–21.
[19] Qin XY, Fukuda T, Yang L, Zaha H, Akanuma H, Zeng Q, et al. Effects of bisphenol A exposure on the proliferation and senescence of normal human mammary epithelial cells. Cancer Biol Ther 2012;13:296–306.
[20] Li S, Washburn KA, Moore R, Uno T, Teng C, Newbold RR, et al. Developmental exposure to diethylstilbestrol elicits demethylation of estrogen-responsive lactoferrin gene in mouse uterus. Cancer Res 1997;57:4356–9.
[21] Li S, Hansman R, Newbold R, Davis B, McLachlan JA, Barrett JC. Neonatal diethylstilbestrol exposure induces persistent elevation of c-fos expression and hypomethylation in its exon-4 in mouse uterus. Mol Carcinog 2003;38:78–84.
[22] Bromer JG, Wu J, Zhou Y, Taylor HS. Hypermethylation of homeobox A10 by in utero diethylstilbestrol exposure: an epigenetic mechanism for altered developmental programming. Endocrinology 2009;150:3376–82.
[23] Tang WY, Newbold R, Mardilovich K, Jefferson W, Cheng RY, Medvedovic M, et al. Persistent hypomethylation in the promoter of nucleosomal binding protein 1 (Nsbp1) correlates with overexpression of Nsbp1 in mouse uteri neonatally exposed to diethylstilbestrol or genistein. Endocrinology 2008;149:5922–31.
[24] Sato K, Fukata H, Kogo Y, Ohgane J, Shiota K, Mori C. Neonatal exposure to diethylstilbestrol alters expression of DNA methyltransferases and methylation of genomic DNA in the mouse uterus. Endocr J 2009;56:131–9.
[25] Hossain MB, Vahter M, Concha G, Broberg K. Low-level environmental cadmium exposure is associated with DNA hypomethylation in Argentinean women. Environ Health Perspect 2012;120:879–84.
[26] Huang D, Zhang Y, Qi Y, Chen C, Ji W. Global DNA hypomethylation, rather than reactive oxygen species (ROS), a potential facilitator of cadmium-stimulated K562 cell proliferation. Toxicol Lett 2008;179:43–7.
[27] Takiguchi M, Achanzar WE, Qu W, Li G, Waalkes MP. Effects of cadmium on DNA-(Cytosine-5) methyltransferase activity and DNA methylation status during cadmium-induced cellular transformation. Exp Cell Res 2003;286:355–65.
[28] Benbrahim-Tallaa L, Waterland RA, Dill AL, Webber MM, Waalkes MP. Tumor suppressor gene inactivation during cadmium-induced malignant transformation of human prostate cells correlates with overexpression of de novo DNA methyltransferase. Environ Health Perspect 2007;115:1454–9.
[29] Jiang G, Xu L, Song S, Zhu C, Wu Q, Zhang L, et al. Effects of long-term low-dose cadmium exposure on genomic DNA methylation in human embryo lung fibroblast cells. Toxicology 2008;244:49–55.
[30] Zhang J, Fu Y, Li J, Wang J, He B, Xu S. Effects of subchronic cadmium poisoning on DNA methylation in hens. Environ Toxicol Pharmacol 2009;27:345–9.
[31] Zhu H, Li K, Liang J, Zhang J, Wu Q. Changes in the levels of DNA methylation in testis and liver of SD rats neonatally exposed to 5-aza-2′ -deoxycytidine and cadmium. J Appl Toxicol 2011;31:484–95.

[32] Wang B, Li Y, Tan Yi, Miao X, Liu XD, Shao C, et al. Low-dose Cd induces hepatic gene hypermethylation, along with the persistent reduction of cell death and increase of cell proliferation in rats and mice. PLoS One 2012;7:e33853.
[33] Lo PK, Sukumar S. Epigenomics and breast cancer. Pharmacogenomics 2008; 9:1879–902.
[34] Huang TH, Esteller M. Chromatin remodeling in mammary gland differentiation and breast tumorigenesis. Cold Spring Harb Perspect Biol 2010;2:a004515.
[35] Pfeifer GP, Dammann R. Methylation of the tumor suppressor gene RASSF1A in human tumors. Biochemistry (Mosc) 2005;70:576–83.
[36] Dumitrescu RG. Epigenetic markers of early tumor development. Methods Mol Biol 2012;863:3–14.
[37] Xu J, Shetty PB, Feng W, Chenault C, Bast RC, Issa JPJ, et al. Methylation of HIN-1, RASSF1A, RIL and CDH13 in breast cancer is associated with clinical characteristics, but only RASSF1A methylation is associated with outcome. BMC Cancer 2012;12:243.
[38] Rocco JW, Sidransky D. p16(MTS-1/CDKN2/INK4a) in cancer progression. Exp Cell Res 2001;264:42–55.
[39] Xu X, Gammon MD, Zhang Y, Cho YH, Wetmur JG, Bradshaw PT, et al. Gene promoter methylation is associated with increased mortality among women with breast cancer. Breast Cancer Res Treat 2010;121.
[40] Lundstrom E, Kallberg H, Alfredsson L, Klareskog L, Padyukov L. Gene-environment interaction between the DRB1 shared epitope and smoking in the risk of anti-citrullinated protein antibody-positive rheumatoid arthritis: all alleles are important. Arthritis Rheum 2009;60:1597–603.
[41] Breitling LP. Current genetics and epigenetics of smoking/tobacco-related cardiovascular disease. Arterioscler Thromb Vasc Biol 2013;33:1468–72.
[42] Breitling LP, Salzmann K, Rothenbacher D, Burwinkel B, Brenner H. Smoking, F2RL3 methylation, and prognosis in stable coronary heart disease. Eur Heart J 2012;33:2841–8.
[43] Zhang Y, Yang R, Burwinkel B, Breitling LP, Brenner H. Methylation as a Biomarker of Current and Lifetime Smoking Exposures. Environ Health Perspect 2013.
[44] Maccani JZ, Koestler DC, Houseman EA, Marsit CJ, Kelsey KT. Placental DNA methylation alterations associated with maternal tobacco smoking at the RUNX3 gene are also associated with gestational age. Epigenomics 2013;5:619–30.
[45] Yanagawa N, Tamura G, Oizumi H, Takahashi N, Shimazaki Y, Motoyama T. Promoter hypermethylation of tumor suppressor and tumor-related genes in non-small cell lung cancers. Cancer Sci 2003;94:589–92.
[46] Yanagawa N, Tamura G, Oizumi H, Endoh M, Sadahiro M, Motoyama T. Inverse correlation between EGFR mutation and FHIT, RASSF1A and RUNX3 methylation in lung adenocarcinoma: relation with smoking status. Anticancer Res 2011;31:1211–14.
[47] Yu GP, Ji Y, Chen GQ, Huang B, Shen K, Wu S, et al. Application of RUNX3 gene promoter methylation in the diagnosis of non-small cell lung cancer. Oncol Lett 2012;3:159–62.
[48] Supic G, Kozomara R, Jovic N, Zeljic K, Magic Z. Hypermethylation of RUNX3 but not WIF1 gene and its association with stage and nodal status of tongue cancers. Oral Dis 2011;17:794–800.
[49] de Freitas Cordeiro-Silva M, Stur E, Agnostini LP, de Podesta JR, de Oliveira JC, Soares MA, et al. Promoter hypermethylation in primary squamous cell carcinoma of the oral cavity and oropharynx: a study of a Brazilian cohort. Mol Biol Rep 2012;39:10111–19.
[50] Xu Y, Wang K, Gao W, Zhang C, Huang F, Wen S, et al. MicroRNA-106b regulates the tumor suppressor RUNX3 in laryngeal carcinoma cells. FEBS Lett 2013; 587:3166–74.

[51] Novakovic B, Ryan J, Pereira N, Boughton B, Craig JM, Saffery R. Postnatal stability and tissue- and time-specific effects of methylation change in response to maternal smoking throughout pregnancy. Epigenetics 2013;9.

[52] Philibert RA, Beach SR, Lei MK, Brody GH. Changes in DNA methylation at the aryl hydrocarbon receptor repressor may be a new biomarker for smoking. Clin Epigenetics 2013;5:19.

[53] Zudaire E, Cuesta N, Murty V, Woodson K, Adams L, Gonzalez N, et al. The aryl hydrocarbon receptor repressor is a putative tumor suppressor gene in multiple human cancers. J Clin Invest 2008;118:640–50.

[54] Heijmans BT, Tobi EW, Stein AD, Putter H, Blauw GJ, Susser ES, et al. Persistent epigenetic differences associated with prenatal exposure to famine in humans. Proc Natl Acad Sci USA 2008;105:17046–9.

[55] Lisboa PC, de Oliveira E, de Moura EG. Obesity and endocrine dysfunction programmed by maternal smoking in pregnancy and lactation. Front Physiol 2012;3:437.

[56] Rahmouni K, Fath MA, Seo S, Thedens DR, Berry CJ, Weiss R, et al. Leptin resistance contributes to obesity and hypertension in mouse models of Bardet-Biedl syndrome. J Clin Invest 2008;118:1458–67.

[57] Suter M, Abramovici A, Showalter L, Hu M, Shope CD, Varner M, et al. In utero tobacco exposure epigenetically modifies placental CYP1A1 expression. Metabolism 2010;59:1481–90.

[58] Suter M, Ma J, Harris A, Patterson L, Brown KA, Shope C, et al. Maternal tobacco use modestly alters correlated epigenome-wide placental DNA methylation and gene expression. Epigenetics 2011;6:1284–94.

[59] DuBois BN, O' Tierney-Ginn P, Pearson J, Friedman JE, Thornburg K, Cherala G. Maternal obesity alters feto-placental cytochrome P4501A1 activity. Placenta 2012;33.

[60] Ogino S, Lochhead P, Chan AT, Nishihara R, Cho E, Wolpin BM, et al. Molecular pathological epidemiology of epigenetics: emerging integrative science to analyze environment, host, and disease. Mod Pathol 2013;26:465–84.

CHAPTER

6

DNA Methylation and Viral Infections

OUTLINE

M. Neidhart: DNA Methylation and Complex Human Disease.
DOI: http://dx.doi.org/10.1016/B978-0-12-420194-1.00006-3

6.1 INTRODUCTION

Epigenetics of human cancer is becoming an area of emerging research direction due to a growing understanding of specific epigenetic pathways and rapid development of detection technologies (Chapters 2,7). Aberrant promoter hypermethylation is a prevalent phenomenon in human cancers. Tumor suppressor genes are often hypermethylated, in part due to the increased activity or deregulation of DNA methyltransferases (DNMTs). A certain number of viruses have been associated with the development of cancers. One of the first defense lines against viruses is the recruitment of DNMTs and methylation of the viral genome [1]. Viruses are associated with 15–20% of human cancers worldwide. In the previous century, many studies were directed towards elucidating the molecular mechanisms and genetic alterations by which viruses cause cancer. Since the virus genome may cause disruption of the host genome by insertion mutations and chromosomal rearrangements, cells infected by virus can be predisposed to cancer. For example, an insertion of viral promoter sequences adjacent to host proto-oncogenes can cause neoplasia [1]. Some DNA and RNA viruses may even carry their own viral oncogenes with transforming activities. Oncoviruses tend to cause persistent infections because they have developed strategies for evading the host immune response. However, viruses are not sufficient for carcinogenesis, and additional factors, including host immunity and cell mutations, are necessary for a tumoral process to be initiated [2]. This chapter focuses on this defense mechanism, which in turn can result in host promoter hypermethylation and tumor suppressor gene silencing. We will review the signal pathways that are involved, particularly in cancers closely associated with various viruses.

6.2 VIRUS INFECTION AND METHYLATION CHANGES

Virus infection, especially with DNA viruses and retroviruses which may cause insertion of viral DNA sequence into the host genome, often triggers the host defense mechanism, in particular DNA methylation machinery, to cause the methylation of foreign movable viral genome (episome) [1]. DNA methylation of these viral DNAs is an effective method to silence viral gene expression. This process, catalyzed by DNMTs, involves the addition of a methyl group in the cytosine residue primarily onto the viral DNA promoter CpG islands. Methylation labels the viral genome, and this in turn recruits other protein factors, such as methyl CpG binding proteins (MeCPs), to form specific protein complex on methylated CpG island containing promoter. These regions are then inaccessible to transcription factors. To maintain the genome stability, DNA methylation induced gene silencing on the integrated virus DNA is crucial for mammalian cells to subdue the intruder's DNA. Exogenous retroviral DNAs can be inactivated due to *de novo* DNA methylation, and are able to integrate into human genome to become part of it [3]. Furthermore, the human genome contains DNA sequences resembling retroviral long terminal repeats (LTRs) and the transposable elements. Interestingly, viruses can find ways to counteract this cell-based modification and even take advantages of it [4]. DNA methylation is responsible, through the silencing of repetitive genomic sequences, for the inactivation of integrated foreign DNA – that is, retrotransposons such as LINE-1 and ALU elements, proviral sequences from endogenous retroviruses, and other transposable elements [3,5]. In this regard, it has been proposed that DNA methylation may have arisen as a genome defense system to prevent chromosomal instability, translocations, and gene disruption caused by the reactivation of these transposable DNA sequences [3,6]. In addition, eukaryotic cells have developed several defense mechanisms against the uptake, integration, and continued expression of foreign DNA (such as viruses) in which gene-specific sequence methylation has an important role [7]. This *de novo* methylation of foreign genes in eukaryotic genomes can be viewed as an ancient cell defense mechanism against the intrusion of foreign genetic material [7–9] (Figure 6.1).

6.2.1 Epstein-Barr Virus

Epstein-Barr virus (EBV) is a ubiquitous human herpes virus, with a huge double-stranded DNA genome of ~172 kb. EBV is one of the

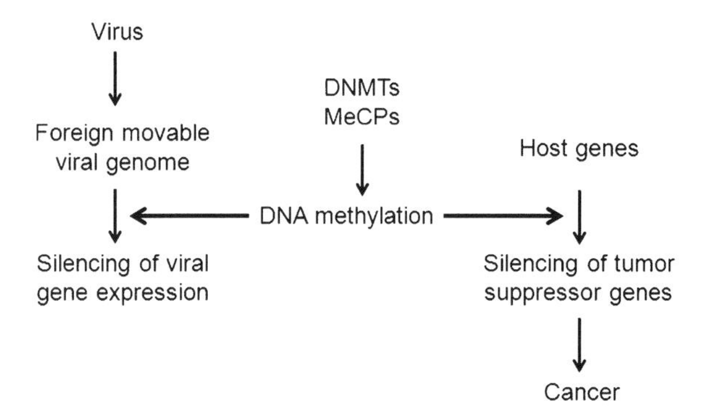

FIGURE 6.1 Viruses induce the recruitment of DNA methyltransferases (DNMTs). The DNA methylation machinery is responsible for silencing viral gene expression. This is done together with methyl binding proteins (MeCPs). The increased activity of DNMTs, however, can also have harmful side effects by silencing host genes. In cases where this affects tumor suppressor genes, the development of cancerous cells is favored.

most common viruses in humans: 90% of the world's adult population is infected by it. The virus persists in most individuals as a lifelong asymptomatic infection of B lymphocytes. When the primary infection occurs during adolescence or early adulthood, it induces infectious mononucleosis in about half of the individuals. EBV is also closely associated with several human malignancies, including nasopharyngeal carcinoma (NPC), Burkitt's lymphoma, T-cell lymphoma, and gastric carcinoma. In NPC, EBV infection is predominantly latent and viral gene expression is under strict control. In addition, EBV infection is involved in the etiology of several lymphoid and epithelial malignancies in immune-compromised humans, such as AIDS and post-transplant patients [9]. DNA methylation in general suppresses viral gene expression, yet methylation of distinct sets of gene promoters in EBV signifies different latency stages (I–III) which are essential for EBV to establish persistent infection and help the virus to evade the host immune system [10]. It is conceivable that once the viral genes are modified, this may also affect the methylation status of the host genome. The epigenetic regulation of viral genes is a very important event in the life cycle of EBV [11]. The expression of latent viral oncogenes, RNAs, and miRNAs is under epigenetic control by DNA methylation and histone modifications that result in the complete silencing of the EBV genome. Although the linear double-stranded EBV genomes packaged into virions are unmethylated, certain viral gene promoters of latent, circular EBV genomes may undergo increased methylation [12]. The EBV genome exists in two different states in the host – lytic and latent – but all phases of the EBV life cycle are associated with human disease. The lytic cycle causes the expression of several viral proteins that

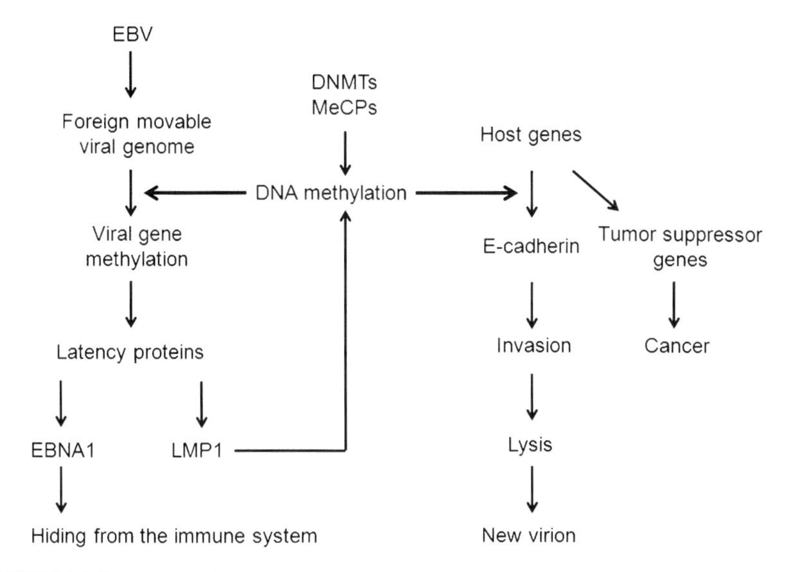

FIGURE 6.2 Upon infection with Epstein-Barr virus (EBV), the viral genome is methylated; however, some latency proteins continue to be produced. The EBN1 gene is incompletely methylated, and the EBN1 protein allows escape from the immune system survey. The LMP1 protein further enhances DNA methylation, resulting in silencing of the E-cadherin gene and thus favoring invasion. As a side effect, tumor suppressor genes can be also affected, triggering the development of cancer cells.

finally produce infectious virions. The latent cycle does not give rise to the production of virions, but a limited number of proteins, known as "latent proteins," are produced in this stage. These latent genes encode the EBV nuclear antigens EBNA1, 2, 3A, 3B, 3C, and LP; the latent membrane proteins (LMP)1, 2A, and 2B; the BARF-1 protein; and the EBER 1, EBER 2, and BART RNA transcripts [9]. In addition, EBV codes for at least 20 miRNAs that are expressed in latently infected cells [13]. EBNA1 and LMP1 play a role in the latency phase, hiding the virus from the immune system and thereby favoring a persistent infection and ongoing invasion (Figure 6.2).

6.2.1.1 EBV Nuclear Antigen 1

Methylation of the EBV genome helps the virus to hide from the host immune system, inhibiting expression of viral latency proteins that are recognized by cytotoxic T cells [14]. The methylation pattern of the EBV genome depends on the stages of EBV latency and the type of tumor. It has been demonstrated that certain viral promoters of latent circular EBV genomes may undergo increased methylation [12]. The DNA methylation level in the EBV genome increases dramatically from an asymptomatic infection to the final neoplastic stages, and has been shown to be involved in regulation of viral gene expression. One of the EBV genes whose expression is epigenetically regulated is EBV nuclear antigen 1 (EBNA1), which plays a crucial function in viral replication and episome maintenance in latency. The methylation profile of the promoter gene coding for EBNA1 is the most extensively analyzed epigenetic mark in this virus. Four promoters (Cp, Wp, Qp, and Fp) that drive EBNA1 expression have been described [15], and there is evidence that DNA methylation regulates the expression of this gene and ultimately defines the type of latency stage. CpG methylation switches off gene expression and induces the alternative transcription of EBNA1 from various promoters in the different latency stages that, at the same time, are associated with the pathology that the virus induces, from a simple infection to a lymphoma or carcinoma [12]. Although Wp, Cp, and Fp have been found methylated in a specific latency type, Qp remains unmethylated, irrespective of its activity, and is thought to be regulated by a putative repressor protein and specific histone modifications [10,14,15].

6.2.1.2 Latent Membrane Protein 1

Another of the viral genes, the latent membrane protein 1 (LMP1), is expressed in ~70% of NPCs. LMP1 is an oncoprotein with transforming capability [16]. It has been demonstrated that LMP1 induces the activation of DNMT1, leading to an increase in the methylation of tumor suppressor gene promoters in nasopharyngeal carcinoma cells [17]. Furthermore, human epithelial cells expressing LMP1 have been found to have higher invasive ability, in correlation with reduced E-cadherin expression [18]. This is the result of LMP1-induced hypermethylation of the E-cadherin gene promoter through activation of DNMT1, 3A, and 3B [19]. Addition of the DNMT inhibitor, 5′-Aza-2′-cytidine, in LMP1-expressing epithelial cells restores expression of E-cadherin. This suggests that LMP1 regulates the expression of specific cellular gene via epigenetic modification. Besides E-cadherin promoter, other tumor suppressor gene promoters, such as RASSF1, retinoic acid receptor, and $p16^{INK4}$, are also hypermethylated in nasopharyngeal carcinoma cells [20]. Similarly, LMP2A protein intermediates in the activation of DNMT1 that leads to downregulation of PTEN gene expression in gastric carcinoma cells [21].

6.2.2 Hepatitis B Virus

Hepatitis B virus (HBV) infection is characterized by acute and chronic infections of the liver, resulting in hepatitis B, cirrhosis, and hepatocellular carcinoma (HCC). It is a member of the hepadnaviruses, and contains a double-stranded circular DNA genome of 3.2 kb. It occurs throughout the world and is characterized by acute and chronic infections of the liver, resulting in hepatitis B, cirrhosis, and tumors. Thus, integrated HBV genome is frequently detected in chronic virus infection and hepatocellular carcinoma (HCC) patients [1]. This has been associated with increased DNA methylation that first targets the viral genome but thereafter also silences host genes, in great part due to the interaction of a viral protein with DNMTs (Figure 6.3).

6.2.2.1 *P16^{INK4}*

P16^{INK4} is frequently inactivated in HBV-related HCC tumor specimens due to extensive CpG methylation within the promoter region [22]. The gene product p16^{INK4} acts as an inhibitor of cyclin-dependent kinases (CDKs) 4 and 6, and inhibits the kinase activity of CDK 4/6 on retinoblastoma (RB) protein to cause G_1-arrest. Inactivation of p16^{INK4} in neoplastic cells promotes uncontrolled cell proliferation. Methylated p16^{INK4} is even present in chronic hepatitis and cirrhosis but with relatively lower frequency [23], indicating that inactivation of p16^{INK4} may occur in the early precancerous stage.

6.2.2.2 *Glutathione S-transferase*

In hepatitis B virus (HBV)-associated HCC tissues, the promoter region of S-transferase (glutathione S-transferase [GSTP1]), encoding an enzyme that protects cells from oxidation damage and electrophilic

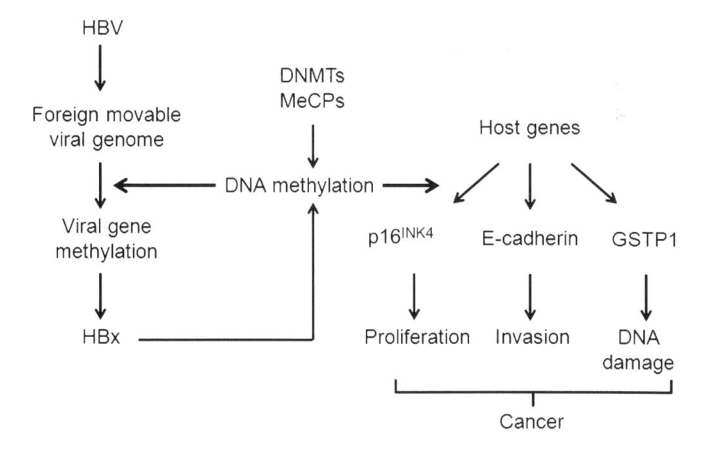

FIGURE 6.3 Upon infection with hepatitis B virus (HBV), the viral genome is methylated; however, the HBx gene escapes silencing. It further enhances DNA methylation, resulting in silencing of the p16^{INK4}, E-cadherin, and GSTP1 genes, resulting in uncontrolled proliferation and invasion, and incomplete DNA repair. This results in the development of cancer cells.

carcinogens, is commonly hypermethylated [24]. In those cases, GSTP1 protein is absent. Thus, HBV-related HCC is highly linked to the hypermethylation of cellular genes p16^{INK4} and GSTP1. However, it is still not known which HBV viral gene is responsible for CpG hypermethylation.

6.2.2.3 HBx Protein

Hepatitis B virus codes for several proteins from four known genes (P, S, X, and C). The P gene codes for DNA polymerase, the S gene for three polypeptides (HBs) of different sizes (large, medium, and small), the X for an HBx protein that is thought to be involved in hepatocarcinogenesis, and the C for the core protein. HBx is a key factor initiating epigenetic alterations induced by this virus. HBx interacts with DNMTs to produce alterations in the expression of key genes through methylation of their promoters [11,25,26]. DNMT1 expression can be activated by HBx, and this increase can then inhibit the expression of tumor-suppressor genes such as P16^{INK4} or E-cadherin [25]. HBx can also regulate the expression of DNMT3a and DNMT3b or interact directly with them, resulting in the hypermethylation or hypomethylation of the promoter regions modulating gene expression [11,26].

6.2.3 Simian Virus 40

Simian virus 40 (SV40) is a member of the polyomaviruses, with a small double-stranded, circular DNA genome of 5 kb. SV40 is a tumor virus of primate origin that has spread to human via contaminated poliovirus vaccines. SV40 large T antigen is a viral protein required for viral DNA replication, and a potent transforming protein. It binds and inactivates tumor suppressor gene products, p53, and RB proteins. SV40 infection has been linked to a number of human cancers, including mesotheliomas, bone and brain tumors, and lymphomas (Figure 6.4).

6.2.3.1 Ras-Association Domain Family 1A

Aberrant methylation of the tumor suppressor gene Ras-association domain family 1A (*RASSF1A*) is detected in SV40-associated malignant mesothelioma (MM). By using methylation specific PCR, the methylation percentage of four genes (hyperplastic polyps [*HPP1*], *RASSF1A*, *cyclin D2*, and Ras-related associated with diabetes [*RRAD*]) was significantly higher in SV40-positive MMs than in SV40-negative MMs [27].

6.2.3.2 DNA Methyltransferase 3B

In addition to inactivating the tumor suppressor gene, SV40 T antigen has been shown to act in concert with activated Ras and telomerase

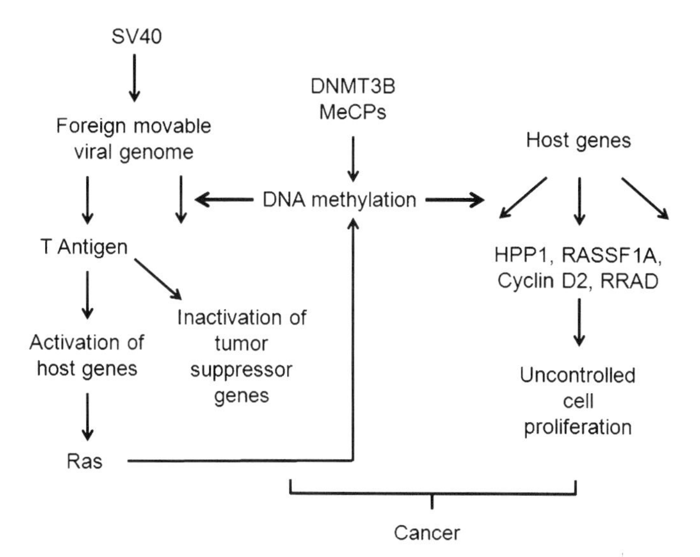

FIGURE 6.4 Upon infection with simian virus 40 (SV40), the viral genome is methylated, but the viral T antigen has already altered host gene expression. On the one hand, SV40 T antigen inactivates tumor suppressor genes; on the other hand, it increased Ras signaling. This results in the recruitment of DNMT3B, a *de novo* DNA methyltransferase, and the methylation of specific promoters that lead to uncontrolled cell proliferation and development of cancer cells.

to form colonies in soft agar assay by using normal human bronchial cells (NHBCs) [28]. The expression of one of the DNA methytransferase 3B (DNMT3B) isoforms was increased in these transformed cells. The introduction of DNMT3B antisense RNA suppressed T antigen-induced transformation. Similarly, mouse embryo fibroblasts expressing T antigen and Ras formed soft agar colonies and tumors, yet fibroblasts from Dnmt3b$^{-/-}$ mice did not grow on soft agar. Furthermore, the transcription of two tumor suppressor genes, fragile histidine triad (FHIT) and tumor suppressor in lung cancer (TSLC1), was reduced in transformed NHBCs, while DNMT3B antisense RNA reactivated the expression of FHIT and TSLC1. These results indicate that T antigen and Ras work together to activate DNMT3B, a *de novo* DNA methyltransferase.

6.2.4 Human Papilloma Virus

Human papilloma viruses are currently believed to be the most common cause of cancers induced by an infectious agent. The HPV genome is divided into three regions: an upstream regulatory region or long control region (LCR); the early region, composed of six open reading frames (ORFs) known as E1, E2, E4, E5, E6, and E7; and the "late region," with two ORFs coding for viral structural proteins L1 and L2. DNA methylation in host cell genes has been investigated as a potential biomarker for early detection of cervical cancer by extensive screening of premalignant and invasive cervical carcinoma samples [29]. Epigenetic alterations such as changes in the DNA methylation pattern of viral and host genomes as well as histone modification are very often associated with HPV infection and cervical carcinogenesis. Methylation of HPV DNA takes place regularly *in vivo* in cervical cells, in clinical samples as well as in cell cultures [30]. The *de novo* methylation is a host defense mechanism for silencing viral replication and transcription that is exploited by the virus to maintain a long-latency infection [14]. Viral DNA hypermethylation is more closely associated with carcinomas than with asymptomatic infections or dysplasia (Figure 6.5).

6.2.4.1 Long Control Region

As the HPV genome does not code for any known gene involved in the DNA methylation machinery, it is believed that the viral genome is methylated by human host cell DNMTs [9]. In HPV16 and HPV18 types, the long control region (LCR) and E6 sequences were commonly found to be unmethylated independent of the stage of neoplastic progression, although the L1 region was densely methylated [31]. In the case of HPV16, the LCR was observed to be methylated in some

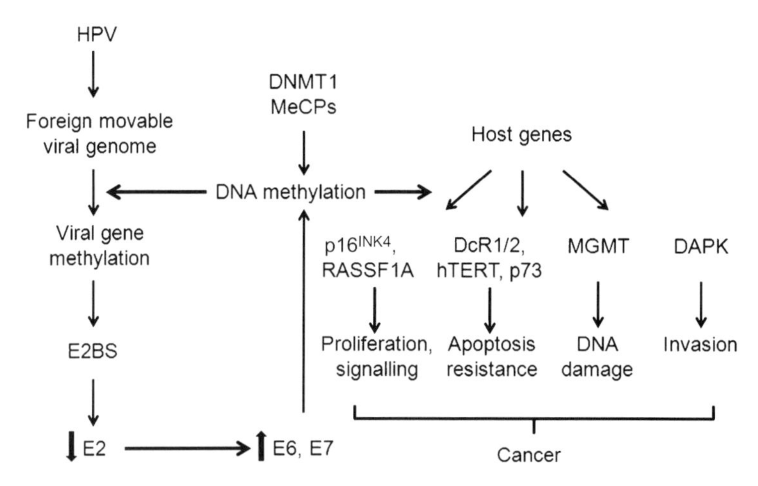

FIGURE 6.5 Upon infection with human papilloma virus (HPV), the viral genome is methylated. Methylation of the E2BS sequence inhibits the binding of E2 protein, which in turn reactivates the E6 and E7 proteins. In particular E7 binds DNMT1, and stimulates its transcription and activity. This results in methylation of specific promoters that lead to uncontrolled cell proliferation; resistance to apoptosis; DNA damage; and invasion and development of cancer cells.

primary cervical carcinomas, especially at E2 binding sites (E2BS, located within the LCR) [14,31]. It has been shown *in vitro* that DNA methylation of the E2BS sequence inhibits the binding of E2, and that this methylation is related to the reactivation of E6 and E7 viral proteins. In this manner, in addition to gene disruption at the E2 locus, the methylation of specific LCR sequences seems to be another mechanism that induces overexpression of the oncogenic proteins E6 and E7 in advanced stages of carcinogenesis induced by HPV16. This is especially interesting because the use of DNA demethylating agents can induce recruitment of E2 to its upstream regulatory region-binding sites and reduce E6 and E7 expression. HPV16 E7 binds DNMT1 and stimulates its enzymatic activity [32], and may activate transcription of DNMT1 as well [33].

6.2.4.2 Changes in Host Phenotype

Changes in the DNA methylation pattern might also be found in the host genome. Several tumor suppressor genes possessing CpG islands in the promoter region are frequently inactivated by hypermethylation in cervical cancer cells [34]. Epigenetic silencing of genes involved in cell cycle regulation [35] (e.g., p16), apoptosis [36–38] (DcR1/DcR2, hTERT, p73), DNA repair [39] (MGMT), development and differentiation [39] (RARβ), hormonal response [40] (ERα), and cellular signaling [41] (RASSF1A), invasion, and metastasis [39] (DAPK) has been detected in cervical cancer cells. However, it is still not clear whether methylation of tumor suppressor genes in cervical cancer cells is induced by HPV viruses or is an effect of carcinogenesis. Difficulties with distinction may result from the fact that almost all cervical cancer cells are HPV positive at diagnosis.

6.2.5 Human Cytomegalovirus

Human cytomegalovirus (HCMV) is the largest herpesvirus (>230 kb) and infects most adults worldwide. DNA methylation in HCMV gene promoters has been associated with *in vitro* and *in vivo* silencing, although only in animal models, in which it is thought to contribute to the modulation of viral latency [42]. Changes in histone acetylation and methylation patterns in HCMV have been found in different stages of infection [43].

6.2.6 Karposi's Sarcoma-Associated Herpesvirus

Karposi's sarcoma-associated herpesvirus (KSHV), also known as human herpesvirus 8 (HHV-8), is associated with Kaposi's sarcoma

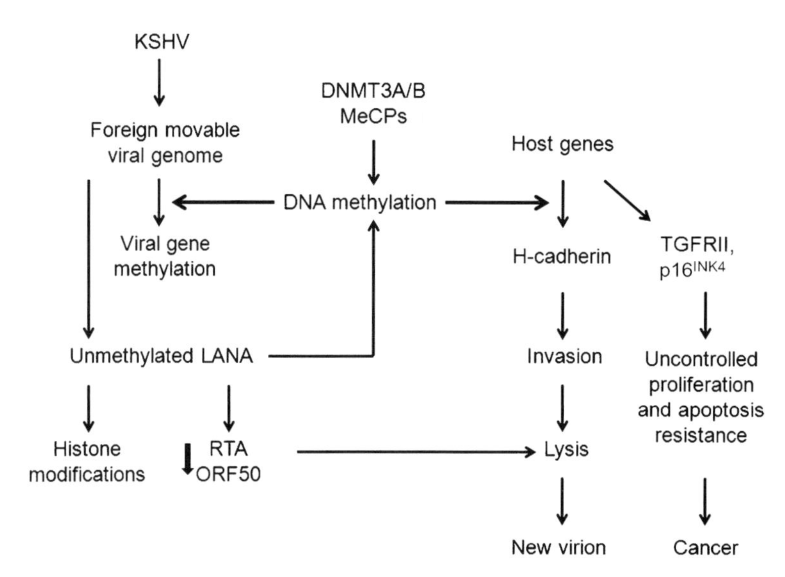

FIGURE 6.6 Upon infection with Karposi's sarcoma-associated herpesvirus (KSHV), the viral genome is methylated, with the exception of the LANA sequence. The viral LANA protein induces histone modifications and inhibits the RTA ORF50 promoter that is responsible for the lytic cycle. LANA also supports the recruitment of *de novo* DNA methyltransferases (DNMT3A/B). This results in methylation of specific promoters that lead to uncontrolled cell proliferation; resistance to apoptosis; and invasion and development of cancer cells.

(one of the commonest cancers in human immune deficiency virus [HIV]-infected individuals) as well as primary effusion lymphoma. There are several KSHV genes closely associated with latency that are potentially tumorigenic, including the latency-associated nuclear protein (LANA). This participates in the maintenance of the episome (i.e., viral material that can be integrated in the host DNA) during latency (Figure 6.6).

6.2.6.1 *Latency-Associated Nuclear Protein*

Latency-associated nuclear protein (LANA) is unmethylated [44] and is known to be associated with histone modifications and subsequent gene expression alterations in host and viral genes that contribute to viral oncogenesis [45–47]. Replication and transcription activator (RTA) is encoded by ORF50 of the viral genome, and is the lytic switch of KSHV. Methylation of RTA (ORF50) promoter is used by virus to maintain the latent cycle. LANA supports maintenance of the latent cycle by the association with ORF50 promoter or binding cellular factors which normally interact with ORF50 [48].

6.2.6.2 *DNMT3A/B and* de novo *Methylation*

It has also been shown that the recruitment of DNMT3A/B causes downregulation of genes by promoter methylation [49,50]. Thus, association and relocalization of DNMT3a induced by LANA has an influence on methylation of the H-cadherin gene promoter. It has also been reported that LANA associates with the TGF-β type II receptor (TβRII) promoter and induces its methylation [50]. Reduction of TβRII expression in primary effusion lymphoma cells results in defective TGF-β signaling pathway, which is important for preventing the development of tumors because it inhibits growth and promotes apoptosis. Another tumor suppressor, $p16^{INK4a}$, is also found to be inactivated by promoter hypermethylation [51]. However, it has not been proved that LANA participates in its downregulation.

6.2.7 Human T-Cell Leukemia Virus Type 1

Human T-cell leukemia virus type 1 (HTLV-1) is a single-stranded RNA retrovirus that infects 20 million people worldwide. HTLV-1 infection is causally associated with a variety of human diseases, including leukemia/lymphoma, myelopathy, uveitis, and arthropathy [52]. Tax protein of HTLV-I, which is considered oncogenic, binds to transcription factors or other cytoplasmic cellular molecules involved in the fundamental cell function and thereby induces cellular changes. Tax is required for transformation, but thereafter can be silenced by methylation [53]. The repression of this viral protein could be associated with the evasion of the immune response, as Tax is the main target of the host's cytotoxic T-lymphocyte response. The next stage can be hypermethylation of the whole incorporated viral sequence, beginning with the 5′-LTR; this often occurs in HTLV-1 induced acute leukemia [54].

6.2.8 Human Immunodeficiency Virus

Human immunodeficiency virus (HIV)-1 is a retrovirus and the causative agent of acquired immunodeficiency syndrome (AIDS). CpG methylation has been implicated in silencing of the integrated provirus genome [55]. *In vitro* studies have shown that DNA methylation suppresses the promoter activity of the HIV-1 long terminal repeat (LTR) [56]. In addition to SV40 and EBV, infection with HIV also causes upregulation of DNMT1 protein expression [57,58]. This activation in turn elevates the global genomic methylation and, conversely, suppresses expression of other genes, such as interferon-gamma (IFN-γ, a cytokine important for immune response during virus infection [57]) and $p16^{INK4}$, a common altered tumor suppressor gene found in cancers.

This aberrant methylation may correlate to AIDS and AIDS-associated malignancies; however, it is not clear which HIV viral gene is involved in such a process.

6.3 VIRUS INFECTIONS, DNA METHYLATION, AND CANCER

Increasing evidence reveals that oncogenic viruses also contribute to the epigenetic changes that are characteristic for cancer cells. Tumor-associated viruses interfere with host epigenetic machinery and cause aberrations of DNA methylation as well as changes in histone modifications. Many studies have shown that viral oncoproteins induce expression and interact with cellular DNMTs [32]. Thus, apart from introducing genetic changes, the presence of the viral genome correlates to the aberrant methylation profile in human cancers. Viruses that establish latent infections need to avoid recognition by the immune system, as this would otherwise eliminate the infection. Different viral evasion strategies have been identified, but all of them are essentially aimed at camouflaging the virus in the host cell, restricting the expression of viral genes and proteins that are indispensable for viral persistency, and avoiding the expression of genes associated with immune response. Viral DNA methylation could be the masking mechanism by which many viruses are able to achieve this [14]. The great majority of the viruses mentioned above, altering the methylome profile, have been associated with cancer development. More importantly, this deregulated methylation can be induced by upregulation of the key methylation enzymes' DNMTs. Abnormal methylation can predispose cells to the precancerous stage through inactivation of tumor suppressor genes, cell cycle regulated genes, and cadherins by promoter hypermethylation.

6.3.1 Global DNA Hypomethylation

Global DNA hypomethylation, promoter methylation, aberrant expression of non-coding RNAs, and dysregulated expression of other epigenetic regulatory genes such as *EZH2* occur upon virus infections, such as HBV [59]. In turn, global hypomethylation leads to aberrant overexpression of oncogenes and genome-wide chromosomal instability, which increases the risk of chromosomal translocations [60,61]. An explanation for global DNA methylation is the deficiency in S-adenosylmethionine, the cell's methyl donor. For example, in hepatocellular carcinoma induced by HBV, methionine adenosyltransferase

(MAT1A) is deregulated. This enzyme converts L-methionine into S-adenosylmethionine. MAT1A is epigenetically regulated, and there are two CpG sites in the first exon, the methylation of which correlates with reduced MAT1A expression. In the liver, deregulation of MAT1A is in turn linked to the overexpression of hepatocyte growth factor (HGF) [62,63]. Downregulation of one or more constituent enzymes of the methionine metabolism may reduce the availability of the labile methyl pool [64], but further studies are necessary.

6.3.2 Recruitment of DNMTs

Viral proteins can produce epigenetic alterations [26,65–67], and it is hypothesized that these are mainly responsible for the carcinogenesis induced by viruses. DNMTs are key enzymes in inactivating transcription by DNA hypermethylation and ensuring proper gene repression. In order to maintain a harmonious internal environment, the protein level or enzymatic activity of DNMTs should be tightly controlled. Nevertheless, disruption of this delicate balance results in chaos. The fact that EBV LMP1, as well as HBV HBx, HPV E7, and KSHV LANA, activates DNMTs highlights the importance of viral proteins as initiators or co-factors to induce epigenetic alterations. In fact, LMP1 is the first viral gene that was identified as having the ability to activate DNMTs via a defined signaling pathway. Upon SV40 infection, the Ras pathways are activated by T antigen: this in turn activates *de novo* DNA methyltransferase 3B [28].

6.3.3 Aberrant DNA Hypermethylation

Alongside global DNA hypomethylation, localized gene promoter hypermethylation frequently occurs. These two phenomena seem to be at odds, and further study is required to elucidate the mechanisms driving these distinct processes. Aberrant DNA hypermethylation induced by viral genes is one of the methods by which the virus contributes to tumor development. Besides E-cadherin promoter, other tumor suppressor gene promoters such as RASSF1A, retinoic acid receptor (*RAR*) beta2, $p16^{INK4}$, and p14 are also hypermethylated in nasopharyngeal carcinoma [68–70] that is induced by CMV. This supports the hypothesis that frequent hypermethylation on cellular promoters may be a common phenomenon in many types of cancer. Similarly, SV40-infected human malignant mesothelial cells showed progressive aberrant methylation in seven genes (RASSF1A, HPP1, Decoy receptor 1 [DcR1], target of methylation-induced silencing [TMS1], cellular retinol-binding protein [CRBP1], hypermethylated in cancer [HIC-1], and RRAD), especially after 50 serial cell passages, suggesting the accumulation of epigenetic errors over time or during DNA replication [27]. Similarly, DNMTs are overexpressed in hepatocellular carcinoma (HCC) induced by HBV, compared to non-cancerous liver samples [71]. HCC is characterized as a CpG island methylator phenotype-positive (CIMP+) cancer [72]. CIMP+ is a term used to describe cancers exhibiting DNA hypermethylation at CpG islands, where the number of promoters that become hypermethylated may increase throughout tumor development. Studies of gene promoter methylation have identified increasing DNA methylation in liver tissues from patients with HCC. The number of methylated gene promoters increases between cirrhotic and primary HCC tumor samples, and between primary HCC and metastatic tumor samples. This is important, as it implies that worsening DNA hypermethylation can drive tumor progression [73]. Genes often found to be hypermethylated in HCC include those involved in gene expression (transcription factors), growth regulation, cell cycle progression ($P16^{INK4}$), invasion (E-cadherin), DNA repair (GSTP1), and apoptosis [23–25,74,75]. Obviously, upon infection with such viruses, inactivation of tumor suppressor genes due to promoter hypermethylation is one of the major reasons leading to tumor progression.

6.3.4 Altered Signaling Pathways

6.3.4.1 AP-1/JNK

By using different LMP1 deletion mutants, it has been determined that the C-terminal activation region 2 (CTAR2) of LMP1 is the domain required for DNMT activation. CTAR2 can activate the AP-1/JNK signaling pathway [76]. A JNK inhibitor (SP60012) has been used to show that DNMT is indeed the downstream target of JNK. In addition, mutation of the AP-1 site on *DNMT1* promoter, and *c-Jun* dominant-negative mutant blocked the LMP1-induced DNMT activation. Chromatin immunoprecipitation (ChIP) assay further demonstrated that DNMT1, 3A, 3B, MeCP2, and HDAC1 formed protein complexes on the E-cadherin promoter in the LMP1-positive cells but not in the control cells. Taken together, a single viral oncoprotein in EBV has been identified, which is capable of activating DNMTs via AP-1/JNK signaling pathway.

6.3.4.2 NF-κB

Upon ENV infection, LMP1 also activates NF-κB, inducing the expression of various genes that encode anti-apoptotic proteins and cytokines [77]. Regarding HBV infection, HBx protein and various HBs envelope proteins seem to be responsible for the alteration of major signaling pathways, including the NF-κB and Wnt/β-catenin pathways, which deregulate normal cell processes [78,79].

6.4 ENDOGENOUS RETROVIRUSES

Retroelements constitute approximately 45% of the human genome [80]. Long interspersed nuclear element (LINE) autonomous retrotransposons are predominantly represented by LINE-1, non-autonomous small interspersed nuclear elements (SINEs) are primarily represented by ALUs, and LTR retrotransposons are represented by several families of human endogenous retroviruses (HERVs). The vast majority of LINE and HERV elements are densely methylated in normal somatic cells and are contained in inactive chromatin. Methylation and the chromatin structure together ensure a stable equilibrium between retroelements and their host. Hypomethylation and expression in developing germ cells opens a "window of opportunity" for retrotransposition and recombination that contributes to evolution [81], but also inherited disease [82]. In somatic cells, the presence of retroelements may be exploited to organize the genome into active and inactive regions, to separate domains and functional regions within one chromatin domain, to suppress transcriptional noise, and to regulate transcript stability. Retroelements, particularly ALUs, may also fulfill physiological roles during responses to stress and infections. Reactivation and hypomethylation of LINEs and HERVs may be important in the pathophysiology of cancer [83] and various autoimmune diseases [84,85], contributing to chromosomal instability and chronically aberrant immune responses.

6.5 CONCLUSION

It is now apparent that viral oncoproteins target the elements of cellular epigenetic machinery, changing their expression and/or activity and thus leading to alterations in the epigenetic state of the host cell. Viral-encoded oncoproteins exploit specific epigenetic processes to force normal quiescent cells to replicate, as well as to regulate viral gene expression during infections. DNA methylation in viral promoters modulates viral gene expression and is the mechanism used by many oncoviruses to avoid detection by the host immune system. Viral proteins are involved either directly or indirectly in the activation of DNMTs, especially the DNMT3A/B genes required for *de novo* DNA methylation. Thus aberrant methylation or epigenetic silencing of the key regulators of cell growth, signaling pathways, and tumor suppressor genes are the main causes of tumor formation. Interestingly, the epigenetic changes observed in pathological conditions such as cancer or autoimmunity could be translated into an effect on the activation of some of the retroelements present in our genome, which ultimately could have a direct or indirect role on the initiation and clinical evolution of certain chronic diseases.

References

[1] Li HP, Leu YW, Chang YS. Epigenetic changes in virus-associated human cancers. Cell Res 2005;15:262–71.
[2] McLaughlin-Drubin ME, Munger K. Viruses associated with human cancer. Biochim Biophys Acta 2008;1782:127–50.
[3] Yoder JA, Walsh CP, Bestor TH. Cytosine methylation and the ecology of intragenomic parasites. Trends Genet 1997;13:335–40.
[4] Verma M. Viral genes and methylation. Ann NY Acad Sci 2003;983:170–80.
[5] Colot V, Rossignol JL. Eukaryotic DNA methylation as an evolutionary device. Bioessays 1999;21:402–11.
[6] Rollins RA, Haghighi F, Edwards JR, Das R, Zhang MQ, Ju J, et al. Large-scale structure of genomic methylation patterns. Genome Res 2006;16:157–63.
[7] Doerfler W. Patterns of DNA methylation – evolutionary vestiges of foreign DNA inactivation as a host defense mechanism. A proposal. Biol Chem Hoppe Seyler 1991;372:557–64.
[8] Doerfler W. A new concept in (adenoviral) oncogenesis: integration of foreign DNA and its consequences. Biochim Biophys Acta 1996;1288:F79–99.
[9] Fernandez AF, Esteller M. Viral epigenomes in human tumorigenesis. Oncogene 2010;29:1405–20.
[10] Tao Q, Robertson KD. Stealth technology: how Epstein-Barr virus utilizes DNA methylation to cloak itself from immune detection. Clin Immunol 2003;109:53–63.
[11] Park JH, Jeon JP, Shim SM, Nam HY, Kim JW, Han BG, et al. Wp specific methylation of highly proliferated LCLs. Biochem Biophys Res Commun 2007;358:513–20.
[12] Niller HH, Wolf H, Minarovits J. Regulation and dysregulation of Epstein-Barr virus latency: implications for the development of autoimmune diseases. Autoimmunity 2008;41:298–328.
[13] Bornkamm GW. Epstein-Barr virus and the pathogenesis of Burkitt's lymphoma: more questions than answers. Int J Cancer 2009;124:1745–55.
[14] Fernandez AF, Rosales C, Lopez-Nieva P, Grana O, Ballestar E, Ropero S, et al. The dynamic DNA methylomes of double-stranded DNA viruses associated with human cancer. Genome Res 2009;19:438–51.
[15] Tao Q, Robertson KD, Manns A, Hildesheim A, Ambinder RF. The Epstein-Barr virus major latent promoter Qp is constitutively active, hypomethylated, and methylation sensitive. J Virol 1998;72:7075–83.
[16] Fahraeus R, Rymo L, Rhim JS, Klein G. Morphological transformation of human keratinocytes expressing the LMP gene of Epstein-Barr virus. Nature 1990;345:447–9.
[17] Niemhom S, Kitazawa S, Kitazawa R, Maeda S, Leopairat J. Hypermethylation of epithelial-cadherin gene promoter is associated with Epstein-Barr virus in nasopharyngeal carcinoma. Cancer Detect Prev 2008;32:127–34.
[18] Fahraeus R, Chen W, Trivedi P, Klein G, Obrink B. Decreased expression of E-cadherin and increased invasive capacity in EBV-LMP-transfected human epithelial and murine adenocarcinoma cells. Int J Cancer 1992;52:834–8.
[19] Tsai CN, Tsai CL, Tse KP, Chang HY, Chang YS. The Epstein-Barr virus oncogene product, latent membrane protein 1, induces the downregulation of E-cadherin gene

expression via activation of DNA methyltransferases. Proc Natl Acad Sci USA 2002;99:10084–9.

[20] Pai S, O'Sullivan B, Abdul-Jabbar I, Peng J, Connoly G, Khanna R, et al. Nasopharyngeal carcinoma-associated Epstein-Barr virus-encoded oncogene latent membrane protein 1 potentiates regulatory T-cell function. Immunol Cell Biol 2007;85:370–7.

[21] Hino R, Uozaki H, Murakami N, Ushiku T, Shinozaki A, Ishikawa S, et al. Activation of DNA methyltransferase 1 by EBV latent membrane protein 2A leads to promoter hypermethylation of PTEN gene in gastric carcinoma. Cancer Res 2009;69:2766–74.

[22] Shim YH, Yoon GS, Choi HJ, Chung YH, Yu E. p16 Hypermethylation in the early stage of hepatitis B virus-associated hepatocarcinogenesis. Cancer Lett 2003;190:213–19.

[23] Li X, Hui AM, Sun L, Hasegawa K, Torzilli G, Minagawa M, et al. p16INK4A hypermethylation is associated with hepatitis virus infection, age, and gender in hepatocellular carcinoma. Clin Cancer Res 2004;10:7484–9.

[24] Zhong S, Tang MW, Yeo W, Liu C, Lo YM, Johnson PJ. Silencing of GSTP1 gene by CpG island DNA hypermethylation in HBV-associated hepatocellular carcinomas. Clin Cancer Res 2002;8:1087–92.

[25] Jung JK, Arora P, Pagano JS, Jang KL. Expression of DNA methyltransferase 1 is activated by hepatitis B virus X protein via a regulatory circuit involving the p16INK4a-cyclin D1-CDK 4/6-pRb-E2F1 pathway. Cancer Res 2007;67:5771–8.

[26] Zheng DL, Zhang L, Cheng N, Xu X, Deng Q, Teng XM, et al. Epigenetic modification induced by hepatitis B virus X protein via interaction with de novo DNA methyltransferase DNMT3A. J Hepatol 2009;50:377–87.

[27] Suzuki M, Toyooka S, Shivapurkar N, Shigematsu H, Miyajima K, Takahashi T, et al. Aberrant methylation profile of human malignant mesotheliomas and its relationship to SV40 infection. Oncogene 2005;24:1302–8.

[28] Soejima K, Fang W, Rollins BJ. DNA methyltransferase 3b contributes to oncogenic transformation induced by SV40T antigen and activated Ras. Oncogene 2003;22:4723–33.

[29] Wentzensen N, Sherman ME, Schiffman M, Wang SS. Utility of methylation markers in cervical cancer early detection: appraisal of the state-of-the-science. Gynecol Oncol 2009;112:293–9.

[30] Turan T, Kalantari M, Callejas-Macias IE, Cubie HA, Cuschieri K, Villa LL, et al. Methylation of the human papillomavirus-18 L1 gene: a biomarker of neoplastic progression?. Virology 2006;349:175–83.

[31] Hublarova P, Hrstka R, Rotterova P, Rotter L, Coupkova M, Badal V, et al. Prediction of human papillomavirus 16 e6 gene expression and cervical intraepithelial neoplasia progression by methylation status. Int J Gynecol Cancer 2009;19:321–5.

[32] Burgers WA, Blanchon L, Pradhan S, de Launoit Y, Kouzarides T, Fuks F. Viral oncoproteins target the DNA methyltransferases. Oncogene 2007;26:1650–5.

[33] Woodman CB, Collins SI, Young LS. The natural history of cervical HPV infection: unresolved issues. Nat Rev Cancer 2007;7:11–22.

[34] Szalmas A, Konya J. Epigenetic alterations in cervical carcinogenesis. Semin Cancer Biol 2009;19:144–52.

[35] Nakashima R, Fujita M, Enomoto T, Haba T, Yoshino K, Wada H, et al. Alteration of p16 and p15 genes in human uterine tumours. Br J Cancer 1999;80:458–67.

[36] Shivapurkar N, Toyooka S, Toyooka KO, Reddy J, Miyajima K, Suzuki M, et al. Aberrant methylation of trail decoy receptor genes is frequent in multiple tumor types. Int J Cancer 2004;109:786–92.

[37] Widschwendter A, Gattringer C, Ivarsson L, Fiegl H, Schneitter A, Ramoni A, et al. Analysis of aberrant DNA methylation and human papillomavirus DNA in

cervicovaginal specimens to detect invasive cervical cancer and its precursors. Clin Cancer Res 2004;10:3396–400.

[38] Liu SS, Leung RC, Chan KY, Chiu PM, Cheung AN, Tam KF, et al. p73 expression is associated with the cellular radiosensitivity in cervical cancer after radiotherapy. Clin Cancer Res 2004;10:3309–16.

[39] Narayan G, Arias-Pulido H, Koul S, Vargas H, Zhang FF, Viellella J, et al. Frequent promoter methylation of CDH1, DAPK, RARB, and HIC1 genes in carcinoma of cervix uteri: its relationship to clinical outcome. Mol Cancer 2003;2:24.

[40] Zambrano P, Segura-Pacheco B, Perez-Cardenas E, Cetina L, Revilla-Vazquez A, Taja-Chayeb L, et al. A phase I study of hydralazine to demethylate and reactivate the expression of tumor suppressor genes. BMC Cancer 2005;5:44.

[41] Cohen Y, Singer G, Lavie O, Dong SM, Beller U, Sidransky D. The RASSF1A tumor suppressor gene is commonly inactivated in adenocarcinoma of the uterine cervix. Clin Cancer Res 2003;9:2981–4.

[42] Brooks AR, Harkins RN, Wang P, Qian HS, Liu P, Rubanyi GM. Transcriptional silencing is associated with extensive methylation of the CMV promoter following adenoviral gene delivery to muscle. J Gene Med 2004;6:395–404.

[43] Cuevas-Bennett C, Shenk T. Dynamic histone H3 acetylation and methylation at human cytomegalovirus promoters during replication in fibroblasts. J Virol 2008;82:9525–36.

[44] Chen J, Ueda K, Sakakibara S, Okuno T, Parravicini C, Corbellino M, et al. Activation of latent Kaposi's sarcoma-associated herpesvirus by demethylation of the promoter of the lytic transactivator. Proc Natl Acad Sci USA 2001;98:4119–24.

[45] Sakakibara S, Ueda K, Nishimura K, Do E, Ohsaki E, Okuno T, et al. Accumulation of heterochromatin components on the terminal repeat sequence of Kaposi's sarcoma-associated herpesvirus mediated by the latency-associated nuclear antigen. J Virol 2004;78:7299–310.

[46] Lu F, Day L, Gao SJ, Lieberman PM. Acetylation of the latency-associated nuclear antigen regulates repression of Kaposi's sarcoma-associated herpesvirus lytic transcription. J Virol 2006;80:5273–82.

[47] Stuber G, Mattsson K, Flaberg E, Kati E, Markasz L, Sheldon JA, et al. HHV-8 encoded LANA-1 alters the higher organization of the cell nucleus. Mol Cancer 2007;6:28.

[48] Pantry SN, Medveczky PG. Epigenetic regulation of Kaposi's sarcoma-associated herpesvirus replication. Semin Cancer Biol 2009;19:153–7.

[49] Shamay M, Krithivas A, Zhang J, Hayward SD. Recruitment of the de novo DNA methyltransferase Dnmt3a by Kaposi's sarcoma-associated herpesvirus LANA. Proc Natl Acad Sci USA 2006;103:14554–9.

[50] Di Bartolo DL, Cannon M, Liu YF, Renne R, Chadburn A, Boshoff C, et al. KSHV LANA inhibits TGF-beta signaling through epigenetic silencing of the TGF-beta type II receptor. Blood 2008;111:4731–40.

[51] Poreba E, Broniarczyk JK, Gozdzicka-Jozefiak A. Epigenetic mechanisms in virus-induced tumorigenesis. Clin Epigenetics 2011;2:233–47.

[52] Uchiyama T. Human T cell leukemia virus type I (HTLV-I) and human diseases. Annu Rev Immunol 1997;15:15–37.

[53] Matsuoka M, Jeang KT. Human T-cell leukaemia virus type 1 (HTLV-1) infectivity and cellular transformation. Nat Rev Cancer 2007;7:270–80.

[54] Taniguchi Y, Nosaka K, Yasunaga JI, Maeda M, Mueller N, Okayama A, et al. Silencing of human T-cell leukemia virus type I gene transcription by epigenetic mechanisms. Retrovirology 2005;2:64.

[55] Kauder SE, Bosque A, Lindqvist A, Planelles V, Verdin E. Epigenetic regulation of HIV-1 latency by cytosine methylation. PLoS Pathog 2009;5:e1000495.

[56] Smith SM. Valproic acid and HIV-1 latency: beyond the sound bite. Retrovirology 2005;2:56.
[57] Mikovits JA, Young HA, Vertino P, Issa JP, Pitha PM, Turcoski-Corrales S, et al. Infection with human immunodeficiency virus type 1 upregulates DNA methyltransferase, resulting in de novo methylation of the gamma interferon (IFN-gamma) promoter and subsequent downregulation of IFN-gamma production. Mol Cell Biol 1998;18:5166–77.
[58] Fang JY, Mikovits JA, Bagni R, Petrow-Sadowski CL, Ruscetti FW. Infection of lymphoid cells by integration-defective human immunodeficiency virus type 1 increases *de novo* methylation. J Virol 2001;75:9753–61.
[59] Ozen C, Yildiz G, Dagcan AT, Cevik D, Ors A, Keles U, et al. Genetics and epigenetics of liver cancer. N Biotechnol 2013;30:381–4.
[60] Herath NI, Leggett BA, MacDonald GA. Review of genetic and epigenetic alterations in hepatocarcinogenesis. J Gastroenterol Hepatol 2006;21:15–21.
[61] Calvisi DF, Ladu S, Gorden A, Farina M, Lee JS, Conner EA, et al. Mechanistic and prognostic significance of aberrant methylation in the molecular pathogenesis of human hepatocellular carcinoma. J Clin Invest 2007;117:2713–22.
[62] Kaelin Jr. WG. Cancer and altered metabolism: potential importance of hypoxia-inducible factor and 2-oxoglutarate-dependent dioxygenases. Cold Spring Harb Symp Quant Biol 2011;76:335–45.
[63] Avila MA, Berasain C, Torres L, Martin-Duce A, Corrales FJ, Yang H, et al. Reduced mRNA abundance of the main enzymes involved in methionine metabolism in human liver cirrhosis and hepatocellular carcinoma. J Hepatol 2000;33:907–14.
[64] Karouzakis E, Gay RE, Gay S, Neidhart M. Increased recycling of polyamines is associated with global DNA hypomethylation in rheumatoid arthritis synovial fibroblasts. Arthritis Rheum 2012;64:1809–17.
[65] Tischoff I, Tannapfe A. DNA methylation in hepatocellular carcinoma. World J Gastroenterol 2008;14:1741–8.
[66] Herceg Z, Paliwal A. HBV protein as a double-barrel shot-gun targets epigenetic landscape in liver cancer. J Hepatol 2009;50:252–5.
[67] Shon JK, Shon BH, Park IY, Lee SU, Fa L, Chang KY, et al. Hepatitis B virus-X protein recruits histone deacetylase 1 to repress insulin-like growth factor binding protein 3 transcription. Virus Res 2009;139:14–21.
[68] Lo KW, Kwong J, Hui ABY, Chan SYY, To KF, Chan ASC, et al. High frequency of promoter hypermethylation of RASSF1A in nasopharyngeal carcinoma. Cancer Res 2001;61:3877–81.
[69] Kwong J, Lo KW, To KF, Teo PM, Johnson PJ, Huang DP. Promoter hypermethylation of multiple genes in nasopharyngeal carcinoma. Clin Cancer Res 2002;8:131–7.
[70] Tong JH, Tsang RK, Lo KW, Woo JK, Kwong J, Chan MW, et al. Quantitative Epstein-Barr virus DNA analysis and detection of gene promoter hypermethylation in nasopharyngeal (NP) brushing samples from patients with NP carcinoma. Clin Cancer Res 2002;8:2612–19.
[71] Choi MS, Shim YH, Hwa JY, Lee SK, Ro JY, Kim JS, et al. Expression of DNA methyltransferases in multistep hepatocarcinogenesis. Hum Pathol 2003;34:11–17.
[72] Li B, Liu W, Wang L, Li M, Wang J, Huang L, et al. CpG island methylator phenotype associated with tumor recurrence in tumor-node-metastasis stage I hepatocellular carcinoma. Ann Surg Oncol 2010;17:1917–26.
[73] Puszyk WM, Trinh TL, Chapple SJ, Liu C. Linking metabolism and epigenetic regulation in development of hepatocellular carcinoma. Lab Invest 2013;93:983–90.
[74] Um TH, Kim H, Oh BK, Kim MS, Kim KS, Jung G, et al. Aberrant CpG island hypermethylation in dysplastic nodules and early HCC of hepatitis B virus-related human multistep hepatocarcinogenesis. J Hepatol 2011;54:939–47.

[75] Liu H, Dong H, Robertson K, Liu C. DNA methylation suppresses expression of the urea cycle enzyme carbamoyl phosphate synthetase 1 (CPS1) in human hepatocellular carcinoma. Am J Pathol 2011;178:652–61.
[76] Kieser A, Kaiser C, Hammerschmidt W. LMP1 signal transduction differs substantially from TNF receptor 1 signaling in the molecular functions of TRADD and TRAF2. EMBO J 1999;18:2511–21.
[77] Young LS, Rickinson AB. Epstein-Barr virus: 40 years on. Nat Rev Cancer 2004;4:757–68.
[78] Lupberger J, Hildt E. Hepatitis B virus-induced oncogenesis. World J Gastroenterol 2007;13:74–81.
[79] Gurtsevitch VE. Human oncogenic viruses: hepatitis B and hepatitis C viruses and their role in hepatocarcinogenesis. Biochemistry (Mosc) 2008;73:504–13.
[80] Schulz WA, Steinhoff C, Florl AR. Methylation of endogenous human retroelements in health and disease. Curr Top Microbiol Immunol 2006;310:211–50.
[81] Kaneko-Ishino T, Ishino F. The role of genes domesticated from LTR retrotransposons and retroviruses in mammals. Front Microbiol 2012;3:262.
[82] Hohn O, Hanke K, Bannert N. HERV-K(HML-2), the Best Preserved Family of HERVs: Endogenization, Expression, and Implications in Health and Disease. Front Oncol 2013;3:246.
[83] Carreira PE, Richardson SR, Faulkner GJ. L1 retrotransposons, cancer stem cells and oncogenesis. FEBS J 2014;281:63–73.
[84] Neidhart M, Rethage J, Kuchen S, Kunzler P, Crowl RM, Billingham ME, et al. Retrotransposable L1 elements expressed in rheumatoid arthritis synovial tissue: association with genomic DNA hypomethylation and influence on gene expression. Arthritis Rheum 2000;43:2634–47.
[85] Nakkuntod J, Avihingsanon Y, Mutirangura A, Hirankarn N. Hypomethylation of LINE-1 but not Alu in lymphocyte subsets of systemic lupus erythematosus patients. Clin Chim Acta 2011;412:1457–61.

CHAPTER

7

DNA Methylation and Cancer

OUTLINE

M. Neidhart: *DNA Methylation and Complex Human Disease.*
DOI: http://dx.doi.org/10.1016/B978-0-12-420194-1.00007-5

7.1 INTRODUCTION

Cancer develops through successive disruptions to the controls of cellular proliferation and programmed cell death (Figure 7.1), as well as migration and angiogenesis. This process requires new abilities to be stably expressed so that they can accumulate in a clonal manner [1]. Mechanisms of genetic mutations (i.e., insertion, deletion, and recombination), are involved in persistent phenotypic changes. Therefore, cancer has long been regarded as a disease based mainly on genetics. However, genetic mutations occur at low frequency, and are thus not particularly efficient in the development of cancer cells.

The development of cancer can be viewed as an evolutionary problem [2]. Cancer cells expand by natural selection and genetic drift, regardless of any negative effects on the whole organism. The fitness of a cancer cell is shaped by its interactions with cells and other factors in its microenvironment. Clonal evolution selects for increased proliferation and survival, and can lead to invasion and metastasis. Some cancer cells acquired DNA repair defects, thus favoring the mutation rate. Alternatively, mechanisms of epigenetic control – for example, DNA methylation and histone modifications – represent another possibility to acquire stable pathological abilities [2,3]. It is important to distinguish patients with cancer caused by mutations and those related to epigenetic silencing of so-called "tumor suppressor genes," because the therapeutic strategy is different.

In general, however, the development of cancer cells involves at least two hits (i.e., mutations or epimutations). Genetic and epigenetic mechanisms influence each other and work cooperatively to enable the acquisition of the cancer phenotype [4]. This chapter focuses on the relations between DNA methylation/hydroxymethylation and cancer development, although it is important to note that other epigenetic alterations, such as histone modifications and microRNAs (miRNAs), are also involved. We will also discuss DNA methylation in the context of interplays with other epigenetic mechanisms.

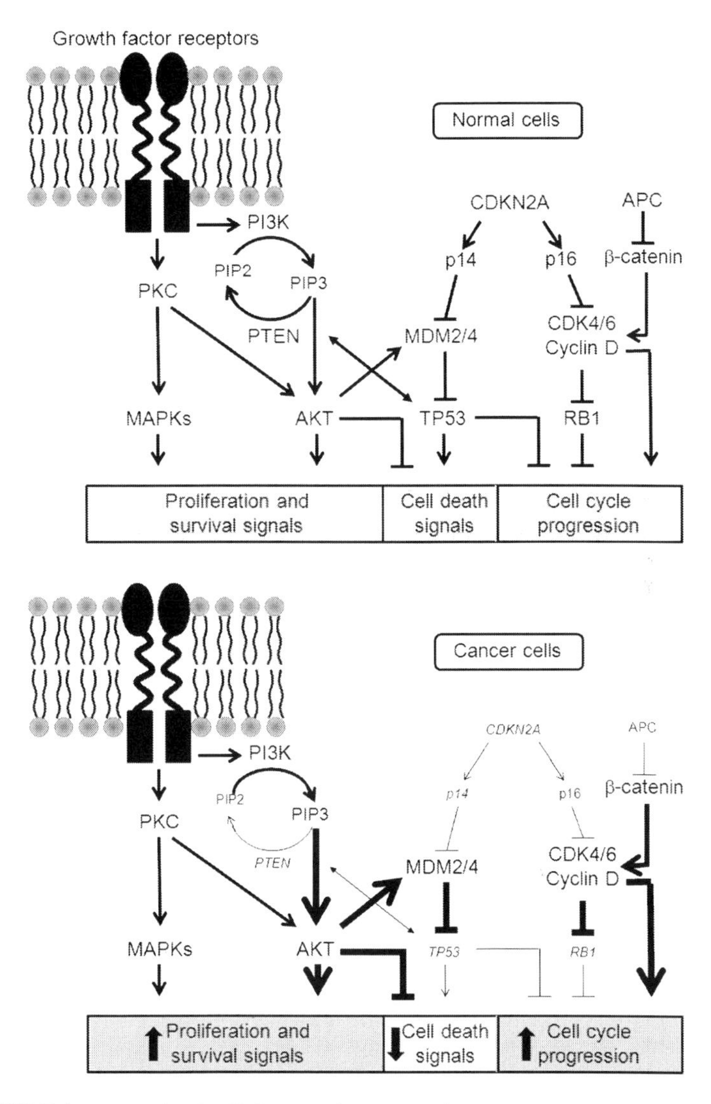

FIGURE 7.1 **Controls of cellular proliferation and programmed cell death in normal cells and cancer cells.** Mutations or epigenetic silencing of specific genes (e.g., PTEN, TP53, CDKN2A, RB1, and APC) increased proliferation and survival of cancer cells. In general, the development of cancer cells involves at least two hits (i.e., mutations or epimutations).

7.2 MAINTENANCE AND PLASTICITY OF THE EPIGENOME

Epigenetic changes refer to alterations in gene expression where the DNA sequence remains unaltered. Being transmitted from mother to

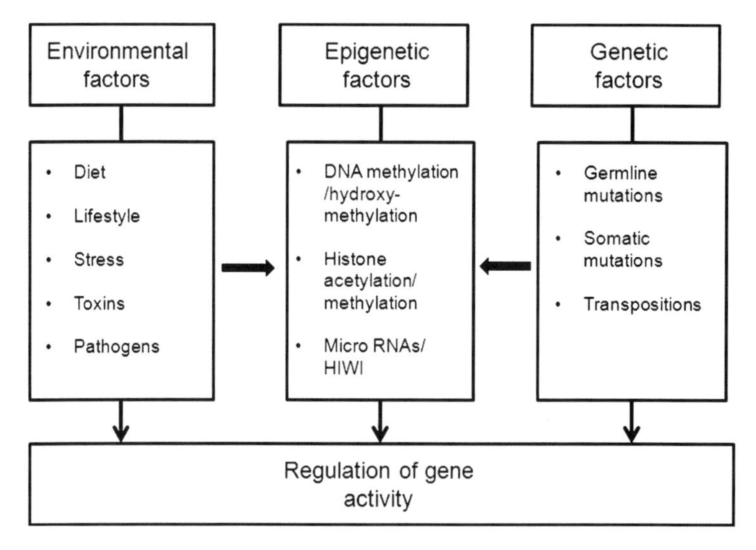

FIGURE 7.2 **Interactions between environmental, epigenetic and genetic factors in the regulation of gene activity.** Epigenetic mechanisms are influenced by both environmental and genetic factors. The integration of all these factors determines the degree of gene expression.

daughter cells, and sometimes even from parents to child, epigenetic changes are candidates to explain aberrant phenotypes in the absence of any structural alterations of the DNA. Highly significant epigenetic differences in dizygotic twins have been shown to occur, as compared with those in monozygotic twins [5]. Twin-based DNA methylation levels showed a mean genome-wide heritability estimate of 18% only [6]. This suggests that both genetic and environmental factors influence the regulation of gene activities. Indeed, numerous studies have shown that the impact of environmental factors can be acquired through the epigenome [7–11], which is one of the current topics in cancer and complex disease studies (Figure 7.2).

7.2.1 General Epigenetic Mechanisms

Epigenetics allows genetically identical cells to achieve diverse stable phenotypes by controlling the transcriptional availability of various parts of the genome through differential chromatin marking and packaging. These mechanisms include DNA methylation [12], DNA hydroxylation [13], histone modifications [14], interacting proteins [15], and microRNA-based regulations [16,17]. Besides covalent modifications of histone tails, the structure of chromatin and the consequent expression of genes located in each region are also regulated by complexes that remodel nucleosomes in an ATP-dependent fashion [18]. Plasticity is an essential feature of epigenetic regulation. These epigenetic marks do not act in isolation but form a network of mutually reinforcing or counteracting signals. Genome-scale projects charting the human epigenome are rapidly extending our understanding of epigenetic marks and how they interact [19–21]. Such new methodologies are presented in Chapters 28–30.

7.2.2 Writers, Readers, and Erasers

A key facet of epigenetics is that these marks can be stably maintained yet adapt to changing developmental or environmental needs. This delicate task is accomplished by initiators, such as long non-coding RNAs; writers, which establish the epigenetic marks; readers, which interpret the epigenetic marks; erasers, which remove the epigenetic marks; remodelers, which can reposition nucleosomes; and insulators, which form boundaries between epigenetic domains. Epigenetic writers are directed to their target locations by sequence context, existing chromatin marks and bound proteins, non-coding RNAs, and/or nuclear architecture. Those marks are then recognized by reader proteins to convey information for various cellular functions. The establishment, maintenance, and change of epigenetic marks are intricately regulated, with crosstalk among the marks and writers to help guide changes to the epigenetic landscape. Regarding DNA methylation, DNA methyltransferases (DNMTs) are the writers, methyl-binding domain proteins (MBDs) the readers, 5-methylcytosine hydroxylases (TETs) the erasers, and CCCTC-binding factor (CTCF) is an insulator.

7.3 DNA METHYLATION, CpG ISLANDS AND CpG SHORES

Methylation of cytosine residues at CpG dinucleotides is a major epigenetic modification in human cells, allowing, for example, cell differentiation and X-chromosome inactivation. The human haploid genome contains about 30 million CpG dinucleotides which form the DNA methylome [22]. This was presented in Chapter 1.

Briefly, much of the mammalian genome consists of large DNA sequences containing sparsely distributed but heavily methylated CpG dinucleotides. Normally, 70–80% of all CpG dinucleotides, and especially those that are part of transposable elements including long interspersed elements (LINEs), are methylated [23]. The genome is punctuated by short regions with unmethylated CpGs occurring at

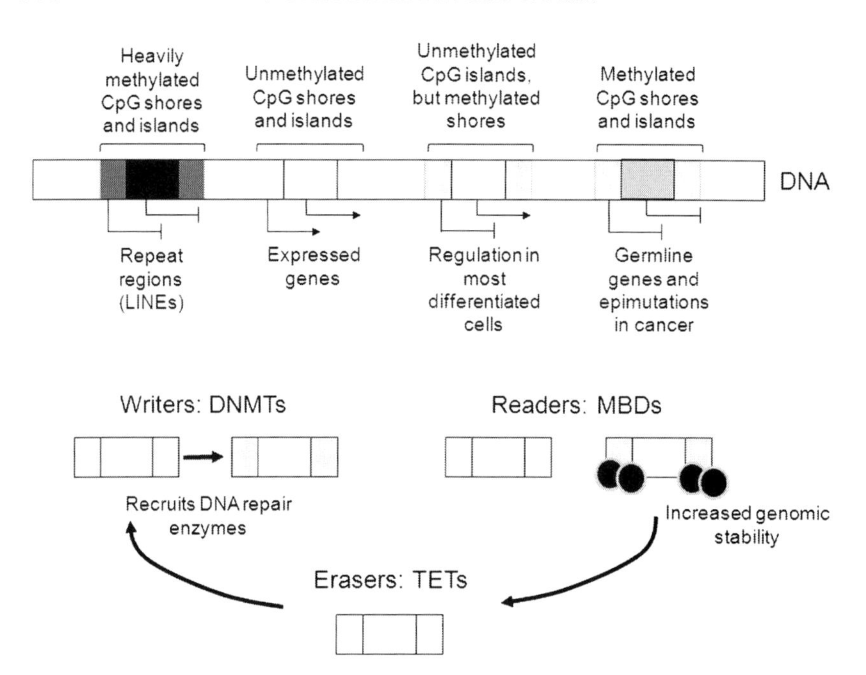

FIGURE 7.3 **DNA methylation homeostasis.** DNA can be divided into CpG islands, shores, and oceans in between. Repeat regions, coding for (for example) endogenous retroviruses (ERVs), are heavily methylated, while most of the other CpG islands are unmethylated. The regulation of gene activity in most tissues occurs through methylation of CpG shores in promoter regions. Silencing of germline genes and epimutations in cancer are reprogramming processes affecting various regions, including CpG shores and islands, both in promoters and in exons. These processes are controlled by DNMTs that copy the methylation pattern from one DNA strand to the other, by MBDs that silence the translation machinery, and by TETs that can erase the methylation marks (for example, in germline cells). This mechanism is part of a system that ensure genomic stability. Dyfunctions, as they occur in cancer, lead not only to aberrant DNA methylation patterns, but also to genomic instability.

higher density, forming distinct CpG islands [24] (Figure 7.3). These are around 1-kb stretches of DNA with high GC content enriched in promoter regions of genes. CpG islands span the transcription start sites of about half of the genes in the human genome. The non-repetitive portion of the human genome contains 27,000 CpG islands [25]. Mapping of DNA methylation in various cell types has confirmed the lack of methylation in the majority of CpG islands, but also uncovered numerous cases of differentially methylated or even constitutively methylated regions [26]. CpG shores are regions of relatively low CpG density that flank traditional CpG islands (up to 2 kb distant). These transition zones display more tissue-specific variation in DNA methylation [27], suggesting an involvement in tissue differentiation.

The enzymes responsible for introducing the methyl group into cytosine are the DNA methyltransferases (DNMTs). Three major proteins with DNMT activity have been identified in mammals: DNMT1, DNMT3A, and DNMT3B [28]. Sites of DNA methylation are occupied by various proteins, including methyl-CpG binding domain (MBD) proteins which recruit the enzymatic machinery to establish silent chromatin [29]. The levels and patterns of DNA methylation are the results of the opposing actions of the methylating and demethylating machineries. The ten–eleven translocation (TET) family of proteins was discovered as 5-methylcytosine dihydroxygenases that convert 5-methylcytosine to 5-hydroxymethylcytosine. TETs have important roles in epigenetic reprogramming in early embryos and primordial germ cells.

7.3.1 Writers: DNA Methyltransferases

DNA methyltransferase 1 (DNMT1) is widely expressed, recognizes hemimethylated DNA, and is responsible for maintaining the existing methylation patterns after DNA replication. In contrast, DNMT3 enzymes are *de novo* DNMTs that introduce methyl groups into previously unmethylated cytosines [28,30]. These enzymes introduce a methyl group into the genome, but this writing must be read by MBDs to silence genes efficiently so that they can't be activated by the translation machinery (i.e., transcription factors, DNA polymerase II, etc.). DNA methylation patterns are guided in part by the primary DNA sequence [31]. The synthesis and physiology of DNMTs are presented in more detail in Chapter 1. Aberrant expression of DNMTs and disruption of DNA methylation patterns are closely associated with many forms of cancer, although the exact mechanisms underlying this link remain elusive [32].

An interesting link between aberrant DNMT expression is their relations with the DNA damage repair systems [32]. These mechanism have evolved to act as a genome-wide surveillance mechanism to maintain chromosome integrity by recognizing and repairing both exogenous and endogenous DNA insults [33]. Impairment of DNA repair systems gives rise to mutations and directly contributes to the development of cancer. Depletion of DNMTs causes an increased microsatellite instability [34], destabilization of repeats [35], and dramatically increased telomere length, as well as telomeric recombination [36]. Microsatellite instability is the condition of genetic hypermutability that results from impaired DNA mismatch repair. Thus, the effects of DNMT depletion appear to be mediated in part by a drop in DNA repair proteins as part of the DNA damage response [37]. The DNMT1 protein has been shown to be recruited to areas of irradiation-induced DNA damage, possibly to facilitate repair of epigenetic information following DNA repair [38]. It is increasingly recognized that chromatin can serve as a cellular sensor for DNA damage and other genomic events [39].

7.3.2 Readers: Methyl-CpG Binding Domain Proteins

As mentioned above, methylated DNA can be specifically recognized by a set of protein readers, which belong to three different structural families in mammals: the methyl-CpG binding domain protein (MBD) family, and the Kaiso and Kaiso-like proteins. Methylated DNA is recognized by MBD or C2H2 zinc finger proteins. The MBD-containing DNA methylation readers include MBD1, MBD2, MBD4, and MeCP2 (methyl-CpG binding protein 2), whereas Kaiso and Kaiso-like proteins use zinc fingers to bind methylated DNA. The involvement of MBDs in DNA methylation is reviewed in more detail in Chapter 1. Once bound to methylated DNA, MBPs translate the DNA methylation silencing signal into appropriate functional states through interactions with diverse partners [40]. MBDs and the Kaiso proteins also participate in DNA methylation-mediated transcriptional repression of tumor suppressor genes after promoter DNA methylation. MBDs function in transcriptional repression and long-range interactions in chromatin, and they also appear to play a role in genomic stability, neural signaling, and transcriptional activation [29]. They are recruited to methylated DNA, and in turn facilitate the recruitment of histone modifiers and chromatin-remodeling complexes [41,42].

Mutations in the MBD family member MeCP2 are the cause of Rett syndrome [43], a severe neurodevelopmental disorder. This has also been shown to have a role in cancerogenesis, in the context of genes becoming silenced through hypermethylation of their promoter. It has been shown to target genes *in vivo* in myeloma, hematological malignancies, and in breast, colorectal, lung, liver, and prostate cancer [44]. Furthermore, polymorphisms in *MBD1* have been shown to be particularly associated with lung cancer risk. In relation to its transcriptional roles, most of the genes silenced by MBD1 in cancer have been identified *in vitro*. These targets have been identified in acute promyelocytic leukemia and in pancreatic, prostate, and colon cancer cell lines [44]. Similarly, MBD2 has been shown to silence genes in various cancers, especially colorectal cancer [45].

7.3.3 Erasers: 5-Methylcytosine Hydroxylases

DNA demethylation can be reversed by the conversion of methylcytosine to hydroxymethylcytosine by the T5-methylcytosine hydroxylases (TET) family of enzymes [46]. TET1 is bound to CpG islands in embryonic stem cells, suggesting that it maintains the fidelity of DNA methylation patterns in cells by maintaining CpG islands in a hypomethylated state [47]. TET2 is important for the rapid re-expression of pluripotency-associated genes in embryonal stem cells. In the context of cancer, it is important to retain that TETs can demethylate the DNA, while TET deficiencies can favor excessive DNA methylation.

7.4 ABERRANT DNA METHYLATION IN CANCER

The DNA methylome of cancer cells can exhibit two striking differences from normal cells: (1) a reduction of global DNA methylation levels, and (2) an aberrant hypermethylation of some sequences, particularly CpG islands that are normally unmethylated. The widespread loss of DNA methylation contrasts with the hypermethylation of CpG islands in cancer [48], including promoter CpG islands that restrict the expression of genes often involved in cell cycle and/or programmed cell death regulation – the so-called tumor-suppressor genes [49]. These seemingly contradictory findings have been widely reported for many types of cancer [50]. The hypermethylation of tumor suppressor genes directly drives the carcinogenic process [51] (Figure 7.4).

7.4.1 Global Hypomethylation

Loss of 5-methylcytosine in cancer cells was discussed more than 30 years ago [52], with global DNA hypomethylation reported in numerous cancer cell lines [23,53] and reduced levels of DNA methylation found at selected genes in primary human tumors compared to normal tissues [54,55]. The underlying causes of the DNA hypomethylation are unknown as yet, but the loss can be mainly localized to particular types of repetitive elements [56] or chromosomal domains [57]. As a result of DNA hypomethylation, repeated regions in the genome have increased frequencies of recombination; this has profound implications for genomic stability [36].

DNA hypomethylation can be a consequence of an active process. The DNA methylation pattern is erased in the early embryo and then re-established in each individual at approximately the time of implantation [58]. In mammals, global DNA demethylation occurs when the unipotent primordial germ cells migrate to the future gonads. At time of this migration, the paternal and maternal genomes undergo an active genome-wide demethylation by TET1 and TET2. Thus, germline cells are normally hypomethylated, allowing the expression of specific germline genes [59]. Surprisingly, DNA hypomethylation in cancer was

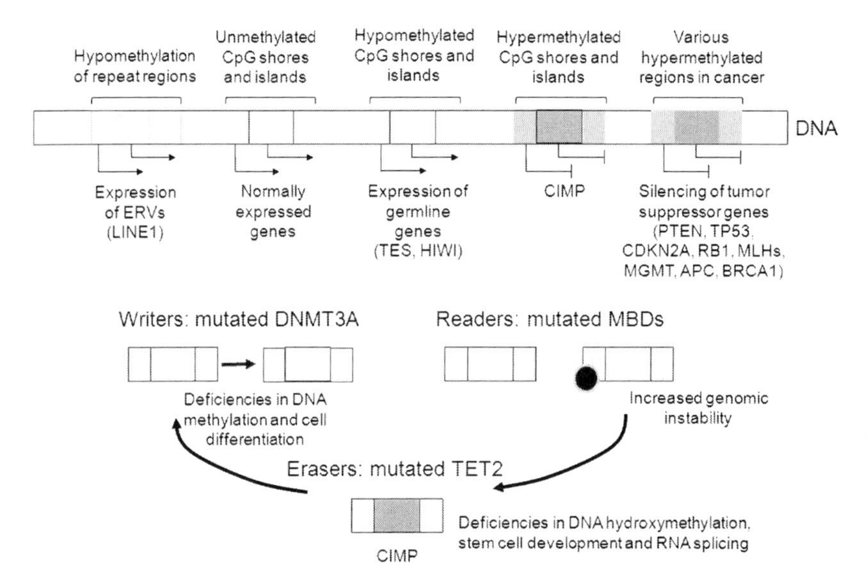

FIGURE 7.4 DNA methylation in cancer. In cancer cells, repeat regions of DNA are hypomethylated; this is reflected by a global decrease of 5-methylcytosine content. Genes that are normally silenced can be expressed, particularly germline-specific genes. Paradoxically, large regions of DNA can be hypermethylated; this is called the CpG island methylator phenotype (CIMP). Transformation into cancer cells occurs through mutations or epimutations (DNA hypermethylation) of specific genes. Mutations of epigenetic genes also can be involved in cancerogenesis; for example, mutated DNMT3A leads to decreased *de novo* DNA methylation and can block cell differentiation; mutated MBDs increase genomic instability; and mutated TET2 can lead to CIMP and altered stem cell development.

found to cause the aberrant activation of only a limited group of genes. Most of these are normally expressed exclusively in germline cells, and therefore were grouped under the term cancer-germline genes [60]. Activation of these genes in tumor cells raises the possibility that their proteins might have oncogenic activities. The concerted expression of germline genes in cancer would correspond to the activation of a gametogenic program, thereby bestowing tumor cells with germ-cell properties, including the capacity to self-renew (a feature of spermatogonial stem cells) and increased motility (a feature of sperm cells). Several MAGE proteins were found to inhibit p53 transactivation function, thereby exerting antiapoptotic properties. GAGE proteins were also shown to render cells resistant to apoptosis. Other studies reported that MAGEA11 serves as a co-stimulator for the androgen receptor, and might therefore contribute to the development of prostate tumors that have become independent of the presence of androgen for their growth. Furthermore, HIWI (human homolog of PIWI [P-lement induced wimpy testis]) and piRNAs are normally expressed in germline cells, but also can be detected in various cancers [60,61]. Its promoter is regulated by DNA methylation and usually silenced in somatic cells. Proteins of the HIWI family favored DNA methylation and decreased LINE1 retrotransposon expression. HIWI-mediated DNA methylation can also be associated with tumor suppressor gene silencing [61].

7.4.2 Decreased DNA Hydroxymethylation

It might be expected that global DNA hypomethylation would be associated with increased DNA hydroxymethylation. In cancer, however, this is not the case. In fact, conversely, a decrease in global hydroxymethylcytosine occurs [62,63], and disruptions of TET enzyme function have been linked to CpG island hypermethylation [64]. This is due to an altered methylation eraser function of TETs. In addition, somatic mutations of isocitrate dehydrogenases (IDHs), enzymes of the Krebs cycle, can be linked to the development of CpG island hypermethylation. Thus, global DNA hypomethylation is accompanied by hypermethylation of CpG islands that can affect CpG islands of specific gene promoters. Cancerogenesis can occur when such promoters belong to genes controlling survival and/or cell cycle progression.

7.4.3 CpG Island Methylator Phenotype

Calcitonin (*CALCA*) was the first gene reported to become hypermethylated in cancer [48]. This was initially viewed as a spontaneous or stochastic event with selection for functionally relevant silencing events. However, the discovery of cases of colorectal cancer with an exceptionally high frequency of CpG island hypermethylation suggested a coordinated event, possibly attributable to an epigenetic control defect. This phenomenon was referred as CpG island methylator phenotype (CIMP) [65]. As mentioned above, excessive DNA methylation can occur because of IDHs or TET mutations.

CpG island hypermethylation may constitute an epigenetic field defect which increases the vulnerability of the bowel mucosa to cancer along with increasing age [66]. The existence of CIMP subsets of cancer is clearly documented for some types of cancer, such as colorectal cancer [67,68] and glioblastoma [69], but not for others, such as serous ovarian cancer [70].

The most distinct examples of CIMP show exceptionally strong associations with other molecular or pathological features of the cancers, lending further validity to the biological relevance to this classification. For example, colorectal cancer CIMP is very tightly associated with a specific mutation of the B-raf (*BRAF*) oncogene [71], whereas glioma CIMP is exceptionally tightly associated with mutation of the isocitrate dehydrogenase-1 (*IDH1*) gene [69]. In such cases, *IDH1* mutation appears to be a causal contributor to the phenotype [72].

Disruption of TET activities can lead to DNA hypermethylation. Indeed, mutations abrogating TET2 enyzmatic activity are frequent in acute myeloid leukemia [73] and correlate with CIMP [74]. Similarly, IDH1 or -2 mutations are associated with acute myeloid leukemia, because a metabolite produced by the mutated enzymes inhibits TET activity [75]. Thus, both TET2 mutation and *IDH1/2* mutations correlate with a similar CIMP, in acute myeloid leukemia [74] or glioblastoma [69]. Such observations underline the importance of IDHs mutations and TETs dysfunctions, associated with preferential hypermethylation of specific CpG islands, in carcinogenesis. The mechanisms by which IDH1 and TET2 are linked to cancer is explained in more detail in Section 7.6.2, below.

7.4.4 Hypermethylation of Tumor Suppressor Genes

Epimutation is defined as abnormal transcriptional repression of active genes and/or abnormal activation of usually repressed genes caused by errors in epigenetic gene repression. Hypermethylation of cytosine residues in promoters is tightly linked with transcriptional repression of the affected gene, and many promoters initially shown to be aberrantly hypermethylated in cancer correspond to known tumor suppressor genes [76]. Aberrant promoter hypermethylation has, therefore, been viewed as an epimutation causing the silencing of such genes [77]. Thus, the current view is that the aberrant hypermethylation of cytosine residues can drive carcinogenesis and cancer progression, and that epimutational events even might outnumber mutations in cancer. The majority of genes that are aberrantly hypermethylated in cancer are in fact already repressed in precancerous cells. Such genes have been termed "epigenetic drivers," and the bulk of hypermethylated genes, which are repressed in normal untransformed tissue, are hence termed "passengers" [50,78]. The notion of driver genes is much more restricted than that of tumor suppressor genes, because it should be shown that they are silenced or activated in precancerous or very early stages of cancer.

7.4.5 Hypermethylation of Driver Genes

Silencing of genes occurs as a result of normal development and differentiation. The aberrant hypermethylation of driver genes in cancer, however, represents an epigenetic reprogramming event. This can be triggered by an already present mutation in the driver gene itself or other genes, and contribute to a growth advantage. Inherited mutations and epimutations often overlap. Driver genes do not necessarily directly control the cell cycle or apoptosis; as shown in the case of *IDH2*, it can involve the whole metabolism. Recently, the HIWI germline protein has been suggested as a driver for cancerogenesis [60,61,79]. Possibly, driver genes also block cell differentiation or restrict epigenetic plasticity and adaptive potential.

7.5 EPIMUTATIONS, HERITABLE ALTERATIONS IN CANCER

It is clear that the cancer genome and epigenome influence each other in a multitude of ways (Figure 7.2). They offer complementary mechanisms to achieve similar results, such as the inactivation of tumor-suppressor genes by either deletion or epigenetic silencing, and they can work cooperatively in, for example, colorectal cancer, where CIMP appears to create a permissive context for further mutation as early as in the precursor lesion [67].

Importantly, a clear mechanism has been described for the inheritance of DNA methylation patterns across cellular generations [80] and, therefore, abnormal DNA methylation states fit the definition of epimutations as heritable alterations. The aberrant hypermethylation of genes such as Retinoblastoma-1 (*RB1*) [81], MutL homolog 1 (*MLH1*) [82,83], and breast cancer 1 (*BRCA1*) [84], whose mutation is associated with inherited cancer predisposition [85], can be regarded as particularly significant.

7.5.1 Inherited Mutations and Epimutations Overlap

Genetics can shed light on the identity of epigenetic drivers by revealing mutual exclusivity with genetic aberrations in the same gene or pathway. *BRCA1* mutations predispose specifically to breast and ovarian cancer, and hypermethylation is limited to cancer of these tissues [86]. Epigenetics may also provide insight into genetic drivers in a similar fashion. For instance, in sporadic cancer, tissue specificity of tumor suppressor genes hypermethylation overlaps with the tissue-specific predispositions caused by inherited mutations in these same genes. Thus, inherited *MLH1* mutations predispose to colorectal cancer and *MLH1* hypermethylation is largely limited to colorectal tumors [86]. Within particular tissues, the phenotypes of cancers that have hypermethylated particular tumor suppressor genes can overlap with the specific phenotypes of cases associated with inherited mutations in the same gene. For example, the same phenotype occurs for mutated and hypermethylated *RB1* retinoblastomas [87]. RB1 is a negative regulator of cell cycle progression; therefore, a dysfunction results in increased proliferation (Figure 7.1). Similarly, colorectal tumors with either mutated or hypermethylated *MLH1* both show microsatellite instability [82,88]. In addition, the understanding of epigenetic networks provides a framework to interpret the functional significance of driver genes.

7.5.2 Epimutations of DNA Mismatch Repair Genes

Epigenetic silencing of DNA repair genes can boost mutation rates and promote genomic instability in cancer cells [89]. Lynch syndrome, often called hereditary non-polyposis colorectal cancer, is an inherited disorder that increases the risk of many types of cancer, particularly cancers of the colon (large intestine) and rectum, which are collectively referred to as colorectal cancer. This syndrome is characterized by microsatellite instability resulting from germline mutations in mismatch repair genes, primarily MutS homologs 1 and 2 (*MLH1* and *MSH2*) (Figure 7.5).

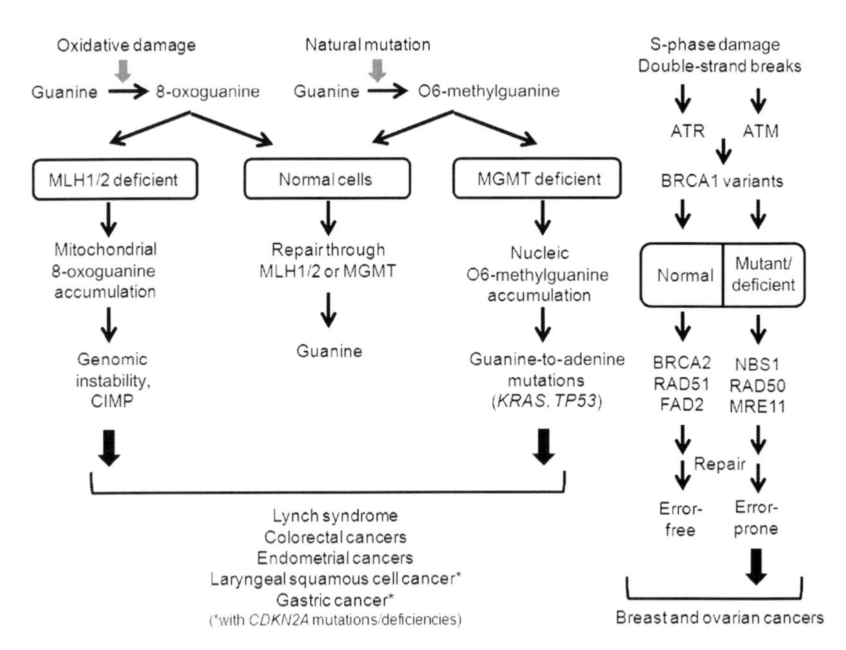

FIGURE 7.5 Dysfunctions of DNA repair and cancerogenesis. MLH1/2- and MGMT-deficient cells (through mutation or epimutation) show accumulation of mutated guanine, leading to genomic instability, CIMP, and/or mutations of various tumor suppressor genes, in particular TP53. The mechanism related to BRCA1 is different. After DNA damage, configuration changes of BRCA1 occur, leading to the activation of DNA repair complexes. The BRCA2 complex is error-free, but the alternative complex is error-prone. Mutant or deficient BRCA1 leads to more error-prone repairs.

Colorectal cancer is one of the major causes of mortality and morbidity, and is the third most common cancer in men and the second most common cancer in women worldwide. The incidence of colorectal cancer shows considerable variation among racially or ethnically defined populations in multiracial/ethnic countries. The cancerogenesis is due to chromosomal instability, or microsatellite instability, or involvement of various proto-oncogenes and driver genes, and also epigenetic changes in the DNA [90]. Approximately 15% of sporadic cases of colorectal cancer display microsatellite instability as a consequence of epigenetic silencing of the *MLH1* mismatch repair gene by promoter CpG island hypermethylation [82], and this in the context of CIMP [65,71] (Figure 7.5).

Microsatellite instability caused by epigenetic silencing of *MLH1* has also been reported in other types of cancer, including 25% of sporadic endometrial cancers [91]. Germline variants of *MLH1* and *MSH2* can predispose to extensive somatic epigenetic silencing of these genes and thereby increase cancer risk [88]. Such familial cases of systemic epigenetic abnormalities can be regarded as a germline transmission of epigenetic defects.

The O6-methylguanine DNA methyltransferase (*MGMT*) enzyme repairs O6-alkylated guanine residues in genomic DNA. O6-methylguanine pairs with thymine and would lead to a G-to-A transition during DNA replication if left unrepaired. *MGMT* promoter methylation in colorectal cancer is associated with G-to-A mutations in *KRAS* [92] and in *TP53* [93] (Figure 7.5). Alkylating agents such as temozolomide are the current standard of care for malignant glioblastoma, but are counteracted by MGMT-mediated repair of the alkylation damage. Epigenetic silencing of *MGMT* by promoter CpG island hypermethylation inactivates this repair pathway and renders the tumor more sensitive to temozolomide treatment [94,95]. DNA methylation and drug susceptibility or resistance are the topics of Chapter 9.

The mechanism related to BRCA1 is different. DNA damage signals through the ataxia telangiectasia mutated (ATM) gene and ATM-related kinase (ATR) in post-translational modifications of BRCA1 that affect its function in terms of DNA repair and cell cycle checkpoints. BRCA1 mutation or epimutation leads to a deficiency and to more repair through the second error-prone mechanism [96]. BRCA1 variants either activate a BRCA2 complex that performs an error-free repair, or another complex that is error-prone (Figure 7.5). The second complex is then associated with an increased risk for breast and ovarian cancers.

Sporadic breast cancers with *TP53* mutations or epigenetic silencing (by hypermethylation) resemble phenotypically *BRCA1* mutated cancers termed *BRCAness* – i.e., those with no *BRCA* mutations but with a dysfunction of the DNA repair system [94]. The loss of *TP53* confers such a large fitness advantage, by enabling cells to survive better and divide more, that the clone is able to tolerate many deleterious mutations and still have a fitness advantage over *TP53* wild-type clones [2] (Figure 7.1). Genetic and epigenetic mechanisms in breast cancer are presented in more detail in Chapter 8.

7.5.3 Epimutations, Cell Cycle, and Apoptosis Genes

The adenomatous polyposis coli (*APC*) gene product is an integral part of the Wnt-signaling mechanism; it plays also a role in cell–cell adhesion, stability of the microtubular cytoskeleton, cell cycle regulation, and apoptosis. It indirectly regulates transcription of a number of critical cell proliferation genes, through its interaction with the transcription factor β-catenin. APC binding to β-catenin leads to ubiquitin-mediated β-catenin destruction; loss of APC function increases transcription of β-catenin targets. These targets include cyclin D (Figure 7.1), c-myc, ephrins, and caspases. In addition to the mutational inactivation, hypermethylation of the *APC* gene promoter is an important mechanism associated with silencing of this driver gene [97]. In many cancers the hypermethylation of CpG islands in the *APC* promoters has been found to be a frequent epigenetic change, and is usually associated with the loss of transcription into APC protein [98–100]. For example, in colorectal cancer, hypermethylation of the APC gene promoters has been reported to be present in about 20–48% of cases.

The cyclin-dependent kinase inhibitor 2A (*CDKN2A*, p16$^{\mathrm{Ink4A}}$) gene functions as an important driver in various cancers. *CDKN2A* promoter methylation is a frequent epigenetic event and an important mechanism leading to silencing and dysfunction of the *CDKN2A* gene, which further results in uncontrolled cell proliferation and cancer development (Figure 7.1). The reactivation of *CDKN2A* prevents carcinogenesis through induction of cell growth arrest and senescence. Hypermethylation of *CDKN2A* has been observed alongside inherited germline mutations [101]. Thus, *CDKN2A* hypermethylation can directly substitute for genetic loss of heterozygosity, as this second hit completely disables its activity [102].

Phosphatase and tensin homolog (*PTEN*) counterbalance phosphoinositide-3-kinase (PI3K) in AKT (protein kinase B) activation. AKT1 is involved in cellular survival pathways by inhibiting apoptotic processes (Figure 7.1). Therefore, *PTEN* is a tumor suppressor gene that plays a critical role in controlling cell growth and survival. Promoter hypermethylation of *PTEN* has been reported in numerous cancers [103]. Alternatively, the increased survival attributed to DNA hypermethylation in this region could be associated with the silencing of a gene product called *KLLN*, which shares the PTEN CpG island and is transcribed from the negative DNA strand in the opposite direction. The *KLLN* gene is necessary for TP53-induced apoptosis, and may therefore possess a driver function.

7.5.4 Epimutations and Metastasis

Metastasis requires that cells leave the primary tumor, but few such cells successfully colonize a distant organ. Colonizing individuals often have high fitness because they can escape from deteriorating local conditions caused by population growth and the overconsumption of resources [2]. Only a few mutations/epimutations have been described to favor metastasis – for example, Tensin-3 (*TNS3*), which is a cytoskeletal regulatory protein inhibiting cell motility. Thus, downregulation of the gene encoding *TNS3* through hypermethylation of the promoter, in human renal cell carcinoma, contributes to cell metastatic behavior [104].

7.5.5 Multiple Mutations/Epimutations

The current "multiple hits" theory of cancerogenesis implies, in general, at least two mutations/epimutations, of germline or somatic origins. It is possible that different mutations give rise to similar phenotypes, or that similar epimutations results in different phenotypes depending on the genetic background. For example, in both laryngeal squamous cell carcinoma and gastric cancer [83], aberrant hypermethylation of *CDKN2A*, *MGMT*, and *MLH1* promoters appear to have predictive values. In Barrett's esophagus, an abnormal precancerous change, inactivation of *TP53* is almost always observed after inactivation of *CDKN2A*. The mutation that is selectively advantageous on its own (i.e., *CDKN2A*) initiates a clonal expansion that creates many opportunities for the other mutation/epimutations (in *TP53*) to occur, sparking a second clonal expansion within the first [2]. Such genetic dependencies lead to regularities in the order in which mutations and epimutations appear.

7.6 EPIMUTATIONS OF EPIGENETIC REGULATORS

Epigenetic states are flexible and persist through multiple cell divisions. Although cancer cells have long been known to undergo epigenetic changes, only recently have wide-scale genomic and epigenomic analyses revealed the widespread occurrence of mutations in epigenetic regulators and the breadth of alterations to the epigenome in cancer cells [3].

DNMT1$^{+/-}$ mice develop fewer precancerous intestinal lesions than *DNMT1* wild-type animals when bred with APC multiple intestinal neoplasia (*Min*) animals predisposed to this neoplasia. Thus, an early hypothesis was that aberrant CpG island hypermethylation in cancer results from the overexpression or increased activity of DNMTs. Such increases were initially reported, but are likely to be attributable to the regulation of DNMTs during the cell cycle [105,106] and an increased number of cycling cells in cancer. Nevertheless, a recent analysis reports that hypermethylation at some genes correlates with increased DNTM3B levels in colorectal tumors [107]. DNMT3B is required for the active suppression of genes and for the survival of cancer cells.

Furthermore, it has become obvious that deficiency of certain epigenetic regulators can be associated with carcinogenesis, including factors involved in DNA methylation, histone modification, or miRNA regulation. In fact, it has been shown that mutations of epigenetic regulator genes may even be present in preneoplastic lesions already [108], in particular mutated DNMTs or TETs, suggesting that they can be involved in early events of cancer development (Figure 7.4).

7.6.1 Mutated DNMT3A

Next-generation sequencing of various cancers has revealed frequent mutations in genes coding for proteins responsible for adding or removing covalent modifications in DNA or histones, as well as in genes encoding subunits of chromatin remodeling complexes, showing that all these genes too can act as drivers in cancer development. Regarding DNA methylation, these regulators include DNMT3A and TET2.

Mutations in coding exons of the *DNMT1* gene are rare, and deletions were detected in 7% of the colorectal cancers [109]. In contrast, the *DNMT3A* gene is often mutated in acute myeloid leukemia [110,111], as well as T-cell lymphoma [112] and other hematological cancer genomes [110,113] (up to 60%); such mutations reduce DNMT3A enzymatic activity [111]. Mutations observed in cancers, in general, are somatic, predominantly truncating mutations with evidence of biallelic inactivation. *DNMT3A* mutations in acute myeloid leukemia constitute an exception, being mainly of the missense type and not showing biallelic inactivation [111]. Such epigenetic perturbation can lead to differentiation block and subsequent cancer development, probably in the context of other dysfunctions.

7.6.2 Mutated TET2

Mutations in the DNA methylation eraser *TET2* gene have also been identified in hematological cancers [73,114,115]. Bone marrow from

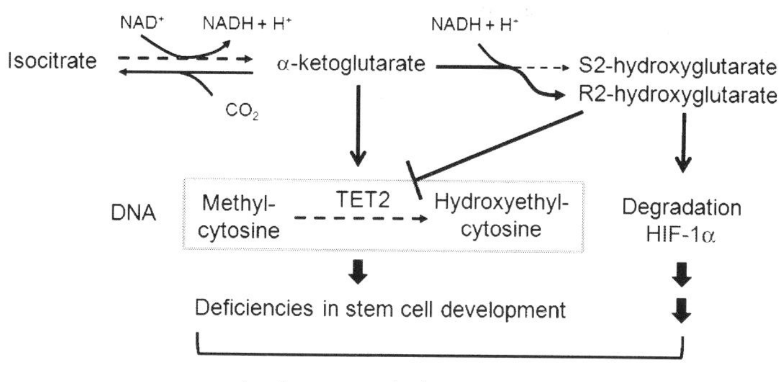

FIGURE 7.6 **IDHs/TET2-related cancerogenesis.** IDH1/2 catalyzes the productions of α-ketoglutarate and 2-hydroxyglutarate. Mutated IDH1/2 shows decreased enzyme activity, and α-ketoglutarate needed for the activity of TET2 is reduced. In addition, mutated IDH1/2 favors the synthesis of the R-enantiomer of 2-hydroxyglutarate, which inhibits TET2 activity, producing CIMP, and increases degradation of HIF-1α – a protein involved in the response to hypoxia. Together, decreased TET2 activity and HIF-1α deficiency can play a role in the development of hematological cancers. A similar phenotype is obtained in cases of TET2 deficiencies or mutations, due to downregulated IDH2.

patients with *TET2* mutations shows reduced levels of 5-hydroxymethylcytosine [116]. In acute myeloid leukemia, *TET2* mutations are mutually exclusive with socitrate dehydrogenase (*IDH1/2*) mutations [117], both however being associated with CIMP. *IDH1/2*, which converts isocitrate to α-ketoglutarate in the Krebs cycle, are the most frequently mutated metabolic genes in cancer, commonly observed in glioblastomas and acute myeloid leukemia. The mutation of IDH1/2 results in increased production of R2-hydroxyglutarate, which inhibits TET2 activity and favors the degradation of HIF-1α. In turn, this suppresses hemopoietic differentiation and can lead to cancerogenesis (Figure 7.6).

In *TET2* mutated cells, the expression of IDH2 is markedly decreased. This enzyme normally produces the S-enantiomer of 2-hydroxyglutarate, which is an important co-factor for TET-mediated 5-methylcytosine hydroxylation, in opposition to the R-enantiomer oncometabolite that inhibits TETs. IDH2 downregulation results in a loss of the remaining TET activity. Importantly, the TET proteins seem to be more than DNA demethylases and transcription regulators; they are multitask proteins involved not only in pluripotency and embryonic development, but also in RNA splicing. With proteins so essential and so versatile, it comes as no surprise that their loss can lead to cancer.

However, it is not known how mutations that affect epigenetic genes cause cancer in the first place, and especially a particular type of cancer.

In the current relatively early era of next-generation sequencing, part of apparent tumor-specificity may reflect the fact that efforts have been restricted to certain types of cancers. Theoretically, possible tissue-specific expression or function of the epigenetic genes themselves or their target genes may play a role. In some instances, known targets of epigenetic gene products may help outline a possible mechanism of cancerogenesis [117].

7.7 EPIGENETIC INTERPLAYS

The N-terminal tails of histones are subjected to several types of post-translational modifications, including acetylation, methylation, phosphorylation, and ubiquitination. The combination of these modifications determines chromatin structure and transcriptional activation or repression of genes. Establishment of the basic DNA methylation profile in the embryo is mediated through histone modifications [118]. A further example of how *de novo* DNA methylation might be linked to histone modifications in early development is the process of heterochromatinization (i.e., the transformation of genetically active euchromatin into genetically inactive heterochromatin), specifically of pericentromeric satellite repeats [119]. DNA hypomethylation on such regions induces chromosomal instability through heterochromatin decondensation and enhancement of chromosomal recombination; this may be a factor involved in the development and progression of certain cancers [120].

In particular, histone acetylation and methylation are known to be linked with DNA methylation (Figure 7.7), but also histone demethylation and the Polycomb repressive complexes show interactions. Therefore, there are situations where histone modifications can provide a mechanism of driver gene silencing which is independent of promoter methylation [121] and may affect neighboring genes across entire chromosome bands, thus mimicking cytogenetic aberrations in cancer [122].

7.7.1 DNA Methylation and Histone Acetylation

Although DNA methylation and histone modification are carried out by different chemical reactions and require different sets of enzymes, there seems to be a biological relationship between the two systems that plays a part in modulating gene repression programming in the organism [123]. Thus, inhibiting cytosine methylation induces histone acetylation, whereas inhibiting histone deacetylation causes the loss of cytosine methylation [124]. Mutation and/or aberrant expression of various histone deacetylases (HDACs) have often been

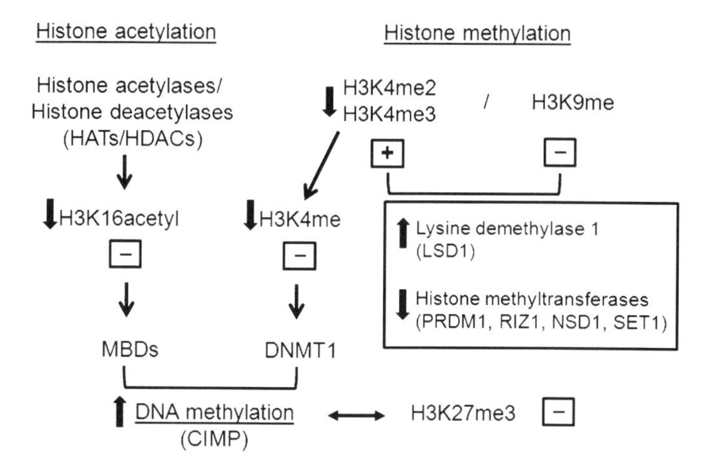

FIGURE 7.7 Interactions between histone modifications and DNA methylation in cancer. Decreased H3K16 acetylation and H3K4 methylation allows binding of MDBs and increased activity of DNMT1, favoring DNA methylation and possibly the development of CIMP. Decreased H3K4 methylation can be caused by increased expression of LSD1 and epimutations of histone methyltransferase genes. In addition, in pre-cancerous stem cells, DNA methylation is associated with H3K27 methylation.

observed in human disease, in particular cancer, making them important therapeutic targets for many human cancers. Therefore, the global pattern of histone acetylation is deregulated in cancer. Indeed, cancer cells undergo a loss of acetylation of histone H4 at lysine 16 (H4K16), indicating that HDAC activity is critical in establishing the tumor phenotype [125]. There is also evidence that DNA methylation is important for maintaining patterns of histone acetylation through cell division. This relationship might be partially mediated through methyl-DNA binding proteins (MBDs) that are capable of recruiting HDACs to the methylated region [41].

The dominant mode of recognition of acetylated lysine residues is by bromodomains present on several human proteins. In particular, the bromodomain and extraterminal domain (BET) family (BRD2, BRD3, BRD4, and BRDT) is important in regulating transcription, epigenetic memory, and cell growth. Currently, targeting of MYCN by BET bromodomain inhibitors appears promising in neuroblastoma [126].

7.7.2 DNA Methylation and Histone Methylation

In contrast to other histone modifications, the importance of histone methylations is highlighted because of their enormously specific dynamics with respect to gene regulation. Over the past decade a tremendous amount of progress has led to the characterization of methyl modifications, as well as the lysine methyltransferases and lysine demethylases that regulate them [127].

The epigenetic stem cell theory of cancerogenesis postulates that chromatin modifications occur even prior to aberrant DNA methylation in cancer [128]. According to this theory, normal tissue stem cells possess a bivalent pattern of activating (H3K4me2) and repressive (H3K27me3) marks. Tumor stem cells express additional repressive marks (H3K9me2/3) which, coupled with DNA hypermethylation, contribute to the abnormal differentiation in tumors.

Histone lysine residues on histone H3 and H4 can become mono-, di- or trimethylated. These modifications are regulated by two classes of enzymes with opposing activities: histone methyltransferases, and histone lysine demethylases. The presence of DNA methylation directs the dimethylation histone 3 at lysine 9 (H3K9me2), which is a mark of repressive chromatin, perhaps through the interaction of the DNA methyltransferase DNMT1 with the replication complex [128,129]. There is also evidence that DNA methylation inhibits the methylation of histone 3 at lysin 4 (H3K4me) [130,131], which is associated with active transcription.

Trimethylation of histone 3 at lysine 9 (H3K9me3) and DNA methylation are present together on individual chromatin fragments in embryonic stem cells [132]. The majority of the H3K9me3 marks require methylated cytosine for their placement. In normal cells, this is not the case for trimethylation of histone 3 at lysine 27 (H3K27m3). However, upon immortalization or tumorigenic transformation of fibroblasts, DNA methylation is required for complete H3K27me3 placement. Importantly, in human promyelocytic cells, H3K27me3 is also dependent on DNA methylation. Because aberrant placement of gene silencing marks at tumor suppressor genes contributes to tumor progression, the improper dependency of H3K27me3 by DNA methylation is likely to be fundamental to cancer. Interestingly, recent studies [133,134] on pediatric glioblastoma have identified missense mutations in genes encoding histone H3. These mutations specifically altered the di- and trimethylation of H3K27. This may reprogram the epigenetic landscape and change gene expression.

Not only genetic mutations but also epimutations can alter the activity of epigenetic genes. For example, the histone methyltransferase gene *PRDM1* (PR domain zinc finger protein 1) may be activated by promoter hypomethylation in lymphoma [135]. *RIZ1* (Retinoblastoma-interacting zinc-finger protein 1), another gene from the same histone methyltransferase gene family, shows frequent inactivation by promoter methylation in gastric and other cancers [136]. The *NSD1* (nuclear receptor SET-domain-containing protein 1) gene, encoding a histone methyltransferase as well, is commonly inactivated by hypermethylation in neuroblastomas and gliomas but only rarely in other types of tumors [137].

7.7.3 DNA Methylation and Histone Demethylases

H3K4 trimethylation is intimately associated with CpG island methylation due to the presence of CXXC finger protein 1 (*CFP1*), which recruits the SET domain histone methyltransferase-1 (*SET1*) [138]. On the one hand, knockdown of certain SET histone methyltransferases causes a decrease in DNA methylation in specific genomic regions [119]. In addition to being implicated in developmental defects, a growing number of studies link alterations in the SET family of proteins to a variety of cancers [139]. On the other hand, aberrant overexpression of LSD1 (lysine-specific demethylase 1), a histone lysine demethylase, has been observed in many types of cancers [140]. LSD1 is functionally linked to DNA methylation and subsequent gene silencing. Thus, DNMTs can interact with the histone H3 tail only when H3K4 is unmethylated [141]. Given that LSD1 demethylates methylated H3K4 to the unmethylated state, it seems to play an important role in *de novo* DNA methylation by generating demethylated H3K4.

7.7.4 DNA Methylation and Polycomb Repressive Complex

Further evidence of synergism between histone modifications and DNA methylation includes observations of Polycomb target genes (which are associated with trimethylation of H3K27) being several times more likely to have cancer-specific promoter DNA hypermethylation than non-targets [140].

Occasionally, opposite changes in the same gene product can be tumorigenic; for example, deficiency of histone-lysine N-methyltransferase (EZH2) deficiency – a member of the Polycomb-group family – is associated with lymphoma, whereas overexpression is typical of breast, prostate, colon, and other cancers [142,143]. In glioblastomas, H3 mutations that alter H3K27me also inhibit the enzymatic activity of the Polycomb repressive complex 2 through interaction with the EZH2 subunit [134]. While it is obvious that epigenetic alterations are no less important than genetic changes in cancer, there is clearly a long and highly exciting road ahead before the exact roles of the individual epigenetic players in cancerogenesis can be fully understood.

7.8 REACTIVATION OF SILENCED TUMOR SUPPRESSOR GENES

The final strong piece of evidence underpinning the causative role of aberrant DNA hypermethylation in silencing tumor suppressor genes in cancer is that they can be reactivated when methylation is removed from their promoters. This is most often achieved by treatment with the drug 5′-aza-2′-deoxycytidine (5-Aza), which inhibits DNMT1 [144]. For example, treatment of cancer cell lines with 5-Aza causes the reactivation of hypermethylated *MLH1* [82,145] and *BRCA1* [146]. Also, genetic knockout or RNA-mediated knockdown of DNMT results in activation of previously hypermethylated *CDNK2A* in a colorectal cancer cell line [147]. These observations demonstrate that gene activation can occur upon ablation of promoter hypermethylation. Epigenetic therapies using demethylation agents are presented in more detail in Chapter 26.

7.9 CONCLUSION

Every cell of an individual has one genotype, but several epigenotypes, depending on its epigenetic marks, that is appropriate for a specific tissue, developmental, or environmental condition [148]. It should be borne in mind that histone modifications and DNA methylation act coordinately in gene expression, and that changes in one will lead to changes in the other [149]. The analysis of cancer genomes and methylomes can help to refine our definition of the type of gene affected by aberrant promoter hypermethylation and generate new hypotheses as to the molecular defect underpinning this epigenetic reprogramming, but many questions remain to be answered. Dissection of this mechanism is also likely to lead to new insights regarding the biology of CpG islands, the most abundant promoter type in our genome. The reinterpretation of cancer-associated CpG island hypermethylation that has occurred as a result of the advent of genome-scale data sets should also be considered as potential epimutations associated with non-neoplastic diseases [150] in, for example, endocrinology (Chapters 10–12), pediatrics (Chapters 13–14), cardiology (Chapter 15), neurology (Chapters 16–18), autoimmunity (Chapters 19–21), and rheumatology (Chapters 22–24). Finally, cancer epigenomes are potentially a rich source of biomarkers, and specific epigenetic defects may be exploitable therapeutic targets [151]. These avenues of research are discussed in Chapter 2. The global understanding of cancer-associated CpG hypermethylation should be used to guide these efforts.

References

[1] Heppner GH, Miller FR. The cellular basis of tumor progression. Int Rev Cytol 1998;177:1–56.

[2] Merlo LM, Pepper JW, Reid BJ, Maley CC. Cancer as an evolutionary and ecological process. Nat Rev Cancer 2006;6:924–35.

[3] You JS, Jones PA. Cancer genetics and epigenetics: two sides of the same coin? Cancer Cell 2012;22:9–20.

[4] Hanahan D, Weinberg RA. Hallmarks of cancer: the next generation. Cell 2011;144:646–74.

[5] Kaminsky ZA, Tang T, Wang SC, Ptak C, Oh GHT, Wong AHC, et al. DNA methylation profiles in monozygotic and dizygotic twins. Nat Genet 2009;41:240–5.

[6] Bell JT, Tsai PC, Yang TP, Pidsley R, Nisbet J, Glass D, et al. Epigenome-wide scans identify differentially methylated regions for age and age-related phenotypes in a healthy ageing population. PLoS Genet 2012;8:e1002629.

[7] Fraga MF, Ballestar E, Paz MF, Ropero S, Setien F, Ballestar ML, et al. Epigenetic differences arise during the lifetime of monozygotic twins. Proc Natl Acad Sci USA 2005;102:10604–9.

[8] Wong AH, Gottesman II, Petronis A. Phenotypic differences in genetically identical organisms: the epigenetic perspective. Hum Mol Genet 2005;14(Spec No 1):R11–18.

[9] Poulsen P, Esteller M, Vaag A, Fraga MF. The epigenetic basis of twin discordance in age-related diseases. Pediatr Res 2007;61:38R–42R.

[10] Szyf M, McGowan P, Meaney MJ. The social environment and the epigenome. Environ Mol Mutagen 2008;49:46–60.

[11] Petronis A. Epigenetics as a unifying principle in the aetiology of complex traits and diseases. Nature 2010;465:721–7.

[12] Jones PA. Functions of DNA methylation: islands, start sites, gene bodies and beyond. Nat Rev Genet 2012;13:484–92.

[13] Ito S, Shen L, Dai Q, Wu SC, Collins LB, Swenberg JA, et al. Tet proteins can convert 5-methylcytosine to 5-formylcytosine and 5-carboxylcytosine. Science 2011; 333:1300–3.

[14] Tan M, Luo H, Lee S, Jin F, Yang JS, Montellier E, et al. Identification of 67 histone marks and histone lysine crotonylation as a new type of histone modification. Cell 2011;146:1016–28.

[15] Ram O, Goren A, Amit I, Shoresh N, Yosef N, Ernst J, et al. Combinatorial patterning of chromatin regulators uncovered by genome-wide location analysis in human cells. Cell 2011;147:1628–39.

[16] Wong JJ, Hawkins NJ, Ward RL. Colorectal cancer: a model for epigenetic tumorigenesis. Gut 2007;56:140–8.

[17] Fabbri M, Calin GA. Epigenetics and miRNAs in human cancer. Adv Genet 2010;70:87–99.

[18] Wilson BG, Roberts CW. SWI/SNF nucleosome remodellers and cancer. Nat Rev Cancer 2011;11:481–92.

[19] Adams D, Altucci L, Antonarakis SE, Ballesteros J, Beck S, Bird A, et al. BLUEPRINT to decode the epigenetic signature written in blood. Nat Biotechnol 2012;30:224–6.

[20] Dunham I, Birney E, Herrero J, Wilder SP, Keefe D, Beal K, et al. An integrated encyclopedia of DNA elements in the human genome. Nature 2012;489:57–74.

[21] Ernst J, Kheradpour P, Mikkelsen TS, Shoresh N, Ward LD, Epstein CB, et al. Mapping and analysis of chromatin state dynamics in nine human cell types. Nature 2011;473:43–9.

[22] Fouse SD, Nagarajan RO, Costello JF. Genome-scale DNA methylation analysis. Epigenomics 2010;2:105–17.

[23] Ehrlich M, Gama-Sosa MA, Huang LH, Midgett RM, Kuo KC, McCune RA, et al. Amount and distribution of 5-methylcytosine in human DNA from different types of tissues of cells. Nucleic Acids Res 1982;10:2709–21.
[24] Rollins RA, Haghighi F, Edwards JR, Das R, Zhang MQ, Ju J, et al. Large-scale structure of genomic methylation patterns. Genome Res 2006;16:157–63.
[25] Han L, Su B, Li WH, Zhao Z. CpG island density and its correlations with genomic features in mammalian genomes. Genome Biol 2008;9:R79.
[26] Lister R, Pelizzola M, Dowen RH, Hawkins RD, Hon G, Tonti-Filippini J, et al. Human DNA methylomes at base resolution show widespread epigenomic differences. Nature 2009;462:315–22.
[27] Irizarry RA, Ladd-Acosta C, Wen B, Wu Z, Montano C, Onyango P, et al. The human colon cancer methylome shows similar hypo- and hypermethylation at conserved tissue-specific CpG island shores. Nat Genet 2009;41:178–86.
[28] Berdasco M, Esteller M. Genetic syndromes caused by mutations in epigenetic genes. Hum Genet 2013;132:359–83.
[29] Bogdanovic O, Veenstra GJ. DNA methylation and methyl-CpG binding proteins: developmental requirements and function. Chromosoma 2009;118:549–65.
[30] Jones PA, Liang G. Rethinking how DNA methylation patterns are maintained. Nat Rev Genet 2009;10:805–11.
[31] Cedar H, Bergman Y. Programming of DNA methylation patterns. Annu Rev Biochem 2012;81:97–117.
[32] Jin B, Robertson KD. DNA methyltransferases, DNA damage repair, and cancer. Adv Exp Med Biol 2013;754:3–29.
[33] Maley CC, Galipeau PC, Finley JC, Wongsurawat VJ, Li X, Sanchez CA, et al. Genetic clonal diversity predicts progression to esophageal adenocarcinoma. Nat Genet 2006;38.
[34] Guo G, Wang W, Bradley A. Mismatch repair genes identified using genetic screens in Blm-deficient embryonic stem cells. Nature 2004;429.
[35] Dion V, Lin Y, Hubert Jr. L, Waterland RA, Wilson JH. Dnmt1 deficiency promotes CAG repeat expansion in the mouse germline. Hum Mol Genet 2008;17:1306–17.
[36] Gonzalo S, Jaco I, Fraga MF, Chen T, Li E, Esteller M, et al. DNA methyltransferases control telomere length and telomere recombination in mammalian cells. Nat Cell Biol 2006;8:416–24.
[37] Loughery JE, Dunne PD, O'Neill KM, Meehan RR, McDaid JR, Walsh CP. DNMT1 deficiency triggers mismatch repair defects in human cells through depletion of repair protein levels in a process involving the DNA damage response. Hum Mol Genet 2011;20:3241–55.
[38] Mortusewicz O, Schermelleh L, Walter J, Cardoso MC, Leonhardt H. Recruitment of DNA methyltransferase I to DNA repair sites. Proc Natl Acad Sci USA 2005;102:8905–9.
[39] Johnson DG, Dent SY. Chromatin: receiver and quarterback for cellular signals. Cell 2013;152:685–9.
[40] Fournier A, Sasai N, Nakao M, Defossez PA. The role of methyl-binding proteins in chromatin organization and epigenome maintenance. Brief Funct Genomics 2012;11:251–64.
[41] Nan X, Ng HH, Johnson CA, Laherty CD, Turner BM, Eisenman RN, et al. Transcriptional repression by the methyl-CpG-binding protein MeCP2 involves a histone deacetylase complex. Nature 1998;393:386–9.
[42] Portela A, Esteller M. Epigenetic modifications and human disease. Nat Biotechnol 2010;28:1057–68.
[43] Amir RE, Van den Veyver IB, Wan M, Tran CQ, Francke U, Zoghbi HY. Rett syndrome is caused by mutations in X-linked MECP2, encoding methyl-CpG-binding protein 2. Nat Genet 1999;23:185–8.

[44] Parry L, Clarke AR. The Roles of the Methyl-CpG Binding Proteins in Cancer. Genes Cancer 2011;2:618–30.
[45] Martin V, Jorgensen HF, Chaubert AS, Berger J, Barr H, Shaw P, et al. MBD2-mediated transcriptional repression of the p14ARF tumor suppressor gene in human colon cancer cells. Pathobiology 2008;75:281–7.
[46] Tahiliani M, Joh KP, Shen Y, Pastor WA, Bandukwala H, Brudno Y, et al. Conversion of 5-methylcytosine to 5-hydroxymethylcytosine in mammalian DNA by MLL partner TET1. Science 2009;324:930–5.
[47] Williams K, Christensen J, Helin K. DNA methylation: TET proteins-guardians of CpG islands?. EMBO Rep 2012;13:28–35.
[48] Baylin SB, Hoppener JW, de Bustros A, Steenbergh PH, Lips CJ, Nelkin BD. DNA methylation patterns of the calcitonin gene in human lung cancers and lymphomas. Cancer Res 1986;46:2917–22.
[49] Jones PA, Baylin SB. The fundamental role of epigenetic events in cancer. Nat Rev Genet 2002;3:415–28.
[50] Baylin SB, Jones PA. A decade of exploring the cancer epigenome - biological and translational implications. Nat Rev Cancer 2011;11:726–34.
[51] Sproul D, Meehan RR. Genomic insights into cancer-associated aberrant CpG island hypermethylation. Brief Funct Genomics 2013;12:174–90.
[52] Ehrlich M, Wang RY. 5-Methylcytosine in eukaryotic DNA. Science 1981;212:1350–7.
[53] Diala ES, Hoffman RM. Hypomethylation of HeLa cell DNA and the absence of 5-methylcytosine in SV40 and adenovirus (type 2) DNA: analysis by HPLC. Biochem Biophys Res Commun 1982;107:19–26.
[54] Feinberg AP, Vogelstein B. Hypomethylation distinguishes genes of some human cancers from their normal counterparts. Nature 1983;301:89–92.
[55] Ehrlich M. DNA hypomethylation in cancer cells. Epigenomics 2009;1:239–59.
[56] Wild L, Flanagan JM. Genome-wide hypomethylation in cancer may be a passive consequence of transformation. Biochim Biophys Acta 2010;1806:50–7.
[57] Hon GC, Hawkins RD, Caballero OL, Lo C, Lister R, Pelizzola M, et al. Global DNA hypomethylation coupled to repressive chromatin domain formation and gene silencing in breast cancer. Genome Res 2012;22:246–58.
[58] Kafri T, Ariel M, Brandeis M, Shemer R, Urven L, McCarrey J, et al. Developmental pattern of gene-specific DNA methylation in the mouse embryo and germ line. Genes Dev 1992;6:705–14.
[59] Shi H, Wang MX, Caldwell CW. CpG islands: their potential as biomarkers for cancer. Expert Rev Mol Diagn 2007;7:519–31.
[60] De Smet C, Loriot A. DNA hypomethylation and activation of germline-specific genes in cancer. Adv Exp Med Biol 2013;754:149–66.
[61] Siddiqi S, Matushansky I. DNA hypomethylation and activation of germline-specific genes in cancer. J Cell Biochem 2012;113:373–80.
[62] Nestor CE, Ottaviano R, Reddington J, Sproul D, Reinhardt D, Duncan D, et al. Tissue type is a major modifier of the 5-hydroxymethylcytosine content of human genes. Genome Res 2012;22:467–77.
[63] Lian CG, Xu Y, Ceol C, Wu F, Larson A, Dresser K, et al. Loss of 5-hydroxymethylcytosine is an epigenetic hallmark of melanoma. Cell 2012; 150:1135–46.
[64] Berman BP, Weisenberger DJ, Aman JF, Hinoue T, Ramjan Z, Liu Y, et al. Regions of focal DNA hypermethylation and long-range hypomethylation in colorectal cancer coincide with nuclear lamina-associated domains. Nat Genet 2012;44:40–6.
[65] Toyota M, Ahuja N, Ohe-Toyota M, Herman JG, Baylin SB, Issa JP. CpG island methylator phenotype in colorectal cancer. Proc Natl Acad Sci USA 1999; 96:8681–6.

[66] Ahuja N, Li Q, Mohan AL, Baylin SB, Issa JP. Aging and DNA methylation in colorectal mucosa and cancer. Cancer Res 1998;58:5489–94.
[67] Hinoue T, Weisenberger DJ, Lange CP, Shen H, Byun HM, Van Den Berg D, et al. Genome-scale analysis of aberrant DNA methylation in colorectal cancer. Genome Res 2012;22:271–82.
[68] Comprehensive molecular characterization of human colon and rectal cancer. Nature 2012;487:330–37.
[69] Noushmehr H, Weisenberger DJ, Diefes K, Phillips HS, Pujara K, Berman BP, et al. Identification of a CpG island methylator phenotype that defines a distinct subgroup of glioma. Cancer Cell 2010;17:510–22.
[70] Bell D, Berchuck A, Birrer M, Chien J, Cramer DW, Dao F, et al. Integrated genomic analyses of ovarian carcinoma. Nature 2011;474:609–15.
[71] Weisenberger DJ, Siegmund KD, Campan M, Young J, Long TI, Faasse MA, et al. CpG island methylator phenotype underlies sporadic microsatellite instability and is tightly associated with BRAF mutation in colorectal cancer. Nat Genet 2006;38:787–93.
[72] Turcan S, Rohle D, Goenka A, Walsha LA, Fang F, Yilmaz E, et al. IDH1 mutation is sufficient to establish the glioma hypermethylator phenotype. Nature 2012;483:479–83.
[73] Abdel-Wahab O, Mullally A, Hedvat C, Garcia-Manero G, Patel J, Wadleigh M, et al. Genetic characterization of TET1, TET2, and TET3 alterations in myeloid malignancies. Blood 2009;114:144–7.
[74] Figueroa ME, Abdel-Wahab O, Lu C, Ward PS, Patel J, Shih A, et al. Leukemic IDH1 and IDH2 mutations result in a hypermethylation phenotype, disrupt TET2 function, and impair hematopoietic differentiation. Cancer Cell 2010;18.
[75] Xu W, Yang H, Liu Y, Yang Y, Wang P, Kim SH, et al. Oncometabolite 2-hydroxyglutarate is a competitive inhibitor of alpha-ketoglutarate-dependent dioxygenases. Cancer Cell 2011;19:17–30.
[76] Feinberg AP, Tycko B. The history of cancer epigenetics. Nat Rev Cancer 2004;4:143–53.
[77] Herman JG, Baylin SB. Gene silencing in cancer in association with promoter hypermethylation. N Engl J Med 2003;349:2042–54.
[78] Kalari S, Pfeifer GP. Identification of driver and passenger DNA methylation in cancer by epigenomic analysis. Adv Genet 2010;70:277–308.
[79] Siddiqi S, Terry M, Matushansky I. Hiwi mediated tumorigenesis is associated with DNA hypermethylation. PLoS One 2012;7:e33711.
[80] Bird A. Perceptions of epigenetics. Nature 2007;447:396–8.
[81] Greger V, Debus N, Lohmann D, Hopping W, Passarge E, Horsthemke B. Frequency and parental origin of hypermethylated RB1 alleles in retinoblastoma. Hum Genet 1994;94:491–6.
[82] Herman JG, Umar A, Polyak K, Graff JR, Ahuja N, Issa JPJ, et al. Incidence and functional consequences of hMLH1 promoter hypermethylation in colorectal carcinoma. Proc Natl Acad Sci USA 1998;95:6870–5.
[83] Xiong HL, Liu XQ, Sun AH, He Y, Li J, Xia Y. Aberrant DNA Methylation of P16, MGMT, hMLH1 and hMSH2 Genes in Combination with the MTHFR C677T Genetic Polymorphism in Gastric Cancer. Asian Pac J Cancer Prev 2013;14:3139–42.
[84] Esteller M, Silva JM, Domingurez G, Bonilla F, Matias-Guiu X, Lerma E, et al. Promoter hypermethylation and BRCA1 inactivation in sporadic breast and ovarian tumors. J Natl Cancer Inst 2000;92:564–9.
[85] Vogelstein B, Kinzler KW. Cancer genes and the pathways they control. Nat Med 2004;10:789–99.
[86] Sproul D, Kitchen RR, Nestor CE, Dixon JM, Sims AH, Harrison DJ, et al. Tissue of origin determines cancer-associated CpG island promoter hypermethylation patterns. Genome Biol 2012;13:R84.

[87] Ohtani-Fujita N, Dryja TP, Rapaport JM, Fujita T, Matsumura S, Ozasa K, et al. Hypermethylation in the retinoblastoma gene is associated with unilateral, sporadic retinoblastoma. Cancer Genet Cytogenet 1997;98:43–9.
[88] Hitchins MP, Rapkins RW, Kwok CT, Srivastava S, Wong JJ, Khachigian LM, et al. Dominantly inherited constitutional epigenetic silencing of MLH1 in a cancer-affected family is linked to a single nucleotide variant within the 5′UTR. Cancer Cell 2011;20:200–13.
[89] Toyota M, Suzuki H. Epigenetic drivers of genetic alterations. Adv Genet 2010;70:309–23.
[90] Sameer AS. Colorectal cancer: molecular mutations and polymorphisms. Front Oncol 2013;3:114.
[91] Simpkins SB, Bocker T, Swisher EM, Mutch DG, Gersell DJ, Kovatich AJ, et al. MLH1 promoter methylation and gene silencing is the primary cause of microsatellite instability in sporadic endometrial cancers. Hum Mol Genet 1999; 8:661–6.
[92] Esteller M, Toyota M, Sanchez-Cespedes M, Capella G, Peinado MA, Watkins DN, et al. Inactivation of the DNA repair gene O6-methylguanine-DNA methyltransferase by promoter hypermethylation is associated with G to A mutations in K-ras in colorectal tumorigenesis. Cancer Res 2000;60:2368–71.
[93] Esteller M, Risques RA, Toyota M, Capella G, Moreno V, Peinado MA, et al. Promoter hypermethylation of the DNA repair gene O(6)-methylguanine-DNA methyltransferase is associated with the presence of G:C to A:T transition mutations in p53 in human colorectal tumorigenesis. Cancer Res 2001; 61:4689–92.
[94] Esteller M, Garcia-Foncillas J, Andion E, Goodman SN, Hidalgo OF, Vanaclocha V, et al. Inactivation of the DNA-repair gene MGMT and the clinical response of gliomas to alkylating agents. N Engl J Med 2000;343:1350–4.
[95] Hegi ME, Diserens AC, Gorila T, Hamou MF, de Tribolet N, Weller M, et al. MGMT gene silencing and benefit from temozolomide in glioblastoma. N Engl J Med 2005;352:997–1003.
[96] Kobayashi H, Ohno S, Sasaki Y, Matsuura M. Hereditary breast and ovarian cancer susceptibility genes (Review). Oncol Rep 2013;30:1019–29.
[97] Chen J, Rocken C, Lofton-Day C, Schulz HU, Muller O, Kutzner N, et al. Molecular analysis of APC promoter methylation and protein expression in colorectal cancer metastasis. Carcinogenesis 2005;26:37–43.
[98] Moreno-Bueno G, Hardisson D, Sanchez C, Sarrio D, Cassia R, Garcia-Rostan G, et al. Abnormalities of the APC/beta-catenin pathway in endometrial cancer. Oncogene 2002;21:7981–90.
[99] Schmidt KN, Leung B, Kwong M, Zarember KA, Satyal S, Navas TA, et al. APC-independent activation of NK cells by the Toll-like receptor 3 agonist double-stranded RNA. J Immunol 2004;172:138–43.
[100] Zhou XL, Eriksson U, Werelius B, Kressner U, Sun XF, Lindblom A. Definition of candidate low risk APC alleles in a Swedish population. Int J Cancer 2004;110:550–7.
[101] Esteller M, Fraga MF, Guo M, Garcia-Foncillas J, Hedenfalk I, Godwin AK, et al. DNA methylation patterns in hereditary human cancers mimic sporadic tumorigenesis. Hum Mol Genet 2001;10:3001–7.
[102] Myohanen SK, Baylin SB, Herman JG. Hypermethylation can selectively silence individual p16ink4A alleles in neoplasia. Cancer Res 1998;58:591–3.
[103] Alyasiri NS, Ali A, Kazim Z, Gupta S, Mandal AK, Singh I, et al. Aberrant promoter methylation of PTEN gene among Indian patients with oral squamous cell carcinoma. Int J Biol Markers 2013;28:298–302.

[104] Carter JA, Gorecki DC, Mein CA, Ljungberg B, Hafizi S. CpG dinucleotide-specific hypermethylation of the TNS3 gene promoter in human renal cell carcinoma. Epigenetics 2013;8:739–47.
[105] Eads CA, Danenberg KD, Kawakami K, Saltz LB, Danenberg PV, Laird PW. CpG island hypermethylation in human colorectal tumors is not associated with DNA methyltransferase overexpression. Cancer Res 1999;59:2302–6.
[106] Robertson KD, Keyomarsi K, Gonzales FA, Velicescu M, Jones PA. Differential mRNA expression of the human DNA methyltransferases (DNMTs) 1, 3a and 3b during the G(0)/G(1) to S phase transition in normal and tumor cells. Nucleic Acids Res 2000;28:2108–13.
[107] Ibrahim AE, Arends MJ, Silva AL, Wyllie AH, Greger L, Ito Y, et al. Sequential DNA methylation changes are associated with DNMT3B overexpression in colorectal neoplastic progression. Gut 2011;60:499–508.
[108] Wiegand KC, Shah SP, Al-Agha OM, Zhao Y, Tse K, Zeng T, et al. ARID1A mutations in endometriosis-associated ovarian carcinomas. N Engl J Med 2010;363:1532–43.
[109] Kanai Y, Ushijima S, Nakanishi Y, Sakamoto M, Hirohashi S. Mutation of the DNA methyltransferase (DNMT) 1 gene in human colorectal cancers. Cancer Lett 2003;192:75–82.
[110] Ley TJ, Ding L, Walter MJ, McLellan MD, Lamprecht T, Larson DE, et al. DNMT3A mutations in acute myeloid leukemia. N Engl J Med 2010;363:2424–33.
[111] Yan XJ, Su J, Gu ZH, Pan CM, Lu G, Shen Y, et al. Exome sequencing identifies somatic mutations of DNA methyltransferase gene DNMT3A in acute monocytic leukemia. Nat Genet 2011;43:309–15.
[112] Couronne L, Bastard C, Bernard OA. TET2 and DNMT3A mutations in human T-cell lymphoma. N Engl J Med 2012;366:95–6.
[113] Walter MJ, Ding L, Shen D, Shao J, Grillot M, McLellan M, et al. Recurrent DNMT3A mutations in patients with myelodysplastic syndromes. Leukemia 2011;25.
[114] Langemeijer SM, Kuiper RP, Berends M, Knops R, Aslanyan MG, Massop M, et al. Acquired mutations in TET2 are common in myelodysplastic syndromes. Nat Genet 2009;41:838–42.
[115] Quivoron C, Couronne L, Valle VD, Lopez CK, Plo I, Wagner-Ballon O, et al. TET2 inactivation results in pleiotropic hematopoietic abnormalities in mouse and is a recurrent event during human lymphomagenesis. Cancer Cell 2011;20:25–38.
[116] Ko M, Huang Y, Jankowska AM, Pape UJ, Tahiliani M, Bandukwala HS, et al. Impaired hydroxylation of 5-methylcytosine in myeloid cancers with mutant TET2. Nature 2010;468:839–43.
[117] Weissmann S, Alpermann T, Grossmann V, Kowarsch A, Nadarajah N, Eder C, et al. Landscape of TET2 mutations in acute myeloid leukemia. Leukemia 2012;26:934–42.
[118] Ooi SK, Qiu C, Bernstein E, Li K, Jia D, Yang Z, et al. DNMT3L connects unmethylated lysine 4 of histone H3 to de novo methylation of DNA. Nature 2007;448:714–17.
[119] Lehnertz B, Ueda Y, Derijck AA, Braunschweig U, Perez-Burgos L, Kubicek S, et al. Suv39h-mediated histone H3 lysine 9 methylation directs DNA methylation to major satellite repeats at pericentric heterochromatin. Curr Biol 2003;13:1192–200.
[120] Nakagawa T, Kanai Y, Ushijima S, Kitamura T, Kakizoe T, Hirohashi S. DNA hypomethylation on pericentromeric satellite regions significantly correlates with loss of heterozygosity on chromosome 9 in urothelial carcinomas. J Urol 2005;173:243–6.
[121] Kondo Y, Shen L, Cheng AS, Ahmed S, Boumber Y, Charo C, et al. Gene silencing in cancer by histone H3 lysine 27 trimethylation independent of promoter DNA methylation. Nat Genet 2008;40:741–50.
[122] Frigola J, Song J, Stirzaker C, Hinshelwood RA, Peinado MA, Clark SJ. Epigenetic remodeling in colorectal cancer results in coordinate gene suppression across an entire chromosome band. Nat Genet 2006;38:540–9.

[123] Cedar H, Bergman Y. Linking DNA methylation and histone modification: patterns and paradigms. Nat Rev Genet 2009;10:295–304.

[124] Lawrence RJ, Earley K, Pontes O, Silva M, Chen ZJ, Neves N, et al. A concerted DNA methylation/histone methylation switch regulates rRNA gene dosage control and nucleolar dominance. Mol Cell 2004;13:599–609.

[125] Fraga MF, Ballestar E, Villar-Garea A, Boix-Chornet M, Espada J, Schotta G, et al. Loss of acetylation at Lys16 and trimethylation at Lys20 of histone H4 is a common hallmark of human cancer. Nat Genet 2005;37:391–400.

[126] Puissant A, Frumm SM, Alexe G, Bassil CF, Qi J, Chanthery YH, et al. Targeting MYCN in neuroblastoma by BET bromodomain inhibition. Cancer Discov 2013; 3:308–23.

[127] Black JC, Van Rechem C, Whetstine JR. Histone lysine methylation dynamics: establishment, regulation, and biological impact. Mol Cell 2012;8:491–507.

[128] Balch C, Nephew KP, Huang TH, Bapat SA. Epigenetic "bivalently marked" process of cancer stem cell-driven tumorigenesis. Bioessays 2007;29:842–5.

[129] Fuks F, Burgers WA, Brehm A, Hughes-Davies L, Kouzarides T. DNA methyltransferase Dnmt1 associates with histone deacetylase activity. Nat Genet 2000;24:88–91.

[130] Hashimshony T, Zhang J, Keshet I, Bustin M, Cedar H. The role of DNA methylation in setting up chromatin structure during development. Nat Genet 2003;34:187–92.

[131] Lande-Diner L, Zhang J, Ben-Porath I, Amariglio N, Keshet I, Hecht M, et al. Role of DNA methylation in stable gene repression. J Biol Chem 2007;282:12194–200.

[132] Murphy PJ, Cipriany BR, Wallin CB, Ju CY, Szeto K, Hagarman JA, et al. Single-molecule analysis of combinatorial epigenomic states in normal and tumor cells. Proc Natl Acad Sci USA 2013;110:7772–7.

[133] Chang KM, Fang D, Gan H, Hashizume R, Yu C, Schroeder M, et al. The histone H3.3K27M mutation in pediatric glioma reprograms H3K27 methylation and gene expression. Genes Dev 2013;27:985–90.

[134] Lewis PW, Muller MM, Koletsky MS, Cordero F, Lin S, Banaszynski LA, et al. Inhibition of PRC2 activity by a gain-of-function H3 mutation found in pediatric glioblastoma. Science 2013;340:857–61.

[135] Zhang YW, Xie HQ, Chen Y, Jiao B, Shen ZX, Chen SJ, et al. Loss of promoter methylation contributes to the expression of functionally impaired PRDM1beta isoform in diffuse large B-cell lymphoma. Int J Hematol 2010;92:439–44.

[136] Oshimo Y, Oue N, Mitani Y, Nakayama H, Kitadai Y, Yoshida K, et al. Frequent epigenetic inactivation of RIZ1 by promoter hypermethylation in human gastric carcinoma. Int J Cancer 2004;110:212–18.

[137] Berdasco M, Ropero S, Setien F, Fraga MF, Lapunzina P, Losson R, et al. Epigenetic inactivation of the Sotos overgrowth syndrome gene histone methyltransferase NSD1 in human neuroblastoma and glioma. Proc Natl Acad Sci USA 2009;106:21830–5.

[138] Thomson JP, Skene PJ, Selfridge J, Clouaire T, Guy J, Webb S, et al. CpG islands influence chromatin structure via the CpG-binding protein Cfp1. Nature 2010;464:1082–6.

[139] Morishita M, di Luccio E. Cancers and the NSD family of histone lysine methyltransferases. Biochim Biophys Acta 2011;1816:158–63.

[140] Lim S, Metzger E, Schule R, Kirfel J, Buettner R. Epigenetic regulation of cancer growth by histone demethylases. Int J Cancer 2010;127:1991–8.

[141] Widschwendter M, Fiegl H, Egle D, Mueller-Holzner E, Spizzo G, Marth C, et al. Epigenetic stem cell signature in cancer. Nat Genet 2007;39:157–8.

[142] Nikoloski G, Langemeijer SMC, Kuiper RP, Knops R, Massop M, Tonnissen ERLTM, et al. Somatic mutations of the histone methyltransferase gene EZH2 in myelodysplastic syndromes. Nat Genet 2010;42:665–7.

[143] Martinez-Garcia E, Licht JD. Deregulation of H3K27 methylation in cancer. Nat Genet 2010;42:100–1.

[144] Patel K, Dickson J, Din S, Macleod K, Jodrell D, Ramsahoye B. Targeting of 5-aza-2′-deoxycytidine residues by chromatin-associated DNMT1 induces proteasomal degradation of the free enzyme. Nucleic Acids Res 2010;38:4313–24.

[145] Veigl ML, Kasturi L, Olechnowicz J, Ma AH, Lutterbaugh JD, Periyasamy S, et al. Biallelic inactivation of hMLH1 by epigenetic gene silencing, a novel mechanism causing human MSI cancers. Proc Natl Acad Sci USA 1998;95:8698–702.

[146] Veeck J, Ropero S, Setien F, Gonzalez-Suarez E, Osorio A, Benitez J, et al. BRCA1 CpG island hypermethylation predicts sensitivity to poly(adenosine diphosphate)-ribose polymerase inhibitors. J Clin Oncol 2010;28:e563–4.

[147] Robert MF, Morin S, Beaulieu N, Gauthier F, Chute IC, Barsalou A, et al. DNMT1 is required to maintain CpG methylation and aberrant gene silencing in human cancer cells. Nat Genet 2003;33:61–5.

[148] Feinberg AP. Methylation meets genomics. Nat Genet 2001;27:9–10.

[149] Hashimoto H, Vertino PM, Cheng X. Molecular coupling of DNA methylation and histone methylation. Epigenomics 2010;2:657–69.

[150] Rakyan VK, Down TA, Balding DJ, Beck S. Epigenome-wide association studies for common human diseases. Nat Rev Genet 2011;12:529–41.

[151] Heyn H, Esteller M. DNA methylation profiling in the clinic: applications and challenges. Nat Rev Genet 2012;13:679–92.

CHAPTER

8

DNA Methylation in Breast and Ovarian Carcinomas

OUTLINE

8.1 INTRODUCTION

Usually breast cancer begins either in the cells of the lobules, which are the milk-producing glands, or in the ducts, the passages that drain milk from the lobules to the nipple. Less commonly, breast cancer can begin in the stromal tissues, which include the fatty and fibrous connective tissues of the breast. Over time, cancer cells can invade nearby healthy breast

M. Neidhart: DNA Methylation and Complex Human Disease.
DOI: http://dx.doi.org/10.1016/B978-0-12-420194-1.00008-7

tissue and make their way into the underarm lymph nodes, small organs that filter out foreign substances in the body. If cancer cells get into the lymph nodes, they then have a pathway into other parts of the body. The breast cancer's stage refers to how far the cancer cells have spread beyond the original tumor. The great majority of breast cancers are due to genetic and/or epigenetic abnormalities that happen as a result of the aging process and the "wear and tear" of life in general.

The term "ovarian cancer" includes several different types of cancer that all arise from cells of the ovary. Most commonly, tumors arise from the epithelium, or lining cells, of the ovary. These include epithelial ovarian (from the cells on the surface of the ovary), fallopian tube, and primary peritoneal (the lining inside the abdomen that coats many abdominal structures) cancer. These are all considered to be one disease process. There is also an entity called borderline ovarian tumors, which has the microscopic appearance of a cancer but tends not to spread much. However, there are also less common forms of ovarian cancer that come from within the ovary itself, including germ cell tumors and sex cord-stromal tumors. Epithelial ovarian cancer accounts for about 70% of all ovarian cancers. It is generally thought of as one of the three types of cancer (ovarian, fallopian tube, and primary peritoneal) that all behave and are treated the same way, depending on the type of cell that causes the cancer. The four most common cell types of epithelial ovarian cancer are serous, mucinous, clear cell, and endometrioid. These cancers arise due to genetic and epigenetic changes in the DNA that lead to the development of tumors. Ovarian cancer is the most lethal gynecological tumor. Due to few early symptoms and a lack of early detection strategies, most patients are diagnosed with advanced stage disease.

Both breast and ovarian cancers have a strong hormonal component. Most of these patients, although initially responsive, eventually develop drug resistance. In this chapter, we describe the changes in DNA methylation in breast [1] and ovarian [2] cancer.

8.2 HORMONAL RECEPTORS

Pathological estrogens have been associated with a higher risk for breast and endometrial cancer, and hormone dependence of breast cancers is correlated with tumor progression and patient prognosis [1]. Most breast cancers are initially positive for estrogen receptors (ERs), and their growth can be stimulated by estrogens and inhibited by antiestrogens. DNA methylation of the ER and progesterone receptor (PGR) promoters has been proposed as a mechanism for the development of ER-negative tumors in cell lines as well as primary tumors [3–6]. Hypermethylation has been discussed as a possible cause of ER loss due to the findings of an early study [7], which demonstrated that ER-negative breast cancer cells are devoid of ER mRNA. Further, ER gene expression can be reactivated in ER-negative cells by inhibition of methylation [8]. However, clinical data remains contradictory. One study [9] found hypermethylation of the ER promoter region in tumors, but another group [10] detected no correlation between gene methylation pattern and ER gene expression in breast tumors. In ovarian cancer, the ER gene is found hypermethylated in some cases [11]. In summary, current evidence suggests that there is no clear link between ESR1 (the gene for ER) methylation and ER status, while PGR methylation is significantly linked to progesterone receptor expression and PGR methylation status might be a predictor for ER status [4].

8.3 HYPOMETHYLATED GENES

Although breast tumors are frequently hypomethylated on a genome-wide scale, the number of genes reported as hypomethylated in breast cancer is relatively small (Table 8.1). This is probably due to the positioning of hypomethylated DNA to regions of pericentromeric DNA and gene-poor regions of the genome, but also to the fact that the focus on DNA methylation in cancer has been on hypermethylation of CpG islands and most techniques will only detect hypermethylated regions. Genes that are hypomethylated in primary breast tumors include the endonuclease *FEN1* [12], the N-acetyltransferase *NAT1* [13], and the P-cadherin *CDH2* [14]. Genes that have been found hypomethylated in breast cancer cell lines but where evidence for hypomethylation

TABLE 8.1 Hypomethylated Genes in Breast Cancer Cells [1]

Gene	Function	Reference(s)
ER	Estrogen receptors (ER-positive breast cancer)	[3]
FEN1	Endonuclease involved in DNA replication and repair	[12]
NAT1	Enzyme metabolizing drugs and other xenobiotics	[13]
CDH3/P-cadherin	Cell motility, aggressiveness	[14]
PLAU/UPA/ urokinase	Serine protease involved in matrix degradation	[15]
BSCG1/γ-synuclein	Cell motility, invasiveness, and metastasis	[16,17]
IGF2	Growth-regulating, insulin-like and mitogenic activities	[18]
CAV1,2/caveolin-1,2	Negative regulators of the RhoC GTPase cascade	[19]

in primary tumors is weak include the metastasis gene *PLAU* [15] and the breast cancer-specific gene 1 (*BCSG1*) [16,17]. The only imprinted gene that has been reported hypomethylated in breast cancer so far is the insulin-like growth factor II (*IGF2*) gene [18]. Hypomethylation of *CAV1,2* and overexpression of caveolin-1/2 have been described in inflammatory breast cancer [19]. In addition, high resolution analysis of DNA hypomethylation in breast cancer identified a large number of hypomethylated sites with around 1500 regions hypomethylated in a cancer-specific manner [20,21]. It is likely that many of these regions contain genes or regulatory sequences that play important roles in cancer development.

8.4 HYPERMETHYLATED GENES

More than 100 genes have been reported to be hypermethylated in breast tumors or breast cancer cell lines [22,23]. Many of the genes aberrantly methylated play important roles in hormone signaling, cell-cycle regulation, tissue invasion and metastasis, DNA repair, cell differentiation, and apoptosis [22–24].

8.4.1 Genes Involved in Hormonal Controls

Women's breast cancer risk is clearly affected by their reproductive history. The hormonal milieu also influences the course of the disease. The female reproductive hormones, estrogens, progesterone, and prolactin, have a major impact on normal mammary gland development. In addition, pathological estrogens have been associated with a higher risk for breast and endometrial cancer, and hormone dependence of breast cancers is correlated with tumor progression and patient prognosis [1]. Most breast cancers initially positive for ER lose it after a certain time. Three distinct tumor subtypes can be considered [25]: ER/progesterone receptor (PGR) positive, tumors overexpressing the human epidermal receptor 2 protein (HER2+), and triple negative breast cancer, which lacks the three markers. Table 8.2 presents the genes that could be hypermethylated and involved in hormonal control of breast cancer cells. DNA methylation of the ER and PGR promoters has been proposed as a

TABLE 8.2 Hypermethylated Genes in Breast Cancer Cells [1] – Hormonal Receptors

Gene	Function	Reference(s)
ER/ESR1,2	Estrogen receptors (ER-negative breast cancer)	[3–6]
PGR	Progesterone receptor	[4]

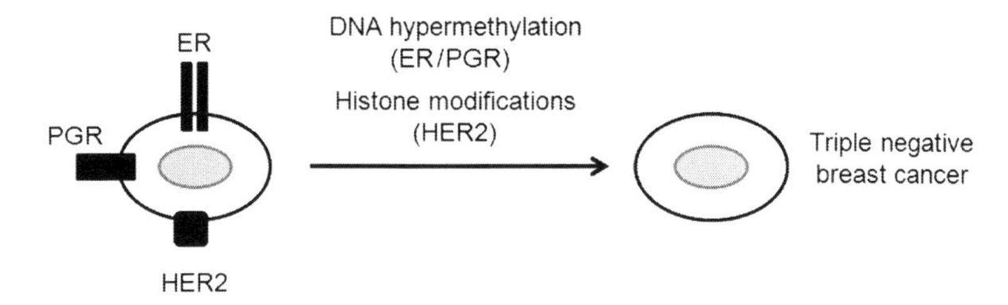

FIGURE 8.1 Appearance of triple negative cancer cells through disruption of epigenetic control mechanisms.

mechanism for the development of ER-negative tumors in cell lines as well as primary tumors [3–6]. Hypermethylation has been discussed as a possible cause of ER loss due to the findings of an early study [7], which demonstrated that ER-negative breast cancer cells are devoid of ER mRNA. Further, ER gene expression can be reactivated in ER-negative cells by inhibition of methylation [8]. Two genes (ESR1 and ESR2) and two ER subunits (α1 and β1) exist. As expected, *ESR1* gene methylation showed an inverse association with ERα1 [4]. Similarly, *ESR2* gene encoding ERβ1 and PGR encoding progesterone receptor can be silenced by methylation in breast cancer [4,6]. Loss of ER expression appeared closely linked to thrombospondin 3 (TSP3) mutations [26]. TSP3 is an adhesive glycoprotein that mediates cell-to-cell and cell-to-matrix interactions. However, clinical data remain contradictory. One study [9] found hypermethylation of the ER promoter region in tumors, but another group [10] detected no correlation between gene methylation pattern and ER gene expression in breast tumors. Current evidence suggests that there is a link between ESR1 methylation and ER status, as well as PGR methylation and progesterone receptor expression: PGR methylation status might be a predictor for ER status [4]. Thus, most recent findings clearly show that DNA hypermethylation is a factor causing the loss of ER and PGR in breast cancer. HER2 expression is controlled by HDAC2 and histone modifications [27] (Figure 8.1).

8.4.2 Genes Involved in the Cell Cycle and Uncontrolled Proliferation

The loss or deregulation of proteins involved in such diverse processes as cellular proliferation, cell cycle and checkpoint control, DNA repair, and cell death is a consistent feature of cancer cells. Cancer cells arise through a process of cellular evolution due to the accumulation of genetic changes. As cancers arise and progress, there is selection for those genetic changes that give the cancer cell a proliferative advantage over normal cells. Table 8.3 presents the prototype genes that are

TABLE 8.3 Hypermethylated Genes in Breast Cancer Cells [1] – Cell Cycle and Proliferation

Gene	Function	Reference(s)
ARH1	Negative regulator of cell proliferation	[28]
CCND2/cyclin D2	Regulator of cell cycle	[29]
CDKN2A/p16^{ink4A}	Negative regulator of cell proliferation	[30,31]
CDKN1C/p57KIP2	Negative regulator of cell proliferation	[32]
GPC3/glypican-3	Cell division and growth regulation	[33]
HIC1	Transcriptional repressor	[34]
HIN1	Negative regulator of cell growth	[35]
IGFBP3	Negative regulator of cell growth	[36]
PPP2R2B	Negative regulator of cell growth and proliferation	[1,37]
PTEN	Negative regulator of cell growth and proliferation	[1,37]
RAR-β2	Retinoic acid receptor beta 2	[38–40]
RASSF1A	DNA repair and negative regulator of cell cycle	[38,41,42]

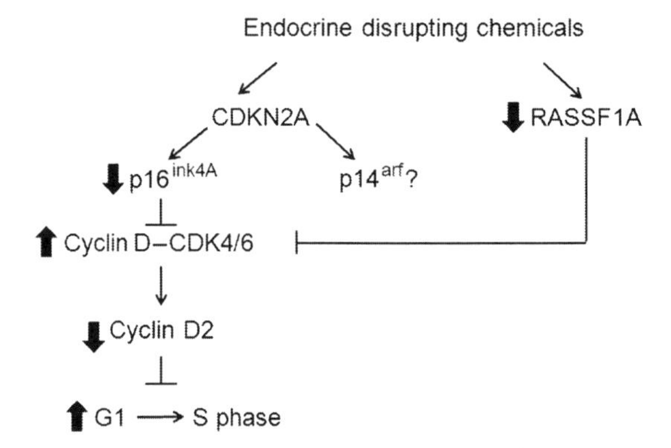

FIGURE 8.2 Disruption of genes limiting cell proliferation can be caused by mutations and/or epigenetic silencing. The best examples are the CDKN2A/RASSF1A/cyclin D pathways in breast cancer. Silencing of CDKN2A and/or RASSF1 decreases the availability of cyclin D2, which limits cell proliferation. Cyclin D2 itself also can be methylated and silenced.

hypermethylated and involved in the uncontrolled proliferation of breast cancer cells. Figure 8.2 shows an overview of the mechanisms that are affected by such genes. Chapter 5 presents the epidemiological evidence that hypermethylation of CDKN2A/p16^{ink4A} and RASSF1A in breast cancer could be due to endocrine-disrupting chemicals.

CDKN2A/p16^{ink4A} is frequently methylated in many human cancers, including breast cancer [30]. CDKN2A inactivation associated with DNA methylation has been observed in human mammary epithelial cells (HMECs) when the cultured cells escape senescence and acquire telomere crisis and chromosomal abnormalities similar to those observed in early neoplastic lesions [31]. CDKN2A methylation has been found in disease-free breast tissue, and it is speculated that this methylation originates from a subpopulation of cells in normal epithelia [43]. CDKN2A/p16^{ink4A} inhibits the G_1 to S transition by inhibiting binding of cyclin D to CDK4 and CDK6. CDKN2A/p14^{arf} is an alternative reading frame from the same locus, also involved in cell cycle regulation. In ovarian cancer, both p16^{ink4A} and p14^{arf} pathways are altered by DNA methylation [44].

RASSF1A also is frequently methylated in breast cancer [41]. CDKN2A and RASSF1A methylation are early epigenetic events in breast cancer and are found in in ductal carcinomas *in situ* [38,42]. Hypermethylated CDKN2A [45] and RASSF1A [11,46] genes also were described in ovarian cancer. RASSF1A inhibits the phosphorylation of JUN kinase that favors formation of the cyclin D–CDK4/6 complex.

Another cell cycle regulator attracting a lot of interest is cyclin D2 itself (CCND2 gene). It is an essential regulator of the cell cycle, and overexpression inhibits the transition between the G_1 and S phases. CCND2 has frequently been found methylated in breast cancer and is also methylated in ductal carcinomas *in situ*, suggesting it to be an early event in tumorigenesis [29]. Similarly, disruption of cyclin-dependent kinase inhibitor 1C (CDKN1C/p57KIP2), which binds cyclins, can be responsible for several hallmarks of cancer, but the mechanism is as yet unclear. CDKN1C is generally not mutated in cancer, but its expression is downregulated through epigenetic changes such as DNA methylation and repressive histone marks at the promoter [32,47]. Disruption of cyclin D1 is involved in ovarian cancer, but in an indirect way; transcription elongation factor (SII)-like 7 (TEFAL7), which transcriptionally represses cyclin D1, can be hypermethylated in ovarian cancer [48].

Functional disruption of negative regulators of cell growth and proliferation is an important feature in this type of cancer. In breast cancer, many other examples of such genes have been reported. For instance, PPP2R2B is a serine/threonine-protein phosphatase implicated in the negative control of cell growth and division. The PPP2R2B gene is often hypermethylated in early invasive breast cancers [37]. Similar findings are reported for PTEN (phosphatase and tensin homolog) [37], which negatively regulates intracellular levels of phosphatidylinositol-3,4, 5-triphosphate in cells and functions as a tumor suppressor by negatively regulating the Akt/PKB signaling pathway. Another methylated regulator of proliferation in breast cancer is the tumor-suppressor RAR-β [38,39]. RAR-β methylation is an early epigenetic event in breast cancer, and is found in *in situ* lesions from both lobular and ductal cancers [38]. In particular, RAR-β2 expression is lower in the breast cancer compared to normal tissue and fibroadenoma [40]. The methylation rate of RAR-β2 in breast cancer and precancerous lesions of breast cancer is higher than that in normal tissues. The HIC1 (hypermethylated in cancer-1) gene functions as a growth regulatory and tumor repressor gene. Hypermethylation or deletion of the region of this gene has been associated with tumors and this gave it its name. Hypermethylation of HIC1 and associated loss of expression is common in primary breast cancer [34]. Furthermore, the HIC1 gene is densely methylated in approximately one-half of the alleles in normal breast epithelium, which may predispose this tissue to inactivation of this gene by loss of heterozygosity. Another gene, HIN1, which is an inhibitor of cell growth, migration and invasion, is frequently also silenced by DNA methylation in breast cancer [35] and ovarian clear cell adenocarcinoma [49]. Glypican-3 (GPC3) is a membrane-bound heparan sulfate proteoglycan that is involved in cell division and growth regulation and is often silenced by DNA methylation in ovarian and breast cancer [33]. Finally, a proper control of protein ADP-ribosylation levels by ADP-ribosylarginine hydrolase (ARH1) is essential to avoid cancerogenesis [50]. ARH1 is expressed in normal ovarian and breast epithelial cells, but its expression is lost in a majority of ovarian and breast cancers [28] due to promoter hypermethylation. Again, this gene is a negative regulator of cell proliferation. In ovarian cancer, ADP-ribosylation factor-like protein 11 (ARL1/ARLTS1) has been also reported to be methylated [51], but alteration in this gene confers apoptosis resistance, not increased proliferation.

Insulin-like growth factor binding protein-3 (IGFBP-3) is a mediator of growth suppression signals. This gene can be hypermethylated in various tumors, including breast [36] and, more often, ovarian cancers [52]. Other regulators of the cell cycle and/or favoring proliferation that are hypermethylated in ovarian cancer include DLEC1 [53], SOCS1 [54], ANGPTL2 [55], and CTCF [56].

8.4.3 Genes Involved in Cell Adhesion, Cell Motility, and Metastasis

Cell–cell adhesion determines the polarity of cells and participates in the maintenance of the cell societies called tissues. Cell–cell adhesiveness is generally reduced in human cancers [57]. Reduced intercellular adhesiveness allows cancer cells to avoid the histological structure, which is the morphological hallmark of malignant tumors. Reduced intercellular adhesiveness is also indispensable for cancer invasion and metastasis. Cadherins and their undercoat proteins, catenins, which connect cadherins to actin filaments, are located at the lateral borders,

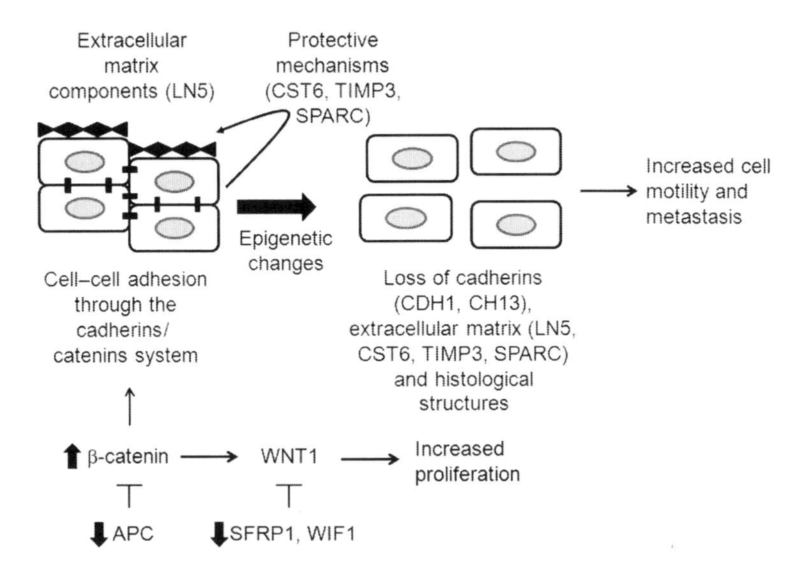

FIGURE 8.3 Disruption of genes allowing tissue formation can be caused by mutations and/or epigenetic silencing. The best examples are the cadherins/catenins in breast cancer. In addition, alteration of the extracellular matrix is caused by silencing of LN5 and protective mechanisms, such as CST6, TIMP3, and SPARC. This results in increased motility and metastasis. The catenin system is also linked to the Wnt pathway and increased proliferation, in which the silencing of APC, SFRP1, and/or WIF1 can play a role.

concentrating on adherens junctions, of epithelial cells and establish firm cell–cell adhesion. The cadherin cell adhesion system in cancer cells is inactivated by various mechanisms that reflect the morphological and biological characteristics of the tumor (Figure 8.3). Table 8.4 presents the genes that could be hypermethylated and involved in the decreased adhesion, increased cell motility, and metastasis of breast cancer cells.

CDH1/E-cadherin is a cell adhesion molecule frequently silenced in breast carcinomas by DNA methylation, and this silencing might be important for tumor cell invasion and metastasis [59,60]. This is also the case in ovarian cancer [73,74]. Similarly, CH13/H-cadherin gene hypermethylation in breast and ovarian cancer is correlated with hypomethylation of the SAT2 gene [61]; this combination shows a poor prognosis. Interestingly, similar to findings in colorectal cancers, it has been suggested that disruption of the adenomatous polyposis coli (APC)/β-catenin pathway is involved in breast carcinogenesis [58]. The regulation of β-catenin by APC prevents genes that stimulate cell division from being turned on too often, and prevents cell overgrowth. β-Catenin is a subunit of the cadherin protein complex, and acts as an intracellular signal transducer in the Wnt signaling pathway. However, in contrast to colorectal cancer, somatic mutations in APC and/or β-catenin are not

TABLE 8.4 Hypermethylated Genes in Breast Cancer Cells [1] – Adhesion and Motility

Gene	Function	Reference(s)
APC	Control of cell division, adhesion and organization into tissues	[58]
CDH1/E-cadherin	Cell–cell adhesion	[59,60]
CDH13/H-cadherin	Cell–cell adhesion	[61,62]
CST6/cystatin M	Cystein protease inhibitor	[63]
LAMA3/α3-laminin	Organization of cells into tissues	[64]
OPCML	Cell adhesion and negative regulator of proliferation	[65]
SFRP1	Negative regulator of Wnt signaling	[66]
SPARC/osteonectin	Cell–matrix interactions, regulation of metalloproteinases	[67]
ROBO1/DUTTI1	Adhesion molecule	[68]
RUNX3	Transcriptional activator	[69]
SYK	Tyrosine kinase	[70]
TIMP3	Inhibitor of metalloproteinases	[71]
WIF1	Negative regulator of Wnt signaling	[72]

found in breast cancer. On the other hand, hypermethylation of the APC promoter can be detected in one-third of primary breast cancers (and not in normal breast tissues). It is likely that hypermethylation of the APC gene disrupts regulation of the APC/β-catenin pathway. This gene can also be hypermethylated in invasive ovarian cancer cells [75].

Wnt (wingless type) signaling is involved in a variety of mammalian developmental processes, including cell proliferation, differentiation and epithelial–mesenchymal interactions, through which they contribute to the development of tissues and organs [76], as well as in the involution of mammary gland following lactation [77]. Wnts are secreted ligands that control cell processes via at least two pathways, one of which, the canonical Wnt signaling pathway, operates through the cytosolic stabilization of a transcriptional co-factor, β-catenin [76]. Wnt proteins bind to "frizzled" receptors, which leads to downstream activation of gene transcription by β-catenin [77]. Recent evidence suggests the role of Wnt signaling in human breast cancer involves elevated levels of nuclear and/or cytoplasmic β-catenin, and overexpression or downregulation of specific Wnt proteins, in particular downregulation of SFRP1 (secreted frizzled protein 1) and WIF1 (Wnt inhibitory factor 1) [77]. SFRP1 acts as a counter-regulator of Wnt signaling. The SFRP1 gene is hypermethylated in the great majority of breast cancers [66]. SFRP1, 2, 4, and 5 were found all to be hypermethylated in ovarian cancer [49,78–81]. Similarly to SFRPs, WIF1 binds to WNT1 and inhibits the growth and invasion of tumors through induction of G_1 arrest. WIF1 expression is commonly diminished in breast tumors, when compared with normal tissue, and this correlates with WIF1 promoter hypermethylation [72].

DUTT1/ROBO1 (human homolog of the Drosophila Roundabout gene) is a member of the NCAM family of receptors. Many breast cancer cell line showed complete hypermethylation of CpG sites within the promoter region of the DUTT1 gene [68]. The same region was also found to be hypermethylated in one fifth of primary invasive breast carcinomas. In ovarian carcinoma, ICAM1 appears hypermethylated [82].

OPCML is an opioid binding cell adhesion molecule. Its gene is hypermethylated in many tumors, including breast [65] and ovarian cancer [83]. Elevation of the RAS signaling pathway plays an important role in epigenetic inactivation of this gene [84]. Ecotopic expression of OPCML leads to inhibition of both anchorage-dependent and -independent growth of carcinoma cells [65]. In ovarian cancer, OPCML promoter methylation is associated with an older age of the patients, an advanced pathological stage, and poor overall survival [85].

Disruption of the histological structure can also be achieved by silencing of genes of the extracellular matrix. Downregulation of Laminin-5 (LN5)-encoding genes (LAMA3, LAMB3, and LAMC2) is reported in various human cancers. In breast cancer, LAMA3 promoter methylation frequency is associated with increased tumor stage and size [64].

Disruption of the extracellular matrix can be favored by silencing a protective mechanism. Cystatin M (CST6) is a lysosomal cysteine proteinase inhibitor found in a variety of human fluids and secretions, where it appears to provide protective functions. Cystatin M is downregulated in metastatic breast tumor cells as compared to primary tumor cells. Loss of expression is likely associated with the progression of a primary tumor to a metastatic phenotype. This occurs by increased CST6 gene methylation [63].

TIMP3 is an inhibitor of matrix metalloproteinases, and this gene has been found to be methylated in some invasive breast [71] and ovarian [11,74] cancer cells. Furthermore, the SPARC gene encodes osteonectin, which is an acidic extracellar matrix glycoprotein that plays an important role in cell–matrix interactions and collagen binding. Osteonectin also increases the production and activity of matrix metalloproteinases, a function important to invading cancer cells. Additional functions of osteonectin beneficial to tumor cells include angiogenesis, proliferation, and migration. Overexpression of osteonectin is reported in many human cancers, such as bone, breast, ovarian, prostate, and colon. SPARC is aberrantly methylated in breast [67] and ovarian cancers [86] that are associated with an increased activity of DNMT3A [86], a *de novo* DNA methyltransferase.

Metastasis can also be produced by very complex interactions upon silencing of specific transcription factors. SYK is a tyrosine kinase and RUNX3 is a transcriptional regulator, the silencing of which in breast cancer has been associated with tumor invasiveness [69,70]. The SYK and RUNX3 genes are hypermethylated in about one-third and one-half of breast cancers, respectively.

8.4.4 Genes Involved in DNA Repair

While large numbers of epigenetic alterations are found in cancers, the epigenetic alterations in DNA repair genes, causing reduced expression of DNA repair proteins, appear to be particularly important (Figure 8.4). Such alterations are thought to occur early in progression to cancer and to be a likely cause of the genetic instability characteristic of cancers. In breast cancer (Table 8.5), specific promoter hypermethylation of the DNA mismatch repair gene hMLH1, the DNA alkyl-repair

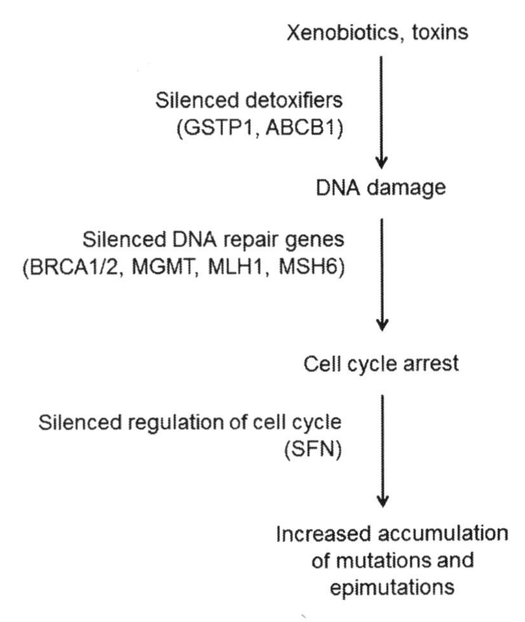

FIGURE 8.4 Disruption of detoxifier genes can be caused by mutations and/or epigenetic silencing. In turn, accumulation of toxic products can cause DNA damage. In the case that DNA repair genes also are defective or silenced, the cell cycle should be arrested and the cell undergoes apoptosis. If again this is not the case, the cell will continue to accumulate mutations and epimutations until it undergoes transformation to a cancer cell.

TABLE 8.5 Hypermethylated Genes in Breast Cancer Cells [1] – DNA Repair

Gene	Function	Reference(s)
ABCB1/Pgp1/CD243	Xenobiotics pump	[1,37]
BRCA1	DNA repair	[87–90]
BRCA2	DNA repair	[62]
GSTP1	Detoxification	[62,87]
MGMT	DNA alkyl repair	[87,91]
MLH1	DNA repair	[92]
MSH6	DNA repair	[62]
SFN/14-3-sigma	Regulator of cell cycle following DNA damage	[93]

gene O(6)-methylguanine-DNA methyltransferase (MGMT) [87,91], MutL homolog 1 (MLH1) [92], MutS homolog 6 (MSH6) [62], the detoxifiers glutathione S-transferase P1 (GSTP1) [87] and ABCB1 [37], as well as the familial breast cancer genes BRCA1 [87] and BRCA2 [62] may lead to specific genetic lesions, microsatellite instability, G to A transitions, steroid-related adducts, and double-strand breaks in DNA [87]. Hypermethylated MLH1 has also been described in ovarian cancer [94].

Chapter 7 discussed the relationship between dysfunction of DNA repair and cancerogenesis. BRCA1 and BRCA2 are normally expressed in the cells of breast and other tissues, where they help to repair damaged DNA or destroy cells if DNA cannot be repaired. They are involved in the repair of chromosomal damage with an important role in the error-free repair of DNA double-strand breaks. Upon DNA damage, configuration change of BRCA1 occurs, leading to activation of DNA repair complexes. The BRCA2 complex is error-free, but the alternative complex is error-prone. Therefore, mutant or deficient BRCA1 leads to more error-prone repairs. Given its important role in familial breast cancer and the fact that no BRCA1 mutations have been detected in sporadic breast cancers, DNA methylation-induced silencing is an attractive mechanism for BRCA1 silencing in these tumors. BRCA1 methylation has been found in sporadic breast cancers but is not a frequent event [87–89], and it is possible that BRCA1 methylation is most common in rare subtypes of basal-like origin [90]. Sporadic and inherited breast tumors have overall similar methylation profiles, but BRCA1 tumors have reduced methylation of certain non-familial genes and have a phenotype associated with basal-like carcinomas [95]. Similarly, BRCA2 gene hypermethylation can occur as an early event in breast cancer [62]. Hypermethylated BRCA1 has also been described in ovarian cancer [46,96].

GSTP1 and ABCB1 both play a role in cellular detoxification, by metabolizing xenobiotics or pumping them out of the cell, respectively. Both genes can be hypermethylated and silenced in breast cancer [1,37,62,87]. This may favor DNA damage. Finally, expression of SFN/14-3-sigma should be induced in response to DNA damage and cause cells to arrest in G2. Remarkably, its transcript is undetectable in most breast cancer biopsies [93], which is associated with hypermethylation of the SFN gene.

8.4.5 Genes Involved in Development and Differentiation

Improper regulation of development genes, such as FOX, HOX or PAX, may result in cancer (Figure 8.5). Table 8.6 presents the genes that could be hypermethylated and involved in differentiation of breast cancer cells.

The FOX gene family provides instructions for making proteins that play a critical role in the formation of many organs and tissues during embryogenesis. The FOX proteins are transcription factors. FOXA1 (forkhead box A1), also known as HNF3α (hepatocyte nuclear factor

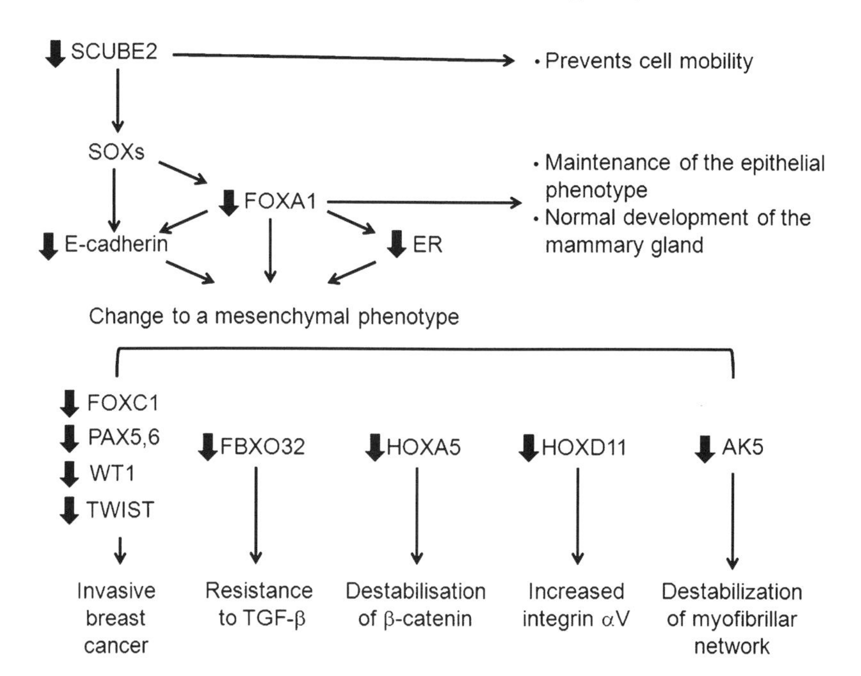

FIGURE 8.5 Disruption of developmental genes can be caused by mutations and/or epigenetic silencing. Central to the normal development of the mammary gland is FOXA1. Its silencing allows the change from an epithelial to a mesenchymal phenotype. With the silencing of other developmental genes, definite phenotypes appeared. Most importantly, FOXC1, PAX5/6, WT1, and TWIST silencing are associated with invasive behavior.

TABLE 8.6 Hypermethylated Genes in Breast Cancer Cells [1] – Development and Differentiation

Gene	Function	Reference(s)
AK5	ATP metabolic process	[22]
FOXA2/HNF3α	Transcriptional activator	[22]
FOXC1	Transcription factor, regulation of embryonic development	[1,37]
HOXA5	Development, differentiation and adhesion	[97]
HOXD11	Development, morphogenesis and adhesion	[22]
PAX5	Development	[62]
PAX6	Development	[62]
SCUBE2	Negative regulator of cell mobility	[98]
TWIST	Transcription factor involved in cell differentiation	[38]
WT1	Transcription factor	[62,99]

3a), is required for postnatal survival due to its essential role in controlling pancreatic and renal function. In addition to regulating a variety of tissues during embryogenesis and early life, rescue experiments have revealed a specific role for FOXA1 in the postnatal development of the mammary gland. Activity of the nuclear estrogen receptor ER1 is also required for proper development of the mammary gland. FOXA1 modulates ER function in breast and prostate cancer cells [100], supporting the postulate that it is involved in ER signaling under normal conditions, and that some carcinogenic processes in these tissues stem from hormonally regulated developmental pathways gone awry [101]. In breast cancer, the promoter of FOXA1 can be hypermethylated [22]. FOXA1 is an important antagonist of the epithelial-to-mesenchymal transition in cancer through its positive regulation of E-cadherin and maintenance of the epithelial phenotype [102]. FOXA1 itself is part of the DNA methylation machinery [103]. Furthermore, FOXC1 (forkhead box C1) has shown a significant increase in the methylation frequency in invasive tumors [1,37]. Low FOXC1 gene expression in both methylated and unmethylated invasive ductal carcinoma indicates that the loss of its expression is an early event during breast cancer progression [37].

SOX genes encode a family of transcription factors that bind to the minor groove of DNA, and belong to a superfamily of genes characterized by a homologous sequence called the HMG (high mobility group) box. This HMG box is a DNA binding domain that is highly conserved throughout eukaryotic species. SOX genes interact with β-catenin and induce both FOXA1 and E-cadherin, which are important for the formation of tissues. This mechanism is enhanced by SCUBE2 (signal peptide-CUB-EGF domain-containing protein 2) that therefore prevents cell mobility. In breast cancer, SCUBE2 is hypermethylated; this is associated with cancer cell mobility and invasiveness [98]. Furthermore, in ovarian cancer, methylation of the SOX1 gene correlates with recurrence and poor survival [79].

Resistance to TGF-β is frequently observed in ovarian cancer, and disrupted TGF-β/SMAD4 signaling results in the aberrant expression of downstream target genes in the disease. This is due to hypermethylation of the FBXO32 gene [104], a member of the F-box genes that play a role in protein ubiquitination and degradation in proteasomes.

HOX genes encode transcription factors governing complex developmental processes in several organs. The evolutionary conserved island of HOXA5 has shown promoter hypomethylation in a few breast cancers and complete methylation in the majority of others [97]. HOXA5 stabilizes adherent junctions through β-catenin retention [105]. In addition, in breast cancer, promoter hypermethylation was reported for HOXD11 [22], a gene involved in embryonic development. It might be a negative regulator of alpha V integrin [106]. In ovarian cancer, HOXA10 [107] and HOXA11 [108] are affected by DNA hypermethylation. HOXA10 is involved in cell differentiation and interacts with PTPN6, which functions as an important regulator of multiple signaling pathways in hematopoietic cells and is down-regulated in various cancers.

The PAX proteins also are important regulators in early development, and alterations in the expression of their genes are thought to contribute to neoplastic transformation. The PAX5 gene is known to play a role in B-cell differentiation, and PAX6 in neurogenesis. They are often hypermethylated in invasive breast cancer [62]. WT1 (Wilm's tumore gene) encodes a transcription factor involved in cell growth differentiation and apoptosis. As with FOXC1, PAX5 and PAX6, WT1 seems often to be hypermethylated in invasive breast cancer [62,99]. The same is true for TWIST-related protein 1 (TWIST1), which is a transcription factor that has been implicated in cell lineage determination and differentiation. Thus, the TWIST1 gene is hypermethylated in breast cancer, but more in invasive cancer than in invasive lobular carcinoma [38].

LMX1A and LMX1B encode two closely related members of the LIM homeobox family of transcription factors. These genes play significant, and frequently overlapping, roles in the development of many structures in the nervous system, but little is known about other roles. The LMX1A gene has been found to be hypermethylated in ovarian cancer [79]. Normalization of LMX1A expression inhibits cell proliferation, migration, invasion and colony formation *in vitro* [109].

TABLE 8.7 Hypermethylated Genes in Breast Cancer Cells [1] – Apoptosis Resistance

Gene	Function	Reference(s)
BCL-2	Negative regulator of apoptosis	[5]
DAPK1	Mediator of γ-interferon induced programmed cell death	[42]
DCC	Negative regulator of tissue growth (pro-apoptotic)	[22]
LOT1	Negative regulator of tissue growth (pro-apoptotic)	[111]
TMS1/PYCARD	Positive regulator of apoptosis	[112]

Energetic and metabolic circuits orchestrate cell differentiation. For example, adenylate kinase (AK) and associated AMP-activated protein kinase (AMPK) constitute a major metabolic signaling axis involved in stem cell cardiac differentiation [110]. Knockdown of AK1, AK2, and AK5 activities with siRNA, or suppression by hyperglycemia, disrupted cardiogenesis, compromising mitochondrial and myofibrillar network formation and contractile performance. In breast cancer, the AK5 gene is hypermethylated [22].

8.4.6 Genes Involved in Apoptosis Resistance

Cancer occurs as the result of a disturbance in the homeostatic balance between cell growth and cell death. Overexpression of antiapoptotic genes, and underexpression of pro-apoptotic genes, can result in the lack of cell death that is characteristic of cancer. Table 8.7 presents the genes that could be hypermethylated and involved in apoptosis resistance of breast cancer cells.

BCL-2 (B cell lymphoma-2) is specifically considered to be an important antiapoptotic protein. Damage to the BCL-2 gene has been identified as a cause of a number of cancers. BCL-2 has been found to be hypermethylated in the MCF-7 breast cancer cell line [5].

DAPK1 (death-associated protein kinase 1) is a calmodulin-dependent serine-threonine kinase that positively mediates γ-interferon induced programmed cell death. DAPK1 gene methylation is often observed in invasive lobular breast cancer [42]. DAPK1 hypermethylation correlates with loss of transcripts, ER positivity, and the absence of p53 overexpression. Methylation of CpG islands in the promoter region of the DAPK1 gene is also common in peripheral blood DNA and tissue samples of patients with ovarian cancer [113].

DCC (deleted in colorectal carcinoma) is a transmembrane receptor for netrin-1. It is a conditional tumor suppressor gene, meaning that it normally prevents cell growth in the absence of netrin-1. *DCC* elimination is not believed to be a key genetic or epigenetic change [22] in tumor formation, but one of many alterations that can promote existing tumor growth. Similarly, LOT1 (lost on transformation 1) plays a significant role as a transcription factor modulating growth suppression through mitogenic signaling pathways, and has pro-apoptotic activity. The LOT1 gene can be hypermethylated in breast cancer [111], but the phenomenon is more obvious in ovarian cancer [111,114].

TMS1/PYCARD encodes for a protein that mediates the assembly of large signaling complexes in the inflammatory and apoptotic signaling pathways via the activation of caspases. TMS1 is aberrantly methylated and silenced in breast cancer cells and in some breast cancers *in situ* [112], as well as in ovarian cancer [115]. Silencing these genes confers a growth advantage that favors the development of other mutations or epimutations.

Many other pro-apoptotic genes have to be found hypermethylated in ovarian cancer – for example, ARL11 [51] and PAR-4/F2RL3 [116]. Interestingly, blood DNA PAR-4 methylation is associated with smoking and is a strong predictor of mortality, including cardiovascular disease and cancer [117].

8.5 CONCLUSION

It is important to distinguish between breast cancer caused by somatic or germline mutations and that caused by DNA hypermethylation. In the first case, surgery and chemotherapy is essential, while in the second case treatment with DNA hypomethylators can be considered. The appearance of triple negative cancer cells (ER−, PGR−, HER2−) occurs through disruption of epigenetic control mechanisms. This can affect the success of therapy. More about this topic (multidrug resistance) is presented in Chapter 9. The first noticeable symptom of breast cancer is typically a lump that feels different from the rest of the breast tissue. Disruption of genes limiting cell proliferation can be caused by mutations and/or epigenetic silencing. Some of these phenomena can be associated with the detrimental effect of endocrine-disrupting chemicals. However, it can also have many other causes. In particular, when detoxifier genes themselves are disrupted by mutations and/or epigenetic silencing, this leads to an accumulation of toxic products that cause DNA damage. The development of cancer always required that multiple control mechanisms fail. Thus, in many breast cancers, DNA repair genes also are defective or silenced, the most known being BRCA1. Even in this case, the cell cycle should be arrested and the cells undergo apoptosis. If this does not occur because the last break is missing, the cell will continue to accumulate mutations and epimutations until its transformation to an invasive cancer cell. The metastatic behavior occurs after disruption of genes maintaining the tissue integrity – i.e., cadherins and laminins. The extracellular matrix can be damaged due to the lack of protective mechanisms limiting, for example, the action of matrix metalloproteinases. Genes that are involved in the normal development of the mammary gland are important to maintain the integrity of this tissue. Central to normal development is FOXA1. Its silencing allows the change from an epithelial to a mesenchymal phenotype. With the silencing of other developmental genes, definite phenotypes appeared. Most importantly, FOXC1, PAX5/6, WT1, and TWIST silencing are associated with an invasive behavior. In addition to increased proliferation, the breast cancer cells also may acquire resistance to apoptosis. Interestingly, the development of ovarian cancer follows a similar scenario, with the same and other genes but with similar functions. Also, in some of these cases, the possibility of epigenetic therapy can be considered. This will be discussed in Chapter 26.

References

[1] Jovanovic J, Ronneberg JA, Tost J, Kristensen V. The epigenetics of breast cancer. Mol Oncol 2010;4:242–54.
[2] Seeber LM, van Diest PJ. Epigenetics in ovarian cancer. Methods Mol Biol 2012;863:253–69.
[3] Piva R, Rimondi AP, Hanau S, Maestri I, Alvisi A, Kumar VL, et al. Different methylation of oestrogen receptor DNA in human breast carcinomas with and without oestrogen receptor. Br J Cancer 1990;61:270–5.
[4] Gaudet MM, Campan M, Figueroa JD, Yang XR, Lissowska J, Peplonska B, et al. DNA hypermethylation of ESR1 and PGR in breast cancer: pathologic and epidemiologic associations. Cancer Epidemiol Biomarkers Prev 2009;18:3036–43.
[5] Chekhun VF, Lukyanova NY, Kovalchuk O, Tryndyak VP, Pogribny IP. Epigenetic profiling of multidrug-resistant human MCF-7 breast adenocarcinoma cells reveals novel hyper- and hypomethylated targets. Mol Cancer Ther 2007;6:1089–98.
[6] Al-Nakhle H, Smith L, Bell SM, Burns PA, Cummings M, Hanby AM, et al. Regulation of estrogen receptor beta1 expression in breast cancer by epigenetic modification of the 5′ regulatory region. Int J Oncol 2013;43:2039–45.
[7] Weigel RJ, deConinck EC. Transcriptional control of estrogen receptor in estrogen receptor-negative breast carcinoma. Cancer Res 1993;53:3472–4.
[8] Ferguson AT, Lapidus RG, Baylin SB, Davidson NE. Demethylation of the estrogen receptor gene in estrogen receptor-negative breast cancer cells can reactivate estrogen receptor gene expression. Cancer Res 1995;55:2279–83.

[9] Lapidus RG, Ferguson AT, Ottaviano YL, Parl FF, Smith HS, Weitzman SA, et al. Methylation of estrogen and progesterone receptor gene 5′ CpG islands correlates with lack of estrogen and progesterone receptor gene expression in breast tumors. Clin Cancer Res 1996;2:805–10.
[10] Hori M, Iwasaki M, Yoshimi F, Asato Y, Itabashi M. Hypermethylation of the Estrogen Receptor Alpha Gene Is Not Related to Lack of Receptor Protein in Human Breast Cancer. Breast Cancer 1999;6:79–86.
[11] Imura M, Yamashita S, Cai LY, Furuta JI, Wakabayashi M, Yasugi T, et al. Methylation and expression analysis of 15 genes and three normally-methylated genes in 13 Ovarian cancer cell lines. Cancer Lett 2006;241:213–20.
[12] Singh P, Yang M, Dai H, Yu D, Huang Q, Tan W, et al. Overexpression and hypomethylation of flap endonuclease 1 gene in breast and other cancers. Mol Cancer Res 2008;6:1710–17.
[13] Kim SJ, Kang HS, Chang HL, Jung YC, Sim HB, Lee KS, et al. Promoter hypomethylation of the N-acetyltransferase 1 gene in breast cancer. Oncol Rep 2008;19:663–8.
[14] Paredes J, Albergaria A, Oliveira JT, Jeronimo C, Milanezi F, Schmitt FC. P-cadherin overexpression is an indicator of clinical outcome in invasive breast carcinomas and is associated with CDH3 promoter hypomethylation. Clin Cancer Res 2005;11:5869–77.
[15] Pakneshan P, Szyf M, Farias-Eisner R, Rabbani SA. Reversal of the hypomethylation status of urokinase (uPA) promoter blocks breast cancer growth and metastasis. J Biol Chem 2004;279:31735–44.
[16] Jia T, Liu YE, Liu J, Shi YE. Stimulation of breast cancer invasion and metastasis by synuclein gamma. Cancer Res 1999;59:742–7.
[17] Gupta A, Godwin AK, Vanderveer L, Lu A, Liu J. Hypomethylation of the synuclein gamma gene CpG island promotes its aberrant expression in breast carcinoma and ovarian carcinoma. Cancer Res 2003;63:664–73.
[18] Ito Y, Koessler T, Ibrahim AE, Rai S, Vowler SL, Abu-Amero S, et al. Somatically acquired hypomethylation of IGF2 in breast and colorectal cancer. Hum Mol Genet 2008;17:2633–43.
[19] Van den Eynden GG, Van Laere SJV, Van der Auwera I, Merajver SD, Van Marck EA, van Dam P, et al. Overexpression of caveolin-1 and -2 in cell lines and in human samples of inflammatory breast cancer. Breast Cancer Res Treat 2006;95:219–28.
[20] Novak P, Jensen T, Oshiro MM, Watts GS, Kim CJ, Futscher BW. Agglomerative epigenetic aberrations are a common event in human breast cancer. Cancer Res 2008;68:8616–25.
[21] Shann YJ, Cheng C, Chiao CH, Chen DT, Li PH, Hsu MT. Genome-wide mapping and characterization of hypomethylated sites in human tissues and breast cancer cell lines. Genome Res 2008;18:791–801.
[22] Miyamoto K, Fukutomi T, Akashi-Tanaka S, Hasegawa T, Asahara T, Sugimura T, et al. Identification of 20 genes aberrantly methylated in human breast cancers. Int J Cancer 2005;116:407–14.
[23] Hinshelwood RA, Clark SJ. Breast cancer epigenetics: normal human mammary epithelial cells as a model system. J Mol Med (Berl) 2008;86:1315–28.
[24] Widschwendter M, Jones PA. DNA methylation and breast carcinogenesis. Oncogene 2002;21:5462–82.
[25] Anderson KN, Schwab RB, Martinez ME. Reproductive risk factors and breast cancer subtypes: a review of the literature. Breast Cancer Res Treat 2014;144:1–10.
[26] Feng W, Shen L, Wen S, Rosen DG, Jelinek J, Hu X, et al. Correlation between CpG methylation profiles and hormone receptor status in breast cancers. Breast Cancer Res 2007;9:R57.
[27] Muller BM, Jana L, Kasajima A, Lehmann A, Prinzler J, Budczies J, et al. Differential expression of histone deacetylases HDAC1, 2 and 3 in human breast

cancer-overexpression of HDAC2 and HDAC3 is associated with clinicopathological indicators of disease progression. BMC Cancer 2013;13:215.

[28] Yuan J, Luo RZ, Fujii S, Wang L, Hu W, Andreeff M, et al. Aberrant methylation and silencing of ARHI, an imprinted tumor suppressor gene in which the function is lost in breast cancers. Cancer Res 2003;63:4174–80.

[29] Evron E, Umbricht CB, Korz D, Raman V, Loeb DM, Niranjan B, et al. Loss of cyclin D2 expression in the majority of breast cancers is associated with promoter hypermethylation. Cancer Res 2001;61:2782–7.

[30] Herman JG, Merlo A, Mao L, Lapidus RG, Issa JP, Davidson NE, et al. Inactivation of the CDKN2/p16/MTS1 gene is frequently associated with aberrant DNA methylation in all common human cancers. Cancer Res 1995;55:4525–30.

[31] Romanov SR, Kozakiewicz BK, Holst CR, Stampfer MR, Haupt MR, Tlsty TD. Normal human mammary epithelial cells spontaneously escape senescence and acquire genomic changes. Nature 2001;409:633–7.

[32] Kobatake T, Yano M, Toyooka S, Tsukuda K, Dote H, Kikuchi T, et al. Aberrant methylation of p57KIP2 gene in lung and breast cancers and malignant mesotheliomas. Oncol Rep 2004;12:1087–92.

[33] Xiang YY, Ladeda V, Filmus J. Glypican-3 expression is silenced in human breast cancer. Oncogene 2001;20:7408–12.

[34] Fujii H, Biel MA, Zhou W, Weitzman SA, Baylin SB, Gabrielson E. Methylation of the HIC-1 candidate tumor suppressor gene in human breast cancer. Oncogene 1998;16:2159–64.

[35] Krop I, Parker MT, Bloushtain-Qimron N, Porter D, Gelman R, Sasaki H, et al. HIN-1, an inhibitor of cell growth, invasion, and AKT activation. Cancer Res 2005;65:9659–69.

[36] Tomii K, Tsukuda K, Toyooka S, Dote H, Hanafusa T, Asano H, et al. Aberrant promoter methylation of insulin-like growth factor binding protein-3 gene in human cancers. Int J Cancer 2007;120:566–73.

[37] Muggerud AA, Ronneberg JA, Warnberg F, Botling J, Busato Fl, Jovanovic J, et al. Frequent aberrant DNA methylation of ABCB1, FOXC1, PPP2R2B and PTEN in ductal carcinoma in situ and early invasive breast cancer. Breast Cancer Res 2010;12:R3.

[38] Fackler MJ, McVeigh M, Evron E, Garrett E, Mehrotra J, Polyak K, et al. DNA methylation of RASSF1A, HIN-1, RAR-beta, Cyclin D2 and Twist in in situ and invasive lobular breast carcinoma. Int J Cancer 2003;107:970–5.

[39] Widschwendter M, Berger J, Hermann M, Muller HM, Amberger A, Zeschnigk M, et al. Methylation and silencing of the retinoic acid receptor-beta2 gene in breast cancer. J Natl Cancer Inst 2000;92:826–32.

[40] Sun J, Xu X, Liu J, Liu H, Fu L, Gu L. Epigenetic regulation of retinoic acid receptor beta2 gene in the initiation of breast cancer. Med Oncol 2011;28:1311–18.

[41] Dammann R, Yang G, Pfeifer GP. Hypermethylation of the cpG island of Ras association domain family 1A (RASSF1A), a putative tumor suppressor gene from the 3p21.3 locus, occurs in a large percentage of human breast cancers. Cancer Res 2001;61:3105–9.

[42] Lehmann U, Celikkaya G, Hasemeier B, Langer F, Kreipe H. Promoter hypermethylation of the death-associated protein kinase gene in breast cancer is associated with the invasive lobular subtype. Cancer Res 2002;62:6634–8.

[43] Holst CR, Nuovo GJ, Esteller M, Chew K, Baylin SB, Herman JG, et al. Methylation of p16(INK4a) promoters occurs in vivo in histologically normal human mammary epithelia. Cancer Res 2003;63:1596–601.

[44] Hashiguchi Y, Tsuda H, Yamamoto K, Inoue T, Ishiko O, Ogita S. Combined analysis of p53 and RB pathways in epithelial ovarian cancer. Hum Pathol 2001;32:988–96.

[45] Milde-Langosch K, Ocon E, Becker G, Loning T. p16/MTS1 inactivation in ovarian carcinomas: high frequency of reduced protein expression associated with hypermethylation or mutation in endometrioid and mucinous tumors. Int J Cancer 1998;79:61–5.
[46] Ibanez de Caceres I, Battagli C, Esteller M, Herman JG, Dulaimi E, Edelson MI, et al. Tumor cell-specific BRCA1 and RASSF1A hypermethylation in serum, plasma, and peritoneal fluid from ovarian cancer patients. Cancer Res 2004;64:6476–81.
[47] Kavanagh E, Joseph B. The hallmarks of CDKN1C (p57, KIP2) in cancer. Biochim Biophys Acta 2011;1816:50–6.
[48] Chien J, Staub J, Avula R, Zhang H, Liu W, Hartmann LC, et al. Epigenetic silencing of TCEAL7 (Bex4) in ovarian cancer. Oncogene 2005;24:5089–100.
[49] Ho CM, Huang CJ, Huang CY, Wu YY, Chang SF, Cheng WF. Promoter methylation status of HIN-1 associated with outcomes of ovarian clear cell adenocarcinoma. Mol Cancer 2012;11:53.
[50] Kato J, Zhu J, Liu C, Stylianou M, Hoffmann V, Lizak MJ, et al. ADP-ribosylarginine hydrolase regulates cell proliferation and tumorigenesis. Cancer Res 2011;71:5327–35.
[51] Petrocca F, Iliopoulos D, Qin HR, Nicoloso MS, Yendamuri S, Wojcik SE, et al. Alterations of the tumor suppressor gene ARLTS1 in ovarian cancer. Cancer Res 2006;66:10287–91.
[52] Wiley A, Katsaros D, Fracchioli S, Yu H. Methylation of the insulin-like growth factor binding protein-3 gene and prognosis of epithelial ovarian cancer. Int J Gynecol Cancer 2006;16:210–18.
[53] Kwong J, Lee JY, Wong KK, Zhou X, Wong DT, Lou KW, et al. Candidate tumor-suppressor gene DLEC1 is frequently downregulated by promoter hypermethylation and histone hypoacetylation in human epithelial ovarian cancer. Neoplasia 2006;8:268–78.
[54] Sutherland KD, Lindeman GJ, Choong DY, Wittlin S, Brentzell L, Phillips W, et al. Differential hypermethylation of SOCS genes in ovarian and breast carcinomas. Oncogene 2004;23:7726–33.
[55] Kikuchi R, Tsuda H, Kozaki K, Kanai Y, Kasamatsu T, Sengoku K, et al. Frequent inactivation of a putative tumor suppressor, angiopoietin-like protein 2, in ovarian cancer. Cancer Res 2008;68:5067–75.
[56] Kikuchi R, Tsuda H, Kanai Y, Kasamatsu T, Sengoku K, Hirohashi S, et al. Promoter hypermethylation contributes to frequent inactivation of a putative conditional tumor suppressor gene connective tissue growth factor in ovarian cancer. Cancer Res 2007;67:7095–105.
[57] Hirohashi S, Kanai Y. Cell adhesion system and human cancer morphogenesis. Cancer Sci 2003;94:575–81.
[58] Jin Z, Tamura G, Tsuchiya T, Sakata K, Kashiwaba M, Osakabe M, et al. Adenomatous polyposis coli (APC) gene promoter hypermethylation in primary breast cancers. Br J Cancer 2001;85:69–73.
[59] Graff JR, Herman JG, Lapidus RG, Chopra H, Xu R, Jarrard DF, et al. E-cadherin expression is silenced by DNA hypermethylation in human breast and prostate carcinomas. Cancer Res 1995;55:5195–9.
[60] Narayan A, Ji W, Zhang XY, Marrogi A, Graff JR, Baylin SB, et al. Hypomethylation of pericentromeric DNA in breast adenocarcinomas. Int J Cancer 1998;77:833–8.
[61] Widschwendter M, Jiang G, Woods C, Muller HM, Fiegl H, Goebel G, et al. DNA hypomethylation and ovarian cancer biology. Cancer Res 2004;64:4472–80.
[62] Murata H, Khattar NH, Kang Y, Gu L, Li GM. Genetic and epigenetic modification of mismatch repair genes hMSH2 and hMLH1 in sporadic breast cancer with microsatellite instability. Oncogene 2002;21:5696–703.

[63] Ai L, Kim WJ, Kim TY, Fields CR, Massoll NA, Robertson KD, et al. Epigenetic silencing of the tumor suppressor cystatin M occurs during breast cancer progression. Cancer Res 2006;66:7899–909.
[64] Sathyanarayana UG, Padar A, Huang CX, Suzuki M, Shigematsu H, Bekele BN, et al. Aberrant promoter methylation and silencing of laminin-5-encoding genes in breast carcinoma. Clin Cancer Res 2003;9:6389–94.
[65] Cui Y, Ying Y, van Hasselt A, Ng KM, Yu J, Zhang Q, et al. OPCML is a broad tumor suppressor for multiple carcinomas and lymphomas with frequently epigenetic inactivation. PLoS One 2008;3:e2990.
[66] Lo PK, Mehrotra J, D'Costa A, Fackler MJ, Garrett-Mayer E, Argani P, et al. Epigenetic suppression of secreted frizzled related protein 1 (SFRP1) expression in human breast cancer. Cancer Biol Ther 2006;5:281–6.
[67] Heitzer E, Artl M, Filipits M, Resel M, Graf R, Weibenbacher B, et al. Differential survival trends of stage II colorectal cancer patients relate to promoter methylation status of PCDH10, SPARC, and UCHL1. Mod Pathol 2013.
[68] Dallol A, Forgacs E, Martinez A, Sekido Y, Walker R, Kishida T, et al. Tumour specific promoter region methylation of the human homologue of the Drosophila Roundabout gene DUTT1 (ROBO1) in human cancers. Oncogene 2002;21:3020–8.
[69] Lau QC, Raja E, Salto-Tellez M, Liu Q, Ito K, Inoue M, et al. RUNX3 is frequently inactivated by dual mechanisms of protein mislocalization and promoter hypermethylation in breast cancer. Cancer Res 2006;66:6512–20.
[70] Yuan Y, Mendez R, Sahin A, Dai JL. Hypermethylation leads to silencing of the SYK gene in human breast cancer. Cancer Res 2001;61:5558–61.
[71] Lui EL, Loo WT, Zhu L, Cheung MN, Chow LW. DNA hypermethylation of TIMP3 gene in invasive breast ductal carcinoma. Biomed Pharmacother 2005;59(Suppl. 2): S363–5.
[72] Ai L, Tao Q, Zhong S, Fields CR, Kim WJ, Lee MW, et al. Inactivation of Wnt inhibitory factor-1 (WIF1) expression by epigenetic silencing is a common event in breast cancer. Carcinogenesis 2006;27:1341–8.
[73] Yuecheng Y, Hongmei L, Xiaoyan X. Clinical evaluation of E-cadherin expression and its regulation mechanism in epithelial ovarian cancer. Clin Exp Metastasis 2006;23:65–74.
[74] Lyu T, Jia N, Wang J, Yan X, Yu Y, Lu Z, et al. Expression and epigenetic regulation of angiogenesis-related factors during dormancy and recurrent growth of ovarian carcinoma. Epigenetics 2013;8:1330–46.
[75] Makarla PB, Saboorian MH, Ashfaq R, Toyooka KI, Toyooka S, Minna JD, et al. Promoter hypermethylation profile of ovarian epithelial neoplasms. Clin Cancer Res 2005;11:5365–9.
[76] Smalley MJ, Dale TC. Wnt signalling in mammalian development and cancer. Cancer Metastasis Rev 1999;18:215–30.
[77] Turashvili G, Bouchal J, Burkadze G, Kolar Z. Wnt signaling pathway in mammary gland development and carcinogenesis. Pathobiology 2006;73:213–23.
[78] Takada T, Yagi Y, Maekita T, Imura M, Nakagawa S, Tsao SW, et al. Methylation-associated silencing of the Wnt antagonist SFRP1 gene in human ovarian cancers. Cancer Sci 2004;95:741–4.
[79] Su HY, Lai HC, Lin YW, Chou YC, Liu CY, Yu MH. An epigenetic marker panel for screening and prognostic prediction of ovarian cancer. Int J Cancer 2009;124:387–93.
[80] Shih YL, Hsieh CB, Yan MD, Tsao CM, Hsieh TY, Liu CH, et al. Frequent concomitant epigenetic silencing of SOX1 and secreted frizzled-related proteins (SFRPs) in human hepatocellular carcinoma. J Gastroenterol Hepatol 2013;28:551–9.

[81] Su HY, Lai HC, Lin YW, Liu CY, Chen CK, Chou YC, et al. Epigenetic silencing of SFRP5 is related to malignant phenotype and chemoresistance of ovarian cancer through Wnt signaling pathway. Int J Cancer 2010;127:555–67.

[82] Arnold JM, Cummings M, Purdie D, Chenevix-Trench G. Reduced expression of intercellular adhesion molecule-1 in ovarian adenocarcinomas. Br J Cancer 2001;85:1351–8.

[83] Sellar GC, Watt KP, Rabiasz GJ, Stronach EA, Li L, Miller EP, et al. OPCML at 11q25 is epigenetically inactivated and has tumor-suppressor function in epithelial ovarian cancer. Nat Genet 2003;34:337–43.

[84] Mei FC, Young TW, Liu J, Cheng X. RAS-mediated epigenetic inactivation of OPCML in oncogenic transformation of human ovarian surface epithelial cells. FASEB J 2006;20:497–9.

[85] Zhou F, Tao G, Chen X, Xie W, Liu M, Cao X. Methylation of OPCML promoter in ovarian cancer tissues predicts poor patient survival. Clin Chem Lab Med 2013;1–8.

[86] Socha MJ, Said N, Dai Y, Kwong J, Ramalingam P, Trieu V, et al. Aberrant promoter methylation of SPARC in ovarian cancer. Neoplasia 2009;11:126–35.

[87] Moelans CB, Verschuur-Maes AH, van Diest PJ. Frequent promoter hypermethylation of BRCA2, CDH13, MSH6, PAX5, PAX6 and WT1 in ductal carcinoma in situ and invasive breast cancer. J Pathol 2011;225:222–31.

[88] Dobrovic A, Simpfendorfer D. Methylation of the BRCA1 gene in sporadic breast cancer. Cancer Res 1997;57:3347–50.

[89] Mancini DN, Rodenhiser DI, Ainsworth PJ, O'Malley FP, Singh SM, Xing W, et al. CpG methylation within the 5′ regulatory region of the BRCA1 gene is tumor specific and includes a putative CREB binding site. Oncogene 1998;16:1161–9.

[90] Turner NC, Reis-Filho JS, Russell AM, Springall RJ, Ryder K, Steele D, et al. BRCA1 dysfunction in sporadic basal-like breast cancer. Oncogene 2007;26:2126–32.

[91] Esteller M, Silva JM, Dominguez G, Bonilla F, Matias-Guiu X, Lerma E, et al. Promoter hypermethylation and BRCA1 inactivation in sporadic breast and ovarian tumors. J Natl Cancer Inst 2000;92:564–9.

[92] Esteller M. Epigenetic lesions causing genetic lesions in human cancer: promoter hypermethylation of DNA repair genes. Eur J Cancer 2000;36:2294–300.

[93] Ferguson AT, Evron E, Umbricht CB, Pandita TK, Chan TA, Hermeking K, et al. High frequency of hypermethylation at the 14-3-3 sigma locus leads to gene silencing in breast cancer. Proc Natl Acad Sci U S A 2000;97:6049–54.

[94] Balch C, Huang TH, Brown R, Nephew KP. The epigenetics of ovarian cancer drug resistance and resensitization. Am J Obstet Gynecol 2004;191:1552–72.

[95] Esteller M, Fraga MF, Guo M, Garcia-Foncillas J, Hedenfalk I, Godwin AK, et al. DNA methylation patterns in hereditary human cancers mimic sporadic tumorigenesis. Hum Mol Genet 2001;10:3001–7.

[96] Press JZ, De Luca A, Boyd N, Young S, Troussard A, Ridge Y, et al. Ovarian carcinomas with genetic and epigenetic BRCA1 loss have distinct molecular abnormalities. BMC Cancer 2008;8:17.

[97] Piotrowski A, Benetkiewicz M, Menzel U, Diaz de Stahl T, Mantripragada K, Grigelionis G, et al. Microarray-based survey of CpG islands identifies concurrent hyper- and hypomethylation patterns in tissues derived from patients with breast cancer. Genes Chromosomes Cancer 2006;45:656–67.

[98] Lin YC, Lee YC, Li LH, Cheng CJ, Yang RB. Tumor suppressor SCUBE2 inhibits breast-cancer cell migration and invasion through the reversal of epithelial-mesenchymal transition. J Cell Sci 2014;127:85–100.

[99] Laux DE, Curran EM, Welshons WV, Lubahn DB, Huang TH. Hypermethylation of the Wilms' tumor suppressor gene CpG island in human breast carcinomas. Breast Cancer Res Treat 1999;56:35–43.

[100] Ijichi N, Ikeda K, Horie-Inoue K, Inoue S. FOXP1 and estrogen signaling in breast cancer. Vitam Horm 2013;93:203–12.
[101] Bernardo GM, Keri RA. FOXA1: a transcription factor with parallel functions in development and cancer. Biosci Rep 2012;32:113–30.
[102] Song Y, Washington MK, Crawford HC. Loss of FOXA1/2 is essential for the epithelial-to-mesenchymal transition in pancreatic cancer. Cancer Res 2010;70:2115–25.
[103] Serandour AA, Avner S, Percevault F, Demay F, Bizot M, Lucchetti-Miganeh C, et al. Epigenetic switch involved in activation of pioneer factor FOXA1-dependent enhancers. Genome Res 2011;21:555–65.
[104] Chou JL, Su HY, Chen LY, Liao YP, Hartman-Frey C, Lai YH, et al. Promoter hypermethylation of FBXO32, a novel TGF-beta/SMAD4 target gene and tumor suppressor, is associated with poor prognosis in human ovarian cancer. Lab Invest 2010;90:414–25.
[105] Kachgal S, Mace KA, Boudreau NJ. The dual roles of homeobox genes in vascularization and wound healing. Cell Adh Migr 2012;6:457–70.
[106] Rossi Degl'Innocenti D, Castiglione F, Buccoliero AM, Bechi P, Taddei GL, Freschi G, et al. Quantitative expression of the homeobox and integrin genes in human gastric carcinoma. Int J Mol Med 2007;20:621–9.
[107] Cheng W, Jiang Y, Liu C, Shen O, Tang W, Wang X. Identification of aberrant promoter hypomethylation of HOXA10 in ovarian cancer. J Cancer Res Clin Oncol 2010;136:1221–7.
[108] Fiegl H, Windbichler G, Mueller-Holzner E, Goebel G, Lechner M, Jacobs IJ, et al. HOXA11 DNA methylation--a novel prognostic biomarker in ovarian cancer. Int J Cancer 2008;123:725–9.
[109] Chao TK, Yo YT, Liao YP, Wang YC, Su PH, Huang TS, et al. LIM-homeobox transcription factor 1, alpha (LMX1A) inhibits tumourigenesis, epithelial-mesenchymal transition and stem-like properties of epithelial ovarian cancer. Gynecol Oncol 2013;128:475–82.
[110] Dzeja PP, Chung S, Faustino RS, Behfar A, Terzic A. Developmental enhancement of adenylate kinase-AMPK metabolic signaling axis supports stem cell cardiac differentiation. PLoS One 2011;6:e19300.
[111] Abdollahi A, Pisarcik D, Roberts D, Weinstein J, Cairns P, Hamilton TC. LOT1 (PLAGL1/ZAC1), the candidate tumor suppressor gene at chromosome 6q24-25, is epigenetically regulated in cancer. J Biol Chem 2003;278:6041–9.
[112] Conway KE, McConnell BB, Bowring CE, Donald CD, Warren ST, Vertino PM. TMS1, a novel proapoptotic caspase recruitment domain protein, is a target of methylation-induced gene silencing in human breast cancers. Cancer Res 2000;60:6236–42.
[113] Collins Y, Dicioccio R, Keitz B, Lele S, Odunsi K. Methylation of death-associated protein kinase in ovarian carcinomas. Int J Gynecol Cancer 2006;16(Suppl. 1):195–9.
[114] Cvetkovic D, Pisarcik D, Lee C, Hamilton TC, Abdollahi A. Altered expression and loss of heterozygosity of the LOT1 gene in ovarian cancer. Gynecol Oncol 2004;95:449–55.
[115] Terasawa K, Sagae S, Toyota M, Tsukada K, Ogi K, Satoh A, et al. Epigenetic inactivation of TMS1/ASC in ovarian cancer. Clin Cancer Res 2004;10:2000–6.
[116] Pruitt K, Ulku AS, Frantz K, Rojas RJ, Muniz-Medina VM, Rangnekar VM, et al. Ras-mediated loss of the pro-apoptotic response protein Par-4 is mediated by DNA hypermethylation through Raf-independent and Raf-dependent signaling cascades in epithelial cells. J Biol Chem 2005;280:23363–70.
[117] Zhang Y, Yang R, Burwinkel B, Breitling LP, Holleczek B, Schottker B, et al. F2RL3 methylation in blood DNA is a strong predictor of mortality. Int J Epidemiol 2014;43(4):1215–25.

CHAPTER

9

DNA Methylation in Acquired Drug Resistance

OUTLINE

9.1 INTRODUCTION

Resistance acquired upon drug therapy ("acquired drug resistance") is a major problem in the treatment of many diseases, including cancer. An estimated 7.5 million cancer patients die worldwide each year, many of them due to failed anticancer therapies as a consequence of acquired resistance to cytotoxic chemotherapeutics and targeted drugs [1] Thus, understanding the mechanisms causing

M. Neidhart: DNA Methylation and Complex Human Disease.
DOI: http://dx.doi.org/10.1016/B978-0-12-420194-1.00009-9

unresponsiveness to anticancer drugs will dramatically improve the design of therapies aimed at preventing the selection of drug-resistant tumor cells. Consequently, such therapies may significantly reduce cancer mortality rates. With the rise of targeted drug therapies it has become increasingly evident that genetic mutations are a critical component of acquired drug resistance. However, genetics is not sufficient in explaining the relatively rapid appearance or the reversibility of non-responsiveness to drug treatment. In addition, the lack of genetic mutations in drug targets and activated parallel pathways suggest a role for non-genetic mechanisms in acquired drug resistance. This chapter will focus on the role of DNA methylation in non-genetic acquired drug resistance.

9.2 DNA METHYLATION AND ACQUIRED DRUG RESISTANCE

Acquired drug resistance has been associated with selection of cells displaying hypomethylation of drug efflux gene promoters, hypermethylation of promoter regions of proapoptotic genes, or altered promoter methylation patterns of DNA repair genes [1]. In addition, the poor prognosis profile of myelodysplastic syndromes and acute myeloid leukemia patients carrying somatic DNMT3A mutations may attest to increased acquired drug resistance due to DNA methylation changes that generate cells with a drug resistance-favorable epigenome (Figure 9.1). Therefore, characterization of a drug resistance-associated DNA methylome may guide the design of therapeutics aimed at targeting DNA methylation.

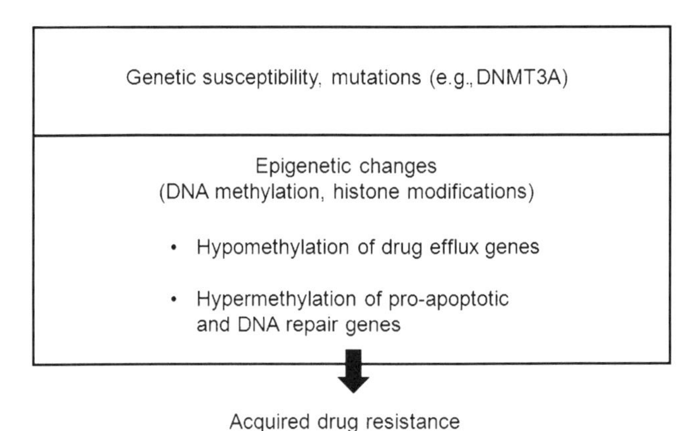

FIGURE 9.1 Genetic and non-genetic factors contribute to the development of acquired drug resistance in cancer.

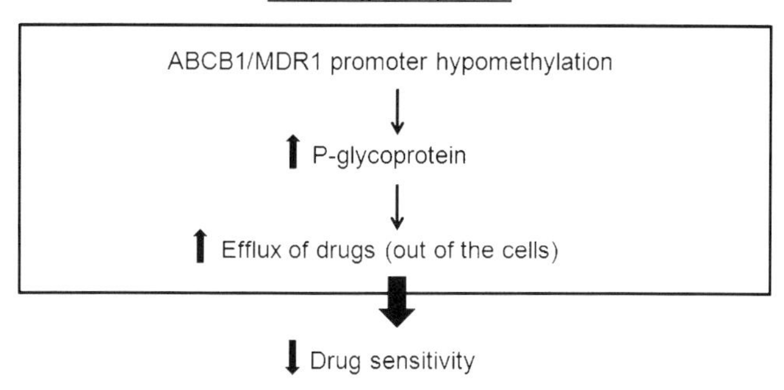

FIGURE 9.2 Decreased drug sensitivity after hypomethylation of the ABCB1 gene in cancer cells.

9.2.1 Hypomethylated Genes

9.2.1.1 Drug Efflux Promoters

Increased expression of P-glycoprotein (Pgp/ABCB1) is a well-known mechanism involved in acquired drug resistance [2]. ABCB1 is also called multidrug resistance gene 1 (MDR1). Pgp belongs to the ATP binding cassette (ABC) transporter superfamily, which facilitates increased efflux of chemotherapeutic drugs from tumor cells resulting in drug insensitivity to various agents (Figure 9.2). Tumors with intrinsically low expression levels of Pgp generally respond well to chemotherapy. Hypo- and hypermethylated DNA has been found at the ABCB1 promoter in different tumorigenic contexts. For instance, while the ABCB1 promoter in normal bladder cells is hypomethylated, early precancerous cells of this tissue usually contain a hypermethylated ABCB1 promoter. Interestingly, upon chemotherapy, methylation of the ABCB1 promoter reverted to its hypomethylated state and correlated with overexpression of the ABCB1 gene in tumors of the bladder; similar observations were made in acute myeloid leukemia patient samples [3]. Hypomethylation of the promoter of the ABCG2/BCRP drug transporter also increased expression in response to chemotherapy, suggesting that this is a general mechanism to regulate ABC transporter expression in response to drug treatment [4,5].

9.2.1.2 Cell Cycle Enhancers

It can be hypothesized that increased proliferation and differentiation rates allow escape from the cytostatic effect of drugs that should induce cell death (Figure 9.3).

Tyrosine-protein phosphatase non-receptor type 6 (PTPN6) is a signaling molecule that regulates a variety of cellular processes,

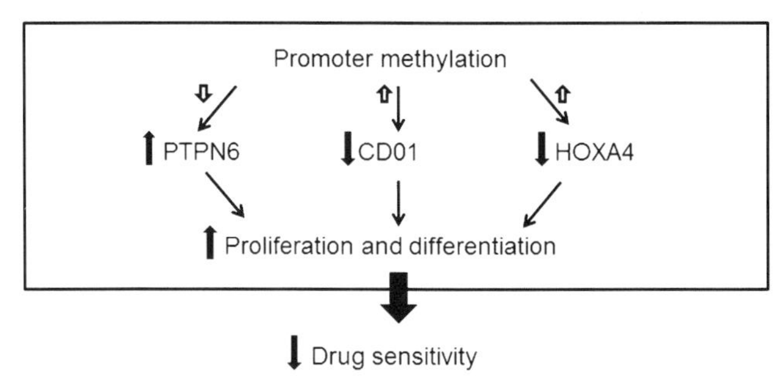

FIGURE 9.3 Increased proliferation and differentiation rates allow escape from the cytostatic effect of drugs that should induce cell death.

including cell growth, differentiation, the mitotic cycle, and oncogenic transformation. PTPN6 expression correlates to poor survival for anaplastic glioma patients [6]. In glioma-derived cell lines, overexpression of PTPN6 caused increased resistance to the chemotherapeutic drugs bortezomib, cisplatin, and melphalan. PTPN6 expression did not affect bortezomib-induced cell cycle arrest, apoptosis, or autophagy. Low PTPN6 promoter methylation correlates with increased protein expression. Thus, increased PTPN6 expression may be a factor contributing to poor survival and influencing the response to chemotherapy.

The human cysteine dioxygenase 1 (CDO1) gene is a non-heme structured, iron-containing metalloenzyme involved in the conversion of cysteine to cysteine sulfinate, and plays a key role in taurine biosynthesis. Promoter hypermethylation of the CDO1 gene, as well as downregulation of CDO1 mRNA and protein levels, is observed in breast, esophagus, lung, bladder, and stomach cancers [7,8]. Restoration of CDO1 expression in human cancer cells markedly decreases the tumor cell growth [7], as well as leading to sensitization to anthracycline treatment [8], a drug that binds DNA and induces cell death.

The human Homeobox (HOX) gene network encodes master regulators in hematopoiesis, and DNA methylation has been implicated to have an important role in aberrant control of HOX gene expression. Inappropriate expression of HOX gene has been implicated in the development of hematopoietic malignancies. Methylation of a HOX gene, HOXA4, has been strongly associated with progression to blast crisis and poor response to treatment in leukemia patients [9]. For example, in chronic myeloid leukemia, promoter hypermethylation of HOXA4 is an epigenetic mechanism mediating the resistance to imatinib mesylate [10], a tyrosine-kinase inhibitor.

9.2.2 Hypermethylated Genes

9.2.2.1 Pro-Apoptotic Gene Promoters

Drug treatment in anticancer therapy often leads to tumor cell death, i.e., apoptosis. Genetic or epigenetic perturbations resulting in a defective execution of an apoptotic response could potentially result in drug-tolerant tumor cells. Indeed, genetic mutations as well as epigenetic changes in proapoptotic genes are a hallmark of human malignancies [11]. Although it remains controversial whether these mutations contribute to drug resistance [12], stochastic epigenetic silencing of proapoptotic genes in a fraction of the total tumor cell population may allow selection of these cells upon drug treatment (Figure 9.4). Indeed, several proapoptotic genes were found to be silenced by promoter methylation upon drug treatment, including death-associated protein kinase 1 (DAPK1) and apoptotic peptidase activating factor 1 (APAF-1) [13].

Reduced expression or inactivation of the proapoptotic DAPK1 by genetic or epigenetic mechanisms is found in human malignancies [14], including breast cancer [15] (Chapter 8). Loss of DAPK1 expression in tumors is caused by homozygous gene deletion, loss of heterozygosity, or hypermethylation of a CpG region in the 5′ UTR [16]. DAPK1 promoter hypermethylation has been reported in a wide range of human tumor types in which DAPK1 silencing correlated with tumor progression, histo-pathological staging, increased metastasis, and high tumor recurrence [16–22]. DAPK1 expression could be restored in tumor cells by demethylating agent 5-aza-cytidine, indicating that DAPK1 silencing involves DNA methylation [18,23]. DAPK1 is involved in resistance to 5-fluorouracil in endometrial adenocarcinoma cells [24], to antiepidermal growth factor receptor biologicals in lung cancer cells [25], to gemcitabine in pancreatic cancer [26], and to cisplatin in cervical squamous cancer cells [27].

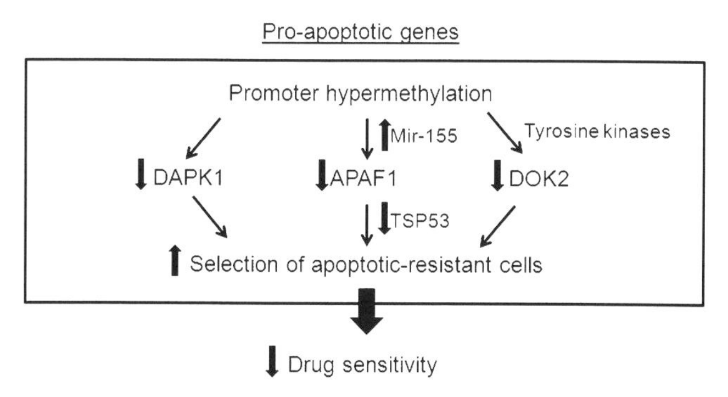

FIGURE 9.4 Increased selection of apoptotic-resistant cells contributes to drug insensitivity in cancer.

APAF1-negative melanoma cells were unable to execute an apoptotic program in response to TP53 activation, resulting in chemotherapy resistance [1]. Loss of APAF1 in metastatic melanoma occurred through genomic loss or epigenetic silencing [28]. In the latter case APAF1 expression was restored by the DNA methylation inhibitor 5-aza-cytidine, resulting in enhanced chemosensitivity and rescue of the apoptotic defects associated with APAF1 loss [13]. Interestingly, miR-155, which is often increased in cancer and autoimmune diseases, targets APAF1 transcripts and inhibits the sensitivity to cisplatin in lung cancer [29].

Docking protein 2 (DOK2), an adapter protein downstream of tyrosine kinase, may modulate cell proliferation induced by cytokines. In ovarian cancer, hypermethylation of DOK2 and the loss of protein expression decreased the level of apoptosis in response to carboplatin [30], a platin-based antineoplastic agent.

9.2.2.2 Mismatch Repair Genes

Inactivation of mismatch repair genes (MMRs) leads to unrepaired deletions in mono- and dinucleotide repeats, resulting in variable repeat lengths, referred to as microsatellite instability. MMR-deficient human cancer cell lines tolerate alkylating agents, suggesting that loss of MMR could cause chemotherapy resistances [31]. The accumulations of somatic mutations due to altered MMR gene expressions (Figure 9.5), as well as altered detoxifiers such as GSTP1, contribute to the development of acquired multidrug resistances. MSI in tumors can only in part be explained by mutations in MMR genes such as MLH1, MSH2, MSH3, and PMS2 [32]. For instance, in less than 10% of all sporadic uterine endometrial carcinomas, MSI was associated with mutations in MMR genes, suggesting that mutations in genes regulating MMR proteins or epigenetics could contribute to inactivation of MMR gene expression.

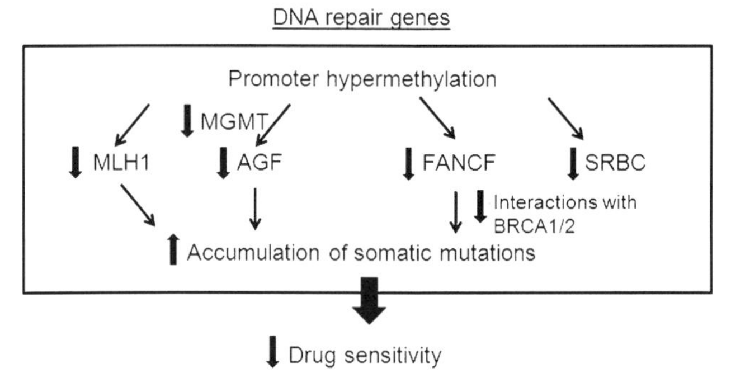

FIGURE 9.5 Increased accumulation of somatic mutations contributes to the development of acquired drug insensitivity in cancer.

Epigenetic-driven inactivation of MMRs is illustrated by the identification of MLH1 promoter hypermethylation in various tumors and as an early event in endometrial cancerogenesis, which correlated with drug resistance and predicted poor overall survival [33–35]. The ability to measure MLH1 promoter methylation status in plasma DNA from ovarian cancer patients provided a non-invasive method to monitor the potential acquisition of drug resistance during drug treatment [35]. Examination of the plasma DNA of patients with epithelial ovarian cancer for methylation of the MLH1 promoter before chemotherapy and at relapse revealed increased MLH1 promoter methylation. Moreover, 25% of the samples from patients with a relapse had MLH1 methylation that was not detected in matched prechemotherapy plasma samples. Acquisition of MLH1 methylation in plasma DNA at relapse predicted poor overall survival of patients.

MGMT encodes the DNA-repairing O6-alkylguanine-DNA-alkyltransferase (AGT), which removes alkylating lesions at position O6 of guanine. DNA damage-induced MGMT expression facilitates DNA repair, and is a possible mechanism to acquire resistance against alkylating agents [1]. Hypermethylation of specific CpG regions in the MGMT promoter resulting in the silencing of MGMT expression has been described in primary human tumors, such as glioma, retinoblastoma, breast and prostate tumors, osteosarcoma, and Hodgkin's lymphoma [5,36–39]. Remarkably, a reduction of MGMT promoter methylation was acquired upon chemotherapy treatment, resulting in MGMT expression and increased AGT activity rendering the cytotoxic drugs ineffective [40,41]. In line with these results, studies in patients with glioma, glioblastoma, non-Hodgkin lymphoma, or male germ cell tumors identified MGMT promoter methylation as a marker for chemotherapy sensitivity: tumor cells with hypermethylated MGMT promoters responded better to chemotherapy, and these patients showed improved survival compared to patients with hypomethylated MGMT promoters [42–44]. As a consequence, MGMT promoter methylation status is an established molecular marker in clinical trials, which is predictive for the response to alkylating chemotherapy and radiotherapy [44].

Another DNA-repair pathway that is linked to chemotherapy resistance involves the Fanconi anemia (FA) group of proteins [1]. FA germline mutations result in chromosomal instability, leading to congenital defects, bone marrow failure, and increased cancer susceptibility [45]. The Fanconi–BRCA1/2 molecular pathway plays an important role in DNA repair, and is necessary for the normal cellular response to interstrand DNA cross-linking agents such as cisplatin, mitomycin C, and diepoxybutane. Somatic mutations or epigenetic silencing of the FA pathway are observed in a variety of tumors. Hypermethylation of the FANCF promoter was found in primary ovarian adenocarcinomas, ovarian granulosa cell tumors, cervical cancer, non-small cell lung cancers, and squamous-cell head and neck cancers [46–48]. In ovarian, glioma, and pancreatic cancer cell lines, FANCF promoter methylation resulted in decreased gene expression and was associated with increased sensitivity to cisplatin and other DNA cross-linking agents [46,49]. Conversely, restoration of FANCF expression by 5-aza-cytidine-mediated promoter hypomethylation resulted in cisplatin resistance [46]. Hypermethylation of other genes is also involved in cisplatin resistance – for example SOX1 [50], GPR56, MT1G, and RASSF1 [51].

Serum deprivation response factor-related gene product that binds to C-kinase (SRBC) was identified as a binding protein of the protein kinase Cδ. Expression of this gene in cultured cell lines is strongly induced by serum starvation. The expression of this protein was found to be downregulated in various cancer cell lines, suggesting the possible tumor suppressor function of this protein [52]. In colorectal cancer, oxaliplatin resistance is associated with SRBC hypermethylation [53]. As for FA proteins, this can also be explained by its interaction with the BRCA1/2 gene products. The BRCA1/2 proteins exerts an important role in DNA double-strand break repair through homologous recombination, so its deficiencies can impair the capacity of cancer cells to repair DNA cross-links caused by chemotherapy drugs such as platinum derivatives.

Glutathione S-transferases (GSTs) are a family of enzymes that play an important role in detoxification. The glutathione S-transferase π gene (GSTP1) is a polymorphic gene encoding active, functionally different GSTP1 variant proteins that are thought to function in xenobiotics metabolism, prevent DNA damage, and play a role in susceptibility to cancer and other diseases. Prostate cancer cells present, almost invariably, hypermethylation of the GSTP1 gene promoter and, as a consequence, low levels of GST-π expression and activity. In these cells, brostallicin (a DNA minor groove binder that shows enhanced antitumor activity in cells with high GST content) shows very little activity. Demethylating agents increased GST expression and restored the response to, for example, brostallicin [54], a DNA minor groove binder.

9.3 MECHANISMS OF ACQUIRED DRUG RESISTANCE

9.3.1 Hypomethylation of Promoters

How chemotherapeutic drugs induce hypomethylation of the ABCB1, PTPN6, CD01, and HOXA1 promoters is still not fully understood, and different mechanisms have been postulated [1,55] (Figure 9.6). For instance, chemotherapeutic drugs could induce

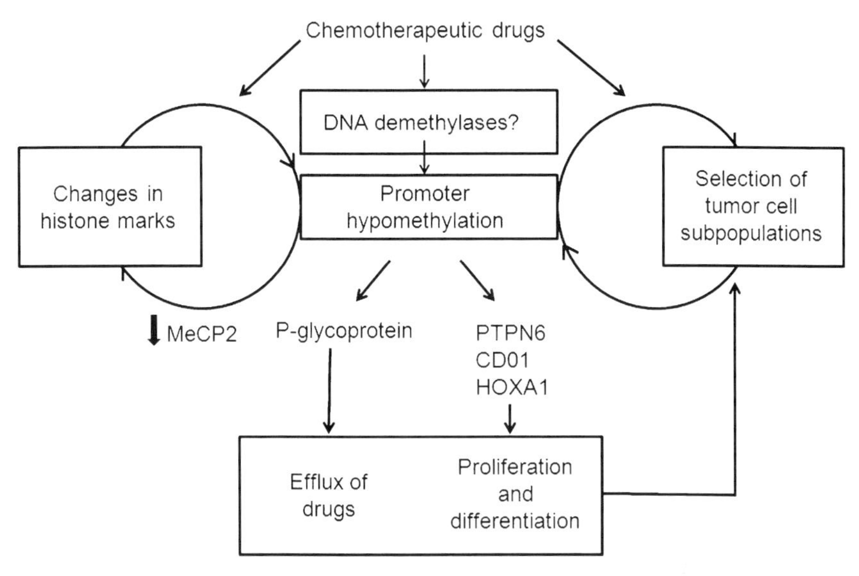

FIGURE 9.6 Promoter hypomethylation of given genes can favor the efflux of drugs, proliferation, and differentiation, and thereby the selection of tumor cell subpopulations that are resistant to chemotherapy.

active DNA demethylation by as yet unknown DNA demethylases. Alternatively, chemotherapeutics could select for a small tumor subpopulation harboring the given hypomethylated promoters. Finally, changes in key chromatin remodeling complex activities could have an impact on the methylation status of such promoters, or lead to alleviated repression at the given promoter, resulting in induced protein expression. There are indications that histone-modifying complexes are recruited to the methylated ABCB1 promoter serving its transcriptional repression. For instance, MeCP2, a well-known methyl CpG-binding protein, binds to hypermethylated DNA at the ABCB1 promoter [56] and may serve as a docking platform for nucleosome modifiers and remodelers such as SWI/SNF, HDAC1, HDAC2, and mSIN3, thereby altering the chromatin state of gene promoters and, subsequently, transcription.

9.3.2 Hypermethylation of Promoters

Similarly, promoter hypermethylation of given genes can favor accumulation of somatic mutation and apoptosis resistance, and thereby the selection of tumor cell subpopulations that are resistant to chemotherapy (Figure 9.7). For instance, the selection favors tumor cells with repressed MMR genes during chemotherapy, resulting in a

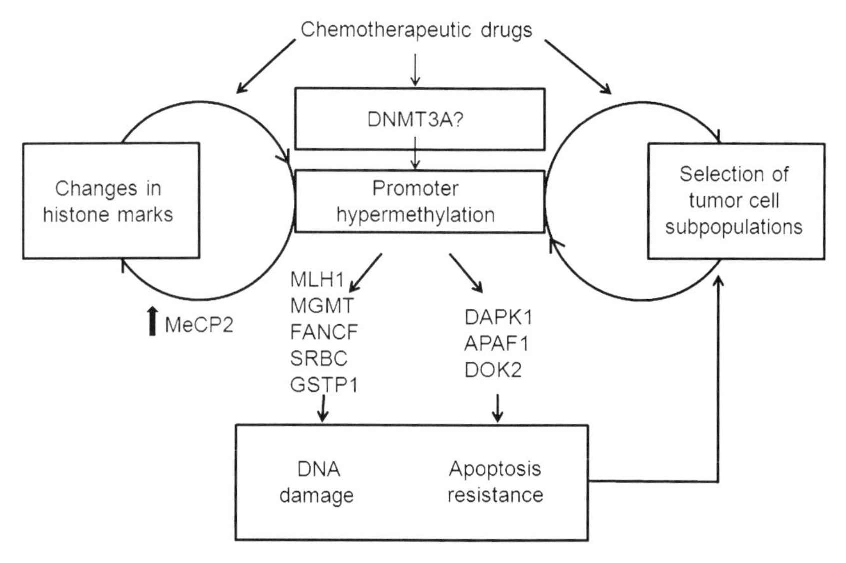

FIGURE 9.7 Promoter hypermethylation of given genes can favor accumulation of somatic mutation and apoptosis resistance, and thereby the selection of tumor cell subpopulations that are resistant to chemotherapy.

drug-resistant tumor, which provides a rational for the use of epigenetic drugs to restore MMR expression and abolish drug resistance [1,35]. Overexpression of *de novo* DNA methyltransferases (DNMT3A/B) plays a central role in the onset of drug resistance to, for example, cisplatin in neuroblastoma cells [57], or docetaxel (an antimitotic drug) in breast cancer cells [58].

9.4 CONCLUSION

Increasing the survival rates of cancer patients will be impossible without improving the efficacy of cytotoxic and targeted anticancer drugs. Since drug resistance is the major cause of treatment failure, it will be vital to avoid non-responsiveness toward current and future treatment modalities. The important contribution of DNA and histone modifications in drug resistance implies the possibility of preventing or abolishing drug resistance by reverting the cancer epigenome of non-responsive cells to a drug responsive state. The development of various "epigenetic" drugs will provide us with the tools for studying the reversibility of drug resistance *in vitro*, in animal models, and in clinical trials [1]. In the near future, anticancer drug regimens will routinely include epigenetic therapies, possibly in conjunction with inhibitors of stemness signal pathways, to effectively reduce the devastating occurrence of cancer chemotherapy resistance [59]. Understanding sensitivity is perhaps even more pressing than understanding resistance. Of course, cancer researchers remain puzzled as to why many therapies work when they do work (why are testicular cancer and pediatric leukemia so curable?), but this is even more pressing for targeted therapies where hypotheses can at least be formulated and tested. Figure 9.8 outlines a number of potential mechanisms by which epigenetic therapy can lead to clonal elimination of neoplasms. It remains to be seen whether these possibilities can be deciphered and distinguished *in vivo*, but if they can, the information will help guide the next generation of clinical trials. For example, if immune modulation is important, then combinations of epigenetic therapy with immunotherapy may be indicated. If a major effect on stem cell renewal is observed, then perhaps epigenetic therapy may have the greatest impact after removing the cancer bulk.

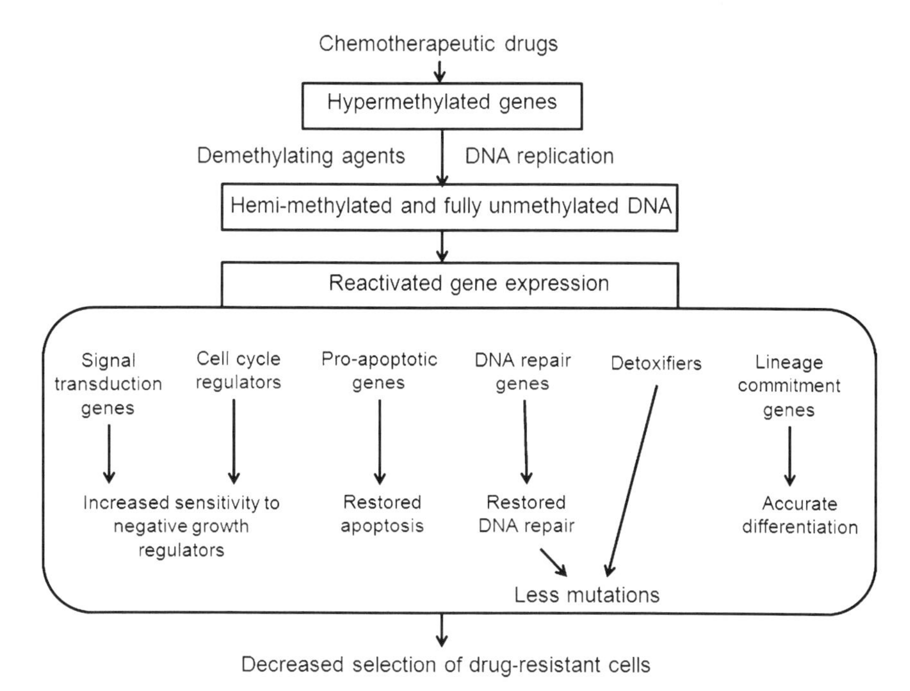

FIGURE 9.8 Demethylating agents can restore gene expression silenced by promoter hypermethylation in cancer; this increases sensitivity to negative growth regulators, apoptosis, and DNA repair. The selection process of mutated drug-resistant cells is then abrogated.

References

[1] Wilting RH, Dannenberg JH. Epigenetic mechanisms in tumorigenesis, tumor cell heterogeneity and drug resistance. Drug Resist Updat 2012;15:21–38.
[2] Sarkadi B, Homolya L, Szakacs G, Varadi A. Human multidrug resistance ABCB and ABCG transporters: participation in a chemoimmunity defense system. Physiol Rev 2006;86:1179–236.
[3] Tada Y, Wada M, Kuroiwa K, Kinugawa N, Harada T, Nagayama J, et al. MDR1 gene overexpression and altered degree of methylation at the promoter region in bladder cancer during chemotherapeutic treatment. Clin Cancer Res 2000;6:4618–27.
[4] Bram EE, Stark M, Raz S, Assaraf YG. Chemotherapeutic drug-induced ABCG2 promoter demethylation as a novel mechanism of acquired multidrug resistance. Neoplasia 2009;11:1359–70.
[5] Oberstadt MC, Bien-Moller S, Weitmann K, Herzog S, Hentschel K, Rimmbach C, et al. Epigenetic modulation of the drug resistance genes MGMT, ABCB1 and ABCG2 in glioblastoma multiforme. BMC Cancer 2013;13:617.
[6] Sooman L, Ekman S, Tsakonas G, Jaiswal A, Navani S, Edgvist PH, et al. PTPN6 expression is epigenetically regulated and influences survival and response to chemotherapy in high-grade gliomas. Tumour Biol 2014;35:4479–88.
[7] Brait M, Ling S, Nagpal JK, Chang X, Park HL, Lee J, et al. Cysteine dioxygenase 1 is a tumor suppressor gene silenced by promoter methylation in multiple human cancers. PLoS One 2012;7:e44951.
[8] Jeschke J, O'Hagan HM, Zhang W, Vatapalli R, Calmon MF, Danilova L, et al. Frequent inactivation of cysteine dioxygenase type 1 contributes to survival of breast cancer cells and resistance to anthracyclines. Clin Cancer Res 2013;19:3201–11.
[9] Strathdee G, Holyoake TL, Sim A, Parker A, Oscier DG, Melo JV, et al. Inactivation of HOXA genes by hypermethylation in myeloid and lymphoid malignancy is frequent and associated with poor prognosis. Clin Cancer Res 2007;13:5048–55.
[10] Elias MH, Baba AA, Husin A, Sulong S, Hassan R, Sim GA, et al. HOXA4 gene promoter hypermethylation as an epigenetic mechanism mediating resistance to imatinib mesylate in chronic myeloid leukemia patients. Biomed Res Int 2013;2013.
[11] Hanahan D, Weinberg RA. Hallmarks of cancer: the next generation. Cell 2011;144:646–74.
[12] Borst P, Borst J, Smets LA. Does resistance to apoptosis affect clinical response to antitumor drugs? Drug Resist Updat 2001;4:129–31.
[13] Soengas MS, Capodieci P, Polsky D, Mora J, Esteller M, Opitz-Araya X, et al. Inactivation of the apoptosis effector Apaf-1 in malignant melanoma. Nature 2001;409:207–11.
[14] Michie AM, McCaig AM, Nakagawa R, Vukovic M. Death-associated protein kinase (DAPK) and signal transduction: regulation in cancer. FEBS J 2010;277:74–80.
[15] Lehmann U, Celikkaya G, Hasemeier B, Langer F, Kreipe H. Promoter hypermethylation of the death-associated protein kinase gene in breast cancer is associated with the invasive lobular subtype. Cancer Res 2002;62:6634–8.
[16] Bialik S, Kimchi A. DAP-kinase as a target for drug design in cancer and diseases associated with accelerated cell death. Semin Cancer Biol 2004;14:283–94.
[17] Tada Y, Wada M, Taguchi K, Mochida Y, Kinugawa N, Tsuneyoshi M, et al. The association of death-associated protein kinase hypermethylation with early recurrence in superficial bladder cancers. Cancer Res 2002;62:4048–53.
[18] Toyooka S, Toyooka KO, Miyajima K, Reddy JL, Toyota M, Sathyanarayana UG, et al. Epigenetic down-regulation of death-associated protein kinase in lung cancers. Clin Cancer Res 2003;9:3034–41.

[19] Raval A, Tanner SM, Byrd JC, Angerman EB, Perko JD, Chen SS, et al. Downregulation of death-associated protein kinase 1 (DAPK1) in chronic lymphocytic leukemia. Cell 2007;129:879–90.
[20] Chaopatchayakul P, Jearanaikoon P, Yuenyao P, Limpaiboon T. Aberrant DNA methylation of apoptotic signaling genes in patients responsive and nonresponsive to therapy for cervical carcinoma. Am J Obstet Gynecol 2010;202:e281–9.
[21] Voso MT, D'Alo F, Greco M, Fabiani E, Criscuolo M, Migliara G, et al. Epigenetic changes in therapy-related MDS/AML. Chem Biol Interact 2010;184:46–9.
[22] Sugita H, Iida S, Inokuchi M, Kato K, Ishiguro M, Ishikawa T, et al. Methylation of BNIP3 and DAPK indicates lower response to chemotherapy and poor prognosis in gastric cancer. Oncol Rep 2011;25:513–18.
[23] Christoph F, Weikert S, Kempkensteffen C, Krause H, Schostak M, Miller K, et al. Regularly methylated novel pro-apoptotic genes associated with recurrence in transitional cell carcinoma of the bladder. Int J Cancer 2006;119:1396–402.
[24] Tanaka T, Bai T, Toujima S, Utsunomiya T, Utsunomiya H, Yukawa K, et al. Impaired death-associated protein kinase-mediated survival signals in 5-fluorouracil-resistant human endometrial adenocarcinoma cells. Oncol Rep 2012;28:330–6.
[25] Ogawa T, Liggett TE, Melnikov AA, Monitto CL, Kusuke D, Shiga K, et al. Methylation of death-associated protein kinase is associated with cetuximab and erlotinib resistance. Cell Cycle 2012;11:1656–63.
[26] Guo Q, Chen Y, Wu Y. Enhancing apoptosis and overcoming resistance of gemcitabine in pancreatic cancer with bortezomib: a role of death-associated protein kinase-related apoptosis-inducing protein kinase 1. Tumori 2009;95:796–803.
[27] Bai T, Tanaka T, Yukawa K, Umesaki N. A novel mechanism for acquired cisplatin-resistance: suppressed translation of death-associated protein kinase mRNA is insensitive to 5-aza-2′-deoxycitidine and trichostatin in cisplatin-resistant cervical squamous cancer cells. Int J Oncol 2006;28:497–508.
[28] Campioni M, Santini D, Tonini G, Murace R, Dragonetti E, Spugnini EP, et al. Role of Apaf-1, a key regulator of apoptosis, in melanoma progression and chemoresistance. Exp Dermatol 2005;14:811–18.
[29] Zang YS, Zhong YF, Fang Z, Li B, An J. MiR-155 inhibits the sensitivity of lung cancer cells to cisplatin via negative regulation of Apaf-1 expression. Cancer Gene Ther 2012;19:773–8.
[30] Lum E, Vigliotti M, Banerjee N, Cutter N, Wrzeszczynski KO, Kahn S, et al. Loss of DOK2 induces carboplatin resistance in ovarian cancer via suppression of apoptosis. Gynecol Oncol 2013;130:369–76.
[31] Anthoney DA, McIlwrath AJ, Gallagher WM, Edlin AR, Brown R. Microsatellite instability, apoptosis, and loss of p53 function in drug-resistant tumor cells. Cancer Res 1996;56:1374–81.
[32] Peltomaki P. Role of DNA mismatch repair defects in the pathogenesis of human cancer. J Clin Oncol 2003;21:1174–9.
[33] Esteller M, Levine R, Baylin SB, Ellenson LH, Herman JG. MLH1 promoter hypermethylation is associated with the microsatellite instability phenotype in sporadic endometrial carcinomas. Oncogene 1998;17:2413–17.
[34] Strathdee G, MacKean MJ, Illand M, Brown R. A role for methylation of the hMLH1 promoter in loss of hMLH1 expression and drug resistance in ovarian cancer. Oncogene 1999;18:2335–41.
[35] Gifford G, Paul J, Vasey PA, Kaye SB, Brown R. The acquisition of hMLH1 methylation in plasma DNA after chemotherapy predicts poor survival for ovarian cancer patients. Clin Cancer Res 2004;10:4420–6.
[36] Watts GS, Pieper RO, Costello JF, Peng YM, Dalton WS, Futscher BW. Methylation of discrete regions of the O6-methylguanine DNA methyltransferase (MGMT) CpG

island is associated with heterochromatinization of the MGMT transcription start site and silencing of the gene. Mol Cell Biol 1997;17:5612–19.
[37] Esteller M, Hamilton SR, Burger PC, Baylin SB, Herman JG. Inactivation of the DNA repair gene O6-methylguanine-DNA methyltransferase by promoter hypermethylation is a common event in primary human neoplasia. Cancer Res 1999;59:793–7.
[38] Kewitz S, Stiefel M, Kramm CM, Staege MS. Impact of O6-methylguanine-DNA methyltransferase (MGMT) promoter methylation and MGMT expression on dacarbazine resistance of Hodgkin's lymphoma cells. Leuk Res 2014;38:138–43.
[39] Guo J, Cui Q, Jiang W, Liu C, Li D, Zeng Y. Research on DNA methylation of human osteosarcoma cell MGMT and its relationship with cell resistance to alkylating agents. Biochem Cell Biol 2013;91:209–13.
[40] Chan CL, Wu Z, Eastman A, Bresnick E. Irradiation-induced expression of O6-methylguanine-DNA methyltransferase in mammalian cells. Cancer Res 1992;52:1804–9.
[41] Fritz G, Kaina B. Stress factors affecting expression of O6-methylguanine-DNA methyltransferase mRNA in rat hepatoma cells. Biochim Biophys Acta 1992;1171:35–40.
[42] Esteller M, Garcia-Foncillas J, Andion E, Goodman SN, Hidalgo OF, Vanaclocha V, et al. Inactivation of the DNA-repair gene MGMT and the clinical response of gliomas to alkylating agents. N Engl J Med 2000;343:1350–4.
[43] Koul S, McKiernan JM, Narayan G, Houldsworth J, Bacik J, Dobrzynski DL, et al. Role of promoter hypermethylation in Cisplatin treatment response of male germ cell tumors. Mol Cancer 2004;3:16.
[44] Weller M, Stupp R, Reifenberger G, Brandes AA, van den Bent MJ, Wick W, et al. MGMT promoter methylation in malignant gliomas: ready for personalized medicine? Nat Rev Neurol 2010;6:39–51.
[45] D'Andrea AD. Susceptibility pathways in Fanconi's anemia and breast cancer. N Engl J Med 2010;362:1909–19.
[46] Taniguchi T, Tischkowitz M, Ameziane N, Hodgson SV, Mathew CG, Joenje H, et al. Disruption of the Fanconi anemia-BRCA pathway in cisplatin-sensitive ovarian tumors. Nat Med 2003;9:568–74.
[47] Marsit CJ, Liu M, Nelson HH, Posner M, Suzuki M, Kelsey KT. Inactivation of the Fanconi anemia/BRCA pathway in lung and oral cancers: implications for treatment and survival. Oncogene 2004;23:1000–4.
[48] Narayan G, Arias-Pulido H, Nandula SV, Basso K, Sugirtharaj DD, Vargas H, et al. Promoter hypermethylation of FANCF: disruption of Fanconi Anemia–BRCA pathway in cervical cancer. Cancer Res 2004;64:2994–7.
[49] Chen CC, Taniguchi T, D'Andrea A. The Fanconi anemia (FA) pathway confers glioma resistance to DNA alkylating agents. J Mol Med (Berl) 2007;85:497–509.
[50] Li N, Li X, Li S, Zhou S, Zhou Q. Cisplatin-induced downregulation of SOX1 increases drug resistance by activating autophagy in non-small cell lung cancer cell. Biochem Biophys Res Commun 2013;439:187–90.
[51] Guo R, Wu G, Li H, Qian P, Han J, Pan F, et al. Promoter methylation profiles between human lung adenocarcinoma multidrug resistant A549/cisplatin (A549/DDP) cells and its progenitor A549 cells. Biol Pharm Bull 2013;36:1310–16.
[52] Xu XL, Wu LC, Du F, Davis A, Peyton M, Tomizawa Y, et al. Inactivation of human SRBC, located within the 11p15.5-p15.4 tumor suppressor region, in breast and lung cancers. Cancer Res 2001;61:7943–9.
[53] Moutinho C, Martinez-Cardus A, Santos C, Navarro-Perez V, Martinez-Balibrea E, Musulen E, et al. Epigenetic inactivation of the BRCA1 interactor SRBC and resistance to oxaliplatin in colorectal cancer. J Natl Cancer Inst 2014;106.

[54] Sabatino MA, Geroni C, Ganzinelli M, Ceruti R, Broggini M. Zebularine partially reverses GST methylation in prostate cancer cells and restores sensitivity to the DNA minor groove binder brostallicin. Epigenetics 2013;8:656–65.

[55] Baker EK, El-Osta A. MDR1, chemotherapy and chromatin remodeling. Cancer Biol Ther 2004;3:819–24.

[56] El-Osta A, Kantharidis P, Zalcberg JR, Wolffe AP. Precipitous release of methyl-CpG binding protein 2 and histone deacetylase 1 from the methylated human multidrug resistance gene (MDR1) on activation. Mol Cell Biol 2002;22:1844–57.

[57] Qiu YY, Mirkin BL, Dwivedi RS. Inhibition of DNA methyltransferase reverses cisplatin induced drug resistance in murine neuroblastoma cells. Cancer Detect Prev 2005;29:456–63.

[58] Kastl L, Brown I, Schofield AC. Altered DNA methylation is associated with docetaxel resistance in human breast cancer cells. Int J Oncol 2010;36:1235–41.

[59] Balch C, Nephew KP. Epigenetic targeting therapies to overcome chemotherapy resistance. Adv Exp Med Biol 2013;754:285–311.

CHAPTER

10

DNA Methylation and Endocrinology

OUTLINE

M. Neidhart: DNA Methylation and Complex Human Disease.

DOI: http://dx.doi.org/10.1016/B978-0-12-420194-1.00010-5

10.1 INTRODUCTION

The endocrine system refers to the glands that secrete hormones directly into the circulation to be carried toward a distant target organ. The major endocrine glands include the pituitary gland, adrenal gland, gonads, thyroid gland, parathyroid gland, and gastrointestinal tract (Figure 10.1). The endocrine system is an information signal system like the nervous system, controlling homeostasis, growth, and development. In vertebrates, the hypothalamus is the neural control center for all endocrine systems. The hypothalamic hormones are referred to as releasing hormones and inhibiting hormones, reflecting their influence on pituitary hormones. The pituitary hormones target various tissues: gonadotropins (FSH, LH) the gonads; prolactin (PRL) and oxytocin (OXY) the breast; adrenocorticotropic hormone (ACTH) the adrenal cortex; growth hormone (GH) the adipose tissue, bone, muscles and liver; thyroid stimulating hormone (TSH) the thyroid gland; and antidiuretic hormone (ADH) the kidney. Some of these target tissues also release hormones, such as estrogen (E), testosterone (T) and progesterone (P) by the gonads, glucocorticoids by the adrenal cortex, somatomedin (S) by the liver, thyroid hormones (T3, T4) by the thyroid gland, and erythropoietin (EPO) by the kidney. The adrenal medulla releases adrenalin, while the pancreas is responsible for the production and release of insulin; together with thyroid hormones, they control the energy metabolism. The glands secrete hormones controlling many

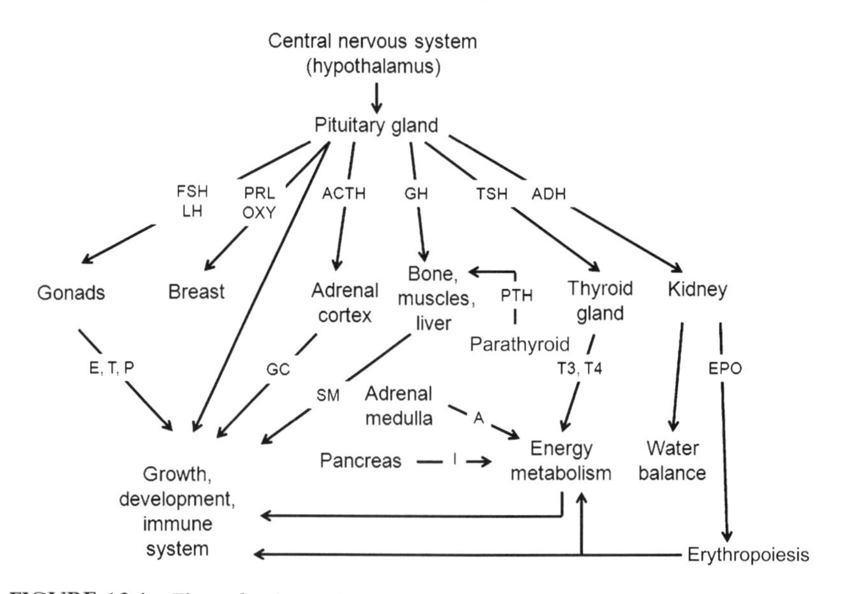

FIGURE 10.1 The endocrine system.

bodily functions, including cell growth and development, mood, sexual functions, and metabolism. Metabolism is a collection of biochemical reactions that takes place in the body's cells to convert the fuel in the food we eat into the energy needed to power everything we do, from growing to moving and thinking. There are specific proteins in the body to control the chemical reactions of metabolism, and each chemical reaction is coordinated with other body functions. In fact, thousands of metabolic reactions happen at the same time, all regulated by the body, to keep our cells healthy and working.

When a cell produces optimum energy it then has the capacity to fulfill its many functions, involving regeneration, detoxification, and its unique genetically and epigenetically programmed role. If cells of like kind have the energy to efficiently fulfill their functions, then the organs or glands they comprise can fulfill their functions. And, if the organs and glands have the energy to efficiently fulfill their functions, then the systems they comprise can efficiently carry out their functions. Without the proper nutritional balance, our bodies are unable to manufacture the energy that is needed for all the life-sustaining processes of metabolism, not just proper immune function. Thus, the endocrine system directly affects metabolism, and metabolism depends on nutrition. Nutrition is the major intrauterine environmental factor that alters expression of the fetal genome and may have lifelong consequences (Figure 10.2). Genetic lesions, including mutations, deletions,

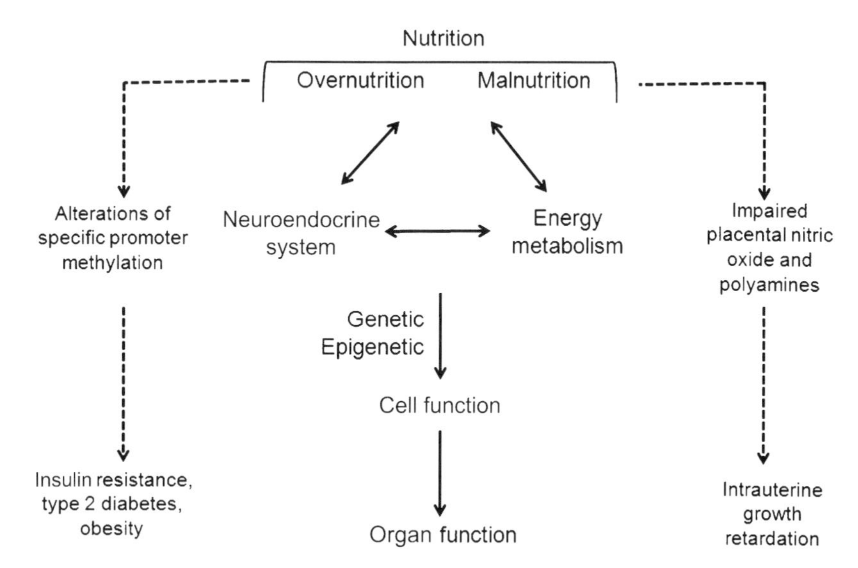

FIGURE 10.2 Influence of nutrition on the neuroendocrine system and energy metabolism, and effects of over- or malnutrition on DNA methylation, fetal growth, and offspring's organ functions.

or breakages, are well known to produce disorders in humans, but there is increasing evidence that diseases can also be caused by epigenetic alterations [1]. Epigenetic changes are not only responsible for normal development, but also involved in diseases. Animal studies show that both maternal undernutrition and overnutrition reduce placental–fetal blood flows and stunt fetal growth. Impaired placental syntheses of nitric oxide (a major vasodilator and angiogenesis factor) and polyamines (key regulators of DNA and protein synthesis) may provide a unified explanation for intrauterine growth retardation in response to the two extremes of nutritional problems with the same pregnancy outcome. There is growing evidence that maternal nutritional status can alter the epigenetic state of the fetal genome. This may provide a molecular mechanism for the impact of maternal nutrition on both fetal programming and genomic imprinting [2]. Changes in DNA methylation can cause silencing of normally active genes, or activation of normally silent genes. This could occur in cancer, in hereditary disorders resulting from DNA methylation defects, and in late-onset diseases caused by the interaction of genomic, epigenetic, and environmental changes (see Chapters 5–7). Similarly, overnutrition can affect DNA methylation of specific promoters, leading to insulin resistance, type 2 diabetes, and obesity. In the present chapter we will introduce the topic of DNA methylation in non-neoplastic endocrine disorders, and particularly what is known from human diseases and animal models. A summary of the changes in DNA methylation observed to date in relation to different endocrine conditions is shown in Table 10.1 [3–29].

10.2 IMPRINTED DISORDERS

Mammals inherit two complete sets of chromosomes, one from the father and one from the mother, and most autosomal genes are expressed from both maternal (M) and paternal (F) alleles. Imprinted genes show expression from only one member of the gene pair (allele), and their expression is determined by the parent during production of the gametes. Imprinted genes represent only a small subset of mammalian genes that are present but not imprinted in other vertebrates. Genomic imprints are erased in both germlines and reset accordingly; thus, they are reversible and this leads to differential expression in the course of development. The Beckwith-Wiedemann syndrome, an overgrowth disorder characterized by an increased risk of cancer and different malformations, is an example of a single-gene epigenetic imprinted disease. Some patients with Beckwith-Wiedemann syndrome show loss of imprinting of insulin-like growth factor 2 (IGF2), leading to an

TABLE 10.1 Endocrine Non-Neoplastic Conditions Associated with DNA Methylation Changes

Conditions	Alterations
Beckwith-Wiedemann syndrome	Loss of imprinting of IGF2 [3]
Pseudohypoparathyroidism type 1A	Differential imprinting of GNAS1 [4]
Type 2 diabetes	Hypomethylated IGFBP1 [5] and IGFBP7 [6]
	High-fat diet decreases IL13RA2 [7] and LEPTIN [8] methylation, but increased NDUFB6 [9] methylation
	Low-protein diet alters PPARα methylation [10]
	Intrauterine growth restriction reduces global DNA methylation [11] and increased PDX1 methylation [12]
	Malnutrition increases HNF4α methylation [13]
	Overfeeding increased PPARγC1α methylation [14]
Obesity	Hypomethylated FTO [15–17]
Behavioral disorders	Glucocorticoid receptor methylation by maternal care [18], in childhood abuse victims [19] and in anxiety [20]
	Arginine vasopressin hypomethylation after infant–mother separation [21]
Cardiovascular diseases	Low-protein diet and AT1B methylation [22,23]
Cushing's syndrome	POMC hypomethylation [24]
Sexual behavior	Estrogen receptor-α methylation [25–28]
	Arginine vasopressin methylation [28]
	REELIN methylation [29]

increase in the levels of this growth factor [3]. A human disorder with multiple hormone resistance, pseudohypoparathyroidism type IA (PHPIA) is also caused by tissue-specific differential imprinting of splice variants of the guanine nucleotide regulatory protein (encoded by GNAS1) [4] (Figure 10.3a).

Other classes of monogenic diseases involves mutation of genes involved themselves in methylation machinery. For instance, Rett syndrome is caused by mutations that affect the methyl-CpG-binding protein MeCP2 [30]. Rett syndrome is a neurodevelopmental disorder that affects mainly females. Among other symptoms, affected patients can have seizures, present intellectual disability with learning difficulties, and have no verbal skills. This phenotype argues for the primacy of MeCP2, as opposed to other MBDs, in the silencing mechanism

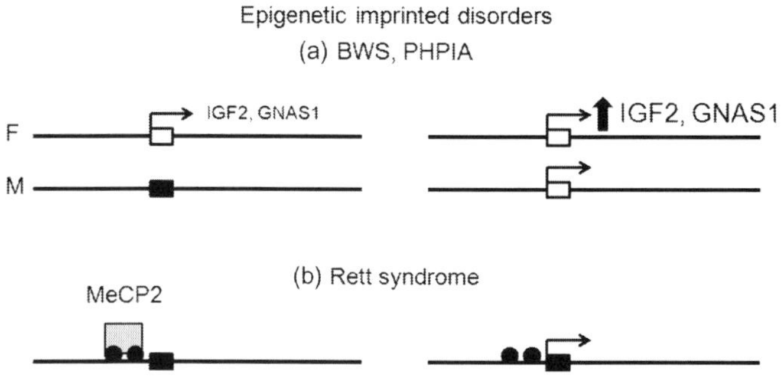

FIGURE 10.3 Epigenetic imprinted disorders with consequences for the endocrine system and homeostasis.

associated with DNA methylation, at least in the central nervous system. In Rett syndrome, DNA methylation is not altered, but gene silencing is abnormal because DNA methylation goes unrecognized (Figure 10.3b).

10.3 RAPID PROMOTER METHYLATION/DEMETHYLATION PROCESSES

10.3.1 Glucocorticoids and Estrogens

DNA methylation is believed to be a slow and stable mechanism, while histone modification is more dynamic. This is not always the case, at least in particular examples reported in endocrinology (Figure 10.4). Thus, although the mechanisms of DNA demethylation are not yet totally understood and remain a controversial issue, data obtained studying the effect of hormones, mainly ligands of nuclear receptors, on transcription of their target genes indicate that they can mediate active DNA demethylation occurring independently of DNA replication. Early evidence was obtained with the glucocorticoid receptor (GR/NR3C1). Glucocorticoids (GCs) cause local DNA demethylation of the tyrosine aminotransferase (TAT) gene around a glucocorticoid response element that is located 2.5 kb upstream of the transcriptional start site [31], showing that demethylation is a consequence of an active mechanism that involves the creation of DNA nicks 3′ to the methylcytidine and the participation of a demethylase initiating a base excision repair.

In the case of the TAT gene, GC-dependent demethylation is slow, being detected after a few hours of stimulation. However, rapid methylation/demethylation at the estrogen receptor target gene PS2 has been observed in breast cancer cells [32]. It had been previously shown that ER is recruited in a cyclic manner to the PS2 promoter and that this is

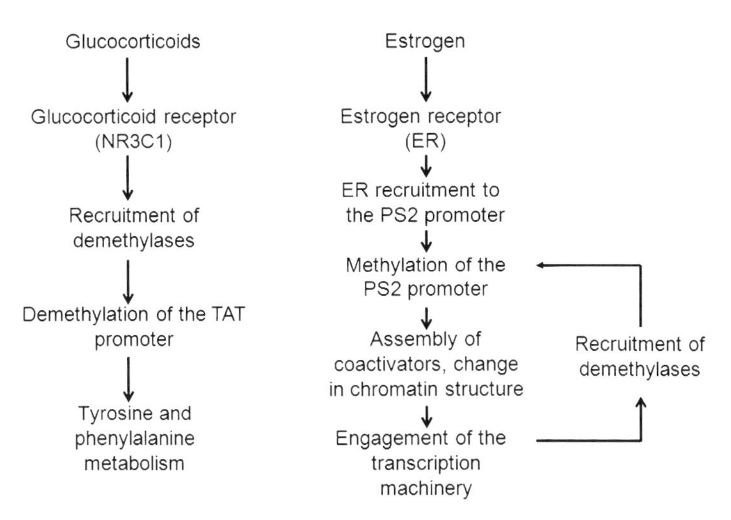

FIGURE 10.4 Examples of rapid promoter methylation/demethylation processes involving steroidal hormones and the recruitment of demethylases.

followed by the assembly of coactivators, which in turn provokes local structural changes in chromatin. This allows engagement of the basal transcription machinery and activation of the RNA polymerase II, which is then followed by the recruitment of co-repressor complexes and initiation of the next transcriptional cycle [33]. Remarkably, these changes are accompanied by changes in the methylation of CpG dinucleotides in the PS2 promoter. Cyclical methylation and demethylation of CpG dinucleotides has a periodicity of around 2 hours, implying the existence of an active demethylation process.

10.3.2 Vitamin D and Parathyroid Hormone

Vitamin D, synthesized in the skin or obtained from the diet, and parathyroid hormone (PTH), secreted by the parathyroid glands, increase serum Ca^{2+} concentrations via actions on the gut, kidney, and bone. These two calcemic hormones regulate cytochrome p450 27B1 (CYP27B1) gene expression. PTH activates CYP27B1 transcription through stimulation of protein kinases A and C (PKA and PKC), whereas vitamin D, through binding to its nuclear receptor, VDR, represses CYP27B1 transcription. Reflecting the vitamin D-mediated repression of the CYP27B1 gene, rapid methylation of CpG sites is induced by vitamin D in this gene promoter [34] (Figure 10.5). This methylation step requires DNMT1 and DNMT3B. Conversely, treatment with PTH causes active CpG demethylation of the CYP27B1 promoter. Thus, these results obtained with the CYP27B1 promoter further suggest that DNA

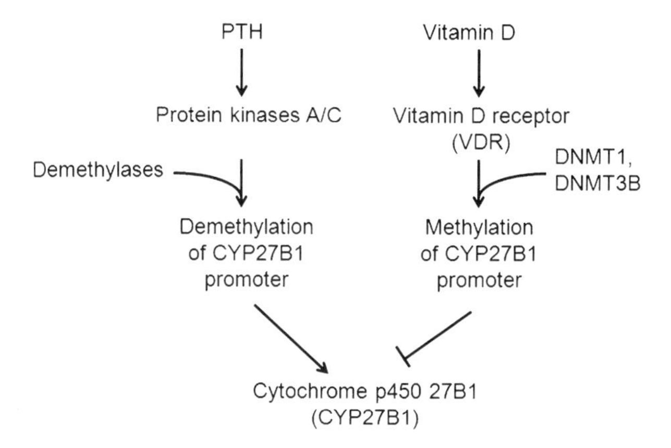

FIGURE 10.5 Example of a rapid promoter methylation/demethylation process involving parathroid hormone (PTH) and vitamin D, and the recruitment of DNA methyltransferases and demethylases.

methylation/demethylation plays an important role in hormonal regulation of transcription.

10.4 EXPOSURE TO STRESS

Stress is a potent stimulus that is considered a vulnerability factor for a number of mental and neurological diseases and is known to significantly alter the epigenome. Likewise, many psychological diseases are inherited abnormally and are controlled by environmental or stochastic events. This points to a role for imprinting and other forms of persistent epigenetic regulation in the etiology of mental illness [35]. The paraventricular nucleus of the hypothalamus corticotropin-releasing hormone (CRH) is the main activator of the hypothalamic–pituitary–adrenal (HPA) axis and is responsible for at least some behavioral changes after stress. In mice subjected to 10 days of social defeat, 4 out of 10 CpG islands of the CRH promoter, including the transcription-activating CRE site, were demethylated in stress-vulnerable but not in stress-resistant mice, and this demethylation co-occurred with increases in CRH transcription [36]. In rats exposed to stress, similarly, CRH promoter methylation is reduced [37]; differences between gender, however, were found regarding the localization – i.e., in the bed nucleus of stria terminalis in both genders, but the amygdala in stressed females only.

Exposure to stress during neurodevelopment has an effect on the quality of physical and mental health. Maternal care might influence HPA function through epigenetic programming of GR expression [18]. In humans, childhood abuse alters HPA stress responses and increases the

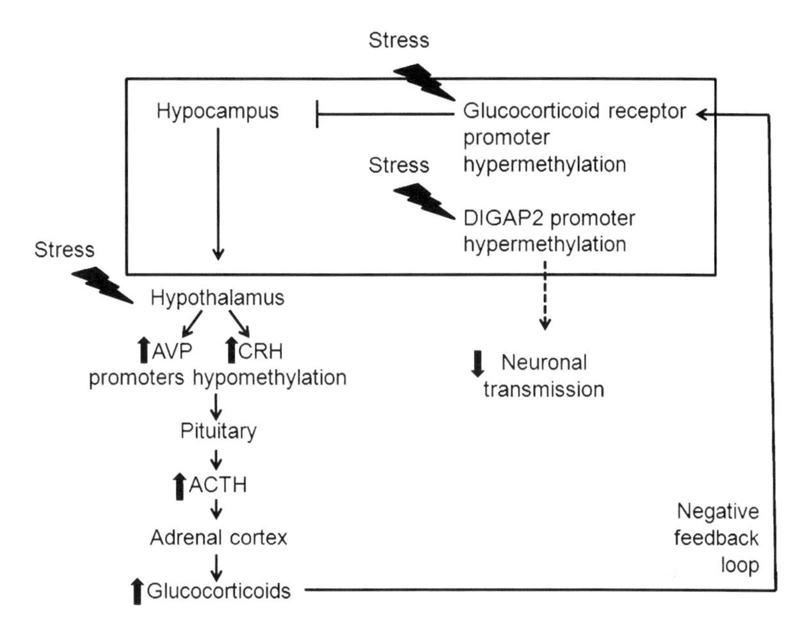

FIGURE 10.6 Chronic activation of the hypothalamo-pituitary–adrenal (HPA) axis by stress. The promoter of hypothalamic secretagogs AVP and CRH becomes demethylated, allowing an increased production of ACTH and glucocorticoids. In turn, these glucocorticoids should decrease the hypocampal activity by a negative feedback loop. This loop is inefficient upon stress because of a glucocorticoid receptor promoter hypermethylation at this site. Stress also reduced neuronal transmission by inducing DIGAP2 promoter hypermethylation.

risk of suicide. Decreased levels of GR mRNA and increased methylation of the GR promoter were found in hippocampus from abused suicide victims that also showed a decrease in NGFI-A transcription factor binding. These findings suggest a common effect of parental care on the epigenetic regulation of hippocampal GR expression [19]. In animals, the most normally used technique to induce early-life stress is periodic infant–mother separation during the neonatal period. This causes an irreversible increase in glucocorticoid secretion with disruption of the activity of the HPA axis and increased sensitivity to stress later in life, which are related to disorders of mood and cognition [38]. Hypothalamic secretagogs regulate HPA axis activity by increasing ACTH expression. Neonatal stress induces lifelong hypomethylation of the arginine vasopressin (AVP) gene, causing increased AVP expression, activation of the HPA axis, and behavioral alterations. Acute predator stress, a rat model of post-traumatic stress disorder, induced genome-wide changes in DNA methylation of the hippocampus [39]. Among the genes reacting to this stress paradigm was DIGAP2, which is involved in synaptic remodeling and neuronal transmission. Figure 10.6 shows a summary of these findings.

10.4.1 Neonatal Development

There is evidence for hypercortisolemia playing a role in the generation of psychiatric symptoms, and for epigenetic variation within HPA axis genes mediating behavioral changes in the adult. When mice are treated with corticosterone, they exhibit anxiety-like behavior together with a significant decrease in the hippocampal mRNA levels of GR and an increase in the stress-related gene FKBP5 [20]. This gene encodes a co-chaperone of HSP90 that binds to GR and promotes cytoplasmic localization of the receptor, regulating GR sensitivity. Differences were seen in FKBP5 methylation in the hippocampus and hypothalamus of glucocorticoid-treated animals. The same occurs in a mouse hippocampal neuronal cell line exposed to corticosterone. This suggests that DNA methylation plays a role in mediating the effects of glucocorticoid exposure on FKBP5 function, with potential consequences for behavior.

Numerous genes involved in the development of the nervous system and pituitary gland are controlled by promoter methylation, such as LHX3 [40] and BDNF [41] (Figure 10.7). Brain-derived neurotrophic factor (BDNF) is a protein implicated in the stress response, long-term memory, Alzheimer's disease, and psychiatric disorders such as bipolar disorder, depression, and schizophrenia [35]. Adult rats, males and females, maltreated during the first week of postnatal life showed hypermethylation of the promoter of the BNDF gene (exon 4) and decreased BDNF transcription in the prefrontal cortex [41]. BDNF also responds to chronic social-predator stress in adult male rats.

Lower weight at birth is associated not only with risk of metabolic syndrome but also with cardiovascular disease and hypertension in adulthood [42]. The renin–angiotensin system appears to play a role in this process, since a maternal low-protein diet results in undermethylation of the AT1B angiotensin receptor promoter and early overexpression of this gene in the adrenal of offspring [22]. Furthermore, maternal glucocorticoids modulate this effect on fetal DNA methylation, since treatment of rat dams with the 11β-hydroxylase inhibitor metyrapone prevents the epigenetic change and hypertension in the offspring. Collectively, these studies suggest that DNA methylation might have an important role in the long-term effects of glucocorticoids in neonatal development.

10.4.2 Weight Control

The pro-opiomelanocortin (POMC) gene plays an important role not only in the regulation of the HPA axis and adrenal development, but also in obesity. POMC serves as a prohormone for adrenocorticotropic hormone (ACTH), a key mediator of the adrenocortical response to stress.

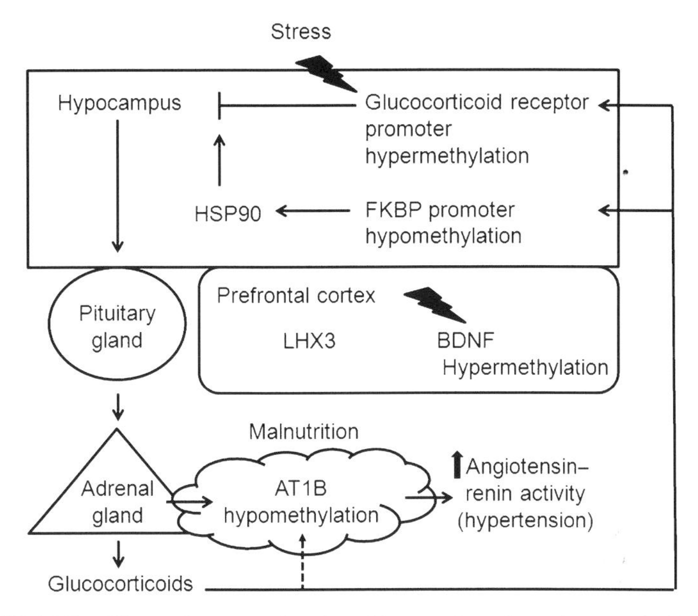

FIGURE 10.7 Effects of stress and malnutrition on the epigenetic control of brain and adrenal development.

POMC expression is retrained by promoter methylation [43]. On the other hand, the POMC gene is intrinsically activated in ACTH-dependent Cushing's syndrome. This disorder may be a consequence of activation of the highly tissue-specific POMC promoter in pituitary and non-pituitary sites. This promoter contains a CpG island, which is methylated in normal non-expressing tissues but is specifically demethylated in expressing tissues and tumors [24]. Methylation near the response element for the tissue-specific POMC activator PTX1 abolishes binding of this transcription factor that plays a key role in pituitary development. The POMC promoter could show different degrees of methylation in POMC-expressing hypothalamic neurons, thus influencing food intake and obesity (Figure 10.8).

10.5 MATERNAL AND SEXUAL BEHAVIOR

10.5.1 Glucocorticoid Receptor

A study [44] compared two groups of rats displaying different levels of maternal behaviors: high or low lick-groom and arched-back nursing. Significant differences in the DNA methylation of exon 1 GR promoter in neurons and glial cells of the hippocampus of adult rats were found.

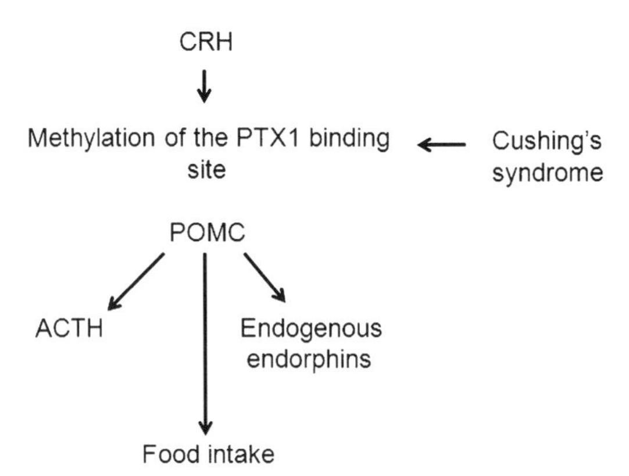

FIGURE 10.8 Differential methylation of the PTX1 binding site in the *POMC* gene promoter influences food intake and body weight, and can be responsible for obesity.

The offspring of arched-back nursing mothers showed lower DNA methylation levels and an increase in the acetylation of histone H3 (H3K9ac). These epigenetic changes correlated with higher hippocampal GR protein content, and decreased fearfulness and post-stress corticosterone concentrations. This effect was primarily dependent on maternal behavior, as cross-fostered pups showed a methylation pattern similar to the foster mother. The exon 1 GR promoter is a site of consensus binding of NGFI-A transcription factor. As suspected, binding of this factor to the exon 1 GR promoter was higher in the offspring of arched-back nursing mothers (Figure 10.9). The opposite effects can be achieved by treating adult rats with methionine, which is a donor of methyl groups and thus an inducer of hypermethylation. The low methylation of the hippocampal GR gene promoter in the adult male offspring of arched-back nursing mothers can be reversed by 7 days of methionine infusions [45]. The capacity for reversal of early epigenetic modifications is therefore present in the adult central nervous system. Similarly, hypermethylation of the GR gene promoter was found in the hippocampus of male suicide victims exposed to child abuse [46] and in blood leukocytes of adults with a history of childhood adversities [47], but not in the hippocampus of suicide victims without an abuse history [46]. It is noticeable that in blood leukocytes the degree of GR promoter methylation correlated with the number of various childhood adversities.

10.5.2 Gonadal Steroids

Many brain sex-specific features arise from the effects of the gonadal steroid hormones that are exerted during the perinatal period. Testosterone is converted into estradiol within developing neurons, and

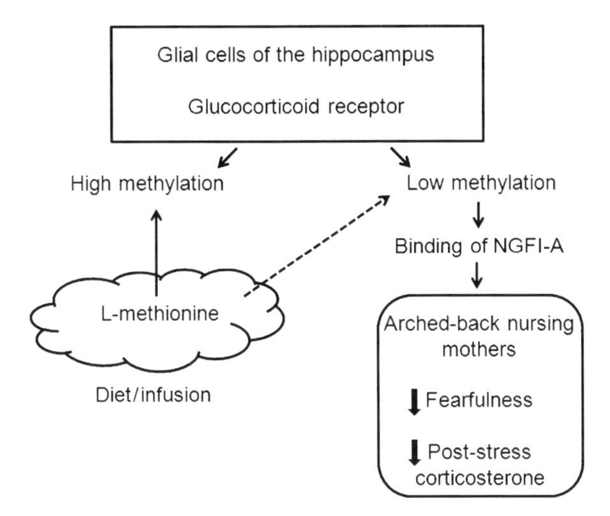

FIGURE 10.9 Epigenetic regulation of the glucocorticoid receptor in arched-back nursing mother mice, and influences of L-methionine.

estradiol mediates several developmental sex differences in cell anatomy and physiology. These differences are responsible for dimorphic regulation of pituitary gonadotropin secretion and for sex-specific behavior [48]. Alterations in DNA methylation of genes essential for sexual brain differentiation, including estrogen receptors (ERα and ERβ) and progesterone receptor (PR), have been described to be sex- and hormonally regulated (Figure 10.10). In rodents, methylation of the ERα promoter within the preoptic area, which is crucial for sexual behavior, is influenced by maternal care during the neonatal period. Thus, maternal licking and grooming has been demonstrated to alter ERα promoter methylation, and consequently ERα expression, in a sex-specific manner [25]. In addition, methylation of the ERα gene in the preoptic area is higher in newborn females than in males or estradiol-treated females, and sex- and hormone-mediated differences in methylation were also observed at later stages [26]. While ERα activation is required for neonatal brain masculinization, ERβ activation appears to be involved in brain defeminization and in the regulation of neuroendocrine functions, since this receptor co-localizes with neuroendocrine hormones such as GnRH, CRH, oxytocin, vasopressin, or prolactin. CpG methylation of the ERβ gene in the preoptic area, hypothalamus, or hippocampus of newborns is not significantly influenced by sex or hormonal treatment. However, there are differences in ERβ methylation in these brain areas in the adult [27]. On the other hand, whereas in hypothalamus of newborn animals no sex differences in PR promoter methylation are detected, significantly lower levels are found in

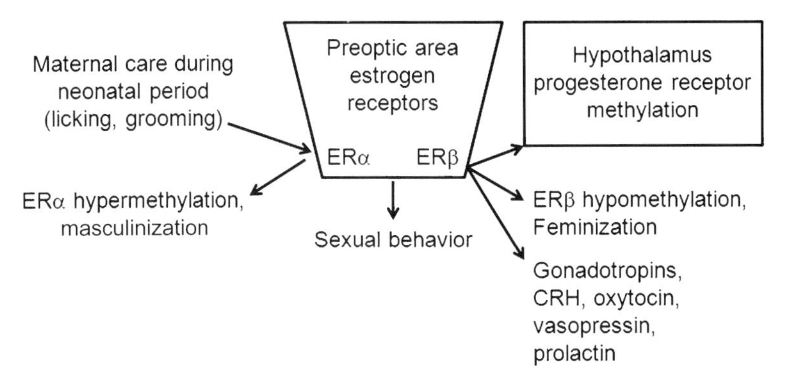

FIGURE 10.10 Differential methylation of estrogen receptors leads to masculinization or feminization of the neonatal brain.

adolescent females than in males. It has been suggested that gonadal female hormones promote PR methylation in the hypothalamus, silencing PR expression during the critical period of sexual differentiation in the male – a process that is essential for masculine behavior in the adult.

10.5.3 Arginine Vasopressin

Whereas DNA methylation can be profoundly influenced by gonadal hormones during brain development, the effect of these hormones in the adult brain has received less attention. However, it has been reported that testosterone regulates expression of arginine vasopressin (AVP) within the bed nucleus of the stria terminalis (BST) in the adult brain (Figure 10.11). Hypothalamic vasopressin is known to potentiate CRH action and stress response. Castration of male rats strongly inhibits AVP expression in this nucleus, and this inhibition is reversed on testosterone treatment. It was found that castration results in AVP promoter methylation at specific CpG sites in the BST. Conversely, castration significantly increased ERα mRNA levels by decreasing ERα promoter methylation [28]. These results suggest that the DNA methylation pattern of some steroid responsive genes is actively regulated by gonadal steroid hormones in the adult brain. The opposing regulation of ERα and AVP promoter methylation in response to changes in steroid hormone levels in the same brain region is intriguing, and how this specificity is regulated remains to be elucidated. In any case, these results indicate that regulation of DNA methylation in the adult brain could play a role in the hormonal control of behavior.

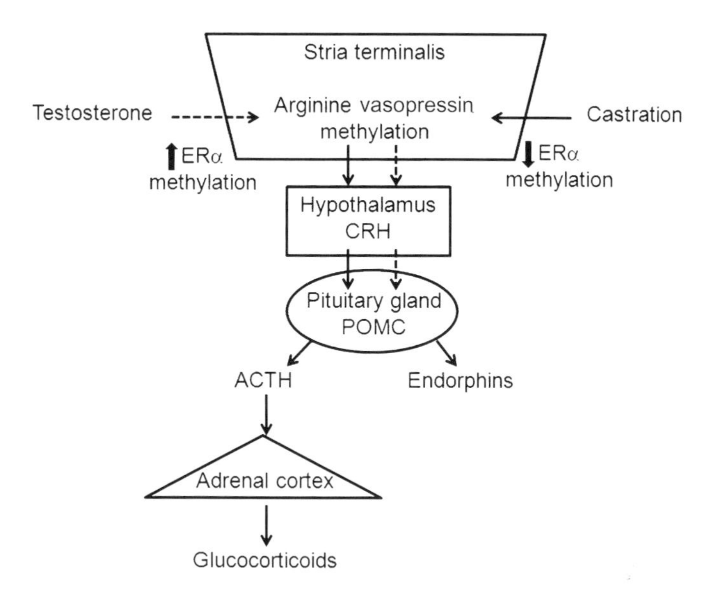

FIGURE 10.11 Epigenetic control of arginine vasopressin in the adult murine brain upon hyper- or hypomethylation of estrogen receptor by testosterone or castration. This could play a role in the hormonal control of behavior.

10.6 EXPOSURE TO ENDOCRINE DISRUPTORS

Epigenetic factors can be altered by the environment (Chapter 4), and increasing evidence suggests that a large variety of environmental and dietary chemicals can interfere with normal endocrine functions and result in adverse consequences (Chapter 5). Initial animal studies demonstrated that compounds with estrogenic activity can disrupt reproductive tract function. However, it is now evident that different chemical compounds, referred to as endocrine disruptors, can either mimic or interfere with the normal actions of hormones, including not only sexual steroids but also thyroid, hypothalamic, and pituitary hormones [49,50] (Table 10.2 [20,51–55]). The effect of endocrine disruptors is particularly strong when exposure occurs during fetal or neonatal periods. If exposure occurs during these critical stages, it can produce permanent effects that may be observed much later in life. For instance, exposure of newborn mice to environmental estrogens causes uterine lesions and uterine tumors in adults.

Mechanistic studies have provided support that estrogens cause both genetic and epigenetic alterations in developing target tissues. Thus, the estrogen-responsive genes lactoferrin and c-fos are permanently upregulated in the uterus after developmental exposure to the estrogen-like

TABLE 10.2 Endocrine Disruptors

Compounds	Actions
Diethylstilbestrol	Hypomethylation of estrogen-responsive genes [20] and changes in NSPB1 promoter methylation [50,51]
Bisphenol-A	Hypomethylation of HOXA10 [52]
Vinclozolin	Increased *de novo* DNA methylatransferases [53,54] and altered methylation in the sperm [55]

compound diethylstilbestrol (DES) due to hypomethylation of the promoter region of these genes after exposure to this chemical [56]. For many years, DES was prescribed to pregnant women to prevent spontaneous abortions. Several genes were identified whose methylation patterns are altered after neonatal treatment with DES or genistein (other estrogenic compound), among them the gene encoding nucleosomal binding protein 1 (NSBP1), a nucleosome core particle binding protein that plays a role in chromatin remodeling [51].

On the other hand, bisphenol-A is a non-steroidal estrogen that is ubiquitous in the environment. Methylation of the HOXA10 gene was decreased in the reproductive tract of mice exposed *in utero* to bisphenol-A [52]. Decreased DNA methylation led to an increase in binding of ERα to the HOXA10 promoter, and to increased estrogen-dependent transcription. Permanent epigenetic alteration of sensitivity to estrogen may then be a mechanism through which endocrine disruptors exert their action. The demonstration that many estrogenic compounds show lifelong effects in animals has raised concern that fetal and neonatal exposure to these compounds in humans could also produce epigenetic changes and impact negatively on human health. For instance, human fetuses can be exposed to high estrogen levels due to unintentional continuation of birth control pill intake by the mother before detection of pregnancy. On the other hand, exposure of the mother to environmental estrogen disruptors or intake of high levels of phytoestrogens can affect the fetus and even the infant during breastfeeding, with possible adverse consequences [57].

10.7 OBESITY AND TYPE 2 DIABETES

There is growing evidence that maternal nutritional status can alter the epigenetic state (stable alterations of gene expression through DNA methylation and histone modifications) of the fetal genome [2]. This may provide a molecular mechanism for the impact of maternal nutrition on both fetal programming and genomic imprinting. Promoting

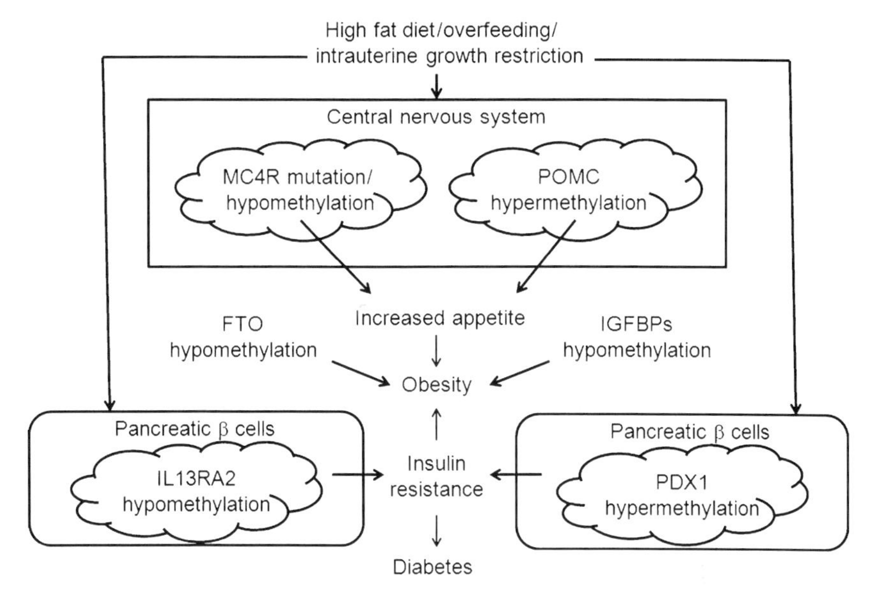

FIGURE 10.12 The altered epigenetic control of appetite and specific gene hypo- or hypermethylation are involved in the development of insulin resistance, type 2 diabetes, and obesity in response to malnutrition or overfeeding.

optimal nutrition will not only ensure optimal fetal development, but also reduce the risk of chronic diseases in adults. In Chapter 3 we mentioned that obesity and/or type 2 diabetes are associated with hypomethylation of IGFBP1 [5], IGFBP7 [6], and FTO [15–17] (Figure 10.12).

10.7.1 Appetite Regulators

A study examining MC4R, mutations of which are associated with obesity, showed that when adult mice were fed a high-fat diet (HFD) this receptor became hypomethylated [58], leading to expression of the gene, and increased appetite and feeding behavior. On the other hand, neonatal mice that were overfed developed hypermethylation of the satiety-mediator POMC in the hypothalamus, whose underexpression is associated with onset of obesity.

10.7.2 Insulin

Insulin plays a key role in metabolic control, and therefore regulation of insulin gene expression has been extensively studied. The insulin promoter is demethylated specifically in pancreatic β cells both in humans and in mice. The insulin gene is methylated in mouse embryonic stem cells, and becomes demethylated on differentiation into insulin-expressing cells. Methylation of a specific CpG located in a cAMP response element (CRE) of the insulin promoter inhibits association of the transcription factors CREB and ATF2 that bind to the CRE, while inducing MeCP2 recruitment, leading to a strong reduction of promoter activity [59]. Therefore, promoter demethylation may play an important role in the specific expression of the insulin gene in pancreatic β cells.

Abnormal nutrition during embryonic development has been shown to influence disease susceptibility in the descendants. The global prevalence of obesity and type 2 diabetes is increasing, and parent obesity is a risk factor for developing obesity in childhood [60]. In rodents it has been shown that when mothers are fed with a high-fat diet (HFD), male offspring exhibit increased body weight and are diabetic and insulin-resistant. Furthermore, the offspring of these males also present insulin resistance, showing that fathers can start intergenerational inheritance of metabolic diseases [61]. Accordingly, paternal HFD alters gene expression in pancreatic β cells of adult female offspring. The IL13RA2 gene, a gene belonging to the JAK–STAT signaling pathway, presented the highest difference in gene expression [7]. An epigenetic mechanism appears to contribute to the altered IL13RA2 expression, since methylation at CpG−960 was reduced in HFD offspring with respect to controls. This CpG is located in a putative recognition site for the transcription factor TCF-1A and for the methylated DNA binding protein NF-X. These results show that paternal HFD could affect metabolism of the offspring by epigenetic regulation of genes important for pancreatic β cell function.

Intrauterine growth restriction (IUGR) also increases susceptibility to age-related diseases, including type 2 diabetes. In a rodent model of IUGR, which develops diabetes in adulthood, global decreases in DNA methylation concomitant with a decrease in DNMT1, MeCP2, and HDAC1 is observed in tissues such as liver or brain [11]. Furthermore, it was found that expression of PDX1, a pancreatic and duodenal homeobox 1 transcription factor critical for β cell function and development, was permanently reduced in IUGR β cells and underwent epigenetic modifications: throughout development there were epigenetic histone modifications, but after the onset of diabetes in adulthood the CpG island in the proximal PDX1 promoter was methylated, resulting in permanent silencing of this locus [12].

10.7.3 Metabolic Regulators

A classic demonstration of epigenetic heritability is the agouti viable yellow (A^{vy}) locus in mice [62]. Expression of this gene, which is

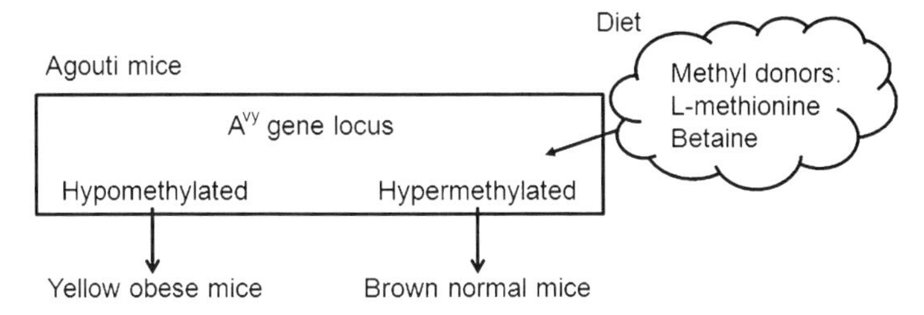

FIGURE 10.13 Epigenetic regulation of the obese phenotype in Agouti mice.

associated with a bright yellow coat and obese phenotype, differs depending on DNA methylation and histone modifications. A diet rich in methyl-donating compounds such as folate and betaine increases methylation of the locus and silences the gene expression. An obese phenotype can be transferred across two generations of mice in response to a high-fat gestational diet [61,63]. In the F3 generation, only females maintained the increased body size phenotype, which was shown to be carried in the paternal lineage, highlighting an imprinting mechanism in the heritability of obesity (Figure 10.13).

Low birth weight and unhealthy diet are risk factors for metabolic diseases. It can even be expected that offspring of males fed a low-protein diet also exhibit metabolic disturbances. Indeed, epigenomic profiling of offspring livers reveals changes in cytosine methylation depending on paternal diet, including reproducible changes in DNA methylation of PPARα, a key lipid regulator [10]. High-fat diets increased plasma triglycerides, trigger insulin resistance, and alter the DNA methylation pattern of the leptin [8] as well as the NDUFB6 promoters [9]. These results indicate that parental diet can affect cholesterol and lipid metabolism in offspring, and define a model system to study environmental reprogramming of the heritable epigenome. Genetic, non-genetic, and epigenetic data propose a role of the key metabolic regulator PPARγ coactivator 1α (PPARγC1α) in the development of type 2 diabetes (Figure 10.14). When challenged with high-fat overfeeding, low birth weight subjects developed insulin resistance and reduced PPARγC1α and OXPHOS gene expression. PPARγC1α methylation was significantly higher in low birth weight subjects during the control diet. However, PPARγC1α methylation increased in only normal birth weight subjects after overfeeding. These changes are reversible, supporting the idea that DNA methylation induced by overfeeding is reversible in humans [14]. Moreover, in a very recent study, it has been shown that the increased risk of type 2 diabetes in the offspring of malnourished mothers is associated with the decreased expression of the orphan nuclear receptor HNF4α, previously linked with this type of diabetes. Specifically, a pancreas-specific enhancer of HNF4α expression is epigenetically inactivated by DNA methylation in the adult

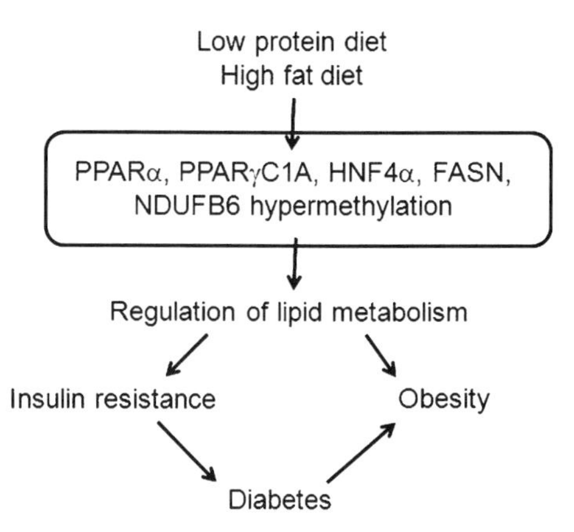

FIGURE 10.14 Altered epigenetic control of metabolic regulators triggered by malnutrition and overfeeding.

offspring of poorly nourished mothers, revealing a novel mechanism by which maternal diet and aging interact through epigenetic processes to determine the risk of age-associated endocrine diseases [13].

10.8 CONCLUSION

Epigenetic mechanisms play a key role in normal endocrine physiology, as well as in the development of endocrine diseases such as type 2 diabetes and obesity. A better knowledge of the association between epigenetic mechanisms and neuroendocrine function should lead to a better understanding of the molecular basis of these disorders, and could help in the development of novel therapeutic strategies. DNA methylation acts at many levels to regulate hormonal actions: during development, in response to environmental factors and endocrine disruptors, in endocrine cancer, endocrine therapies, etc. It is clear that the development of obesity is not a simple overshooting of calories in versus calories out, rather a much more complex sum of many coexisting factors. Substantial evidence exists showing links between epigenetic events and regulation of metabolic pathways leading to obesity and type 2 diabetes (Chapter 11). The significance of the timing of such events cannot be overlooked, particularly in fetal development, in which exposure to an undersupply or oversupply of energy and some nutrients results in altered chromatin structuring and DNA modification, thereby influencing expression of key genes. Although many aspects of endocrine system epigenetics are still unknown, the observations in animals described here shed a new light on the importance of this epigenetic modification in the functioning of the endocrine glands and in the response of target tissues to hormones.

References

[1] Garcia-Carpizo V, Ruiz-Llorente L, Fraga M, Aranda A. The growing role of gene methylation on endocrine function. J Mol Endocrinol 2011;47:E75–89.
[2] Wu G, Bazer FW, Cudd TA, Meininger CJ, Spencer TE. Maternal nutrition and fetal development. J Nutr 2004;134:2169–72.
[3] DeBaun MR, Niemitz EL, McNeil DE, Brandenburg SA, Lee MP, Feinberg AP. Epigenetic alterations of H19 and LIT1 distinguish patients with Beckwith-Wiedemann syndrome with cancer and birth defects. Am J Hum Genet 2002;70:604–11.
[4] Lalande M. Imprints of disease at GNAS1. J Clin Invest 2001;107:793–4.
[5] Gu T, Gu HF, Hilding A, Sjoholm LK, Ostenson CG, Ekstrom TJ, et al. Increased DNA methylation levels of the insulin-like growth factor binding protein 1 gene are associated with type 2 diabetes in Swedish men. Clin Epigenetics 2013;5:21.
[6] Gu HF, Gu T, Hilding A, Zhu Y, Karvestedt L, Ostenson CG, et al. Evaluation of IGFBP-7 DNA methylation changes and serum protein variation in Swedish subjects with and without type 2 diabetes. Clin Epigenetics 2013;5:20.
[7] Ng SF, Lin RCY, Laybutt DR, Barres R, Owens JA, Morris MJ. Chronic high-fat diet in fathers programs beta-cell dysfunction in female rat offspring. Nature 2010;467:963 U103.
[8] Milagro FI, Campion J, Garcia-Diaz DF, Goyenechea E, Paternain L, Martinez JA. High fat diet-induced obesity modifies the methylation pattern of leptin promoter in rats. J Physiol Biochem 2009;65:1–9.
[9] Ling C, Poulsen P, Simonsson S, Rönn T, Holmkvist J, Almgren P, et al. Genetic and epigenetic factors are associated with expression of respiratory chain component NDUFB6 in human skeletal muscle. J Clin Invest 2007;117:3427–35.
[10] Carone BR, Fauquier L, Habib N, Shea JM, Hart CE, Li R, et al. Paternally induced transgenerational environmental reprogramming of metabolic gene expression in mammals. Cell 2010;143:1084–96.
[11] Lillycrop KA. Symposium 1: Nutrition and epigenetics Effect of maternal diet on the epigenome: implications for human metabolic disease. Proc Nutr Soc 2011;70:64–72.
[12] Park JH, Stoffers DA, Nicholls RD, Simmons RA. Development of type 2 diabetes following intrauterine growth retardation in rats is associated with progressive epigenetic silencing of Pdx1. J Clin Invest 2008;118:2316–24.
[13] Sandovici I, Smith NH, Nitert MD, Ackers-Johnson M, Uribe-Lewis S, Ito Y, et al. Maternal diet and aging alter the epigenetic control of a promoter-enhancer interaction at the Hnf4a gene in rat pancreatic islets. Proc Natl Acad Sci USA 2011; 108:5449–54.
[14] Brons C, Jacobsen S, Nilsson E, Ronn T, Jensen CB, Storgaard H, et al. Deoxyribonucleic acid methylation and gene expression of PPARGC1A in human muscle is influenced by high-fat overfeeding in a birth-weight-dependent manner. J Clin Endocrinol Metab 2010;95:3048–56.
[15] Toperoff G, Aran D, Kark JD, Rosenberg M, Dubnikov T, Nissan B, et al. Genome-wide survey reveals predisposing diabetes type 2-related DNA methylation variations in human peripheral blood. Hum Mol Genet 2012;21:371–83.
[16] Bell CG, Finer S, Lindgren CM, Wilson GA, Rakyan VK, Teschendorff AE, et al. Integrated genetic and epigenetic analysis identifies haplotype-specific methylation in the FTO type 2 diabetes and obesity susceptibility locus. PLoS One 2010;5: e14040.
[17] Almen MS, Jacobsson JA, Moschonis G, Benedict C, Chrousos GP, Fredriksson R, et al. Genome wide analysis reveals association of a FTO gene variant with epigenetic changes. Genomics 2012;99:132–7.
[18] Weaver ICG, D'Alessio AC, Brown SE, Hellstrom IC, Dymov S, Sharma S, et al. The transcription factor nerve growth factor-inducible protein A mediates epigenetic

programming: Altering epigenetic marks by immediate-early genes. J Neurosci 2007; 27:1756–68.
[19] McGowan PO, Sasaki A, D'Alessio AC, Dymov S, Labonte B, Szyf M, et al. Epigenetic regulation of the glucocorticoid receptor in human brain associates with childhood abuse. Nat Neurosci 2009;12:342–8.
[20] Lee RS, Tamashiro KL, Yang X, Purcell RH, Harvey A, Wilour VL, et al. Chronic corticosterone exposure increases expression and decreases deoxyribonucleic acid methylation of Fkbp5 in mice. Endocrinology 2010;151:4332–43.
[21] Murgatroyd C, Patchev AV, Wu Y, Micale V, Bockmuhl Y, Fischer D, et al. Dynamic DNA methylation programs persistent adverse effects of early-life stress. Nat Neurosci 2009;12:1559 U1108.
[22] Bogdarina I, Welham S, King PJ, Burns SP, Clark AJL. Epigenetic modification of the renin-angiotensin system in the fetal programming of hypertension. Circ Res 2007;100:520–6.
[23] Bogdarina I, Haase A, Langley-Evans S, Clark AJL. Glucocorticoid effects on the programming of AT1b angiotensin receptor gene methylation and expression in the rat. PLoS One 2010;5:e9237.
[24] Newell-Price J. Proopiomelanocortin gene expression and DNA methylation: implications for Cushing's syndrome and beyond. J Endocrinol 2003;177:365–72.
[25] Champagne FA, Weaver IC, Diorio J, Dymov S, Szyf M, Meaney MJ. Maternal care associated with methylation of the estrogen receptor-alpha1b promoter and estrogen receptor-alpha expression in the medial preoptic area of female offspring. Endocrinology 2006;147:2909–15.
[26] Kurian JR, Olesen KM, Auger AP. Sex differences in epigenetic regulation of the estrogen receptor-alpha promoter within the developing preoptic area. Endocrinology 2010;151:2297–305.
[27] Schwarz JM, Nugent BM, McCarthy MM. Developmental and hormone-induced epigenetic changes to estrogen and progesterone receptor genes in brain are dynamic across the life span. Endocrinology 2010;151:4871–81.
[28] Auger CJ, Coss D, Auger AP, Forbes-Lorman RM. Epigenetic control of vasopressin expression is maintained by steroid hormones in the adult male rat brain. Proc Natl Acad Sci USA 2011;108:4242–7.
[29] Zhao ZR, Fan L, Frick KM. Epigenetic alterations regulate estradiol-induced enhancement of memory consolidation. Proc Natl Acad Sci USA 2010;107:5605–10.
[30] Bienvenu T, Chelly J. Molecular genetics of Rett syndrome: when DNA methylation goes unrecognized (vol 7, pg 415, 2006). Nat Rev Genet 2006;7: 583.
[31] Kress C, Thomassin H, Grange T. Active cytosine dernethylation triggered by a nuclear receptor involves DNA strand breaks. Proc Natl Acad Sci USA 2006;103:11112–17.
[32] Kangaspeska S, Stride B, Metivier R, Polycarpou-Schwarz M, Ibberson D, Carmouche RP, et al. Transient cyclical methylation of promoter DNA. Nature 2008;452:112. U114.
[33] Metivier R, Gallais R, Tiffoche C, Le Peron C, Jurkowska RZ, Carmouche RP, et al. Cyclical DNA methylation of a transcriptionally active promoter. Nature 2008;452:45 U42.
[34] Kim MS, Kondo T, Takada I, Youn MY, Yamamoto Y, Takahashi S, et al. DNA demethylation in hormone-induced transcriptional derepression (Retracted article; see Vol. 486, pp. 280, 2012). Nature 2009;461:1007–12.
[35] Stankiewicz AM, Swiergiel AH, Lisowski P. Epigenetics of stress adaptations in the brain. Brain Res Bull 2013;98:76–92.

[36] Elliott E, Ezra-Nevo G, Regev L, Neufeld-Cohen A, Chen A. Resilience to social stress coincides with functional DNA methylation of the Crf gene in adult mice. Nat Neurosci 2010;13:1351–3.
[37] Sterrenburg L, Gaszner B, Boerrigter J, Santbergen L, Bramini M, Roubos EW, et al. Sex-dependent and differential responses to acute restraint stress of corticotropin-releasing factor-producing neurons in the rat paraventricular nucleus, central amygdala, and bed nucleus of the stria terminalis. J Neurosci Res 2012;90:179–92.
[38] Murayama A, Kim MS, Yanagisawa J, Takeyama K, Kato S. Transrepression by a liganded nuclear receptor via a bHLH activator through co-regulator switching. EMBO J 2004;23:1598–608.
[39] Chertkow-Deutsher Y, Cohen H, Klein E, Ben-Shachar D. DNA methylation in vulnerability to post-traumatic stress in rats: evidence for the role of the post-synaptic density protein Dlgap2. Int J Neuropsychopharmacol 2010;13:347–59.
[40] Malik RE, Rhodes SJ. The role of DNA methylation in regulation of the murine Lhx3 gene. Gene 2014;534:272–81.
[41] Roth TL, Lubin FD, Funk AJ, Sweatt JD. Lasting epigenetic influence of early-life adversity on the BDNF gene. Biol Psychiatry 2009;65:760–9.
[42] Heijmans BT, Tobi EW, Stein AD, Putter H, Blauw GJ, Susser ES, et al. Persistent epigenetic differences associated with prenatal exposure to famine in humans. Proc Natl Acad Sci USA 2008;105:17046–9.
[43] Wu Y, Patchev AV, Daniel G, Almeida OF, Spengler D. Early-life stress reduces DNA methylation of the Pomc gene in male mice. Endocrinology 2014;155:1751–62.
[44] Weaver IC, Cervoni N, Champagne FA, D'Alessio AC, Sharma S, Seckl JR, et al. Epigenetic programming by maternal behavior. Nat Neurosci 2004;7:847–54.
[45] Weaver IC, Champagne FA, Brown SE, Dymov S, Sharma S, Meaney MJ, et al. Reversal of maternal programming of stress responses in adult offspring through methyl supplementation: altering epigenetic marking later in life. J Neurosci 2005;25:11045–54.
[46] McGowan PO, Sasaki A, D'Alessio AC, Dymov S, Labonte B, Szyf M, et al. Epigenetic regulation of the glucocorticoid receptor in human brain associates with childhood abuse. Nat Neurosci 2009;12:342–8.
[47] Tyrka AR, Price LH, Marsit C, Walters OC, Carpenter LL. Childhood adversity and epigenetic modulation of the leukocyte glucocorticoid receptor: preliminary findings in healthy adults. PLoS One 2012;7:e30148.
[48] Mccarthy MM. Estradiol and the developing brain. Physiol Rev 2008;88:91–124.
[49] Newbold RR, Hanson RB, Jefferson WN, Bullock BC, Haseman J, McLachlan JA. Increased tumors but uncompromised fertility in the female descendants of mice exposed developmentally to diethylstilbestrol. Carcinogenesis 1998;19:1655–63.
[50] Newbold RR, Padilla-Banks E, Jefferson WN. Adverse effects of the model environmental estrogen diethylstilbestrol are transmitted to subsequent generations. Endocrinology 2006;147:S11–17.
[51] Tang WY, Newbold R, Mardilovich K, Jefferson W, Cheng RY, Medvedovic M, et al. Persistent hypomethylation in the promoter of nucleosomal binding protein 1 (Nsbp1) correlates with overexpression of Nsbp1 in mouse uteri neonatally exposed to diethylstilbestrol or genistein. Endocrinology 2008;149:5922–31.
[52] Bromer JG, Zhou YP, Taylor MB, Doherty L, Taylor HS. Bisphenol-A exposure *in utero* leads to epigenetic alterations in the developmental programming of uterine estrogen response. FASEB J 2010;24:2273–80.
[53] Anway MD, Skinner MK. Transgenerational effects of the endocrine disruptor vinclozolin on the prostate transcriptome and adult onset disease. Prostate 2008; 68:517–29.

[54] Cowin PA, Gold E, Aleksova J, O'Bryan MK, Foster PM, Scott HS et al. Vinclozolin exposure in utero induces postpubertal prostatitis and reduces sperm production via a reversible hormone-regulated mechanism. Endocrinology **151**: 783–792.

[55] Guerrero-Bosagna C, Settles M, Lucker B, Skinner MK. Epigenetic transgenerational actions of vinclozolin on promoter regions of the sperm epigenome. PLoS One 2010; 5:e13100.

[56] Li SF, Washburn KA, Moore R, Uno T, Teng C, Newbold RR, et al. Developmental exposure to diethylstilbestrol elicits demethylation of estrogen-responsive lactoferrin gene in mouse uterus. Cancer Res 1997;57:4356–9.

[57] Prins GS. Estrogen imprinting: when your epigenetic memories come back to haunt you. Endocrinology 2008;149:5919–21.

[58] Widiker S, Karst S, Wagener A, Brockmann GA. High-fat diet leads to a decreased methylation of the Mc4r gene in the obese BFMI and the lean B6 mouse lines. J Appl Genet 2010;51:193–7.

[59] Kuroda A, Rauch TA, Todorov I, Ku HT, Al-Abdullah IH, Kandeel F, et al. Insulin gene expression is regulated by DNA methylation. PLoS One 2009;4:e6953.

[60] Whitaker RC, Wright JA, Pepe MS, Seidel KD, Dietz WH. Predicting obesity in young adulthood from childhood and parental obesity. New Engl J Med 1997;337:869–73.

[61] Dunn GA, Bale TL. Maternal high-fat diet promotes body length increases and insulin insensitivity in second-generation mice. Endocrinology 2009;150:4999–5009.

[62] Delaney C, Garg SK, Fernandes C, Hoeltzel M, Allen RH, Stabler S, et al. Maternal diet supplemented with methyl-donors protects against atherosclerosis in F1 ApoE (−/−) mice. PLoS One 2013;8:e56253.

[63] Dunn GA, Bale TL. Maternal high-fat diet effects on third-generation female body size via the paternal lineage. Endocrinology 2011;152:2228–36.

CHAPTER

11

DNA Methylation in Metabolic Diseases

OUTLINE

11.1 INTRODUCTION

It is known that the gestational environment has direct implications for the phenotype of the offspring without making changes to the fetal genome [1]. This is supported by data showing that children born to obese women before bariatric surgery are at higher risk for obesity

M. Neidhart: DNA Methylation and Complex Human Disease.
DOI: http://dx.doi.org/10.1016/B978-0-12-420194-1.00011-7

than those born after such surgical intervention. Epidemiological studies in the past two decades in the UK, USA, India, and China have pointed to the phenomenon that both very low and very high birth weights correlate with increased risk of obesity and metabolic syndrome diseases (cardiovascular disease, diabetes, dyslipidemia). The period of embryonic development has been recognized as a critical window in the establishment of the epigenome. In Chapter 10, we presented evidence in animal models that an adverse prenatal and early postnatal environment can increase obesity risk in later life. This has led to the search for nutritional interventions during pregnancy and lactation that have the potential to mitigate or overcome this adverse programming [2]. The importance of the *in utero* environment is also highlighted succinctly in a study of survivors of the Dutch Hunger Winter [1]. Individuals exposed to famine conditions while *in utero* displayed a higher incidence of coronary heart disease and dyslipidemia in later life. Diet and weight loss interventions in obese mothers may lead to a decreased risk of obesity in the offspring, possibly mediated through changes in insulin signaling, fat storage, energy expenditure, or appetite control pathways. Epigenetic mechanisms are likely to have a role in this altered risk profile, and the findings of obesity-associated methylation marks in genes involved in these processes also support this hypothesis. Human studies showing a direct relationship between specific prenatal nutritional exposures on methylation profiles of the offspring and subsequent risk of obesity in later life are scarce. However, there are a number of studies that have assessed differences in methylation of candidate genes in children in relation to maternal/paternal characteristics [3–6] or have explored relationships between epigenetic markers in the cord blood at delivery and obesity/metabolic outcomes in childhood – for example, methylations of CASP10, CDKN1C, EPHA1, HLA-DOB, NID2, MMP9, and MPL, and obesity, at age 10 [7]. Another study [8] compared methylation profiles of siblings born before and after maternal weight loss surgery; it reported differences between the siblings in obesity characteristics and in methylation profiles for genes involved in the regulation of glucose homeostasis and immune function, some of which translated into alterations in gene expression and insulin sensitivity. Although this was a small study, its findings suggest that significant weight loss and, presumably, improved metabolic health profiles in the mother are associated with a distinct epigenome and lower weight and waist circumference in the children. This review also discusses recent progress in understanding how diet and cellular metabolism affects the epigenome, and how epigenetic factors regulate energy metabolism. We will consider the nutritional aspects of lifestyle-associated diseases.

11.2 OBESITY

Obesity is a major health problem that is determined by interactions between lifestyle and environmental and genetic factors. Overall, the available global methylation studies in obesity do not provide consistent evidence for a relationship between global methylation and obesity [9]. Multiple studies, however, have used a hypothesis-driven candidate gene approach, where methylation sites in (or near) known candidate genes for obesity susceptibility have been the subject of investigation.

11.2.1 Multiple Promoter Methylation Alterations

In some cases, the choice of genes has been based on prior analysis of gene expression differences in the same subjects. Candidate gene methylation studies have focused on a range of genes implicated in obesity, appetite control and/or metabolism, insulin signaling, immunity, growth, circadian clock regulation, and imprinted genes, and assessed their relationship with a variety of obesity markers. Collectively, these studies have identified lower methylation of tumor necrosis factor alpha (TNF-α) in peripheral blood leukocytes [10], pyruvate dehydrogenase kinase 4 (PDK4) in muscle [11], and leptin (LEP) in whole blood [12], and increased methylation of pro-opiomelanocortin (POMC) in whole blood [13], PPARγ coactivator 1 alpha (PGC1α) [11] in muscle, and CLOCK as well as aryl hydrocarbon receptor nuclear translocator-like (BMAL1) [14] genes in peripheral blood leukocytes, in obese compared with lean individuals. Associations between body mass index, adiposity, and waist circumference, and methylation in PDK4 in muscle [11], melanin-concentrating hormone receptor 1 (MCHR1) in whole blood [15], and the serotonin transporter (SLC6A4) gene [16], the androgen receptor (AR) [17,11] β-hydroxysteroid dehydrogenase type 2 (HSD2) [18], period circadian clock 2 (PER2) [14] and glucocorticoid receptor (GR) [18] in peripheral blood leukocytes, have also been reported. The most consistently observed epigenetic association has been that of methylation at the IGF2/H19 imprinting region in blood cells with measures of adiposity [18,19]. Studies of offspring born to women during the Dutch famine in the 1940s suggested a link between the fetal environment (including nutrition) and postnatal health, particularly cardiovascular function or dysfunction, in humans [20]. Interestingly, 60 years after birth, offspring with early prenatal experience of the famine exhibited less methylation of the imprinted IGF2 gene as compared to same-gender siblings without exposure to prenatal malnutrition [21]. Furthermore, individuals with periconceptional exposure to the famine had lower methylation of the INSIGF gene but higher methylation levels for several other genes (IL10,

LEP, ABCA1, GNASAS, and MEG3) [22]. These genes are closely linked with nutrient metabolism, cardiovascular function, and inflammation. Interactions between undernutrition and sex also affect methylation of the INSIGF, LEP, and GNASAS genes. Collectively, these studies provide evidence that obesity and other metabolic diseases are associated with altered epigenetic regulation of a number of metabolically important genes, some of which are inherited.

In a separate group of studies, comparison of the methylation profiles of people who successfully lost weight during interventions and those who did not has been used in order to determine whether there may be biomarkers that predict individual responsiveness to weight loss interventions. After weight loss surgery, promoter methylation of PCG1α decreased and of PDK4 increased in obese women, to levels comparable to lean women [11]. Comparing low and high responders to 8-week caloric restriction, differences were found for LEP- and TNFA-promoter methylation [23]. Similarly, comparing the response to 10-week diet and exercise weight loss intervention, important differences were revealed in, for example, AQP9, DUSP22, HIPK3, TNNT1, and TNNI3 [24]. Another study compared methylation at baseline and after 16 weeks of weight loss intervention, and found differences for CLOCK and PER2 [24].

Obesity-associated differentially methylated sites in peripheral blood cells were also detected in genome-wide studies, showing, for example, an association between FTO methylation and obesity [25]. A more recent study [26] has reported significant associations between methylation at three probes targeting specific CpG sites within intron 1 of HIF3A and body mass index in a discovery cohort, and subsequently confirmed them in two independent cohorts. HIF3A encodes a component of the hypoxia inducible transcription factor that mediates the cellular response to hypoxia by regulating expression of many downstream genes. This transcription factor has been previously implicated in metabolism and obesity [27].

11.2.2 Epigenetic Prognostic Biomarkers

It has long been assumed that DNA methylation profiles would remain stable throughout adult life; however, this view is now changing. Interventions such as exercise, diets, and weight loss surgery have been shown to modulate methylation profiles in different tissue types [9,11]. Interestingly, methylation profiles of obese individuals became more similar to those of lean individuals following weight loss surgery [11]. Although this was only demonstrated in a small study, it suggests that methylation profiles of obese individuals can be modified by reductions in body weight/fat mass. This conclusion may imply that some methylation marks are a consequence of the obese phenotype, rather than a programmed mark that predisposes people to become obese. These findings again highlight the importance of studies in which methylation marks are measured early in life before disease manifests, to define which acquired marks become permanent, and thus potential early markers for disease risk, and which are transient and modifiable in later life.

Several studies have explored the association of DNA methylation at birth with adiposity in later life. Methylation variation in the promoter of the retinoid X receptor alpha gene (RxRα) in umbilical cord tissue was found to explain up to 26% of the variation in childhood adiposity [28]. DNA methylation also appears to be important in the regulation of peroxisome proliferator-activated receptors γ (PPARγ) [29] and PPARγ coactivator 1 alpha (PGC1α) [11,30,31], which interact with each other. This system is involved in the epigenome–metabolism crosstalk (see below).

Variation in methylation within tumor-associated calcium signal transducer 2 (TACST2) at birth was also found to correlate with fat mass in later life; however, further analysis including single nuclear polymorphism data of this gene showed that reverse causation or confounding was likely to account for the observed correlation [32]. IGF2 is another example of a gene showing loci-specific variation in methylation at birth, and also at childhood, that is associated with growth characteristics and obesity in later life [18,19,33]. The epigenetic regulation of IGF2 has been of particular interest, given its role in control of fetal growth and development. Differences in the degree of methylation near IGF2 have often been linked to exposure to a suboptimal environment *in utero* [18,21,22,34].

The finding of an association between variation in matrix metallopeptidase 9 (MMP9) methylation levels at birth and childhood adiposity [7] is also of interest, given the critical role that metalloproteinases have in extra cellular matrix remodeling during adipose tissue formation, and coupled with the fact that altered MMP9 plasma levels and gene expression has previously been found in obese individuals [35]. Moreover, variation in methylation near MMP9, and another metalloproteinase called PM20D1, was associated with body mass index in a genome-wide study at two time points 11 years apart [36]. These findings show that these marks are most likely established at an early age and may be associated with adiposity at different stages in later life, which suggests that these methylation changes could be potentially useful to predict obesity risk from an early age.

11.3 TYPE 2 DIABETES

Diabetes is a multifactorial disease, with numerous pathways influencing its progression. In all forms of the disease, the diabetic milieu is characterized by hyperglycemia and a relative or absolute lack of insulin signaling.

Despite strong familial clustering of type 2 diabetes and identification of several common variants, genome-wide association studies provide limited capacity to predict the disease. Thus the complexity of type 2 diabetes cannot be entirely accounted for by genetic changes. The observed familial clustering may be attributable to a significant epigenetic component [37]. For instance, gestational metabolic programming by environmental cues is an important determinant of type 2 diabetes predisposition. This is exemplified by follow-up studies of Pima Native Americans that demonstrate significantly higher predisposition to type 2 diabetes for individuals born to diabetic mothers than in offspring of non-diabetics and, importantly, siblings born prior to maternal diagnosis. Second, childhood and adult lifestyle factors are strongly associated with type 2 diabetes risk, and a shared family environment including dietary factors and exercise could promote common environmentally derived epigenetic changes. Early clinical findings demonstrated a strong correlation between low birth weight and later incidence of the disease.

11.3.1 Pancreatic Islet Dysfunction

In intrauterine growth retardation, expression of pancreatic and duodenal homeobox 1 (PDX1) is silenced by DNA methylation-driven transcriptional repression during transition from IUGR to diabetes in adults [38]. PDX1 is a transcription factor that functions as a critical regulator of β-cell differentiation, growth, and insulin secretion. Changes in PDX1 are linked to type 2 diabetes and β-cell dysfunction in human and animal models [39,40].

Compared to healthy control subjects, DNA hypermethylation at the peroxisome proliferator-activated receptor-γ coactivator 1α (PPARGC1A) promoter and concomitant transcriptional repression was observed in pancreatic islets isolated from patients with type 2 diabetes [31]. Genome-wide promoter methylation analysis of skeletal muscle from patients with type 2 diabetes revealed similar observations [41]. Expression of this gene is linked with islet insulin secretion, and a recently described polymorphism in the coding region is associated with increased risk for type 2 diabetes [42]. Combined with the aforementioned relation of PDX1 promoter hypermethylation to islet dysfunction and development [38], this observation suggests a link between DNA methylation and pancreatic dysfunction in diabetes.

11.3.2 Adipocyte Dysfunction

Dutch and Chinese middle-aged men exposed to maternal starvation at developmentally crucial periods of gestation were found to have much higher rates of type 2 diabetes [37]. As mentioned in the introduction, some evidence suggests that DNA methylation may link prenatal famine to diabetes through decreased methylation of the imprinted IGF2 gene [21] and sex-specific hypermethylation of GNASAS and LEP [43]. GNASA encodes for results in different forms (by altenative splicing) of the stimulatory G-protein alpha subunit, a key element of the classical signal transduction pathway linking receptor–ligand interactions with the activation of adenylyl cyclase (converting ATP into cyclic AMP) and a variety of cellular responses. It is well known that the secretion of leptin by adipose tissue is controlled by epigenetic mechanisms. Specifically, the LEP promoter that contains a CpG island has a high degree of methylation in preadipocytes [44], and the promoter is demethylated in association with induction of leptin expression in differentiated adipocytes [45]. Furthermore, a high-fat diet increases DNA methylation of one CpG site in the LEP promoter of adult adipocytes [46].

11.3.3 Mitochondrial Dysfunction

The reduction of fatty acid β-oxidation is one of the important features of mitochondrial dysfunction in the skeletal muscle, which is closely associated with the insulin resistance in metabolic diseases. Interestingly, methylation of the cytochrome c oxidase polypeptide 7A1 (COX7A1) promoter is increased in skeletal muscle of diabetic animals, and its gene transcription is reduced in association with reduced glucose uptake by muscle [47]. COX7A1 is part of complex 4 of the mitochondrial respiratory chain.

11.4 EPIGENOME-METABOLISM CROSSTALK

Obesity and consequent insulin resistance represent typical examples of a cellular failed energy strategy. It is well documented that nutrient- and hormone-driven transcription regulators greatly contribute to the disrupted metabolic gene expression under caloric excess [48]. In the following section we will describe in detail an example of epigenome-metabolism crosstalk, including links between the methionine cycle, retinoid-related pathway and adipogenesis.

11.4.1 Methionine Cycle and Folate

The methionine cycle is responsible for the synthesis of S-adenosylmethionine (AdoMet). Figure 11.1 presents the relationship between methionine, AdoMet, folate, DNA methylation, and increased production of triglycerides. AdoMet-dependant methylation reactions are required for post-translational modifications of proteins (e.g., histone

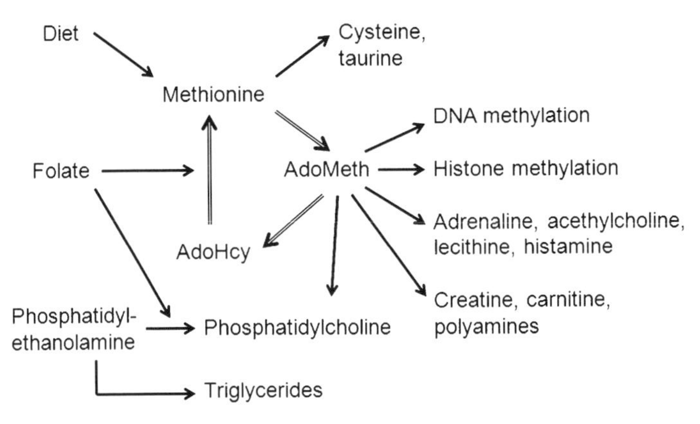

FIGURE 11.1 Relationship between the methionine cycle (with AdoMeth), DNA methylation, folate, biosynthesis of neurotransmitters, small molecules (such as polyamines), and triglycerides.

methylation), methylation of DNA, as well as the synthesis of hormones and other small molecules, including creatine, carnitine, phosphatidylcholine, and polyamines. Impaired methylation capacity can be caused by either a decrease in AdoMet or an increase in S-adenosylhomocysteine (AdoHcy), a competitive inhibitor of many methyltransferase reactions [49]. Importantly, the nutritional and physiological state of the animal will alter the production of these bioactive substances, thus regulating the availability of AdoMeth. When cysteine or taurine is deficient in the diet, their synthesis from methionine will be increased *in vivo*, thus decreasing total AdoMeth availability for DNA or protein methylation. Inadequate synthesis of glycine and serine, coupled with low supplies from the diet, can also impair the methionine cycle. In baboons, a 30% reduction in maternal nutrient intake did not alter fetal weight at stages 0.5 and 0.9 of gestation (full term being 1.0). However, tissue-specific global methylation status [50] and the availability of AdoMeth [51] were altered in the kidneys of the monkeys at both 0.5 and 0.9 of gestation. As a result, such a relatively mild nutrient restriction can induce alterations in gene expression and provide a potential mechanism responsible for the increased incidence of endothelial dysfunction, renal dysfunction, and hypertension in offspring [20]. Furthermore, an important function of folate is in the maintenance of cellular AdoMet and AdoHcy concentrations. Folate deficiency decreases flux through phosphatidylethanolamine N-methyltransferase, an enzyme that synthesizes phosphatidylcholine via the methylation of phosphatidylethanolamine [52]. Thus, folate deficiency reduces *de novo* phosphatidylcholine synthesis, resulting in accumulation of triglycerides. Changes in AdoMet and folate status are observed in obesity and metabolic syndromes [49].

11.4.2 Retinoid X Receptor Alpha

As mentioned above, methylation of retinoid X receptor alpha (RxRα) is a prognostic marker for childhood adiposity [28]. RxRα is a nuclear receptor with a known role in adipogenesis; it forms a heterodimer with the transcription factor PPARγ to activate transcription of genes involved in adipocyte differentiation, glucose metabolism, inflammation, and energy homeostasis. Interestingly, hepatocyte RXRα-deficient mice have significant lower levels of AdoMeth [53]. It is known that retinoid-mediated pathways can regulate the synthesis of AdoMeth [54]. For example, excessive retinol consumption by rats affects transmethylation reactions that utilize AdoMeth.

11.4.3 Peroxisome Proliferator-Activated Receptor γ

All peroxisome proliferator-activated receptors (PPARs) heterodimerize with RxRs and bind to specific regions on the DNA of target genes. These DNA sequences are termed PPREs (peroxisome proliferator hormone response elements). PPARγ is a key regulator of adipogenesis and has been shown to promote lipid storage and contribute to hepatic steatosis. Folic acid supplementation during *in utero* and postnatal periods, or after weaning, decreases the methylation of the PPARγ promoter in liver [55]. Promoter methylation is inversely related to gene expression, which suggests increased PPARγ expression in response to folic acid supplementation. In adipose tissue, PPARγ is nearly tripled upon folic acid supplementation. Therefore, increased body weight in folic acid supplemented mammals may be due to enhanced lipid storage via PPARγ-dependent mechanisms. In addition, AdoMeth also stimulates fatty acid metabolism in muscle satellite cells, as well as PPARγ mRNA and protein synthesis, in a dose-dependent manner [56]. Through its interaction with PGC1α, increased PPARγ is associated with risks for type 2 diabetes and obesity [11,30,31].

11.4.4 PPARγ Coactivator 1 Alpha

Increased DNA methylation on the promoter of PPARγ coactivator 1 alpha (PGC1α) and decreased expression occur in the skeletal muscle of type 2 diabetes mellitus patients [41]. In cardiomyopathies, the increased concentrations of triglycerides and acylcarnitines are associated with a deficit in fatty acid oxidation, decreased AdoMeth, and imbalanced methylation of PCG1α (Figure 11.2) [57]. As PGC1α is an integrative transcriptional regulator of oxidative metabolism, including fatty acid oxidation and mitochondrial biogenesis, DNA methylation-dependent repression of this gene might explain the mechanism of

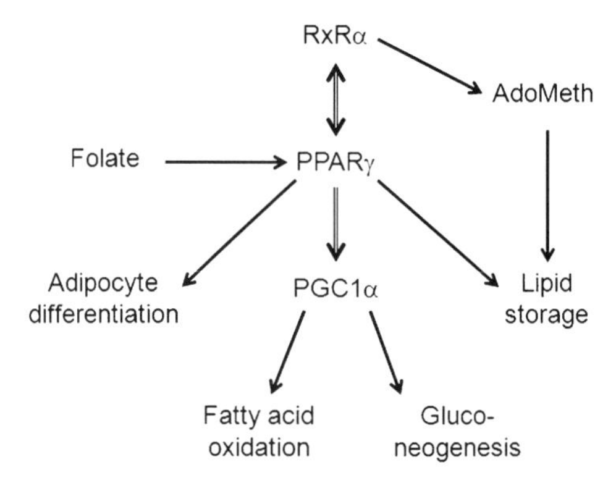

FIGURE 11.2 Relationship between the RxRα, PPARγ, PGC1α axis, folate, AdoMeth, and lipid metabolism. Hypomethylation of RxRα and/or PPARγ, as well as hypermethylation of PCG1α, are associated with obesity.

mitochondrial dysfunction. Another report analyzed PGC1α promoter methylation in young healthy men with low and normal birth weights [58]. The skeletal muscle of low birth weight individuals harbored higher methylation frequencies than normal birth weight individuals, and was more susceptible to insulin resistance after short-term overfeeding. This implies that the perinatal metabolic condition is responsible for the epigenetic changes that lead to an increased sensitivity to caloric excess later in life. Liver is another important source of oxidative metabolism, which contributes to the systemic metabolic homeostasis. Calorie overload leads to the accumulation of hepatic lipids, resulting in the development of non-alcoholic fatty liver disease including non-alcoholic steatohepatitis, which is associated with mitochondrial dysfunction [48]. The decreased mitochondrial DNA content in non-alcoholic fatty liver disease patients was coincident with increased PCG1α promoter methylation, suggesting that an epigenetic event might be involved in mitochondrial dysfunction in the liver [59]. In addition, PGC1α is a crucial regulator of gluconeogenesis in the liver, indicating that hepatic glucose production might also be regulated epigenetically.

11.5 CONCLUSION

Overall, significant progress has been made in the field of epigenetics, obesity, and type 2 diabetes, but there is still much to be learned before we fully understand the role of the epigenome in the development of complex diseases. Epigenetics is a rapidly evolving area of research, and the first steps are already being made in identifying potential biomarkers for obesity that could be detected at birth. Eventually this may help in predicting an individual's obesity risk at a young age, before the phenotype develops, and thus open possibilities for introducing targeted strategies to prevent the condition. It is also now clear that several epigenetic marks are modifiable, not only by changing the exposure *in utero* but also by lifestyle changes in adult life, which implies that there is the potential for interventions to be introduced in postnatal life to modify or rescue unfavorable epigenomic profiles. Testing the possible involvement of nutrient- and/or hormone-driven signaling pathways and transcription factors may help to elucidate this point. Compelling evidence indicates that the fetal and early neonatal periods of development are extremely sensitive to environmental cues, which have long-lasting consequences for postnatal growth, health, and likely athletic performance. Much research is needed with both cell and animal models to understand basic mechanisms responsible for effects mediated by specific nutrients in fetal and neonatal programming. Such new knowledge is expected to aid in designing effective therapeutic means for metabolic abnormalities in offspring that experience an adverse intrauterine environment. In terms of metabolic diseases, we lack knowledge on whether the diet affects the availability of nutrient-derived epigenetic cofactors (for example, methionine, AdoMeth, folate, etc.) in the cells to drive aberrant epigenetic regulation. Investigation on this point would provide mechanistic insights into how specific epigenetic factors might influence the pathogenesis of metabolic diseases.

References

[1] Drummond EM, Gibney ER. Epigenetic regulation in obesity. Curr Opin Clin Nutr Metab Care 2013;16:392–7.

[2] Muhlhausler BS, Gugusheff JR, Ong ZY, Vithayathil MA. Nutritional approaches to breaking the intergenerational cycle of obesity. Can J Physiol Pharmacol 2013;91:421–8.

[3] St-Pierre J, Hivert MF, Perron P, Poirier P, Guay SP, Brisson D, et al. IGF2 DNA methylation is a modulator of newborn's fetal growth and development. Epigenetics 2012;7:1125–32.

[4] Perng W, Rozek LS, Mora-Plazas M, Duchin O, Marin C, Forero T, et al. Micronutrient status and global DNA methylation in school-age children. Epigenetics 2012;7:1133–41.

[5] Soubry A, Schildkraut JM, Murtha A, Wang F, Huang Z, Bernal A, et al. Paternal obesity is associated with IGF2 hypomethylation in newborns: results from a Newborn Epigenetics Study (NEST) cohort. BMC Med 2013;11:29.

[6] Michels KB, Harris HR, Barault L. Birthweight, maternal weight trajectories and global DNA methylation of LINE-1 repetitive elements. PLoS One 2011;6:e25254.

[7] Relton CL, Groom St. A, Pourcain B, Sayers AE, Swan DC, Embleton ND, et al. DNA methylation patterns in cord blood DNA and body size in childhood. PLoS One 2012;7:e31821.

[8] Guenard F, Deshaies Y, Cianflone K, Kral JG, Marceau P, Vohl MC. Differential methylation in glucoregulatory genes of offspring born before vs. after maternal gastrointestinal bypass surgery. Proc Natl Acad Sci USA 2013;110:11439–44.
[9] van Dijk SJ, Molloy PL, Varinli H, Morrison JL, Muhlhausler BS. Epigenetics and human obesity. Int J Obes (Lond) 2014.
[10] Hermsdorff HH, Mansego ML, Campion J, Milagro FL, Zulet MA, Martinez JA. TNF-alpha promoter methylation in peripheral white blood cells: relationship with circulating TNFalpha, truncal fat and n-6 PUFA intake in young women. Cytokine 2013;64:265–71.
[11] Barres R, Kirchner H, Rasmussen M, Yan J, Kantor FR, Krook A, et al. Weight loss after gastric bypass surgery in human obesity remodels promoter methylation. Cell Rep 2013;3:1020–7.
[12] Obermann-Borst SA, Eilers PH, Tobi EW, de Jong FH, Slagboom PE, Heijmans BT, et al. Duration of breastfeeding and gender are associated with methylation of the LEPTIN gene in very young children. Pediatr Res 2013;74:344–9.
[13] Kuehnen P, Mischke M, Wiegand S, Sers C, Horsthemke B, Lau S, et al. An Alu element-associated hypermethylation variant of the POMC gene is associated with childhood obesity. PLoS Genet 2012;8:e1002543.
[14] Milagro FI, Gomez-Abellan P, Campion J, Martinez JA, Ordovas JM, Garaulet M. CLOCK, PER2 and BMAL1 DNA methylation: association with obesity and metabolic syndrome characteristics and monounsaturated fat intake. Chronobiol Int 2012;29:1180–94.
[15] Stepanow S, Reichwald K, Huse K, Gausmann U, Nebel A, Rosenstiel P, et al. Allele-specific, age-dependent and BMI-associated DNA methylation of human MCHR1. PLoS One 2011;6:e17711.
[16] Zhao J, Goldberg J, Vaccarino V. Promoter methylation of serotonin transporter gene is associated with obesity measures: a monozygotic twin study. Int J Obes (Lond) 2013;37:140–5.
[17] Moverare-Skrtic S, Mellstrom D, Vandenput L, Ehrich M, Ohlsson C. Peripheral blood leukocyte distribution and body mass index are associated with the methylation pattern of the androgen receptor promoter. Endocrine 2009;35:204–10.
[18] Drake AJ, McPherson RC, Godfrey KM, Cooper C, Lillycrop KA, Hanson MA, et al. An unbalanced maternal diet in pregnancy associates with offspring epigenetic changes in genes controlling glucocorticoid action and foetal growth. Clin Endocrinol (Oxf) 2012;77:808–15.
[19] Huang RC, Galati JC, Burros S, Beilin LJ, Li X, Pennell CE, et al. DNA methylation of the IGF2/H19 imprinting control region and adiposity distribution in young adults. Clin Epigenetics 2012;4:21.
[20] Wang J, Wu Z, Li D, Li N, Dindot SV, Satterfield MC, et al. Nutrition, epigenetics, and metabolic syndrome. Antioxid Redox Signal 2012;17:282–301.
[21] Heijmans BT, Tobi EW, Stein AD, Putter H, Blauw GJ, Susser ES, et al. Persistent epigenetic differences associated with prenatal exposure to famine in humans. Proc Natl Acad Sci USA 2008;105:17046–9.
[22] Tobi EW, Slagboom PE, van Dongen J, Kremer D, Stein AD, Putter H, et al. Prenatal famine and genetic variation are independently and additively associated with DNA methylation at regulatory loci within IGF2/H19. PLoS One 2012;7:e37933.
[23] Cordero P, Campion J, Milagro FI, Goyenechea E, Steemburgo T, Javierre BM, et al. Leptin and TNF-alpha promoter methylation levels measured by MSP could predict the response to a low-calorie diet. J Physiol Biochem 2011;67:463–70.
[24] Moleres A, Campion J, Milagro FI, Marcos A, Campoy C, Garagorri JM, et al. Differential DNA methylation patterns between high and low responders to a weight

loss intervention in overweight or obese adolescents: the EVASYON study. FASEB J 2013;27:2504–12.

[25] Almen MS, Jacobsson JA, Moschonis G, Benedict C, Chrousos GP, Fredriksson R, et al. Genome wide analysis reveals association of a FTO gene variant with epigenetic changes. Genomics 2012;99:132–7.

[26] Dick KJ, Nelson CP, Tsaprouni L, Sandling JK, Aissi D, Wahl S, et al. DNA methylation and body-mass index: a genome-wide analysis. Lancet 2014;383:1990–8.

[27] Jiang C, Qu A, Matsubara T, Chanturiya T, Jou W, Gavrilova O, et al. Disruption of hypoxia-inducible factor 1 in adipocytes improves insulin sensitivity and decreases adiposity in high-fat diet-fed mice. Diabetes 2011;60:2484–95.

[28] Godfrey KM, Sheppard A, Gluckman PD, Lillycrop KA, Burdge GC, McLean C, et al. Epigenetic gene promoter methylation at birth is associated with child's later adiposity. Diabetes 2011;60:1528–34.

[29] Issa JP. Epigenetic variation and cellular Darwinism. Nat Genet 2011;43:724–6.

[30] Ribel-Madsen R, Fraga MF, Jacobsen S, Bork-Jensen J, Lara E, Valvanese V, et al. Genome-wide analysis of DNA methylation differences in muscle and fat from monozygotic twins discordant for type 2 diabetes. PLoS One 2012;7:e51302.

[31] Ling C, Del Guerra S, Lupi R, Ronn T, Granhall C, Luthman H, et al. Epigenetic regulation of PPARGC1A in human type 2 diabetic islets and effect on insulin secretion. Diabetologia 2008;51:615–22.

[32] Groom A, Potter C, Swan DC, Fatemifar G, Evans DM, Ring SM, et al. Postnatal growth and DNA methylation are associated with differential gene expression of the TACSTD2 gene and childhood fat mass. Diabetes 2012;61:391–400.

[33] Perkins E, Murphy SK, Murtha AP, Schildkraut J, Jirtle RL, Demark-Wahnefried W, et al. Insulin-like growth factor 2/H19 methylation at birth and risk of overweight and obesity in children. J Pediatr 2012;161:31–9.

[34] Cooper WN, Khulan B, Owens S, Elks CE, Seidel V, Prentice AM, et al. DNA methylation profiling at imprinted loci after periconceptional micronutrient supplementation in humans: results of a pilot randomized controlled trial. FASEB J 2012;26:1782–90.

[35] Glowinska-Olszewska B, Urban M. Elevated matrix metalloproteinase 9 and tissue inhibitor of metalloproteinase 1 in obese children and adolescents. Metabolism 2007;56:799–805.

[36] Feinberg AP, Irizarry RA, Fradin D, Aryee MJ, Murakami P, Aspelund T, et al. Personalized epigenomic signatures that are stable over time and covary with body mass index. Sci Transl Med 2010;2:49ra67.

[37] Keating ST, El-Osta A. Epigenetic changes in diabetes. Clin Genet 2013;84:1–10.

[38] Park JH, Stoffers DA, Nicholls RD, Simmons RA. Development of type 2 diabetes following intrauterine growth retardation in rats is associated with progressive epigenetic silencing of Pdx1. J Clin Invest 2008;118:2316–24.

[39] Kulkarni RN, Jhala US, Winnay JN, Krajewski S, Montminy M, Kahn CR. PDX-1 haploinsufficiency limits the compensatory islet hyperplasia that occurs in response to insulin resistance. J Clin Invest 2004;114:828–36.

[40] Brissova M, Blaha M, Spear C, Nicholson W, Radhika A, Shiota M, et al. Reduced PDX-1 expression impairs islet response to insulin resistance and worsens glucose homeostasis. Am J Physiol Endocrinol Metab 2005;288:E707–714.

[41] Barres R, Osler ME, Yan J, Rune A, Fritz T, Caidahl K, et al. Non-CpG methylation of the PGC-1alpha promoter through DNMT3B controls mitochondrial density. Cell Metab 2009;10:189–98.

[42] Ek J, Andersen G, Urhammer SA, Gaede PH, Drivsholm T, Borch-Johnsen K, et al. Mutation analysis of peroxisome proliferator-activated receptor-gamma coactivator-1

(PGC-1) and relationships of identified amino acid polymorphisms to Type II diabetes mellitus. Diabetologia 2001;44:2220–6.
[43] Tobi EW, Lumey LH, Ralens RP, Kremer D, Putter H, Stein AD, et al. DNA methylation differences after exposure to prenatal famine are common and timing- and sex-specific. Hum Mol Genet 2009;18:4046–53.
[44] Melzner I, Scott V, Dorsch K, Fischer P, Wabitsch M, Bruderlein S, et al. Leptin gene expression in human preadipocytes is switched on by maturation-induced demethylation of distinct CpGs in its proximal promoter. J Biol Chem 2002;277:45420–7.
[45] Yokomori N, Tawata M, Onaya T. DNA demethylation modulates mouse leptin promoter activity during the differentiation of 3T3-L1 cells. Diabetologia 2002;45:140–8.
[46] Milagro FI, Campion J, Garcia-Diaz DF, Goyenechea E, Paternain L, Martinez JA. High fat diet-induced obesity modifies the methylation pattern of leptin promoter in rats. J Physiol Biochem 2009;65:1–9.
[47] Ronn T, Poulsen P, Hansson O, Holmkvist J, Almgren P, Nilsson P, et al. Age influences DNA methylation and gene expression of COX7A1 in human skeletal muscle. Diabetologia 2008;51:1159–68.
[48] Hino S, Nagaoka K, Nakao M. Metabolism-epigenome crosstalk in physiology and diseases. J Hum Genet 2013;58:410–15.
[49] da Silva RP, Kelly KB, Al Rajabi A, Jacobs RL. Novel insights on interactions between folate and lipid metabolism. Biofactors 2014;40:277–83.
[50] Unterberger A, Szyf M, Nathanielsz PW, Cox LA. Organ and gestational age effects of maternal nutrient restriction on global methylation in fetal baboons. J Med Primatol 2009;38:219–27.
[51] Farley D, Tejero ME, Comuzzie AG, Higgins PB, Cox L, Werner SL, et al. Feto-placental adaptations to maternal obesity in the baboon. Placenta 2009;30:752–60.
[52] Chew TW, Jiang X, Yan J, Wang W, Lusa AL, Carrier BJ, et al. Folate intake, MTHFR genotype, and sex modulate choline metabolism in mice. J Nutr 2011;141:1475–81.
[53] Dai T, Wu Y, Leng AS, Ao Y, Robel RC, Lu SC, et al. RXRalpha-regulated liver SAMe and GSH levels influence susceptibility to alcohol-induced hepatotoxicity. Exp Mol Pathol 2003;75:194–200.
[54] Fell D, Steele RD. Effect of retinol toxicity on hepatic S-adenosylmethionine-dependent transmethylation in rats. Drug Nutr Interact 1987;5:1–7.
[55] Sie KK, Li J, Ly A, Sohn KJ, Croxford R, Kim YI. Effect of maternal and postweaning folic acid supplementation on global and gene-specific DNA methylation in the liver of the rat offspring. Mol Nutr Food Res 2013;57:677–85.
[56] Yue T, Fang Q, Yin J, Li D, Li W. S-adenosylmethionine stimulates fatty acid metabolism-linked gene expression in porcine muscle satellite cells. Mol Biol Rep 2010;37:3143–9.
[57] Garcia MM, Gueant-Rodriguez RM, Pooya S, Brachet P, Alberto JM, Jeannesson E, et al. Methyl donor deficiency induces cardiomyopathy through altered methylation/acetylation of PGC-1alpha by PRMT1 and SIRT1. J Pathol 2011;225:324–35.
[58] Brons C, Jacobsen S, Nilsson E, Ronn T, Jensen CB, Storgaard H, et al. Deoxyribonucleic acid methylation and gene expression of PPARGC1A in human muscle is influenced by high-fat overfeeding in a birthweight-dependent manner. J Clin Endocrinol Metab 2010;95:3048–56.
[59] Sookoian S, Rosselli MS, Gemma C, Burgueno AL, Fernandez Gianotti T, Castano GO, et al. Epigenetic regulation of insulin resistance in nonalcoholic fatty liver disease: impact of liver methylation of the peroxisome proliferator-activated receptor gamma coactivator 1alpha promoter. Hepatology 2010;52:1992–2000.

CHAPTER

12

DNA Methylation in Pituitary Diseases

OUTLINE

12.1 INTRODUCTION

The limbic–hypothalamic–pituitary–adrenal axis (LHPA) is the primary circuit that initiates, regulates, and terminates a stress response. The same brain areas that control stress also react to stress dynamically and with long-term consequences. One of the biological processes

M. Neidhart: DNA Methylation and Complex Human Disease.
DOI: http://dx.doi.org/10.1016/B978-0-12-420194-1.00012-9

evoking potent adaptive changes in the CNS, such as changes in behavior, gene activity, or synaptic plasticity in the hippocampus, is psychogenic stress [1]. The LHPA axis acts through a number of mediators such as corticotropin-releasing hormone (CRH) and glucocorticosteroids (GC). Stress alters neurotransmission and synaptic plasticity in the brain areas involved in the LHPA axis, such as the prefrontal cortex, hippocampus, or amygdala. Stimuli that are long lasting and intensive can lead to a persistent change in the stress response and mechanisms, and in the function and structure of the brain itself. These changes can lead to cognitive deficits and behavior alterations that can translate to neurodegenerative or mental illnesses. The functioning of the hypothalamic–pituitary–adrenal (HPA) axis and the serotonergic system is known to be intertwined with mood. Alterations in these systems are often associated with depression; however, neither is sufficient to cause depression in and of itself. It is now becoming increasingly clear that the environment plays a crucial role, particularly the perinatal environment [2]. The changes in promoter methylation induced by stress of the glucocorticoid receptor (GC) in the central nervous system were discussed in Chapter 10. GC methylation in offspring explained variations in the stress response that originally depends on maternal behavior [3]. In this chapter we will discuss the role of DNA methylation in pituitary functions, as well as the effect of pituitary hormones on DNA methylation in the periphery.

12.2 HYPOTHALAMO-PITUITARY AXIS

The paraventricular nucleus (PVN) of the hypothalamus CRH is the main activator of the hypothalamo-pituitary (HPA) axis, and is responsible for at least some behavioral changes after stress. In addition, hypothalamic arginine vasopressin (AVP) is known to potentiate CRH action and stress response. CRH and AVP induce the stress-related changes in the ACTH precursor, POMC, gene expression. CRH [4] and AVP [5] expression are both modulated by promoter methylation. POMC is intrinsically upregulated in Cushing's syndrome due to a hypomethylation of its promoter in pituitary and non-pituitary sites [6]. Cushing's disease is characterized by increased secretion of ACTH from the anterior pituitary. This is most often due to excess production of CRH or to a pituitary adenoma.

Many other hypothalamic factors regulate anterior pituitary activities – for example, growth-hormone releasing hormone (GRH) and somatostatin affect growth hormone (GH), dopamine regulating prolactin (PRL), gonadotropin-releasing hormone (GnRH) for luteinizing hormone (LH), etc. Site-specific promoter methylation modulates the action of transcription factors that dictate the tissue-specific expression of PRL and GH [7]; regarding other pituitary hormones, the information is missing. PRL release is mainly under the negative control of dopamine; however, estrogen can stimulate PRL expression in part by reducing the methylation level of the PRL promoter in lactotropic pituitary cells [8]. During lactation, the GH gene is hypermethylated in somatotropic cells, while the PRL gene is hypomethylated in lactotropes [9]. However, normally, the GH gene is unmethylated in somatotropes and mainly controlled by the balance of GRH and somatostatin. This is also the case in pathologies such as acromegaly, where promoter methylation does not play a role [10]. In pituitary adenomas secreting PRL or GH, the situation is different (see below).

In general, GHRH and/or GH gene hypermethylation is not reported in growth retardation. Prader-Willi syndrome is a genetic disorder characterized by low GH, short stature, mild mental retardation, abnormal body composition, muscular hypotonia, and distinctive behavioral features [11]. Excessive eating causes progressive obesity with increased cardiovascular morbidity and mortality. It is caused by loss of one or more normally active paternal genes on chromosome 15q11−13. Russell-Silver syndrome, for example, which is characterized by intrauterine and postnatal growth retardation, relative macrocephaly, a typical triangular face, and asymmetry, is due to a chromosome 11p15 hypomethylation [12]. Beckwith-Wiedemann syndrome is caused by dysregulation of imprinted genes (KvDMR1) on chromosome 11p15.5. The syndrome includes overgrowth, macroglossia, organomegaly, abdominal wall defects, hypoglycemia, and long-term malignancy risk. Some of these symptoms are associated with a loss of imprinting of the IGF2/H19 locus (see below). In cases of hypopituitarism and short stature, therapy with GH appears to be indicated [11,13].

12.3 PITUITARY DEVELOPMENT

Regarding epigenetics, it might be interesting to investigate the genes involved in hypothalamic and pituitary gland development, such as KISS1, LHX3, LHX4, or bone morphogenetic protein 4 (BMP4).

Kisspeptin and its receptor have been implicated as critical regulators of reproductive physiology, with humans and mice without functioning kisspeptin systems displaying severe pubertal and reproductive defects. Alterations in the expression of KISS1 (the gene encoding kisspeptin) over development, along with differences in Kiss1 expression between the sexes in adulthood, may be critical for the maturation and functioning of the neuroendocrine reproductive system, and could possibly contribute to pubertal progression, sex differences in LH secretion, and other facets of reproductive physiology. It is therefore important to understand how KISS1 gene expression develops, and what possible regulatory mechanisms govern the modulation of its expression. A number of recent studies [14], primarily in rodent or cell line models, have focused on the contributions of epigenetic mechanisms to the regulation of KISS1 gene expression.

Furthermore, mutations in the LHX3 gene are associated with pediatric diseases featuring severe hormone deficiencies, hearing loss, developmental delay, rigid cervical spine, and other symptoms. LHX3 is regulated by promoter methylation [15,16]. LHX4, mutations of which can be associated with hypopituitarism, is also regulated by promoter methylation [17]. BMP4 is a key mediator of anterior pituitary organogenesis. However, through inappropriate expression patterns, BMP4 is also pathogenic in pituitary adenomas. In these cases, increased or decreased in BMP4 in lactotrope- and corticotrope-derived adenomas, respectively, is consistent with a bifunctional role for this protein toward either promotion or inhibition of cell proliferation and hormone secretion. BMP4 transcript and protein are differentially expressed and show an increase in the majority of prolactinomas relative to normal pituitary. In pituitary tumor cell lines, reduced BMP4 expression is associated with silencing histone tail marks (H3K27me3) and an increase in CpG island methylation [16].

12.4 PITUITARY ADENOMA

In pituitary adenomas, a majority of the tumor suppressor genes silenced by DNA methylation act via modulation of apoptosis and cell cycle progression [18]. Two major human cell cycle regulator pathways, the CDKN2A/p16^{inka4}/RB1 and P53 pathways, act through these very mechanisms to inhibit cell proliferation and neoplastic growth. The CDKN2A/p16^{inka4}/RB1 pathway normally inhibits the G_1 to S cell cycle phase transition, and the P53 pathway both regulates cell cycle progression and promotes apoptosis. Both pathways limit growth potential and tumorigenesis by decreasing cell line viability and uncontrolled growth. These pathways are commonly affected in many cancer types, including pituitary adenomas [19–21].

In general, pro-oncogenes produce the opposite effects of tumor suppressor genes by facilitating cell-cycle progression, maintaining chromosomal stability, and inducing aneuploidy. Unlike the methylation-induced silencing observed in tumor suppressor genes, these genes show increased expression. For example, melanoma-associated antigen 3 (MAGEA3) increases gene expression through interactions with fibroblast growth factor receptor 2 (FGFR2/CD332) to produce gene promoter hypomethylation, which decreases inhibition of transcription and increases gene expression [22].

Maternal Imprinted Gene 3 (MEG3) is an imprinted gene that is modified in PAs via DNA methylation. In normal pituitary cells, MEG3 retains strong expression, functions as a powerful growth suppressor, and primarily acts by increasing P53 expression and modifying its transcriptional activation. Loss of MEG3 leads to decreased expression of P53, increased cell survival through activated cell cycle progression, and decreased apoptosis. In normal pituitary gland, MEG3 displays approximately 50% methylation rates (consistent with an imprinted gene). The MEG3 promoter regions in pituitary adenomas with MEG3 deficiency showed significantly increased (>50%) methylation of CpG islands [23].

In rat prolactinomas, abnormal expression of the prolactin gene is correlated with hypomethylated status of CpG sites in exons 1, 2, and 4 of the prolactin gene [24]. Similarly, in pituitary adenomas secreting GH, hypomethylation of the GH gene occurs in the whole organ [25].

Perhaps the most important clinical aspect of this research lies in its potential to develop drugs that can modify epimutations. Although the majority of symptomatic non-prolactinoma pituitary adenomas are currently treated via surgical resection, important advances will soon likely facilitate improved medical treatment for these tumors outside of the well-established dopamine analogs for prolactinomas. Unlike gene mutations, epigenetic modifications are typically reversible, and epigenetically altered pituitary adenomas could therefore be potentially cured with drug therapy. Two major strategies exist for targeting epigenetic dysregulation of pituitary adenomas: (1) alteration of genome-scale epigenetic processes; and (2) alteration of methylation or acetylation of specific genes [26]. Several drugs (i.e., DNMT inhibitors) are already known to accomplish the former goal.

12.5 PRO-OPIOMELANOCORTIN GENE

The pro-opiomelanocortin (POMC) gene is recognized as playing an important role in the regulation of the LHPA, adrenal development, and obesity (see also Chapter 11). POMC is activated in ACTH-dependent Cushing's syndrome. The syndrome may occur when the highly tissue-specific 5′ promoter of human POMC is activated in pituitary and non-pituitary sites. Whilst the factors involved in transcription in the corticotropes of the anterior pituitary gland are becoming well delineated, the mechanism of activation in non-pituitary sites is not fully understood [6]. This promoter is embedded within a defined CpG island, and, in contrast to somatically expressed CpG island promoters reported to date, is methylated in normal non-expressing tissues, but is specifically unmethylated in expressing tissues, tumors and the POMC-expressing DMS-79 small-cell lung cancer cell line. Methylation *in vitro* is sufficient for silencing of expression. In particular, methylation near the response element for the tissue-specific POMC activator pituitary homeobox 1 (PTX1) diminishes POMC expression [27]. Sites outside the PTX1 response element may be important for binding, and this may have implications for pituitary development. DMS-79 cells lack POMC-demethylating activity, implying that the methylation and expression patterns are likely to be set early or prior to neoplastic transformation, and that targeted *de novo* methylation might be a potential therapeutic strategy.

It is conceivable that, in POMC neurons of the hypothalamus, the POMC promoter is subject to a variable density of methylation with clear implications for the signaling of satiety and obesity [28]. Thus, leptin treatment during lactation may program methylation of POMC in the hypothalamus of rats fed with high fat diets, with possible protection against the development of obesity [29].

12.6 PROLACTIN AND GROWTH HORMONE

In pituitary lactotropic cells, promoter methylation inversely correlate with prolactin (PRL) expression [7]. In non-pituitary cells, PRL gene is silenced by methylation of both promoter and exons. Estrogen increases PRL secretion in part by hypomethylating the PRL gene promoter [8]. On the other hand, abnormal expressions of the PRL gene in some pituitary adenomas [24] and in the B-lymphoblastoid cell line IM-9-P [30] are associated with altered methylation in exons.

PRL is an important growth modulatory hormone in fetal and adult tissues. It stimulates DNA synthesis and enhances the cell cycle in various tissues. PRL causes global hypomethylation of DNA, for example in the kidney and liver [31], favoring in general gene expression. Puberty in females initiates a branching morphogenesis of the mammary gland, which requires growth hormone (GH) and estrogen, as well as insulin-like growth factor 1, to create a ductal tree that fills the fat pad. Upon pregnancy, the combined actions of progesterone and PRL generate alveoli, which secrete milk during lactation. Hypomethylation of the casein gene occurs during this period in the mammary gland [32]. Specifically, there is an increase in DNA methylation at three CpG dinucleotides of the functional STAT5-binding site of the casein promoter [33]. There is a decline in methylation following a non-milking period. Lack of demand for milk at weaning initiates the process of involution, whereby the gland is remodeled back to its prepregnancy state. These processes require numerous signaling pathways that have distinct regulatory functions at different stages of gland development. Expression of the epithelial-specific ETS transcription factor ELF5 is increased following PRL stimulation. ELF5 is at the origin of a genomic regulatory network responsible for co-ordinating all of the mechanisms that lead to a differentiated breast alveolar compartment [34]. Methylation of the ELF5 promoter has been proposed to act as a lineage gatekeeper during embryonic development. ELF5 promoter methylation is lineage-specific and developmentally regulated in the mammary gland [35].

In pituitary somatotropic cells, DNA methylation can modulate but normally does not control GH gene expression [36]. However, a glucocorticoid receptor binding site (GRBS) exists approximately 100 base pairs downstream from the start of GH transcription. This GRBS is selectively inhibited by methylation of two GC pairs within this site [37], avoiding an enhancing effect of glucocorticoids. The GH gene also has multiple thyroid responsive elements (TRE) that are unmethylated [38], allowing modulation by thyroid hormones [39]. In other cells of the pituitary and in non-pituitary tissues, the GH gene is silenced by methylation of CGs about 140 base pairs from the transcription site [40,41]. In some pituitary adenomas (secreting GH), however, the GH gene is entirely hypomethylated [10,25,42]. In somatotropes, the PRL gene is silenced by promoter hypermethylation [43,44].

GH regulates a broad range of physiological processes, including long bone growth, fatty acid oxidation, glucose uptake, and hepatic steroid and foreign compound metabolism. The liver is its primary target. GH exerts sex-dependent effects on the liver in many species, with many hepatic genes, most notably genes coding for cytochrome P450 enzymes, being transcribed in a sex-dependent manner due to differences in promoter methylation [45]. Downregulation of DNMT3A/B plays a role in age- and sex-dependent, as well as GH-induced, DNA hypomethylation [46]. On the other hand, GH induced DNMT1 in dwarf mice [47]. Thus, GH has different effects on methylation, depending on the physiological and pathological conditions.

Somatomedins (e.g., IGF1/2) is a group of hormones that is produced, when stimulated by GH, to promote cell growth and division. Insulin-like growth factor 1 (IGF1) is a hormone similar in structure to insulin. It plays an important role in childhood growth, and continues to have anabolic effects in adults. Deficiency of either GH or IGF1 results in diminished stature. IGF1 binds to at least two cell surface receptors: the IGF1 receptor (IGF1R), and the insulin receptor. The promoter methylation and expression of IGF1 can be altered in the diabetic skeletal muscle [48].

Insulin-like growth factor 2 (IGF-2) also shares structural similarity to insulin. The major role of IGF-2 is as a growth promoting hormone during gestation. It exerts its effects by binding to the IGF1 receptor. Altered methylation of the IGF2/H19 locus is involved in imprinted development disorders, such as the Beckwith-Wiedemann syndrome [49]. The IGF2/H19 locus was essential in the marsupial and mammalian evolution [50], and has an important role to play in feto-placental development. IGF2 affects the growth, morphology, and nutrient transfer capacity of the placenta, and thereby controls the nutrient supply for fetal growth [51].

12.7 GONADOTROPINS (LUTEINIZING HORMONE, FOLLICLE STIMULATING HORMONE)

In testis, DNA methylation is differentially regulated during development and is controlled by gonadotropic hormones. Follicle stimulating hormone (FSH) decreases the activity of DNMTs in testis [52]. Administration of FSH to immature rats caused hypomethylation of seminiferous tubular DNA, while luteinizing hormone (LH) caused similar effects in Leydig cells [53].

The luteinizing hormone receptor (LHR) is expressed primarily in gonads, where its expression is tightly controlled to mediate LH signals regulating ovarian cyclic changes and testicular function. In addition, LHR gene expression has been observed in several non-gonadal tissues, including placenta, uterus, normal mammary glands, and breast tumor and placenta cell lines. Coordinated histone hyperacetylation and DNA demethylation lead to derepression of the LHR gene expression in these tissues [54]. On the other hand, rapid silencing is mainly due to histone hypoacethylation upon recruitment of histone deacetylases HDAC1/2 at an SP1 site of the LHR promoter [55].

Gonadotropins LH, FSH, and steroids influence the acquisition of developmental competence of the oocyte; they all altered the level of global DNA methylation [56]. This is required for the oocyte to mature correctly. IGF2 promotes granulosa cell proliferation during the follicular phase of the menstrual cycle, acting alongside FSH. After ovulation has occurred, IGF-2 promotes progesterone secretion during the luteal phase of the menstrual cycle, together with LH [57].

12.8 THYROID STIMULATING HORMONE

In thyroid cells, promoter methylation inversely correlates with thyroid stimulating hormone receptor (TSHR) expression [58,59]. TSHR is silenced by promoter methylation in epithelial thyroid cancer [60]. This is associated with decreased or absent TSH-promoted iodine uptake. It can be restored by using DNA demethylation agents [61].

12.9 ARGININE VASOPRESSIN AND ATRIAL NATRIURETIC PEPTIDE

Adverse early life events can induce long-lasting changes in physiology and behavior. Early-life stress causes enduring hypersecretion of glucocorticoids and alterations in memory. This phenotype is accompanied by a persistent increase in arginine vasopressin (AVP) expression in neurons of the hypothalamic paraventricular nucleus (PVN), and is reversed by an AVP receptor antagonist. Altered AVP expression is associated with sustained DNA hypomethylation [5]. AVP is derived from a preprohormone precursor that is synthesized in the PVN and stored in vesicles at the posterior pituitary. AVP can trigger neuroendocrine and behavioral alterations that are frequent features in depression. Early-life stress leads to epigenetic marking of the AVP gene underpinning sustained expression and increased HPA axis activity [62,63]. The relationship between steroid hormones and receptors with AVP promoter methylation in the brain [64] was discussed in Chapter 10.

AVP is also called antidiuretic hormone (ADH). It regulates the body's retention of water by acting to increase water reabsorption in the collecting ducts of the kidney nephron. The syndrome of inappropriate antidiuretic hormone secretion is due to an overproduction of AVP by the posterior pituitary gland. On the other hand, diabetes insipidus can be caused by underproduction of AVP and/or mutation of the AVP receptor. The molecular mechanisms responsible for the development of this syndrome is not well-known, but it is believed that it might be inherent to mutations or depend on epigenetic factors such as DNA methylation [65]. For example, an association exists between incomplete X chromosome inactivation by methylation in woman, AVP type 2 receptor dysfunction, and diabetes insipidus [66]. Disturbances of volume-regulating mechanisms have already been implicated in the pathophysiology of eating disorders such as anorexia or bulimia nervosa, with the peptide hormones AVP and atrial natriuretic peptide (ANP) being of special interest. ANP is a vasodilator via heart muscle cells. It is involved in the homeostatic control of body water, sodium, potassium, and adipose tissue. Lower levels of ANP transcripts are detected in patients with eating disorders [67]. This downregulation is accompanied by a hypermethylation of the ANP gene promoter in the bulimic subgroup, without change in AVP promoter methylation. In these patients, ANP mRNA expression is inversely associated with impaired impulse regulation. In patients with alcohol dependence, AVP appears normal and ANP is increased; however, the AVP promoter appears hypermethylated, while the ANP promoter is hypomethylated [68,69]. Further studies focusing on longitudinal changes of epigenetic regulation and gene expression of both peptides are needed to clarify the pathophysiological role of these findings.

12.10 CONCLUSION

Although genetics determines endocrine phenotypes, it cannot fully explain the great variability and reversibility of the system in response to environmental changes. Evidence now suggests that epigenetics links genetics and environment in shaping endocrine function. Epigenetic mechanisms, including DNA methylation, separate the genome into active and inactive domains based on endogenous and exogenous environmental changes and developmental stages, creating phenotype plasticity that can explain interindividual and population endocrine variability. We reviewed the current understanding of epigenetics in the pituitary field, specifically, the regulation by epigenetics of the release of pituitary hormones, as well as the levels of hormone actions. A three-dimensional model is necessary to explain the phenomena related to progressive changes in endocrine functions with age, the early origin of endocrine disorders, and rapid shifts in disease patterns among populations experiencing major lifestyle changes, including the many endocrine disruptions in contemporary life. In spite of endocrinology and epigenetics being well established disciplines, the links between them are rather incomplete; this is particularly obvious regarding the hypothalamus and pituitary gland, despite this system controlling a great part of body's functions. The key for understanding epigenetics in endocrinology is the identification, through advanced high-throughput screening technologies, of plasticity genes or loci that respond directly to a specific environmental stimulus. Investigations to determine whether epigenetic changes induced by today's lifestyles or environmental exposures can be inherited and are reversible should be an important priority for research.

References

[1] Stankiewicz AM, Swiergiel AH, Lisowski P. Epigenetics of stress adaptations in the brain. Brain Res Bull 2013;98:76–92.

[2] Booij L, Wang D, Levesque ML, Tremblay RE, Szyf M. Looking beyond the DNA sequence: the relevance of DNA methylation processes for the stress-diathesis model of depression. Philos Trans R Soc Lond B Biol Sci 2013;368:20120251.

[3] Szyf M, Weaver IC, Champagne FA, Diorio J, Meaney MJ. Maternal programming of steroid receptor expression and phenotype through DNA methylation in the rat. Front Neuroendocrinol 2005;26:139–62.
[4] Elliott E, Ezra-Nevo G, Regev L, Neufeld-Cohen A, Chen A. Resilience to social stress coincides with functional DNA methylation of the Crf gene in adult mice. Nat Neurosci 2010;13:1351–3.
[5] Murgatroyd C, Patchev AV, Wu Y, Micale V, Bockmuhl Y, Fischer D, et al. Dynamic DNA methylation programs persistent adverse effects of early-life stress. Nat Neurosci 2009;12:1559–66.
[6] Newell-Price J. Proopiomelanocortin gene expression and DNA methylation: implications for Cushing's syndrome and beyond. J Endocrinol 2003;177:365–72.
[7] Ngo V, Gourdji D, Laverriere JN. Site-specific methylation of the rat prolactin and growth hormone promoters correlates with gene expression. Mol Cell Biol 1996;16:3245–54.
[8] Kulig E, Landefeld TD, Lloyd RV. The effects of estrogen on prolactin gene methylation in normal and neoplastic rat pituitary tissues. Am J Pathol 1992;140:207–14.
[9] Kumar V, Biswas DK. Dynamic state of site-specific DNA methylation concurrent to altered prolactin and growth hormone gene expression in the pituitary gland of pregnant and lactating rats. J Biol Chem 1988;263:12645–52.
[10] Adams EF, Buchfelder M, Huttner A, Moreth S, Fahlbusch R. Recent advances in the molecular biology of growth-hormone secreting human pituitary tumours. Exp Clin Endocrinol 1993;101:12–16.
[11] Hoybye C. Endocrine and metabolic aspects of adult Prader-Willi syndrome with special emphasis on the effect of growth hormone treatment. Growth Horm IGF Res 2004;14:1–15.
[12] Eggermann T. Russell-Silver syndrome. Am J Med Genet C Semin Med Genet 2010;154C:355–64.
[13] Baiocchi M, Yousuf FS, Hussain K. Hypopituitarism in a patient with Beckwith-Wiedemann syndrome due to hypomethylation of KvDMR1. Pediatrics 2014;133: e1082–1086.
[14] Semaan SJ, Kauffman AS. Emerging concepts on the epigenetic and transcriptional regulation of the Kiss1 gene. Int J Dev Neurosci 2013;31:452–62.
[15] Malik RE, Rhodes SJ. The role of DNA methylation in regulation of the murine Lhx3 gene. Gene 2014;534:272–81.
[16] Yacqub-Usman K, Duong CV, Clayton RN, Farrell WE. Epigenomic silencing of the BMP-4 gene in pituitary adenomas: a potential target for epidrug-induced re-expression. Endocrinology 2012;153:3603–12.
[17] Rauch T, Li H, Wu X, Pfeifer GP. MIRA-assisted microarray analysis, a new technology for the determination of DNA methylation patterns, identifies frequent methylation of homeodomain-containing genes in lung cancer cells. Cancer Res 2006;66:7939–47.
[18] Pease M, Ling C, Mack WJ, Wang K, Zada G. The role of epigenetic modification in tumorigenesis and progression of pituitary adenomas: a systematic review of the literature. PLoS One 2013;8:e82619.
[19] Baylin SB, Herman JG, Graff JR, Vertino PM, Issa JP. Alterations in DNA methylation: a fundamental aspect of neoplasia. Adv Cancer Res 1998;72:141–96.
[20] Seemann N, Kuhn D, Wrocklage C, Keyvani K, Hackl W, Buchfelder M, et al. CDKN2A/p16 inactivation is related to pituitary adenoma type and size. J Pathol 2001;193:491–7.
[21] Yoshino A, Katayama Y, Ogino A, Wantabe T, Yachi K, Ohta T, et al. Promoter hypermethylation profile of cell cycle regulator genes in pituitary adenomas. J Neurooncol 2007;83:153–62.

[22] Zhu X, Asa SL, Ezzat S. Fibroblast growth factor 2 and estrogen control the balance of histone 3 modifications targeting MAGE-A3 in pituitary neoplasia. Clin Cancer Res 2008;14:1984–96.
[23] Zhao J, Dahle D, Zhou Y, Zhang X, Klibanski A. Hypermethylation of the promoter region is associated with the loss of MEG3 gene expression in human pituitary tumors. J Clin Endocrinol Metab 2005;90:2179–86.
[24] Xu RK, Wu XM, Di AK, Xu JN, Pang CS, Pang SF. Pituitary prolactin-secreting tumor formation: recent developments. Biol Signals Recept 2000;9:1–20.
[25] U HS, Kelley P, Lee WH. Abnormalities of the human growth hormone gene and protooncogenes in some human pituitary adenomas. Mol Endocrinol 1988; 2:85–9.
[26] Yacqub-Usman K, Richardson A, Duong CV, Clayton RN, Farrell WE. The pituitary tumour epigenome: aberrations and prospects for targeted therapy. Nat Rev Endocrinol 2012;8:486–94.
[27] Lamonerie T, Tremblay JJ, Lanctot C, Therrien M, Gauthier Y, Drouin J. Ptx1, a bicoid-related homeo box transcription factor involved in transcription of the pro-opiomelanocortin gene. Genes Dev 1996;10:1284–95.
[28] Plagemann A, Harder T, Brunn M, Harder A, Roepke K, Wittrock-Staar M, et al. Hypothalamic proopiomelanocortin promoter methylation becomes altered by early overfeeding: an epigenetic model of obesity and the metabolic syndrome. J Physiol 2009;587:4963–76.
[29] Palou M, Pico C, McKay JA, Sanchez J, Priego T, Mathers JC, et al. Protective effects of leptin during the suckling period against later obesity may be associated with changes in promoter methylation of the hypothalamic pro-opiomelanocortin gene. Br J Nutr 2011;106:769–78.
[30] Gellersen B, Kempf R. Human prolactin gene expression: positive correlation between site-specific methylation and gene activity in a set of human lymphoid cell lines. Mol Endocrinol 1990;4:1874–86.
[31] Reddy PM, Reddy PR. Effect of prolactin on DNA methylation in the liver and kidney of rat. Mol Cell Biochem 1990;95:43–7.
[32] Platenburg GJ, Vollebregt EJ, Karatzas CN, Kootwijk EP, De Boer HA, Strijker R. Mammary gland-specific hypomethylation of Hpa II sites flanking the bovine alpha S1-casein gene. Transgenic Res 1996;5:421–31.
[33] Singh K, Molenaar AJ, Swanson KM, Gudex B, Arias JA, Erdman RA, et al. Epigenetics: a possible role in acute and transgenerational regulation of dairy cow milk production. Animal 2012;6:375–81.
[34] Oakes SR, Rogers RL, Naylor MJ, Ormandy CJ. Prolactin regulation of mammary gland development. J Mammary Gland Biol Neoplasia 2008;13:13–28.
[35] Lee HJ, Hinshelwood RA, Bouras T, Gallego-Ortega D, Valdes-Mora F, Blazek K, et al. Lineage specific methylation of the Elf5 promoter in mammary epithelial cells. Stem Cells 2011;29:1611–19.
[36] Lan NC. The effects of 5-azacytidine on the expression of the rat growth hormone gene. Methylation modulates but does not control growth hormone gene activity. J Biol Chem 1984;259:11601–6.
[37] Moore DD, Markas AR, Buckley DI, Kapler G, Payvar F, Goodman HM. The first intron of the human growth hormone gene contains a binding site for glucocorticoid receptor. Proc Natl Acad Sci USA 1985;82:699–702.
[38] Norman MF, Lavin TN, Baxter JD, West BL. The rat growth hormone gene contains multiple thyroid response elements. J Biol Chem 1989;264:12063–73.
[39] Darling DS, Gaur NK, Zhu B. A zinc finger homeodomain transcription factor binds specific thyroid hormone response elements. Mol Cell Endocrinol 1998; 139:25–35.

[40] Strobl JS, Dannies PS, Thompson EB. Rat growth hormone gene expression is correlated with an unmethylated CGCG sequence near the transcription initiation site. Biochemistry 1986;25:3640–8.
[41] Gaido ML, Strobl JS. Inhibition of rat growth hormone promoter activity by site-specific DNA methylation. Biochim Biophys Acta 1989;1008:234–42.
[42] Huttner A, Adams EF, Buchfelder M, Fahlbusch R. Growth hormone gene structure in human pituitary somatotrophinomas: promoter region sequence and methylation studies. J Mol Endocrinol 1994;12:167–72.
[43] Zhang ZX, Kumar V, Rivera RT, Pasion SG, Chisholm J, Biswas DK. Suppression of prolactin gene expression in GH cells correlates with site-specific DNA methylation. DNA 1989;8:605–13.
[44] Arnold TE, Farrance IK, Morris J, Ivarie R. Prolactin-deficient GH3B3 cells are defective in the utilization of the endogenous prolactin promoter yet are fully competent to initiate transcription from a transfected prolactin promoter. DNA Cell Biol 1991;10:105–12.
[45] Waxman DJ, O'Connor C. Growth hormone regulation of sex-dependent liver gene expression. Mol Endocrinol 2006;20:2613–29.
[46] Takasugi M, Hayakawa K, Arai D, Shiota K. Age- and sex-dependent DNA hypomethylation controlled by growth hormone in mouse liver. Mech Ageing Dev 2013;134:331–7.
[47] Armstrong VL, Rakoczy S, Rojanathammanee L, Brown-Borg HM. Expression of DNA methyltransferases is influenced by growth hormone in the long-living ames dwarf mouse *in vivo* and *in vitro*. J Gerontol A Biol Sci Med Sci 2013.
[48] Nikoshkov A, Sunkari V, Savu O, Forsberg E, Catrina SB, Brismar K. Epigenetic DNA methylation in the promoters of the Igf1 receptor and insulin receptor genes in db/db mice. Epigenetics 2011;6:405–9.
[49] Soejima H, Higashimoto K. Epigenetic and genetic alterations of the imprinting disorder Beckwith-Wiedemann syndrome and related disorders. J Hum Genet 2013;58:402–9.
[50] Renfree MB, Suzuki S, Kaneko-Ishino T. The origin and evolution of genomic imprinting and viviparity in mammals. Philos Trans R Soc Lond B Biol Sci 2013;368:20120151.
[51] Fowden AL, Sibley C, Reik W, Constancia M. Imprinted genes, placental development and fetal growth. Horm Res 2006;65(Suppl 3):50–8.
[52] Reddy PM, Reddy PR. Regulation of DNA methyltransferase in the testis of rat. Biochem Int 1988;16:543–7.
[53] Reddy PM, Reddy PR. Differential regulation of DNA methylation in rat testis and its regulation by gonadotropic hormones. J Steroid Biochem 1990;35:173–8.
[54] Zhang Y, Fatima N, Dufau ML. Coordinated changes in DNA methylation and histone modifications regulate silencing/derepression of luteinizing hormone receptor gene transcription. Mol Cell Biol 2005;25:7929–39.
[55] Dufau ML, Liao M, Zhang Y. Participation of signaling pathways in the derepression of luteinizing hormone receptor transcription. Mol Cell Endocrinol 2010;314:221–7.
[56] Murray AA, Swales AK, Smith RE, Molinek MD, Hillier SG, Spears N. Follicular growth and oocyte competence in the in vitro cultured mouse follicle: effects of gonadotrophins and steroids. Mol Hum Reprod 2008;14:75–83.
[57] Brogan RS, Mix S, Puttabyatappa M, VandeVoort CA, Chaffin CL. Expression of the insulin-like growth factor and insulin systems in the luteinizing macaque ovarian follicle. Fertil Steril 2010;93:1421–9.
[58] Ikuyama S, Niller HH, Shimura H, Akamizu T, Kohn LD. Characterization of the 5′-flanking region of the rat thyrotropin receptor gene. Mol Endocrinol 1992; 6:793–804.

[59] Yokomori N, Tawata M, Saito T, Shimura H, Onaya T. Regulation of the rat thyrotropin receptor gene by the methylation-sensitive transcription factor GA-binding protein. Mol Endocrinol 1998;12:1241–9.

[60] Xing M, Usadel H, Cohen Y, Yokumaru Y, Guo Z, Westra WB. Methylation of the thyroid-stimulating hormone receptor gene in epithelial thyroid tumors: a marker of malignancy and a cause of gene silencing. Cancer Res 2003;63:2316–21.

[61] Provenzano MJ, Fitzgerald MP, Krager K, Domann FE. Increased iodine uptake in thyroid carcinoma after treatment with sodium butyrate and decitabine (5-Aza-dC). Otolaryngol Head Neck Surg 2007;137:722–8.

[62] Murgatroyd C, Spengler D. Epigenetic programming of the HPA axis: early life decides. Stress 2011;14:581–9.

[63] Murgatroyd C. Epigenetic programming of neuroendocrine systems during early life. Exp Physiol 2014;99:62–5.

[64] Auger CJ, Coss D, Auger AP, Forbes-Lorman RM. Epigenetic control of vasopressin expression is maintained by steroid hormones in the adult male rat brain. Proc Natl Acad Sci USA 2011;108:4242–7.

[65] Forga L, Anda E, Martinez de Esteban JP. Paraneoplastic hormonal syndromes [in Spanish]. An Sist Sanit Navar 2005;28:213–26.

[66] Satoh M, Ogikubo S, Yoshizawa-Ogasawara A. Correlation between clinical phenotypes and X-inactivation patterns in six female carriers with heterozygote vasopressin type 2 receptor gene mutations. Endocr J 2008;55:277–84.

[67] Frieling H, Bleich S, Otten J, Romer KD, Kornhuber J, de Zwaan M, et al. Epigenetic downregulation of atrial natriuretic peptide but not vasopressin mRNA expression in females with eating disorders is related to impulsivity. Neuropsychopharmacology 2008;33:2605–9.

[68] Hillemacher T, Frieling H, Luber K, Yazici A, Muschler MA, Lenz B, et al. Epigenetic regulation and gene expression of vasopressin and atrial natriuretic peptide in alcohol withdrawal. Psychoneuroendocrinology 2009;34:555–60.

[69] Glahn A, Riera Knorrenschild R, Rhein M, Haschemi Nassab M, Groschl M, Heberlein A, et al. Alcohol-induced changes in methylation status of individual CpG sites, and serum levels of vasopressin and atrial natriuretic peptide in alcohol-dependent patients during detoxification treatment. Eur Addict Res 2013;20:143–50.

CHAPTER

13

DNA Methylation and Development

OUTLINE

13.1 INTRODUCTION

In mammals, oocyte-derived mRNAs are degraded shortly after fertilization. Therefore, embryonic genome activation and production of embryo-derived transcript must have to occur early during development. The major activation of transcription occurs at species-specific stages of development – for example, at the end of the second cell cycle in mice, the third cycle in pigs, and fourth cycle in humans. The most marked morphological changes occur within the nucleus, where functional

M. Neidhart: *DNA Methylation and Complex Human Disease.*
DOI: http://dx.doi.org/10.1016/B978-0-12-420194-1.00013-0

ribosome-synthesizing nucleoli develop from inactive precursor bodies. From a molecular point of view, a prerequisite for embryonic genome activation is the epigenetic remodeling of the specialized parental genomes into the totipotent genome of the zygote and initial blastomeres. Mammalian development is characterized by a bimodal DNA methylation reprogramming, with a first round occurring during gametogenesis and second round occurring after fertilization during pre-implantation embryonic development [1–3]. Epigenetic mechanisms contribute to the regulation of gene expression in the early embryo, in particular controlling the expression of specific transcription factors [4] (Figure 13.1).

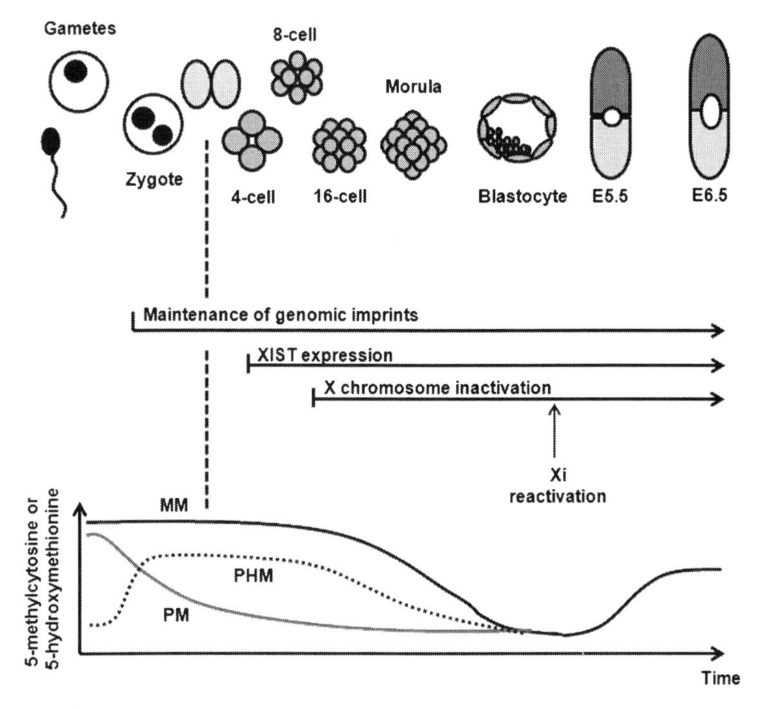

FIGURE 13.1 **Epigenetic regulation in early development.** The trophectoderm lineage is the first to differentiate from cells that have an outside position in morula stage embryos. At the blastocyst stage, the hypoblast is specified. Inner cell-mass cells will give rise to the developing mouse embryo, whereas the TE and hypoblast form extraembryonic tissues. Genomic imprints are parent-of-origin-specific marks that are maintained during embryogenesis and regulate the differential expression of the maternal and paternal copies of imprinted genes. X chromosome inactivation and reactivation is observed during development of female embryos. The diagram illustrates global changes in DNA methylation and DNA hydroxymethylation levels. In cleavage-stage embryos, the paternal methylation (PM), paternal hydroxymethylation (PHM), and maternal (MM, maternal methylation) genomes are differentially marked by 5-methylcytosine and 5-hydroxymethylcytosine. Both decrease during development to the blastocyst stage, after which the 5-methylcytosine increases as the embryonic lineages are formed. E5.5 and E6.5 indicate the weeks of embryonic development in mice.

During the period of early embryonic development, methylation patterns of the germline and somatic cell lineages are established [5]. During the cleavage phase (i.e., the early cell divisions that occur as a fertilized egg begins to develop into an embryo), methylation in the zygote's genome is almost completely removed. After implantation, as the cells produced during the cleavage phase begin to organize themselves (i.e., gastrulation), the organism's methylation patterns are re-established by *de novo* methylation [6]. *De novo* DNA methylation is mediated by DNA methyltrasferases DNMT3a and DNMT3b, and is guided by specialized chromatin structures on the unmethylated allele [7]. Such patterning and repatterning of methylation marks also occurs in trophoblast lineages (i.e., the various specialized cells comprising the placenta). Proper setting and resetting of methylation marks throughout development is crucial for the proper health and development of the embryo. Importantly, gender-specific DNA methylation of particular loci occurs, forming the basis of genomic imprinting. Thus, a number of imprinted genes are expressed from a single parental allele. Genomic imprinting implies that information for expression or repression of a parental allele must be maintained from the maternal and paternal germlines throughout fertilization and development. Since diffusible transcription factors have equal access to both alleles, genomic imprinting is best explained by a mechanism that links gene regulatory information to the DNA or chromatin of the gene locus [8]. Epigenetic regulation is also highlighted by the process of X inactivation in female embryos [9]. Thereby, the X-linked gene dosage is equalized to one active X chromosome between male (XY) and female (XX) cells. Inactivation of the paternally inherited X chromosome is initiated at the four-cell stage. Imprinted inactivation of the paternal X chromosome is maintained in the extraembryonic lineages, whereas reactivation of the inactive X chromosome (Xi) is observed in the cells of the inner cell mass of the blastocyte. In the embryonic lineages, dosage compensation is re-established at the time of gastrulation by random inactivation of either the maternally or the paternally inherited X chromosome. When considering epigenetic research in general, DNA methylation is the most heavily studied mode of epigenetic regulation [10]. Thus, it is not surprising that the majority of published articles on the epigenetics of maternal lifestyle (e.g., nutrition or smoking) during pregnancy describe associations with DNA methylation.

13.2 DIFFERENT EMBRYONIC LINEAGES

During the two-, four-, and eight-cell and the morula stage, no significant differences in DNA methylation seem to occur. Yet, at the

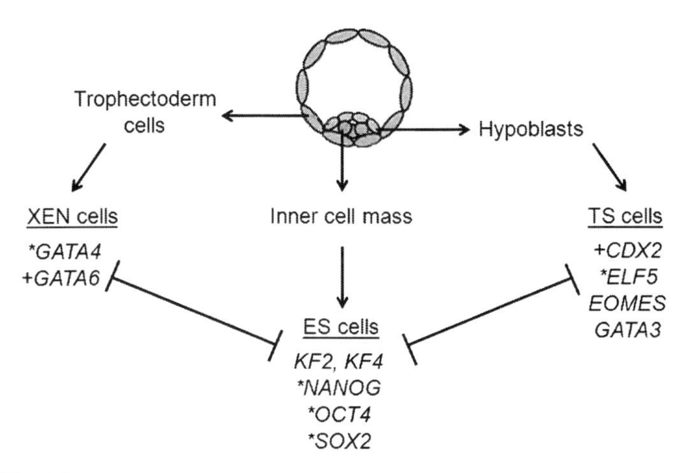

FIGURE 13.2 **Transcriptional control and epigenetic regulation in the lineages of the blastocyst.** The three lineages of the blastocyst can give rise to stem cell lines in culture. Transcription factor networks as observed in trophectoderm stem (TS) cells, extraembryonic endoderm stem (XEN) cells, and embryonic stem (ES) cells are shown, and their mutual antagonistic regulation is indicated. *, regulated by promoter methylation; +, regulated by promoter hydroxymethylation.

blastocyte stage, *de novo* methylation starts preferentially in the inner cell mass cells [11]. The function of epigenetic regulation has been elucidated in different stem cell lines of the early embryo that reflect characteristics of different lineages [8,12] (Figure 13.2). Trophoblast stem (TS) cells and extraembryonic endoderm stem (XEN) cells maintain the capacity to differentiate into the trophoblast and primitive endoderm (PE) lineage, respectively. In both stem cell types, imprinted inactivation of the paternal X chromosome is observed. Embryonic stem (ES) cells are derived from the inner cell mass of the blastocyst. ES cells maintain the potential for differentiating into all cell types of the embryo, which has been impressively demonstrated by using tetraploid aggregation. In contrast to their wide lineage potential, ES cells do normally not contribute to extraembryonic tissues (TE and PE), consistent with characteristics of cells from the inner cell mass of the late blastocyst and early epiblast. A second pluripotent stem cell type can be obtained from the mouse post-implantation epiblast. Similar to ES cells, these epiblast-derived stem cells (EpiSCs) possess a wide lineage differentiation potential in culture. Furthermore, EpiSCs have initiated X inactivation whereas female ES cells possess two active X chromosomes. This indicates that the two pluripotent cell types are distinguished by developmental and epigenetic characteristics. It is interesting to observe that pluripotent stem cells established from most mammals, including humans, resemble the EpiSC type with the notable exception of the rat [13]. This could potentially indicate that the EpiSC cell state is maintained more stably in evolution. The developmentally restricted differentiation potential of the stem cells of the early mouse embryo has been studied for understanding transcriptional and epigenetic regulation in lineage specification.

13.2.1 DNA Methylation in Early Embryogenesis

Gene deletion studies have provided insight into the function of DNA methylation in development. Disruption of DNMT1 results in lethality before E10.5 [14] (weeks of embryonic development in mice). Similarly, loss of DNMT3B is lethal before murine E9.5 [15]. Although deletion of DNMT3A is compatible with embryonic development, DNMT3A-deficient mice die within the first weeks after birth. Interestingly, the specification and function of early embryonic lineages is largely unaffected by loss of DNA methyltransferases. Deficiency for all three DNA methyltransferases in ES cells leads to a complete loss of DNA methylation [16]. These ES cells show increased cell death in differentiation. Notably, the extraembryonic lineages can be established in the absence of DNA methylation, whereas maintenance of differentiated embryonic cell types is dependent on the DNA methylation system [17]. Maintenance of DNA methylation patterns is mediated by the restoration of symmetrical methylation on hemimethylated CpG dinucleotides following DNA replication. This is facilitated by the NP95/UHRF1 protein that recruits DNMT1 to hemimethylated CpG sites [18]. NP95 is required for maintaining DNA methylation patterns, and its disruption in mice leads to lethality after gastrulation. Thus, deletions of NP95 and DNMT1 appear to cause similar phenotypes.

13.2.2 DNA Hydroxymethylation in Early Embryogenesis

Recent studies have reported initial findings suggesting that an additional mode of epigenetic regulation, namely, conversion of 5-methylcytosine (5mC) to 5-hydroxymethylcytosine (5hmC), may play a role in the reprogramming of the zygote [19]. Methylated cytosine in DNA can be modified by the TET family of proteins that catalyze the oxidation of 5mC to 5hmC. TET1 is highly expressed and specific for ES cells [20]. TET1 binds preferentially to CpG-rich sequences at promoters, and its activity has been associated with both activating and repressing functions. It has been suggested that 5hmC could potentially mask the silencing effect of 5mC. In addition, 5hmC appears to be involved in the repression of Polycomb-targeted developmental regulators. In ES cells, TET1-repressed genes include GATA6 and CDX2 (Figure 13.2).

In line with these observations, disruption of TET1 expression predisposes to differentiation into extraembryonic tissues [21]. A recent study [22] of TET1-deficient mice has shown that TET1 is dispensable for maintaining pluripotency, and its loss is compatible with embryonic and postnatal development. This could possibly point towards compensation by other TET1 family proteins. For example, it has been shown that TET3 is required for hydroxymethylation of the paternal genome in zygotes [23]. In mouse cleavage stage embryos, the paternal genome is largely devoid of 5mC but enriched in 5hmC, whereas the maternal genome is marked by 5mC [24]. Loss of maternal TET3 leads to increased developmental failure after implantation. Both 5hmC and 5mC can be further oxidized by the TET dioxygenases to 5-carboxylcytosine (5acC), which in turn is removed by thymine-DNA glycosylase, thereby establishing a mechanism for DNA demethylation [25]. These findings suggest a complex chemistry leading to different modification states of cytosine in DNA, and allowing methyl removal to unmethylated DNA [26]. Thus, similar to histone modifications, DNA methylation has to be understood as a dynamic and reversible modification in development. Further studies will be necessary to understand more completely the implications of these findings on the patterning of methylation marks during development, as well as the effect of exposures on the underlying mechanisms.

13.2.3 Methylation State of Key Transcription Factors

Cell signaling pathways during early development have been well characterized in mice and humans. The activation of GATA4 in embryonic cells is known to drive their differentiation to endoderm. ES cells block their GATA4-induced endoderm differentiation. Transcriptome analysis of the cells' response to GATA4 over time revealed groups of endoderm and mesoderm developmental genes whose expression was induced by GATA4 only when DNA methylation was lost [27], suggesting that DNA methylation restricts the ability of these genes to respond to GATA4, rather than controlling their transcription *per se*. GATA4-binding-site profiles and DNA methylation analyses suggested that DNA methylation modulates the GATA4 response through diverse mechanisms. Thus, epigenetic regulation by DNA methylation functions as a heritable safeguard to prevent transcription factors from activating inappropriate downstream genes, thereby contributing to the restriction of the differentiation potential of somatic cells.

The trophoblast (TR) is the first to differentiate during mammalian embryogenesis and play a pivotal role in the development of the placenta. One key molecular player in the specification of the trophectoderm is CDX2 [28]. CDX2 mutants fail to implant. The promoter of CDX2 is hypomethylated in TS cells compared with blastocysts. In contrast, the DNA methylation status of the promoter regions of OCT4, NANOG, and SOX2 are higher in TS cells compared with blastocysts [29]. CDX2 is co-expressed with OCT4 in the early embryo, but following cell differentiation into the inner cell mass and the trophectoderm, CDX2 expression is restricted to the latter [30]. CDX2 has a conserved role in specifying trophoectoderm specification in mammals [3]. ELF5 is expressed in the human placenta in villous cytotrophoblast cells, but not in post-mitotic syncytiotrophoblast and invasive extravillous cytotrophoblast cells. ELF5 establishes a circuit of mutually interacting transcription factors with CDX2 and EOMES, and the highly proliferative ELF5(+)/CDX2(+) double-positive subset of cytotrophoblast cells demarcates a putative TS cell compartment in the early human placenta. In contrast to placental trophoblast, however, ELF5 is hypermethylated and largely repressed in human ES cells and derived trophoblast cell lines, as well as in induced pluripotent stem cells and murine epiblast stem cells [31]. Thus, these cells exhibit an embryonic lineage-specific epigenetic signature and do not undergo an epigenetic reprogramming to reflect the trophoblast lineage at key loci. Thus, promoter methylation of GATA4, OCT4, NANOG, SOX2, and ELF5 determines the fate of early embryonic stem cells (Figure 13.2).

13.3 MESODERM DEVELOPMENT

In humans and mice, the inner cell mass becomes cavitated forming the amniotic cavity. The innermost cells sealing this cavity form the epiblast and the embryonic disc. The blastocyte enlarges and the embryonic disc develops into an oval shape. This process is also accompanied by the ingression of cells into the space between the epiblast/trophectoderm and the hypoblast. The ingressing cells form mesoderm, which quickly moves into the extra-embryonic region posterior to the embryonic disc. Mesenchymal stem cells (MSCs) are multipotent stem cells of mesodermal origin that can be isolated from various sources and induced into different cell types [32]. Epigenomic changes in DNA methylation and chromatin structure appeared to be critical in the determination of lineage-specific differentiation of MSCs. For instance, TRIP10 locus was identified as the target of Polycomb group protein and modified by DNA methylation during MSC differentiation [32,33]. In human bone marrow-derived MSCs, TRIP10 is hypomethylated in the undifferentiated stage and becomes hypermethylated during MSC-to-liver differentiation, but

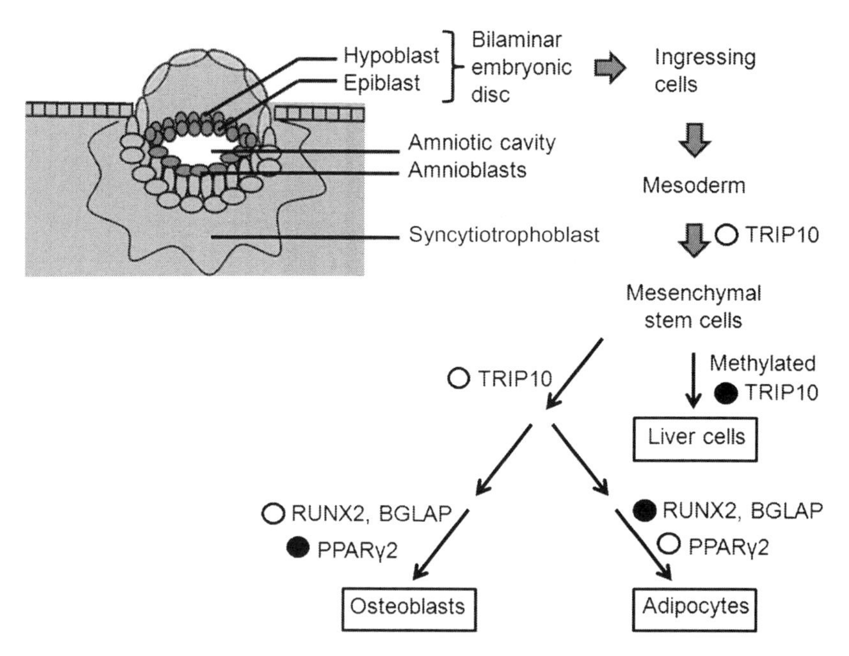

FIGURE 13.3 Mesoderm development and mesenchymal stem cell differentiation (open circle, hypomethylated promoter; closed circle, hypermethylated promoter).

remains hypomethylated during MSC-to-adipocyte differentiation. In addition, there are methylation differences within the promoters of tissue-specific genes between the bone- and adipose-derived MSCs; they correlate with their differences in lineage differentiation potential [32]. For example, osteoblast-specific genes such as RUNX2 and BGLAP are hypermethylated in adipose-derived MSCs as compared to the bone-derived MSCs, whereas PPARγ2, the adipocyte-specific gene, is hypomethylated in adipose-derived MSCs (Figure 13.3).

13.4 ECTODERM DEVELOPMENT

After gastrulation, neural crest cells are specified at the border of the neural plate and the non-neural ectoderm. During neurulation, the borders of the neural plate, also known as the neural folds, converge at the dorsal midline to form the neural tube. Subsequently, neural crest cells from the roof plate of the neural tube undergo an epithelial to mesenchymal transition, delaminating from the neuroepithelium and migrating through the periphery, where they differentiate into varied cell types. Neural crest cells are a transient, multipotent, migratory cell population unique to vertebrates that gives rise to a diverse cell lineage, including melanocytes, craniofacial cartilage and bone, smooth muscle, peripheral and enteric neurons, and glia. The formation of the central nervous system primarily depends on the genetic information stored in the genome. Nevertheless, epigenetic mechanisms through DNA methylation and histone modification are found to be essential for the expression of genes required for various steps during neural development, including neurogenesis and gliogenesis. Neuronal genes are repressed in non-neuronal cells by repressor element 1 silencing transcription factor (REST)-dependent DNA methylation events [34]. The developmental expression of glutamatergic receptor subunits in the post-mitotic neurons is also regulated by REST-dependent DNA methylation [35]. The generation of glias is also regulated by DNA methylation mechanisms. The Janus kinase (JAK)-signal transducer and activator of transcription (STAT) signaling pathway have been shown to be implicated in gliogenesis. The genetic deletion of DNA methyltransferase 1 (DNMT1) leads to premature glia differentiation. Methylation of certain JAK/STAT pathway genes is required for controlling gliogenesis in the correct timing [36].

13.5 ENDODERM DEVELOPMENT

The embryonic endoderm develops into the interior linings of two tubes in the body, the digestive and respiratory tubes. As mentioned above, in early embryos the specification of the pluripotent epiblast and two differentiating lineages (trophectoderm and primitive endoderm) is controlled by transcription factors that are regulated by autoactivating and reciprocal repressive mechanisms as well as by ERK-mediated signaling [37]. Experimental evidence for the involvement of underlying epigenetic modifications (e.g., genomic DNA methylation) in long-term studies has just started to emerge with regard to the gastrointestinal tract. Also, interactions between the microbiota and the host are considered pivotal in the early programming of gut functions [38]. Postnatal decrease of cholecystokinin type A receptor promoter methylation allows cholecystokinin to stimulate pancreatic growth, amylase secretion, and gallbladder contraction [39]. The endogenous glucocorticoid surge in late gestation plays a vital role in the maturation of several organ systems. For this reason, pregnant women at risk of preterm labor are administered synthetic glucocorticoids to promote fetal lung development. Animal studies have shown that fetal synthetic glucocorticoid exposure can cause lifelong changes in endocrine and metabolic function. This is associated with alterations in global DNA methylation [40].

13.6 HEMATOPOIETIC STEM CELLS

Blood is one of the most highly regenerative tissues, with approximately one trillion new cells arising daily in adult human bone marrow (BM). Although adult hematopoietic stem cells (HSCs) remain relatively quiescent, HSCs can enter the cell cycle and either self-renew or differentiate into multipotent progenitors that provide diverse mature blood cells, resulting in a well-known hierarchy of blood cells that maintain homeostasis. Epigenetic regulation is required not only for development but also for tissue homeostasis, which is maintained via the self-renewal and differentiation of somatic stem cells. Accumulating evidence suggests that epigenetic regulators play critical roles in the maintenance of self-renewing HSCs [41]. DNMT3a is a critical epigenetic regulator of HSCs that is required for their efficient differentiation [42]. TET2, on the other hand, limits the expansion and self-renewal activity of HSCs [43].

13.7 CONCLUSION

Reprogramming is important because eggs and sperm develop from specialized cells with stable gene expression profiles. In other words, their genetic information is marked with epigenetic tags. Before the new organism can grow into a healthy embryo, the epigenetic tags must be erased. During embryonic development, cells are directed toward their future lineages, and DNA methylation poses a fundamental epigenetic barrier that guides and restricts differentiation and prevents regression into an undifferentiated state. DNA methylation also plays an important role in sex chromosome dosage compensation, the repression of retrotransposons that threaten genome integrity, the maintenance of genome stability, and the coordinated expression of imprinted genes. However, DNA methylation marks must be globally removed to allow for sexual reproduction and the adoption of the specialized, hypomethylated epigenome of the primordial germ cell and the pre-implantation embryo. Recent technological advances in genome-wide DNA methylation analysis and the functional description of novel enzymatic DNA demethylation/hydroxymethylation pathways have provided significant insights into the molecular processes that prepare the mammalian embryo for normal development.

References

[1] Reik W, Dean W, Walter J. Epigenetic reprogramming in mammalian development. Science 2001;293:1089–93.
[2] Santos F, Dean W. Epigenetic reprogramming during early development in mammals. Reproduction 2004;127:643–51.

[3] Oestrup O, Hall V, Petkov SG, Wolf XA, Hyldig S, Hyttel P. From zygote to implantation: morphological and molecular dynamics during embryo development in the pig. Reprod Domest Anim 2009;44(Suppl. 3):39–49.
[4] Leeb M, Wutz A. Establishment of epigenetic patterns in development. Chromosoma 2012;121:251–62.
[5] Maccani MA, Marsit CJ. Epigenetics in the placenta. Am J Reprod Immunol 2009;62:78–89.
[6] Jaenisch R. DNA methylation and imprinting: why bother? Trends Genet 1997;13:323–9.
[7] Ko YG, Nishino K, Hattori N, Arai Y, Tanaka S, Shiota K. Stage-by-stage change in DNA methylation status of Dnmt1 locus during mouse early development. J Biol Chem 2005;280:9627–34.
[8] Ferguson-Smith AC. Genomic imprinting: the emergence of an epigenetic paradigm. Nat Rev Genet 2011;12:565–75.
[9] Augui S, Nora EP, Heard E. Regulation of X-chromosome inactivation by the X-inactivation centre. Nat Rev Genet 2011;12:429–42.
[10] Bird A. Perceptions of epigenetics. Nature 2007;447:396–8.
[11] Fulka J, Fulka H, Slavik T, Okada K, Fulka Jr. J. DNA methylation pattern in pig in vivo produced embryos. Histochem Cell Biol 2006;126:213–17.
[12] Rossant J. Stem cells and lineage development in the mammalian blastocyst. Reprod Fertil Dev 2007;19:111–18.
[13] Blair K, Wray J, Smith A. The liberation of embryonic stem cells. PLoS Genet 2011;7: e1002019.
[14] Li E, Bestor TH, Jaenisch R. Targeted mutation of the DNA methyltransferase gene results in embryonic lethality. Cell 1992;69:915–26.
[15] Okano M, Bell DW, Haber DA, Li E. DNA methyltransferases Dnmt3a and Dnmt3b are essential for de novo methylation and mammalian development. Cell 1999;99:247–57.
[16] Tsumura A, Hayakawa T, Kumaki Y, Takebayashi S, Sakaue M, Matsuoka C, et al. Maintenance of self-renewal ability of mouse embryonic stem cells in the absence of DNA methyltransferases Dnmt1, Dnmt3a and Dnmt3b. Genes Cells 2006;11:805–14.
[17] Sakaue M, Ohta H, Kumaki Y, Oda M, Sakaide Y, Matsuoka C, et al. DNA methylation is dispensable for the growth and survival of the extraembryonic lineages. Curr Biol 2010;20:1452–7.
[18] Sharif J, Muto M, Takebayashi S, Suetake I, Iwamatsu A, Endo TA, et al. The SRA protein Np95 mediates epigenetic inheritance by recruiting Dnmt1 to methylated DNA. Nature 2007;450:908–12.
[19] Wossidlo M, Nakamura T, Lepikhov K, Marques CJ, Zakhartchenko V, Boiani M, et al. 5-Hydroxymethylcytosine in the mammalian zygote is linked with epigenetic reprogramming. Nat Commun 2011;2:241.
[20] Wu H, D'Alessio AC, Ito S, Xia K, Wang Z, Cui K, et al. Dual functions of Tet1 in transcriptional regulation in mouse embryonic stem cells. Nature 2011;473:389–93.
[21] Ficz G, Branco MR, Seisenberger S, Santos F, Krueger F, Hore TA, et al. Dynamic regulation of 5-hydroxymethylcytosine in mouse ES cells and during differentiation. Nature 2011;473:398–402.
[22] Dawlaty MM, Ganz K, Powell BE, Hu YC, Markoulaki S, Cheng AW, et al. Tet1 is dispensable for maintaining pluripotency and its loss is compatible with embryonic and postnatal development. Cell Stem Cell 2011;9:166–75.
[23] Gu TP, Guo F, Yang H, Wu HP, Xu GF, Liu W, et al. The role of Tet3 DNA dioxygenase in epigenetic reprogramming by oocytes. Nature 2011;477:606–10.
[24] Iqbal K, Jin SG, Pfeifer GP, Szabo PE. Reprogramming of the paternal genome upon fertilization involves genome-wide oxidation of 5-methylcytosine. Proc Natl Acad Sci USA 2011;108:3642–7.

[25] He YF, Li BZ, Li Z, Liu P, Wang Y, Tang Q, et al. Tet-mediated formation of 5-carboxylcytosine and its excision by TDG in mammalian DNA. Science 2011;333:1303–7.
[26] Nabel CS, Kohli RM. Molecular biology. Demystifying DNA demethylation. Science 2011;333:1229–30.
[27] Oda M, Kumaki Y, Shigeta M, Jakt LM, Matsuoka C, Yamagiwa A, et al. DNA methylation restricts lineage-specific functions of transcription factor Gata4 during embryonic stem cell differentiation. PLoS Genet 2013;9:e1003574.
[28] Strumpf D, Mao CA, Yamanaka Y, Ralston A, Chawengsaksophak K, Beck F, et al. Cdx2 is required for correct cell fate specification and differentiation of trophectoderm in the mouse blastocyst. Development 2005;132:2093–102.
[29] Huang X, Han X, Uyunbilig B, Zhang M, Duo S, Zuo Y, et al. Establishment of bovine trophoblast stem-like cells from in vitro-produced blastocyst-stage embryos using two inhibitors. Stem Cells Dev 2014.
[30] Niwa H, Toyooka Y, Shimosato D, Strumpf D, Takahashi K, Yagi R, et al. Interaction between Oct3/4 and Cdx2 determines trophectoderm differentiation. Cell 2005;123:917–29.
[31] Hemberger M, Udayashankar R, Tesar P, Moore H, Burton GJ. ELF5-enforced transcriptional networks define an epigenetically regulated trophoblast stem cell compartment in the human placenta. Hum Mol Genet 2010;19:2456–67.
[32] Leu YW, Huang TH, Hsiao SH. Epigenetic reprogramming of mesenchymal stem cells. Adv Exp Med Biol 2013;754:195–211.
[33] Hsiao SH, Lee KD, Hsu CC, Tseng MJ, Jin VX, Sun WS, et al. DNA methylation of the Trip10 promoter accelerates mesenchymal stem cell lineage determination. Biochem Biophys Res Commun 2010;400:305–12.
[34] Ballas N, Grunseich C, Lu DD, Speh JC, Mandel G. REST and its corepressors mediate plasticity of neuronal gene chromatin throughout neurogenesis. Cell 2005;121:645–57.
[35] Rodenas-Ruano A, Chavez AE, Cossio MJ, Castillo PE, Zukin RS. REST-dependent epigenetic remodeling promotes the developmental switch in synaptic NMDA receptors. Nat Neurosci 2012;15:1382–90.
[36] Fan G, Martinowich K, Chin MH, He F, Fouse SD, Hutnick L, et al. DNA methylation controls the timing of astrogliogenesis through regulation of JAK–STAT signaling. Development 2005;132:3345–56.
[37] Albert M, Peters AH. Genetic and epigenetic control of early mouse development. Curr Opin Genet Dev 2009;19:113–21.
[38] Lalles JP. Long term effects of pre- and early postnatal nutrition and environment on the gut. J Anim Sci 2012;90(Suppl 4):421–9.
[39] Matsusue K, Takiguchi S, Takata Y, Funakoshi A, Miyasaka K, Kono A. Expression of cholecystokinin type A receptor gene correlates with DNA demethylation during postnatal development of rat pancreas. Biochem Biophys Res Commun 1999;264:29–32.
[40] Crudo A, Petropoulos S, Suderman M, Moisiadis VG, Kostaki A, Hallett M, et al. Effects of antenatal synthetic glucocorticoid on glucocorticoid receptor binding, DNA methylation, and genome-wide mRNA levels in the fetal male hippocampus. Endocrinology 2013;154:4170–81.
[41] Sashida G, Iwama A. Epigenetic regulation of hematopoiesis. Int J Hematol 2012;96:405–12.
[42] Challen GA, Sun D, Jeong M, Luo M, Jelinek J, Berg JS, et al. Dnmt3a is essential for hematopoietic stem cell differentiation. Nat Genet 2012;44:23–31.
[43] Moran-Crusio K, Reavie L, Shih A, Abdel-Wahab O, Ndiaye-Lobry D, Lobry C, et al. Tet2 loss leads to increased hematopoietic stem cell self-renewal and myeloid transformation. Cancer Cell 2011;20:11–24.

CHAPTER

14

DNA Methylation in Growth Retardation

OUTLINE

14.1 INTRODUCTION

Dietary protein restriction in animal models and protein malnutrition in humans causes remarkable changes in the methyl transfer *in vivo*. Although the specific consequences of perturbation in maternal and fetal methyl transfer remain to be determined, a profound influence is suggested by the demonstrated relationship between maternal folate and vitamin B12 insufficiency and metabolic programming [1]. Indeed, a growing body of evidence supports the notion that epigenetic changes

M. Neidhart: DNA Methylation and Complex Human Disease.
DOI: http://dx.doi.org/10.1016/B978-0-12-420194-1.00014-2

such as DNA methylation and histone modifications, both involving chromatin remodeling, contribute to fetal metabolic programming. The complex metabolic networks that modulate fetal metabolic programming and putative answers are investigated by systems biology [2]. This picture summarizes the putative molecular mechanisms linking impaired nutrient availability during the fetal period with adult chronic diseases such as metabolic and cardiovascular disorders, including coronary heart disease, type 2 diabetes, and insulin resistance. Epigenetic modifications, such as DNA methylation and covalent posttranslational histone modifications, provide a molecular explanation of how these complex metabolic networks coordinately influence fetal metabolic programming. Genomic imprinting is an epigenetic phenomenon that gives rise to parent-of-origin-specific, monoallelic expression of genes during development [3]. Approximately 120 imprinted genes have been discovered in humans. These unusual genes are organized in evolutionarily conserved gene clusters that can comprise up to several megabases of DNA. The monoallelic expression of imprinted genes is mediated by so-called "imprinting control regions" (ICRs). These essential *cis*-acting sequences of several kilobases in size are CpG-rich and marked by cytosine methylation on one of the two parental alleles only. The monoallelic DNA methylation imprints at ICRs are acquired during male or female gametogenesis. After fertilization, they are maintained somatically throughout development and postnatal life, which confers an epigenetic "memory" to the ICR. Because of their allelic DNA methylation, most ICRs are functionally active on one of the two parental chromosomes only, which explains their parental allele-specific effects on gene expression [3,4] (Figure 14.1). Epigenetic alterations were found to be causally involved in imprinting-related diseases. These so-called epimutations involve gains or losses of DNA methylation at ICRs and at other differentially methylated regions (DMRs), and affect the regulation of one or multiple imprinted domains, although in most cases such epimutations are thought to arise early in development and, in exceptional cases, are due to perturbed imprint acquisition in germ cells. The aberrant DNA methylation profiles at ICRs often result from perturbation of methylation maintenance mechanisms. Several studies in animals and humans have indicated that this intrinsic maintenance process can be readily influenced by environmental cues [4]. For instance, superovulation, embryo culture, and other *in vitro* procedures used in the fertility clinic could be responsible for the reported increased occurrence of imprinting disorders in babies conceived by assisted reproduction [5]. Primary epimutations at imprinted gene clusters involve the ICR, which shows either reduced or increased DNA methylation (hypomethylation and hypermethylation, respectively).

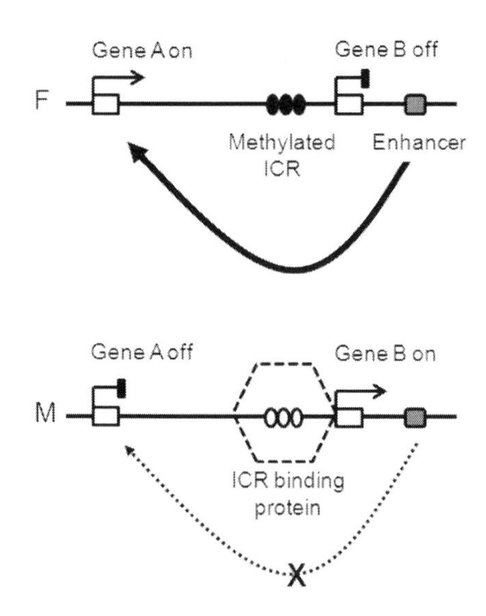

FIGURE 14.1 **Principle of genomic imprinting.** A and B are genes that are expressed from only the paternal (F, father) or maternal (M, mother) chromosome. In this example, B is expressed from the maternal chromosome only; A is expressed only from the paternal chromosome. The two genes share an enhancer region, located downstream of B. The imprinting control region (ICR) in the promoter of B is a boundary element, controlled by DNA methylation. In the upper figure, the paternal ICR is methylated (filled circles), preventing the interaction of ICR binding proteins. This allows the enhancer to contact the promoters of the paternal gene A, allowing it to be transcripted. The paternal gene B is silenced by methylation. In the lower figure, an ICR binding protein binds to the unmethylated maternal ICR. This prevents the promoters located in the A gene from interacting with the enhancer element, resulting in transcriptional silencing of A.

In several imprinting-related disorders, this seems to be the most frequent causal mechanism [6]. Well-studied examples are provided by two imprinted domains at the telomeric region of the short arm of chromosome 11 (11p15.5). These two growth-related gene clusters are involved in Silver-Russell syndrome (SRS), characterized by intrauterine growth restriction (IUGR), and in Beckwith-Wiedemann syndrome (BWS), characterized by fetal overgrowth and high birth weight and height. A third imprinting disease, which is also often caused by primary epimutations, is transient neonatal diabetes mellitus (TNDM), a neonatal type of diabetes genetically linked to chromosome 6q24. In this chapter, we will show how complex interactions of genetic and epigenetic factors are involved in developmental disorders.

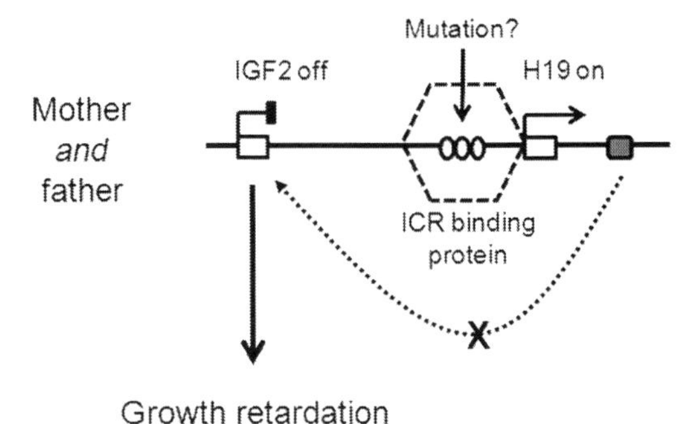

FIGURE 14.2 Loss of imprinting of the IGF2-H19 domain in Silver-Russell syndrome.

14.2 SILVER-RUSSELL SYNDROME

Silver-Russell syndrome is a rare (incidence 1:100,000) and mostly sporadic disorder characterized by intrauterine and postnatal growth retardation, facial dysmorphism, feeding difficulties, and body asymmetry. It is a genetically heterogeneous disorder in which, in rare cases, maternal uniparental disomy of chromosome 7 (UPD7) and sometimes other chromosomal alterations are detected. Epigenetic studies also have linked this syndrome to one of the two imprinted gene clusters on human chromosome 11p15.5, namely the IGF2/H19 domain [7] (Figure 14.2). This growth-related locus comprises the paternally expressed IGF2 and INS (imprinted in the yolk sac only) genes and a growth-regulating non-coding RNA, called H19, expressed from the maternal chromosome. The domain is controlled by an intergenic ICR located several kilobases upstream of H19, called the H19-ICR. More than 50% of SRS cases are caused by hypomethylation at the ICR of the IGF2/H19 locus [7,8], which provides a molecular explanation for this severe growth restriction syndrome [9]. The loss of methylation occurred early in development or could sometimes be caused by a constitutive genetic mutation(s). Whether other imprinted genes also contribute to SRS remains unclear. However, the maternal UPD7 observed in a minority of the patients suggests that a growth-regulating imprinted gene on this chromosome could also be involved. One candidate on this chromosome is the maternally expressed growth factor receptor-bound protein 10 (GRB10) gene [10,11].

14.3 BECKWITH-WIEDEMANN SYNDROME

Contrary to SRS, BWS is characterized by fetal and postnatal overgrowth and also presents an increased risk of childhood cancers

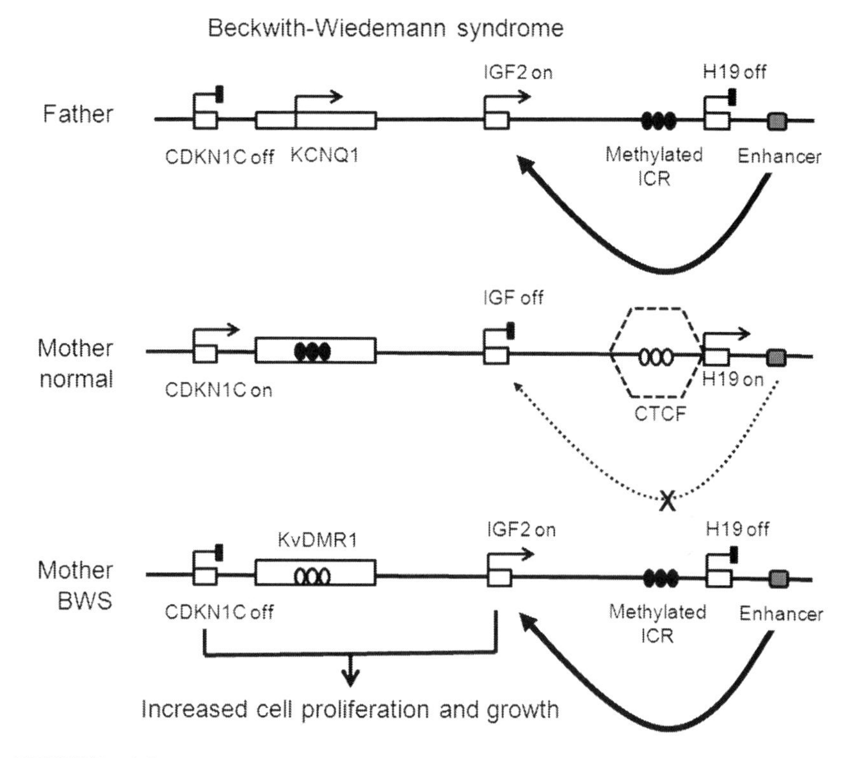

FIGURE 14.3 Loss of imprinting of the IGF2-H19 and KCNQ1-CDKN1C domains in Beckwith-Wiedemann syndrome.

(10% of all patients). A minority of children with Beckwith-Wiedemann syndrome (BWS) develop Wilms' tumor (kidney cancer), and some present with adrenocortical carcinoma. Other clinical features of this rare syndrome (incidence 1:15,000) include macroglossia, midline abdominal wall defects, earlobe creases or ear pits, and neonatal hypoglycemia. BWS maps to chromosome 11p15. Besides the distal IGF2/H19 imprinted domain, this chromosomal region also comprises the large (1-Mb) growth-regulating KCNQ1 imprinting cluster (Figure 14.3). The key gene in this second, proximally located cluster is CDKN1C (p57KIP2), a negative regulator of the cell cycle. A frequently observed loss of CDKN1C expression in cancer suggests that this is also a tumor suppressor gene [12].

Epimutations are responsible for approximately 60–70% of BWS cases. In approximately 10% of the patients there is hypermethylation at the H19-ICR, which leads to biallelic (and therefore increased) IGF2 expression responsible for the observed overgrowth phenotype [9]. Strikingly, this is the exact opposite epigenetic change to that observed in SRS, in which this ICR loses its DNA methylation during early development.

CTCF is thought to be a major DNA-binding protein, which is able to establish insulators. Specifically, CTCF binding can block enhancer function when it occurs between enhancers and promoters. One of the best illustrations of such an enhancer-insulating function is provided by the H19/IGF2 locus above. The differentially methylated ICR of this imprinted locus contains a cluster of CTCF-binding sites, and imprinted H19/IGF2 expression is regulated by selective binding of CTCF to the unmethylated allele. CTCF binding on the maternal chromosome mediates an allele-specific higher-order chromatin configuration at the domain, and thus constitutes an enhancer-blocking transcriptional insulator of the IGF2 gene [13]. By contrast, methylation of the paternal allele prevents CTCF binding and thus prevents establishment of the insulator function, allowing IGF2 expression on this parental chromosome. CTCF collocates with cohesin, and CTCF binding is strictly required to recruit cohesin to the chromatin [14]. Cohesin play a critical role in maintaining CTCF-mediated chromatin conformation. Mutation of all CTCF-binding sites in the H19-ICR leads to aberrant gain of DNA methylation on the maternal allele during post-implantation development [15]. Similarly, in BWS, small maternally inherited ICR deletions (1.4–2 kb) that remove one or more of the CTCF sites were reported to be associated with an aberrant gain of DNA methylation at the remaining target sites, leading to increased expression of IGF2 [16,17].

However, the most commonly observed epimutation in BWS affects imprinting along the KCNQ1 imprinted domain, leading to a strong reduction in CDKN1C expression. In particular, at the domain's ICR (called the KvDMR1) there is loss of DNA methylation early in development. The lack of DNA methylation on both of the alleles of this ICR induces gene silencing along the domain on both the parental chromosomes, including that of CDKN1C, as mentioned above. The resulting loss of CDKN1C expression is thought to be the main determinant in these cases of BWS, since deleterious point mutations in CDKN1C, on their own, also induce an overgrowth syndrome [18]. A striking observation has been that cohorts of BWS babies present an increased frequency of monozygotic twinning. The affected twins are females and are discordant, with only one of the two girls showing the clinical features of the overgrowth syndrome [19]. Methylation studies have shown that there is often loss of methylation at the ICR of the KCNQ1 domain, with the epimutation occurring in the affected twin girl. Methylation changes in the peripheral blood, however, are often observed in both twins. In some monozygotic female twins, hypomethylation was also detected at the H19-ICR and the ICR of the DLK1-DIO3 imprinted domain. Similarly to the H19-ICR, DLK1-DIO3 ICR (type 3 iodothyronine deiodinase gene) is involved in the epigenetic process that causes a subset of genes to be regulated based on their

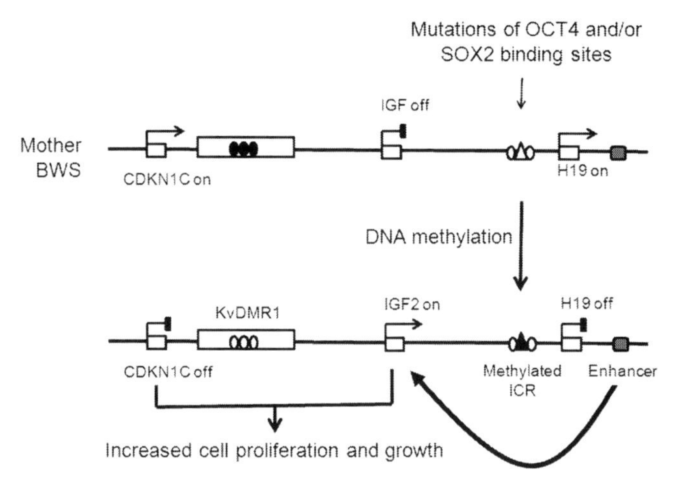

FIGURE 14.4 Mutations in the OCT4 and/or SOX2 binding sites can change the methylation pattern at the H19-ICR. This produced a Beckwith-Wiedemann syndrome phenotype. It is transmitted by the female germline cells only.

parental origin. Significantly, in the overgrowth syndrome, discordant monozygotic female twins have been reported that have hypomethylation at the H19-ICR, again affecting only one of the two girls [20]. One interpretation for this intriguing phenomenon is that the methylation losses somehow trigger the twinning process. Alternatively, female preimplantation embryos are more prone to loss of DNA methylation than early male embryos, and this effect is exacerbated by the increased number of cell divisions due the monozygotic twinning process.

Pluripotency factors could be involved in epigenetic maintenance processes at imprinted loci. Recent genetic studies indicate that this is indeed the case. Octamer-binding protein 4 (OCT4, also known as POU5F1), along with other transcription factors (for example NANOG), is involved in early embryogenesis and stem cell pluripotency through modulation of DNA methylation [21]. Point mutations and small deletions at OCT4- and SOX2-binding sites in BWS patients at the H19-ICR could were found to correlate with acquisition of DNA methylation at this ICR-controlling IGF2 expression [22–24] (Figure 14.4). In the reported cases, the defects were transmitted through the female germline only, with no phenotypic consequences following paternal transmission of the genetic mutations [22,24]. This interesting finding strongly suggests that binding of pluripotency factors to this ICR, on the unmethylated maternal allele, is essential in keeping this parental allele unmethylated through pre-implantation development. SOX2–OCT4 motifs (octamer motifs and their flanking sequences) in the ICR determine the cell type, DNA region, and allele specificity of DNA methylation [25,26].

Such processes can also influence the expression of multiple other genes. In rare cases of BWS that presented loss of DNA methylation at the KvDMR1 ICR (the KCNQ1 domain), it was found that this methylation defect was linked to a frame-shift mutation in exon 6 of NLRP2 [27], revealing a *trans* mechanism in BWS. NLRP2 and NLRP7 (also called NALP2 and NALP7) encode proteins that comprise a nucleotide-binding oligomerization domain, a leucine-rich repeat, and a pyrin domain (NLRP), and belong to a family of cytoplasmic proteins of unknown function comprising 14 members. NALP proteins are characterized by an N-terminal pyrin domain (PYD) and are involved in the activation of caspase-1 by Toll-like receptor 4. They may also be involved in protein complexes that activate pro-inflammatory caspases [28]. NLRP2 is highly homologous to NLRP7, and both proteins are expressed in oocytes and pre-implantation embryos.

14.4 TRANSIENT NEONATAL DIABETES MELLITUS

Transient neonatal diabetes mellitus (TNDM) is a growth deficiency syndrome characterized by IUGR and persistent hyperglycemia during the first 6 weeks of life owing to a lack of insulin. This extremely rare syndrome (incidence 1:100,000) has been linked to paternal uniparental disomy of chromosome 6 (UPD6) and to large duplications of region 6q24 on the paternal chromosome. In approximately 50% of cases, however, the disease seems not to be genetic in origin and is associated with defective DNA methylation at the imprinted transcription factor gene ZAC/PLAG1, localized on chromosome 6q24. ZAC encodes a zinc finger protein that regulates apoptosis and cell cycle arrest. The protein is expressed at very high levels in insulin-producing cells in the human fetal pancreas, but not adult islets. ZAC can also function as a transcriptional activator of CDKN1C. ZAC is believed to control induction of the pituitary adenylate cyclase-activating polypeptide (PACAP), a strong activator of glucose-stimulated insulin secretion. These features of the ZAC gene make this a strong candidate for the pathogenesis of TNDM [29–31].

ZAC and the long non-coding RNA (lncRNA) HYMAI are both paternally expressed from a common promoter that is also the so-called P1/ICR [31,32]. Normally, the maternal alleles are silenced. However, in some tissues ZAC is biallelically expressed from an upstream promoter (P2) [31] (Figure 14.5). In the pancreas and/or pituitary, this is thought to induce TNDM [31,33,34]. As mentioned above, the ZAC transcription factor is involved in cell-cycle control and insulin secretion in the pancreas.

It is possibly through the effects of ZAC on other imprinted loci that TNDM shows clinical overlap with BWS [35]. Mouse studies have shown that ZAC regulates the expression of several other imprinted

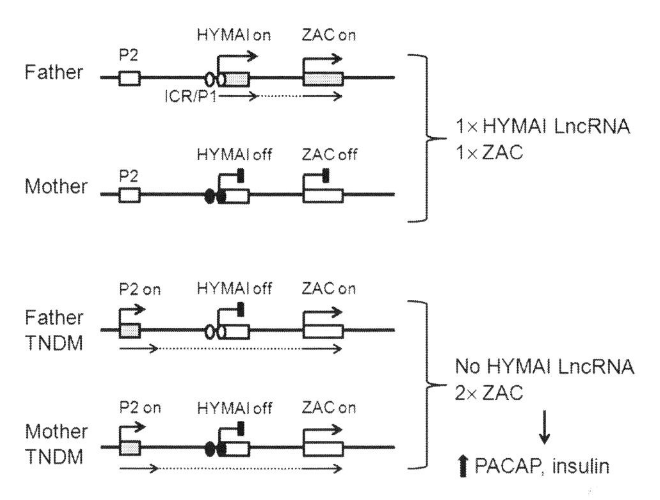

FIGURE 14.5 Imprinting at the ZAC-HYMAI locus and biallelic expression of ZAC in transient neonatal diabetes mellitus.

genes, including IGF2 and CDKN1C, and influences the allelic silencing activity of the ICR of the KCNQ1 imprinted domain [36].

Mutations of zinc finger protein 57 homolog (ZFP57), a transcription repressor, is also associated with a TNDM phenotype. ZFP57 is a nuclear protein, which protects ICRs and other DMRs against loss of methylation in the embryo. Its key role initially emerged from studies on TNDM pedigrees, in which loss of DNA methylation at multiple DMRs was found to be linked to homozygous mutations in the ZFP57 gene [37]. Subsequently, additional loss-of-function mutations were identified at the human ZFP57 gene. A recently discovered 1-bp deletion resulting in premature translation termination and a truncated protein was associated with hypomethylation at multiple maternal-methylated ICRs in TNDM patients [37,38]. Gene-targeting experiments in mice indicated that this protein is required for the early embryonic maintenance of DNA methylation imprints [39], and binds to a methylated hexanucleotide motif (TGCCGC) that is present in most ICRs [40]. This critical transcription factor is expressed in pluripotent cells and maintains the somatic methylation of imprinted domains through the recruitment of KAP1 (also called TRIM28, TIF1β), which recruits repressive chromatin modifiers, including ESET (SETDB1), a lysine methyltransferase that deposits H3K9me3 onto chromatin (Figure 14.6). The latter repressive mark is recognized by the UHRF1 (NP95, ICBP90), which facilitates recruitment of the maintenance DNMTs following DNA replication [41].

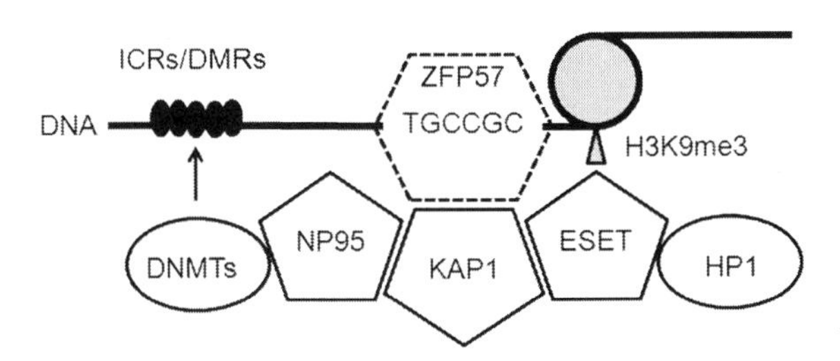

FIGURE 14.6 ZFP57-mediated regulation of allele-specific DNA methylation at multiple ICRs/DMRs (differentially methylated regions). ZFP57 acts as an anchor for KAP1, which in turn recruits other major epigenetic regulators such as ESET/HP1, NP95, and the DNMTs.

Although the precise relationship between these different protein factors remains unclear, combined, these proteins ensure that DNA methylation imprints are efficiently maintained, particularly during the critical preimplantation stages of development [42].

Other mechanisms can be involved in the pathogenesis of similar syndromes. Ciliary neurotrophic factor may regulate mitochondrial function by upregulating transcription factor A, mitochondrial (TFAM) [43]. Altered increased DNA methylation of the TFAM promoter and decreased mitochondrial DNA copy number are found in adolescents with metabolic syndrome components, in particular insulin resistance [44,45].

Furthermore, exploring changes in the pattern of DNA methylation in the human umbilical cord has a tremendous potential because of the pluripotency of this tissue. Promoter methylation of the peroxisome proliferator-activated receptor gamma coactivator 1α (PPARGC1A) gene in umbilical cord of newborns is positively correlated with maternal body mass index [2]. Aberrant DNA methylation of IFG2 in umbilical cord blood was associated with overweight or obesity in a multiethnic cohort [46]. In addition, umbilical cord tissue methylation at particular CpGs of retinoid X receptor-α and endothelial nitric oxide synthase from healthy neonates was associated with childhood fat mass [47].

14.5 PSEUDOHYPOPARATHYROIDISM TYPE 1B (PHP1B)

Pseudohypoparathyroidism represents a heterogeneous group of disorders whose common feature is end-organ resistance to parathyroid hormone, leading to hypocalcemia, hyperphosphatemia, and obesity. Patients have a short stature, a rounded face, and often mild mental retardation, and other characteristic features jointly termed Albright's hereditary osteodystrophy. These syndromes are due to genetic or epigenetic

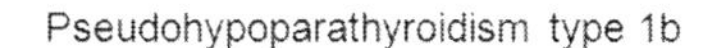

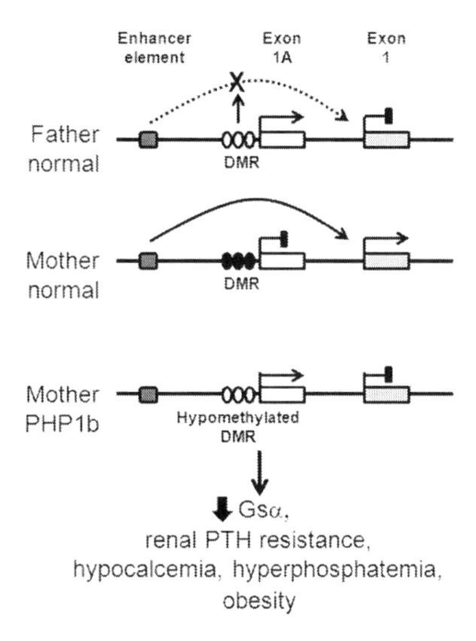

FIGURE 14.7 Tissue-specific (in renal proximal tubules) imprinting of GNSA1 and loss of maternal imprinting in PHP1b.

mutations affecting the GNAS1 locus on chromosome 20q13.11, which encodes the stimulatory G-protein subunit Gsα, necessary for hormone signal transduction. GNAS1 transcription is predominantly biallelic, with selective imprinting and alternative promoter usage in different tissues. Pseudohypoparathyroidism type 1a (PHP1a) arises upon maternal inheritance of molecular defects at GNAS1. PHP1b is associated with a paternal-specific imprinting pattern of exon 1A on both alleles, which leads to decreased Gsα expression in renal proximal tubules [48] (Figure 14.7). These patients have renal parathyroid hormone resistance, decreased cAMP response to parathyroid hormone infusion, and normal erythrocyte Gsα activity, resulting in loss of the maternal allele expression in renal tissue, but without other features of Albright's hereditary osteodystrophy, presumably because activity is maintained in tissues where Gsα is expressed biallelically. Approximately 15–20% of PHP1b patients display a specific loss of methylation at the GNAS1 DMRs, whereas most PHP1b patients (80–85%) display a broad loss of imprinting at the GNAS domain [49,50].

14.6 PRADER-WILLI AND ANGELMAN SYNDROMES

Prader-Willi syndrome (PWS) and Angelman syndrome (AS) are rare neurodevelopmental genetic syndromes (incidence 1:10,000) and were

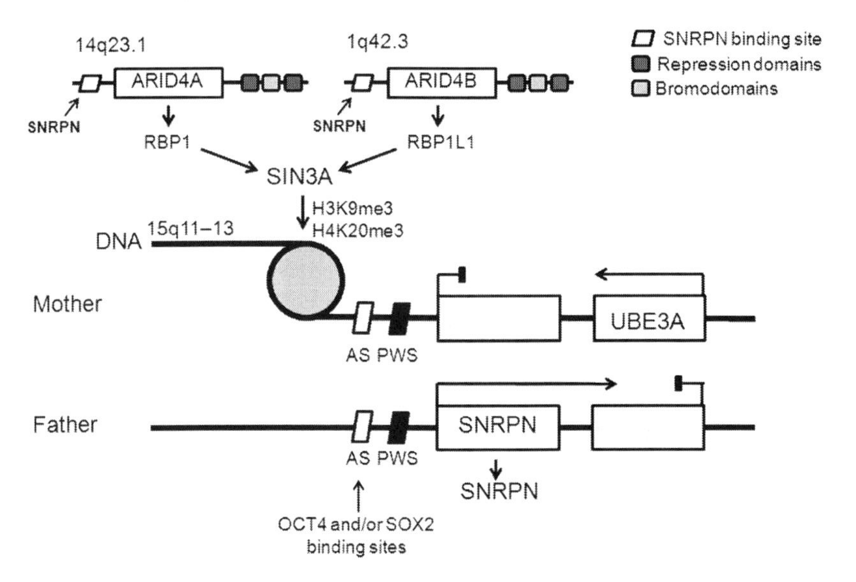

FIGURE 14.8 Disruption of the ARID/SIN3A system leads to loss of the inhibitory histone marks and methylation of the PWS ICR and decreased UBE3A gene expression from the maternal allele.

the first reported imprinting diseases. They both map to human chromosome 15q11–13, a large imprinted domain comprising many genes that are expressed from the paternal chromosome in the brain, and one gene, namely ubiquitin-protein ligase E3A (UBE3A), expressed from the maternal chromosome. Whereas PWS is most often caused by maternal UPD of chromosome 15 (UPD15), AS is caused most frequently by maternal deletion of chromosome 15q11–13. The clinical phenotype of PWS is highly complex and caused by loss of expression of multiple paternally expressed genes on the domain. By contrast, AS is caused by loss of expression of the single maternally expressed gene UBE3A, which maps to 15q11–13. Since PWS and AS are caused almost exclusively by genetic mutations affecting the imprinted domain itself [5,6], it has not been systematically explored in humans whether *trans*-acting factors could also be affected. Mouse studies, however, have identified several proteins involved in the embryonic maintenance of DNA methylation imprint at the domain's AS/PWS ICRs [51].

Important examples are ARID4A and ARID4B, two closely related members of the ARID gene family (Figure 14.8). ARID proteins play important roles in development, tissue-specific gene expression, and cell growth regulation. These genes encode RBP1 and RBP1L1, respectively, which are both members of chromatin-remodeling complexes and could recruit the histone deacetylase-dependent SIN3A transcriptional co-repressor. Targeted deletion of both of these genes in mice resulted in loss of DNA methylation and reduced H3K9me3 and H4K20me3 on the maternal allele at AS/PWS ICRs of the SNRPN (small nuclear ribonucleoprotein-associated protein N) gene [51].

As mentioned for BWS above, a potential role of OCT4 and SOX2 binding was also discovered at the ICR linked to AS. In mechanistic studies in the mouse, mutation of either OCT4- or SOX2-binding sites resulted in a significant maternal allele-specific gain of methylation at this control region in oocytes [52]. However, how this plays a role in the human disease is as yet unclear.

14.7 SYNDROMES MIMICKING UNIPARENTAL DISOMY 14 (UPD14)

Maternal uniparental disomy of chromosome 14 (UPD14) is an extremely rare condition that causes pre- and post-natal growth retardation, congenital hypotonia, joint laxity, motor delay, and mild mental retardation. Paternal UPD14 has a more severe phenotype, with polyhydramnios (excess of amniotic fluid), thoracic and abdominal wall defects, growth retardation, and severe developmental delay. Cases of segmental UPD14 indicate that the distal portion of chromosome 14q is the critical region in these syndromes [53]. The 14q32.2 region contains a large cluster of imprinted genes that include the DLK1, RTL1, and DIO3 genes expressed from the paternal chromosome, and the MEG3 (GTL2) and MEG8 (RIAN), the antisense genes to RTL1 (RTL1as), and micro RNA and small non-coding RNA gene clusters expressed from the maternal chromosome [54]. This imprinted domain is under the control of an ICR called the "intergenic germline DMR" [55], which is methylated on the paternally inherited chromosome. Gene-targeting studies in mice suggest that excessive RTL1 expression (a retrotransposon-derived gene) could explain the phenotype observed in patients with paternal UPD14, while decreased DLK1 and RTL1 gene expression could account for maternal UPD14 [53]. Protein delta homolog 1 (DLK1) is expressed as a transmembrane protein, and a soluble form cleaved off by the protease ADAM17 is active in inhibiting adiopogenesis. This may explain the epigenetic UPD14-like syndromes.

Several recent reports have described deletions and epimutations affecting the imprinted DLK1-DIO3 region in individuals with a phenotype similar to that of maternal UPD14 (but who did not present evidence for the occurrence of UPD) [56–58] (Figure 14.9). In these studies, loss of methylation was detected at the IG-DMR and also at the DMR comprising the MEG3 promoter (a long non-coding RNA associated with regulation of cell proliferation), indicating the occurrence of an epimutation (hypomethylation) affecting both of these paternally methylated DMRs.

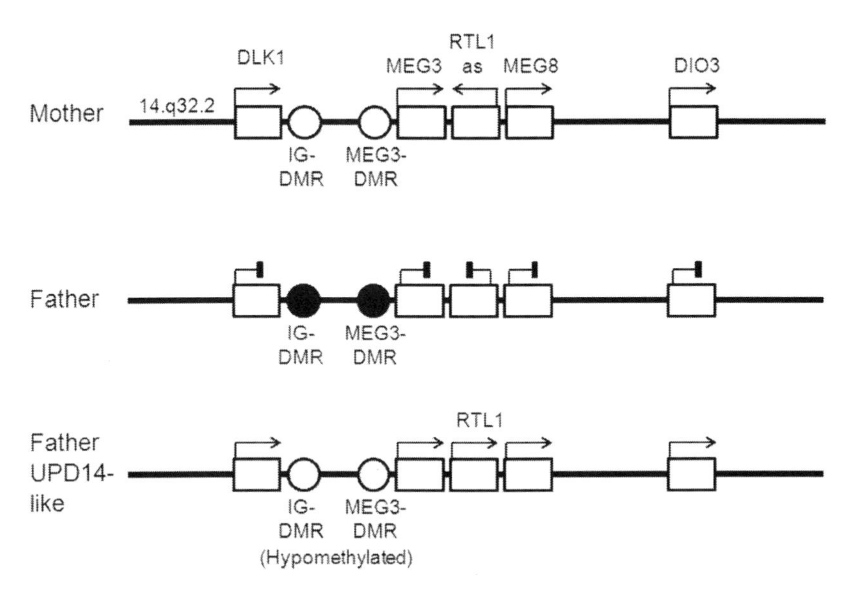

FIGURE 14.9 Genomic imprinting of the DLK1-DIO3 locus and hypomethylation of the IG- and MEG3-DMRs on the paternal allele in UPD14-like syndromes.

14.8 FAMILIAL BIPARENTAL HYDRATIDIFORM MOLE

The lack of a maternal genome mostly results in proliferation of the extraembryonic tissues, while embryonic development fails. A rare familial, biparental form of this extremely severe clinical phenotype, named familial biparental hydatidiform mole (FBHM), has been identified, in which affected women suffer from recurrent abnormal pregnancy with no embryonic development and cystic degeneration of placental villi. In different cases of FBHM, there was a failure to establish maternal imprints in the oocyte [59–61]. The imprint acquisition failure at multiple maternal ICRs suggested that a *trans*-acting factor is genetically affected. Indeed, recessive mutations in NLRP7 were identified as one cause of FBHM [62]. NLRPs (NACHT, LRR, and PYD domain-containing proteins) are constituents of inflammasomes.

However, some FBHM families do not present NLRP7 mutations, showing that there is genetic heterogeneity in familial hydatidiform moles. Thus, mutations in C6ORF221 (ECAT1) were identified as a second genetic cause of FBHM [63]. C6ORF221 is part of a group of four highly similar genes within a 100-kb cluster on human chromosome 6. Members of this gene family display a mostly oocyte and/or early embryonic expression pattern. This highly specific expression pattern, and the fact that genetic mutations in C6ORF221 and NLRP7 induce identical phenotypes (i.e., absence of maternal imprints), make it likely that NLRP7 and C6ORF221 participate in an oocyte-specific protein complex involved in imprint establishment. Indeed, the C6ORF221 and NLRP7 proteins were found to interact [63]. Further studies are required to elucidate the precise molecular mechanisms, and to which extent these factors could also be important in the embryonic maintenance of imprints.

14.9 CONCLUSION

Several novel insights have emerged regarding the genetic and epigenetic origins of imprinting disorders. Possibly the most relevant novel finding is that in some patients, altered DNA methylation affects not only the canonical ICR involved in the disease but also multiple other ICRs/DMRs to different extents. The recent reports on H19-ICR microdeletions and point mutations in BWS raise the question of what percentage of patients presenting small genetic mutations, including single nucleotide changes, could be responsible for the observed epimutations (hypomethylation and hypermethylations) at specific ICRs/DMRs. To address this question, it should be relevant in future studies not only to assess DNA methylation levels at ICRs but also to carefully study their DNA sequences. The occurrence of ICR point mutations and microdeletions highlights the importance of searching for genetic defects in case the methylation change is confined to the ICR associated with the imprinting disorder. In case the methylation change affects multiple imprinted (and non-imprinted) gene loci, however, it should seem relevant to explore whether *trans*-acting factor genes are mutated or altered in their expression. In such genetic studies, it might not be sufficient to include only the known *trans*-acting factors of imprinting disorders; novel candidates emerging from mouse studies could be included in a targeted screen as well. Particularly when a familial component is suspected, it should be worth using broad, high-throughput sequencing approaches for mutation detection (e.g., exome sequencing). Hopefully, in the years to come, it will become clearer how commonly *trans*-acting factors are perturbed in imprinting-related diseases in comparison with other causal mechanisms. The coming years will undoubtedly pinpoint new players involved in imprinting-related diseases, and may shed further light on the extent to which these epigenetic diseases could be genetic in origin.

References

[1] Kalhan SC. One-carbon metabolism, fetal growth and long-term consequences. Nestle Nutr Inst Workshop Ser 2013;74:127–38.
[2] Sookoian S, Gianotti TF, Burgueno AL, Pirola CJ. Fetal metabolic programming and epigenetic modifications: a systems biology approach. Pediatr Res 2013;73:531–42.

[3] Ferguson-Smith AC. Genomic imprinting: the emergence of an epigenetic paradigm. Nat Rev Genet 2011;12:565–75.
[4] Feil R, Fraga MF. Epigenetics and the environment: emerging patterns and implications. Nat Rev Genet 2011;13:97–109.
[5] Arnaud P, Feil R. Epigenetic deregulation of genomic imprinting in human disorders and following assisted reproduction. Birth Defects Res C Embryo Today: Rev 2005;75:81–97.
[6] Hirasawa R, Feil R. Genomic imprinting and human disease. Essays Biochem 2010;48:187–200.
[7] Gicquel C, Rossignol S, Cabrol S, Houang M, Steunou V, Barbu V, et al. Epimutation of the telomeric imprinting center region on chromosome 11p15 in Silver-Russell syndrome. Nat Genet 2005;37:1003–7.
[8] Schonherr N, Meyer E, Eggermann K, Ranke MB, Wollmann HA, Eggermann T. (Epi)mutations in 11p15 significantly contribute to Silver-Russell syndrome: but are they generally involved in growth retardation? Eur J Med Genet 2006;49:414–18.
[9] Feil R. Epigenetic asymmetry in the zygote and mammalian development. Int J Dev Biol 2009;53:191–201.
[10] Butler MG. Genomic imprinting disorders in humans: a mini-review. J Assist Reprod Genet 2009;26:477–86.
[11] Dias RP, Bogdarina I, Cazier JB, Buchanan C, Donaldson MC, Johnston LB, et al. Multiple segmental uniparental disomy associated with abnormal DNA methylation of imprinted loci in Silver-Russell syndrome. J Clin Endocrinol Metab 2012;97:e2188–93.
[12] Uribe-Lewis S, Woodfine K, Stojic L, Murrell A. Molecular mechanisms of genomic imprinting and clinical implications for cancer. Expert Rev Mol Med 2011;13:e2.
[13] Bell AC, Felsenfeld G. Methylation of a CTCF-dependent boundary controls imprinted expression of the Igf2 gene. Nature 2000;405:482–5.
[14] Wendt KS, Yoshida K, Itoh T, Bando M, Koch B, Schirghuber E, et al. Cohesin mediates transcriptional insulation by CCCTC-binding factor. Nature 2008;451:796–801.
[15] Engel N, Thorvaldsen JL, Bartolomei MS. CTCF binding sites promote transcription initiation and prevent DNA methylation on the maternal allele at the imprinted H19/Igf2 locus. Hum Mol Genet 2006;15:2945–54.
[16] Choufani S, Shuman C, Weksberg R. Beckwith-Wiedemann syndrome. Am J Med Genet C Semin Med Genet 2010;154C:343–54.
[17] Beygo J, Citro V, Sparago A, De Crescenzo A, Cerrato F, Heitmann M, et al. The molecular function and clinical phenotype of partial deletions of the IGF2/H19 imprinting control region depends on the spatial arrangement of the remaining CTCF-binding sites. Hum Mol Genet 2013;22:544–57.
[18] Lee MP, DeBaun M, Randhawa G, Reichard BA, Elledge SJ, Feinberg AP. Low frequency of p57KIP2 mutation in Beckwith-Wiedemann syndrome. Am J Hum Genet 1997;61:304–9.
[19] Bliek J, Alders M, Maas SM, Oostra RJ, Mackay DM, van der Lip K, et al. Lessons from BWS twins: complex maternal and paternal hypomethylation and a common source of haematopoietic stem cells. Eur J Hum Genet 2009;17:1625–34.
[20] Yamazawa K, Kagami M, Fukami M, Matsubara K, Ogata T. Monozygotic female twins discordant for Silver-Russell syndrome and hypomethylation of the H19-DMR. J Hum Genet 2008;53:950–5.
[21] You JS, Kelly TK, De Carvalho DD, Taberlay PC, Liang G, Jones PA. OCT4 establishes and maintains nucleosome-depleted regions that provide additional layers of epigenetic regulation of its target genes. Proc Natl Acad Sci USA 2011; 108:14497–502.
[22] Demars J, Shmela ME, Rossignol S, Okabe J, Netchine I, Azzi S, et al. Analysis of the IGF2/H19 imprinting control region uncovers new genetic defects, including

mutations of OCT-binding sequences, in patients with 11p15 fetal growth disorders. Hum Mol Genet 2010;19:803–14.
[23] Berland S, Appelback M, Bruland O, Beygo J, Buiting K, Mackay DJ, et al. Evidence for anticipation in Beckwith-Wiedemann syndrome. Eur J Hum Genet 2013; 21:1344–8.
[24] Poole RL, Leith DJ, Docherty LE, Shmela ME, Gicguel C, Splitt M, et al. Beckwith-Wiedemann syndrome caused by maternally inherited mutation of an OCT-binding motif in the IGF2/H19-imprinting control region, ICR1. Eur J Hum Genet 2012;20:240–3.
[25] Sakaguchi R, Okamura E, Matsuzaki H, Fukamizu A, Tanimoto K. Sox–Oct motifs contribute to maintenance of the unmethylated H19 ICR in YAC transgenic mice. Hum Mol Genet 2013;22:4627–37.
[26] Hori N, Yamane M, Kouno K, Sato K. Induction of DNA demethylation depending on two sets of Sox2 and adjacent Oct3/4 binding sites (Sox–Oct motifs) within the mouse H19/insulin-like growth factor 2 (Igf2) imprinted control region. J Biol Chem 2012;287:44006–16.
[27] Meyer E, Lim D, Pasha S, Tee LJ, Rahman F, Yates JR, et al. Germline mutation in NLRP2 (NALP2) in a familial imprinting disorder (Beckwith-Wiedemann syndrome). PLoS Genet 2009;5.
[28] Tschopp J, Martinon F, Burns K. NALPs: a novel protein family involved in inflammation. Nat Rev Mol Cell Biol 2003;4:95–104.
[29] Temple IK, Shield JP. Transient neonatal diabetes, a disorder of imprinting. J Med Genet 2002;39:872–5.
[30] Arima T, Drewell RA, Arney KL, Inoue J, Makita Y, Hata A, et al. A conserved imprinting control region at the HYMAI/ZAC domain is implicated in transient neonatal diabetes mellitus. Hum Mol Genet 2001;10:1475–83.
[31] Kameswaran V, Kaestner KH. The Missing lnc(RNA) between the pancreatic beta-cell and diabetes. Front Genet 2014;5:200.
[32] Varrault A, Bilanges B, Mackay DJ, Basyuk E, Ahr B, Fernandez C, et al. Characterization of the methylation-sensitive promoter of the imprinted ZAC gene supports its role in transient neonatal diabetes mellitus. J Biol Chem 2001;276:18653–6.
[33] Piras G, El Kharroubi A, Kozlov S, Escalante-Alcalde D, Hernandez L, Copeland NG, et al. Zac1 (Lot1), a potential tumor suppressor gene, and the gene for epsilon-sarcoglycan are maternally imprinted genes: identification by a subtractive screen of novel uniparental fibroblast lines. Mol Cell Biol 2000;20:3308–15.
[34] Kamiya M, Judson H, Okazaki Y, Kusakabe M, Muramatsu M, Takada S, et al. The cell cycle control gene ZAC/PLAGL1 is imprinted – a strong candidate gene for transient neonatal diabetes. Hum Mol Genet 2000;9:453–60.
[35] Mackay DJ, Hahnemann JM, Boonen SE, Poerksen S, Bunyan DJ, White HE, et al. Epimutation of the TNDM locus and the Beckwith-Wiedemann syndrome centromeric locus in individuals with transient neonatal diabetes mellitus. Hum Genet 2006;119:179–84.
[36] Varrault A, Gueydan C, Delalbre A, Bellmann A, Houssami S, Aknin C, et al. Zac1 regulates an imprinted gene network critically involved in the control of embryonic growth. Dev Cell 2006;11:711–22.
[37] Mackay DJ, Callaway JL, Marks SM, White HE, Acerini CL, Boonen SE, et al. Hypomethylation of multiple imprinted loci in individuals with transient neonatal diabetes is associated with mutations in ZFP57. Nat Genet 2008;40:949–51.
[38] Court F, Martin-Trujillo A, Romanelli V, Garin I, Iglesias-Platas I, Salafsky I, et al. Genome-wide allelic methylation analysis reveals disease-specific susceptibility to multiple methylation defects in imprinting syndromes. Hum Mutat 2013; 34:595–602.

[39] Li X, Ito M, Zhou F, Youngson N, Zuo X, Leder P, et al. A maternal-zygotic effect gene, Zfp57, maintains both maternal and paternal imprints. Dev Cell 2008;15:547–57.
[40] Baglivo I, Esposito S, De Cesare L, Sparago A, Anvar Z, Riso V, et al. Genetic and epigenetic mutations affect the DNA binding capability of human ZFP57 in transient neonatal diabetes type 1. FEBS Lett 2013;587:1474–81.
[41] Sharif J, Muto M, Takebayashi S, Suetake I, Iwamatsu A, Endo TA, et al. The SRA protein Np95 mediates epigenetic inheritance by recruiting Dnmt1 to methylated DNA. Nature 2007;450:908–12.
[42] Ideraabdullah FY, Bartolomei MS. ZFP57: KAPturing DNA methylation at imprinted loci. Mol Cell 2011;44:341–2.
[43] Liu QS, Wang QJ, Du GH, Zhu SY, Gao M, Zhang L, et al. Recombinant human ciliary neurotrophic factor reduces weight partly by regulating nuclear respiratory factor 1 and mitochondrial transcription factor A. Eur J Pharmacol 2007;563:77–82.
[44] Gemma C, Sookoian S, Dieuzeide G, Garcia SI, Gianotti TF, Gonzalez CD, et al. Methylation of TFAM gene promoter in peripheral white blood cells is associated with insulin resistance in adolescents. Mol Genet Metab 2010;100:83–7.
[45] Gianotti TF, Sookoian S, Dieuzeide G, Garcia SI, Gemma C, Gonzalez CD, et al. A decreased mitochondrial DNA content is related to insulin resistance in adolescents. Obesity (Silver Spring) 2008;16:1591–5.
[46] Perkins E, Murphy SK, Murtha AP, Schildkraut J, Jirtle RL, Demark-Wahnefried W, et al. Insulin-like growth factor 2/H19 methylation at birth and risk of overweight and obesity in children. J Pediatr 2012;161:31–9.
[47] Godfrey KM, Sheppard A, Gluckman PD, Lillycrop KA, Burdge GC, McLean C, et al. Epigenetic gene promoter methylation at birth is associated with child's later adiposity. Diabetes 2011;60:1528–34.
[48] Liu J, Litman D, Rosenberg MJ, Yu S, Biesecker LG, Weinstein LS. A GNAS1 imprinting defect in pseudohypoparathyroidism type IB. J Clin Invest 2000;106:1167–74.
[49] Bastepe M, Frohlich LF, Linglart A, Abu-Zahra HS, Tojo K, Ward LM, et al. Deletion of the NESP55 differentially methylated region causes loss of maternal GNAS imprints and pseudohypoparathyroidism type Ib. Nat Genet 2005;37:25–7.
[50] Chillambhi S, Turan S, Hwang DY, Chen HC, Juppner H, Bastepe M. Deletion of the noncoding GNAS antisense transcript causes pseudohypoparathyroidism type Ib and biparental defects of GNAS methylation in cis. J Clin Endocrinol Metab 2010;95:3993–4002.
[51] Wu MY, Tsai TF, Beaudet AL. Deficiency of Rbbp1/Arid4a and Rbbp1l1/Arid4b alters epigenetic modifications and suppresses an imprinting defect in the PWS/AS domain. Genes Dev 2006;20:2859–70.
[52] Kaufman Y, Heled M, Perk J, Razin A, Shemer R. Protein-binding elements establish in the oocyte the primary imprint of the Prader-Willi/Angelman syndromes domain. Proc Natl Acad Sci USA 2009;106:10242–7.
[53] Kagami M, Sekita Y, Nishimura G, Irie M, Kato F, Okada M, et al. Deletions and epimutations affecting the human 14q32.2 imprinted region in individuals with paternal and maternal upd(14)-like phenotypes. Nat Genet 2008;40:237–42.
[54] Girardot M, Cavaille J, Feil R. Small regulatory RNAs controlled by genomic imprinting and their contribution to human disease. Epigenetics 2012;7:1341–8.
[55] Lin SP, Youngson N, Takada S, Seitz H, Reik W, Paulsen M, et al. Asymmetric regulation of imprinting on the maternal and paternal chromosomes at the Dlk1-Gtl2 imprinted cluster on mouse chromosome 12. Nat Genet 2003;35:97–102.
[56] Buiting K, Kanber D, Martin-Subero JI, Lieb W, Terhal P, Albrecht B, et al. Clinical features of maternal uniparental disomy 14 in patients with an epimutation and a deletion of the imprinted DLK1/GTL2 gene cluster. Hum Mutat 2008;29:1141–6.

[57] Zechner U, Kohlschmidt N, Rittner G, Damatova N, Beyer V, Haaf T, et al. Epimutation at human chromosome 14q32.2 in a boy with a upd(14)mat-like clinical phenotype. Clin Genet 2009;75:251–8.

[58] Hosoki K, Ogata T, Kagami M, Tanaka T, Saitoh S. Epimutation (hypomethylation) affecting the chromosome 14q32.2 imprinted region in a girl with upd(14)mat-like phenotype. Eur J Hum Genet 2008;16:1019–23.

[59] Hayward BE, De Vos M, Talati N, Abdollahi MR, Taylor GR, Meyer E, et al. Genetic and epigenetic analysis of recurrent hydatidiform mole. Hum Mutat 2009;30:E629–39.

[60] El-Maarri O, Seoud M, Coullin P, Herbiniaux U, Oldenburg J, Rouleau G, et al. Maternal alleles acquiring paternal methylation patterns in biparental complete hydatidiform moles. Hum Mol Genet 2003;12:1405–13.

[61] Judson H, Hayward BE, Sheridan E, Bonthron DT. A global disorder of imprinting in the human female germ line. Nature 2002;416:539–42.

[62] Murdoch S, Djuric U, Mazhar B, Seoud M, Kahn R, Kuick R, et al. Mutations in NALP7 cause recurrent hydatidiform moles and reproductive wastage in humans. Nat Genet 2006;38:300–2.

[63] Parry DA, Logan CV, Hayward BE, Shires M, Landolsi H, Diggle C, et al. Mutations causing familial biparental hydatidiform mole implicate c6orf221 as a possible regulator of genomic imprinting in the human oocyte. Am J Hum Genet 2011;89:451–8.

CHAPTER 15

DNA Methylation in Cardiology

OUTLINE

15.1 INTRODUCTION

A current hypothesis is that cardiovascular risk factors might influence and remodel epigenomic patterns and that cardiovascular biomarkers are associated with epigenetic modifications. Many evidences suggest that epigenetic modifications contribute to subclinical and clinical cardiovascular disease [1]. Participants in the Dutch Hunger Winter Families Study who were exposed *in utero* to the 1944–1945 famine, a condition that has been associated with overweight, impaired glucose homeostasis, and increased cardiovascular risk in adulthood, exhibited hypomethylation of the imprinted insulin growth factor-2 (IGF2) and insulin-induced gene 1 (INSIGF1) and hypermethylation of five other genes, including GNASAS and MEG3 (see Chapter 14), compared to

M. Neidhart: *DNA Methylation and Complex Human Disease*.
DOI: http://dx.doi.org/10.1016/B978-0-12-420194-1.00015-4

unexposed siblings [2]. Epigenomics is inherently interconnected with genetics because epigenetic modifications can alter the expression of genetic variations, and genetic variation is one of the determinants of DNA methylation and histone modifications. Because of its dynamic nature, the epigenome is hypothesized to show signatures associated with cardiovascular risk biomarkers. The individual epigenomic background may determine the levels of these biomarkers or their responses to acquired risk factors. In particular, DNA methylation has been linked to several cardiovascular-related biomarkers, including homocysteine [3] and C-reactive protein [4]. In turn, in cardiovascular diseases, combined vitamin B12 deficiency and hyperhomocysteinemia has the potential to cause global DNA hypomethylation and an altered transcriptional program by lowering synthesis of S-adenosylmethionine [5], the cell's methyl donor. Furthermore, inflammation is a component of cardiovascular diseases and DNA methylation is emerging as a primary regulator of it. Thus, DNA methylation has been shown to control leukocyte functions related to cardiovascular risk, including the expression of soluble mediators and surface molecules that direct margination, adhesion, and migration of blood leukocytes in vascular tissues [6]. The malleability of the epigenome in response to exogenous factors provides a simple and likely hypothesis to be tested that is relevant to metabolic diseases in general: that dietary and environmental risk factors can induce a proatherogenic gene expression program through changes in the epigenome [7]. This idea was first proposed in the late 1990s by Newman [7], who argued that folate or vitamin B6 and B12 deficiency causes atherosclerosis by inducing DNA hypomethylation.

15.2 ATHEROSCLEROSIS

Atherosclerosis is the leading cause of death among western populations. Atherosclerosis is a chronic inflammatory condition that begins with "fatty streak" lesions in the artery walls, due to lipid retention in the intima. Fibrous elements and inflammatory cells accumulate in the inner layer of the walls of arteries. As these deposits (plaques) build up, the lumen of the blood vessel narrows, restricting the passage of blood. The surface of atherosclerotic plaques may erode or rupture, releasing substances that encourage platelets to adhere and a blood clot to form, causing blockage of the artery. Mice deficient in genes coding for methylation enzymes, such as DNA methyltransferases (DNMTs, which establish or replicate DNA methylation) or methylenetetrahydrofolate reductase (MTHFR, related to methyl donor generation), show hypomethylation of their DNA. In MTHFR-deficient mice, DNA hypomethylation has been shown to precede the formation of aortic fatty streaks [8]. Similarly, lower DNA methylation content in peripheral blood leukocytes is reported in patients with atherosclerotic cardiovascular diseases [9].

Additionally, atherosclerosis-prone apoE$^{-/-}$ mice develop specific changes in DNA methylation of transcribed gene sequences as well as in interdispersed elements both in peripheral blood leukocytes and in the aorta before developing vascular lesions [10]. In these mice, changes in DNA methylation in peripheral blood leukocytes have been shown to contribute to the dysregulation of inflammation and promotion of atherosclerosis. Hypomethylation is also characteristic of areas of smooth muscle cell proliferation that has been found in advanced atherosclerotic plaques in human pathology specimens as well as in atheromas of apoE$^{-/-}$ mice [8] and in neointimal thickenings of rabbit aortas [11]. ApoE$^{-/-}$ mouse aortas exhibit a decrease in DNA methylation that can be detected as early as age 4 weeks, thus anticipating any histological changes associated with atherosclerosis [12]. In addition, hypo- and hypermethylation have been demonstrated at genes involved in key cellular functions in atherogenesis: smooth muscle cell proliferation and survival, extracellular matrix remodeling, and cellular redox reactions [13,14]. Some of these changes are induced by or associated with abnormal levels of lipoproteins, thus possibly driven by dietary and/or endogenous lipids.

Additionally, in human atherosclerotic tissues the estrogen receptor-α and -β promoters show increased methylation. Estrogen receptor (ER) promoter methylation has been well demonstrated to increase with age even in normal tissues, and to reach near complete methylation level in the elderly [1]. Decreased ERs limited the beneficial effect of 27-hydroxycholosterol on cell adhesion and production of pro-inflammatory cytokines.

Advanced human atherosclerotic lesions show demethylation of the few normally hypermethylated somatic CpG islands [15]. This demethylation impacts mainly on the expression of a subset of transcription factors known to regulate smooth muscle cell biology and inflammation, such as homeobox (HOX) members PROX1, NOTCH1, and FOXP1. Similarly, lower DNA methylation content in peripheral blood leukocytes is reported in patients with atherosclerotic cardiovascular diseases [9].

Figure 15.1 shows the pathways in atherosclerosis that are modulated by DNA methylation. Risk factors for cardiovascular diseases are high cholesterol and triglycerides, lack of exercise, overweight, diabetes, and smoking. At the present time, however, there is still no clear evidence that exposed individuals develop epigenetic alterations which in turn increase the risk of atherosclerotic disease, and/or whether epigenetic changes in DNA repair are associated with changes in vascular cell functions [16]. Investigating the epigenome during the transition

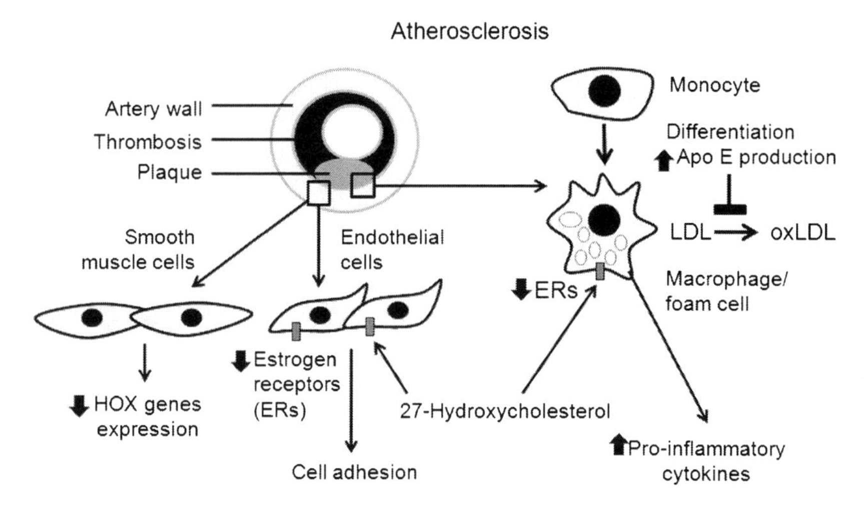

FIGURE 15.1 Pathways in atherosclerosis that are modulated by DNA methylation.

from a healthy vessel to a vessel harboring fatty streaks would be very important, particularly in the light of evidence that aberrant DNA methylation precedes lesion formation in mice [10].

15.3 ARRHYTHMIA

Cardiac arrhythmia is caused by altered conduction properties of the heart, which may arise in response to ischemia, inflammation, fibrosis, aging, or from genetic factors. A first evidence for epigenetic regulation of cardiac rhythm was raised from a study conducting microarrays on heart rhythm determinants on tissue from mice exposed to either intermittent or chronic hypoxia, and untreated wild-type mice. A different environment (hypoxia) profoundly restructured the heart rhythm by changing the hierarchy of the composing genes and by identifying new role players. This was the case for the epigenetic modulators histone deacetylase 5 (HDAC5) and myocyte enhancer factor 2B (MEF2B) [17].

Postural tachycardia syndrome (POTS) has multiple symptoms, one of such being tachycardia. Dysfunction of the norepinephrine transporter (NET) gene has previously been implicated in POTS, with a reported coding mutation in the NET gene (SLC6A2) [18]. NET is a monoamine transporter and is responsible for the Na^+/Cl^--dependent reuptake of extracellular norepinephrine (NE), which is also known as noradrenaline. NET can also reuptake extracellular dopamine. The reuptake of these two neurotransmitters is essential in regulating concentrations in the synaptic cleft. Head-up tilt experiments in POTS patients showed that

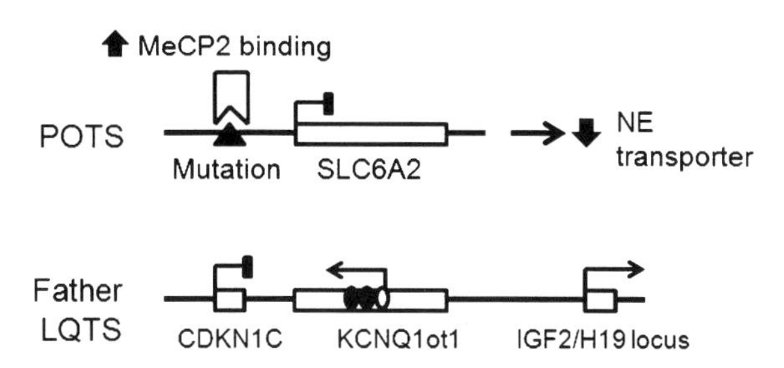

FIGURE 15.2 Pathways in arrhythmia that are modulated by DNA methylation.

the expression of norepinephrine transported is lower in POTS patients compared to healthy subjects. In the absence of an altered SLC6A2 gene sequence or promoter methylation, the observed reduced expression of norepinephrine was directly correlated with chromatin modifications. Changes in expression were attributable to increased binding of the repressive methyl CpG-binding protein 2 (MeCP2) regulatory complex, in association with an altered histone modification composition at the promoter region of the SLC6A2 gene (Figure 15.2).

Long QT syndrome (LQTS) is a rare inherited heart condition in which delayed repolarization of the heart following a heartbeat increases the risk of episodes of torsades de pointes (TDP, a form of irregular heartbeat that originates from the ventricles). These episodes may lead to palpitations, fainting, and sudden death due to ventricular fibrillation. Variable imprinting of the KCNQ1 gene provides a possible explanation for LQTS in the absence of a coding sequence mutation in KCNQ1 (see Chapter 14). Paternal imprinting is probably relieved in cardiac tissue, meaning that during differentiation methylation of the paternal chromosome must occur to block production of the suppressive KCNQot1 transcript [19] (Figure 15.2).

15.4 ISCHEMIC HEART DISEASE

Heart failure is the end stage of several pathological cardiac conditions, including myocardial infarction, cardiac hypertrophy, and hypertension. Various molecular and cellular mechanisms are involved in the development of heart failure. Lower LINE-1 methylation in peripheral blood mononuclear cells (PBMCs) is a predictor of incidence and mortality from ischemic heart disease and stroke [20]. Interestingly, three independent studies consistently demonstrated that exposure to air pollution, an established risk factor for ischemic heart disease and stroke, is also associated with reduced blood methylation of LINE-1 [21]. Using a candidate gene approach, hyper- and hypomethylation of specific genes was shown to be related to air pollutant exposures, including increased CDKN2B methylation and decreased NOS2 and MAGEA1 methylation. Smoking is a well-known risk factor for cardiovascular diseases. A smoking mother is also a risk for the embryo; smoking induces lower Alu and LINE-1 DNA methylation in child buccal cell DNA as well as changes in methylation of specific genes identified through methylation profiling [22].

Comparing left ventricle tissue from heart failure patients and healthy controls, a microarray hybridization study [23] identified three differentially methylated angiogenesis-related loci and correlated to differential expression levels of the corresponding gene. Hypermethylation of the 5′ regulatory region of platelet endothelial cell adhesion molecule 1 (PECAM1/CD31) and hypomethylation of the angiomotin-like protein 2 (AMOTL2) in failing hearts correlated with reduced expression of those genes, while hypermethylation within the Rho GTPase activating protein 24 gene (ARHGAP24) is paradoxically correlated with increased expression of ARHGAP24 in failing hearts. AMOTL2 is a protein that binds angiostatin, a circulating inhibitor of the formation of new blood vessels, whereas ARHGAP24 is a negative regulator of Rho GTPases, which affect various cellular functions, including cytoskeleton rearrangement. Genome-wide methylation profiles allowed the identification of two genes with decreased methylation of promoters in patients compared with control subjects, namely lymphocyte antigen 75 (LY75) and adenosine A2a receptor (ADORA2A) [21,24]. DNA methylation was found to be responsible for the hypermutability of distinct cardiac genes. This is the case for the cardiac isoform of the myosin binding protein C gene (MYBPC3) that has a significantly higher level of exonic methylation of CpG sites than the skeletal isoform (MYBPC2) [25]. This suggests that there are unique aspects of the MYBPC3 gene or its epigenetic environment that are prone to generate genetic mutations. MYBPC3 plays a role in cardiac contraction by modifying the activity of actin-activated myosin ATPase (Figure 15.3).

15.5 HYPERTENSION

In PBMCs of patients with hypertension, recent studies have shown a loss of global genomic methylation content [26]. One of the main issues regarding this is to establish the meaning of DNA methylation status in peripheral blood cells in hypertensive patients, especially when considering that the tissue and cell specificity of the epigenetic phenomena and the relationship between an epigenetic alteration at

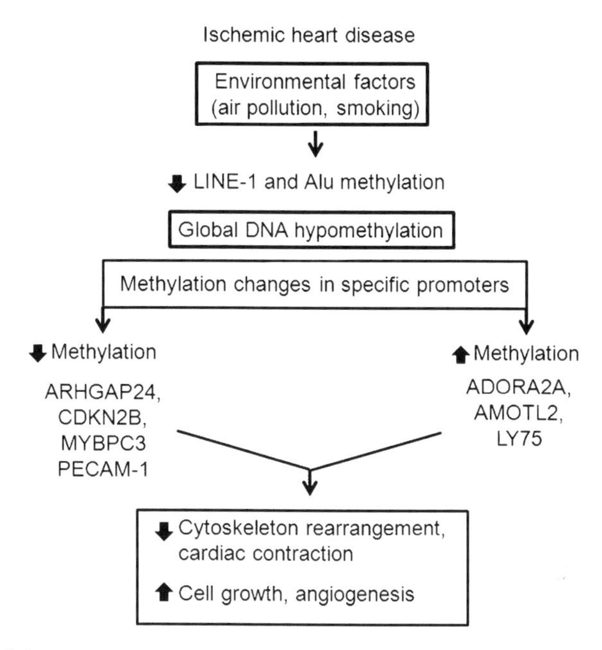

FIGURE 15.3 Pathways in ischemic heart disease that are modulated by DNA methylation.

molecular levels of PBMCs is not clearly associated with the pathogenesis of hypertensive disease, by current knowledge [27].

Considering the importance of mineralocorticoids and glucocorticoids such as aldosterone and cortisol in the onset and progression of hypertension, attention has been devoted to the regulation of a gene known to be involved in the cortisol/cortisone shuttle in affecting the competing binding to mineralocorticoid receptors by either cortisol or aldosterone, depending on either one's concentration in the blood. Special interest has been devoted to the function of 11-β-hydroxysteroid dehydrogenase type 2 (HSD11B2), whose function is that of catalyzing the dehydrogenation of 11-β-hydroxyglucocorticoids to 11-keto-steroids by using NAD^+ as a co-factor. This enzyme catalyzes the conversion of cortisol to cortisone, its inactive metabolite, therefore a loss of 11-β-hydroxysteroid dehydrogenase type 2 activity may lead to the occupancy of mineralocorticoid receptors by excess cortisol, which causes a renal sodium reabsorption that results in a salt-sensitive increase in blood pressure. Inappropriate mineralocorticoid receptor stimulation by cortisol is the pathogenetic mechanism of a rare form of salt-sensitive hypertension caused by mutations in the HSD11B2 gene leading to a substantial or even complete loss of enzymatic activity. Deleterious homozygous mutations in the HSD11B2 gene are associated with apparent

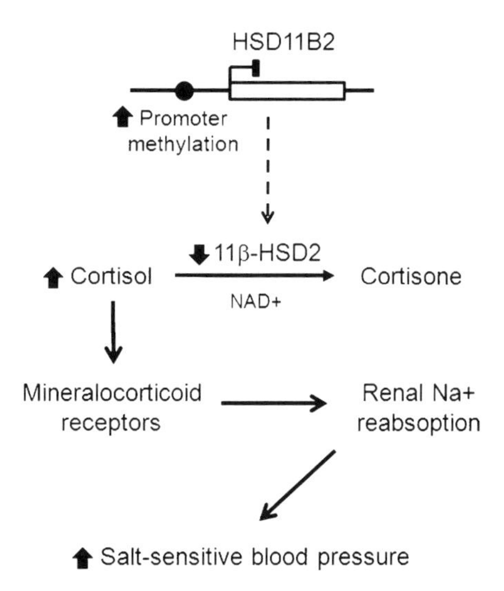

FIGURE 15.4 Increased HSD11B2 promoter methylation and hypertension.

mineralocorticoid excess syndrome and heterozygous alterations have been described as possible causes, although they could only partially explain the impaired enzyme function. Interestingly, in some patients with hypertension, hypermethylation of the HSD11B2 gene has been reported [28], linking epigenetics to blood pressure control (Figure 15.4).

In an intrauterine growth restriction (IUGR) rat model of hypertension, a condition that is known to increase the risk of adult hypertension development and increased promoter methylation at the HSD11B2 site has been observed [29]. In human placentas, methylation levels at HSD11B2 promoter sites were significantly higher in IUGR newborns than in controls, and related to lower HSD11B2 expression [30]. The methylation pattern at different CpG sites correlated also with measures of fetal growth, such as birth weight and ponderal index, therefore showing that promoter methylation at the HSD11B2 site may be linked to the development of IUGR and possibly to the risk of future development of hypertension.

For the known role of alterations in ion flux mechanisms in the pathogenesis of hypertension, the role of membrane transporter genes has been also evaluated – for example, the $Na^{+}/K^{+}/2Cl^{-}$ co-transporter 1 (NKCC1/SLC12A2), a gene encoding a solute carrier that transports sodium, potassium, and chloride through the cellular membrane. The methylation status of NKCC1 promoter was measured in the aorta and hearts of spontaneously hypertensive rats (SHRs) compared with Wistar Kyoto rats used as controls [31]. NKCC1 expression is upregulated by a mechanism induced by gene promoter hypomethylation in this model. Thereafter, it has been confirmed in another rodent model that NKCC1 is epigenetically upregulated during postnatal development of hypertension [32].

Furthermore, the somatic angiotensin-converting enzyme gene (sACE) is a key regulator of blood pressure by catalyzing the conversion of angiotensin I into physiologically active angiotensin II, a substance with potent vasopressive properties. The higher levels of promoter methylation levels of sACE in cultured human endothelial cell lines and in rats *in vivo* were associated with transcriptional repression, therefore highlighting a possible involvement of epigenetic mechanisms through sACE methylation in hypertension [33].

In a human cell culture model using adrenocortical cells, the DNA methylation status was shown to be diminished by stimulation with interleukin-6 at a CCAAT/enhancer-binding protein binding site and a transcription start site causing the expression activation of the angiotensinogen (AGT) gene [34]. AGT is the primary substrate of the renin–angiotensin–aldosterone system, and is therefore one of the major target proteins for the study of hypertension. The demethylation was accompanied by increased chromatin accessibility of the AGT promoter, thereby switching the AGT phenotypic expression from an inactive to an active state. In a rat model, the high salt intake induced a decreased DNA methylation activity in rat visceral adipose tissue. This study shows that excess circulating aldosterone concentrations and a high salt intake act as a nutritional stimulus for the AGT gene expression in adipose tissue both in rats and humans.

Promoter methylation of the α-adducin (ADD1) gene was evaluated for the risk of developing hypertension [35]. Adducin binds with high affinity to Ca^{2+}/calmodulin and is a substrate for protei kinases A and C. Polymorphism in ADD1 has been associated with hypertension. In addition, lower DNA methylation at the ADD1 promoter site was found to be related to an increased essential hypertension risk, whereas a differential power in predicting the risk of hypertension was observed for different sites of specific CpG sites of methylation in males compared with females.

Ambient particulate matter has an effect on blood pressure [36]. In humans, exposure to fine and coarse concentrated ambient particles induces blood hypomethylation of LINE-1 and Alu repeats, as mentioned above, as well as Toll-like receptor 4 (TLR4). Hypomethylation of these factors was found to be associated with increased systolic blood pressure after exposure. This is of great interest, since many epidemiological studies [37] have reported a correlation between particulate-matter exposure, cardiovascular disease, and death, and this may therefore represent a novel mechanism that mediates environmental effects on blood pressure and, indirectly, cardiovascular disease and heart failure.

15.6 CONCLUSION

A growing body of evidence suggests that changes in the population's exposure to diet- and lifestyle-related factors induce epigenetic changes that in turn contribute to the recent epidemics of metabolic diseases such as obesity, diabetes, metabolic syndrome, and vascular complications. Epigenetics holds the promise to clarify at least two pending issues in cardiovascular and epidemiological research. One is to provide molecular mechanisms for environment–gene interactions with the ultimate goal of explaining how environment-related risk factors for cardiovascular diseases act on the cell's transcriptional program. Secondly, epigenetic information should be used to complement genetic analysis with the hope of finally identifying genetic variants that explain a large portion of cardiovascular risk variability.

References

[1] Baccarelli A, Rienstra M, Benjamin EJ. Cardiovascular epigenetics: basic concepts and results from animal and human studies. Circ Cardiovasc Genet 2010;3:567–73.

[2] Tobi EW, Lumey LH, Talens RP, Kremer D, Putter H, Stein AD, et al. DNA methylation differences after exposure to prenatal famine are common and timing- and sex-specific. Hum Mol Genet 2009;18:4046–53.

[3] Ingrosso D, Cimmino A, Perna AF, Masella L, De Santo NG, De Bonis ML, et al. Folate treatment and unbalanced methylation and changes of allelic expression induced by hyperhomocysteinaemia in patients with uraemia. Lancet 2003;361:1693–9.

[4] Fu LH, Cong B, Zhen YF, Li SJ, Ma CL, Ni ZY, et al. [Methylation status of the IL-10 gene promoter in the peripheral blood mononuclear cells of rheumatoid arthritis patients]. Yi Chuan 2007;29:1357–61.

[5] Zaina S, Lund G. Epigenetics: a tool to understand diet-related cardiovascular risk?. J Nutrigenet Nutrigenomics 2011;4:261–74.

[6] Baccarelli A, Tarantini L, Wright RO, Bollati V, Litonjua AA, Zanobetti A, et al. Repetitive element DNA methylation and circulating endothelial and inflammation markers in the VA normative aging study. Epigenetics 2010;5:222–8.

[7] Newman PE. Can reduced folic acid and vitamin B12 levels cause deficient DNA methylation producing mutations which initiate atherosclerosis? Med Hypotheses 1999;53:421–4.

[8] Chen Z, Karaplis AC, Ackerman SL, Pogribny IP, Melnyk S, Lussier-Cacan S, et al. Mice deficient in methylenetetrahydrofolate reductase exhibit hyperhomocysteinemia and decreased methylation capacity, with neuropathology and aortic lipid deposition. Hum Mol Genet 2001;10:433–43.

[9] Castro R, Rivera I, Struys EA, Jansen EE, Ravasco P, Camilo ME, et al. Increased homocysteine and S-adenosylhomocysteine concentrations and DNA hypomethylation in vascular disease. Clin Chem 2003;49:1292–6.

[10] Lund G, Andersson L, Lauria M, Lindholm M, Fraga MF, Villar-Garea A, et al. DNA methylation polymorphisms precede any histological sign of atherosclerosis in mice lacking apolipoprotein E. J Biol Chem 2004;279:29147–54.

[11] Laukkanen MO, Mannermaa S, Hiltunen MO, Aittomaki S, Airenne K, Janne J, et al. Local hypomethylation in atherosclerosis found in rabbit ec-sod gene. Arterioscler Thromb Vasc Biol 1999;19:2171–8.

[12] Turunen MP, Aavik E, Yla-Herttuala S. Epigenetics and atherosclerosis. Biochim Biophys Acta 2009;1790:886–91.
[13] Ordovas JM, Smith CE. Epigenetics and cardiovascular disease. Nat Rev Cardiol 2010;7:510–19.
[14] Zawadzki C, Chatelain N, Delestre M, Susen S, Quesnel B, Juthier F, et al. Tissue factor pathway inhibitor-2 gene methylation is associated with low expression in carotid atherosclerotic plaques. Atherosclerosis 2009;204:e4–14.
[15] Castillo-Diaz SA, Garay-Sevilla ME, Hernandez-Gonzalez MA, Solis-Martinez MO, Zaina S. Extensive demethylation of normally hypermethylated CpG islands occurs in human atherosclerotic arteries. Int J Mol Med 2010;26:691–700.
[16] Borghini A, Cervelli T, Galli A, Andreassi MG. DNA modifications in atherosclerosis: from the past to the future. Atherosclerosis 2013;230:202–9.
[17] Iacobas DA, Iacobas S, Haddad GG. Heart rhythm genomic fabric in hypoxia. Biochem Biophys Res Commun 2010;391:1769–74.
[18] Bayles R, Harikrishnan KN, Lambert E, Baker EK, Agrotis A, Guo L, et al. Epigenetic modification of the norepinephrine transporter gene in postural tachycardia syndrome. Arterioscler Thromb Vasc Biol 2012;32:1910–16.
[19] Bokil NJ, Baisden JM, Radford DJ, Summers KM. Molecular genetics of long QT syndrome. Mol Genet Metab 2010;101:1–8.
[20] Baccarelli A, Wright R, Bollati V, Litonjua A, Zanobetti A, Tarantini L, et al. Ischemic heart disease and stroke in relation to blood DNA methylation. Epidemiology 2010;21:819–28.
[21] Baccarelli A, Bollati V. Epigenetics and environmental chemicals. Curr Opin Pediatr 2009;21:243–51.
[22] Breton CV, Byun HM, Wenten M, Pan F, Yang A, Gilliland FD. Prenatal tobacco smoke exposure affects global and gene-specific DNA methylation. Am J Respir Crit Care Med 2009;180:462–7.
[23] Movassagh M, Choy MK, Goddard M, Bennett MR, Down TA, Foo RSY. Differential DNA methylation correlates with differential expression of angiogenic factors in human heart failure. PLoS One 2010;5:e8564.
[24] Haas J, Frese KS, Park YJ, Keller A, Vogel B, Lindroth AM, et al. Alterations in cardiac DNA methylation in human dilated cardiomyopathy. EMBO Mol Med 2013;5:413–29.
[25] Meurs KM, Kuan M. Differential methylation of CpG sites in two isoforms of myosin binding protein C, an important hypertrophic cardiomyopathy gene. Environ Mol Mutagen 2011;52:161–4.
[26] Smolarek I, Wyszko E, Barciszewska AM, Nowak S, Gawronska I, Jablecka A, et al. Global DNA methylation changes in blood of patients with essential hypertension. Med Sci Monit 2010;16:CR149–155.
[27] Friso S, Carvajal CA, Fardella CE, Olivieri O. Epigenetics and arterial hypertension: the challenge of emerging evidence. Transl Res 2015;165(1):154–65.
[28] Friso S, Pizzolo F, Choi SW, Guarini P, Castagna A, Ravagnani V, et al. Epigenetic control of 11 beta-hydroxysteroid dehydrogenase 2 gene promoter is related to human hypertension. Atherosclerosis 2008;199:323–7.
[29] Baserga M, Kaur R, Hale MA, Bares A, Yu X, Callaway CW, et al. Fetal growth restriction alters transcription factor binding and epigenetic mechanisms of renal 11beta-hydroxysteroid dehydrogenase type 2 in a sex-specific manner. Am J Physiol Regul Integr Comp Physiol 2010;299:R334–342.
[30] Zhao Y, Gong X, Chen L, Li L, Liang Y, Chen S, et al. Site-specific methylation of placental HSD11B2 gene promoter is related to intrauterine growth restriction. Eur J Hum Genet 2014;22:734–40.

[31] Lee HA, Baek I, Seok YM, Yang E, Cho HM, Lee DY, et al. Promoter hypomethylation upregulates $Na^+-K^+-2Cl^-$ cotransporter 1 in spontaneously hypertensive rats. Biochem Biophys Res Commun 2010;396:252–7.
[32] Cho HM, Lee HA, Kim HY, Han HS, Kim IK. Expression of $Na^+-K^+-2Cl^-$ cotransporter 1 is epigenetically regulated during postnatal development of hypertension. Am J Hypertens 2011;24:1286–93.
[33] Riviere G, Lienhard D, Andrieu T, Vieau D, Frey BM, Frey FJ. Epigenetic regulation of somatic angiotensin-converting enzyme by DNA methylation and histone acetylation. Epigenetics 2011;6:478–89.
[34] Wang F, Demura M, Cheng Y, Zhu A, Karashima S, Yoneda T, et al. Dynamic CCAAT/enhancer binding protein-associated changes of DNA methylation in the angiotensinogen gene. Hypertension 2014;63:281–8.
[35] Zhang LN, Liu PP, Wang L, Yuan F, Xu L, Xin Y, et al. Lower ADD1 gene promoter DNA methylation increases the risk of essential hypertension. PLoS One 2013;8:e63455.
[36] Bellavia A, Urch B, Speck M, Brook RD, Scott JA, Albetti B, et al. DNA hypomethylation, ambient particulate matter, and increased blood pressure: findings from controlled human exposure experiments. J Am Heart Assoc 2013;2:e000212.
[37] Brook RD, Rajagopalan S, Pope 3rd CA, Brook JR, Bhatnagar A, Diez-Roux AV, et al. Particulate matter air pollution and cardiovascular disease: an update to the scientific statement from the American Heart Association. Circulation 2010;121:2331–78.

CHAPTER

16

DNA Methylation and Neurology

OUTLINE

16.1 INTRODUCTION

Epigenetics pervades all aspects of development, including the neurons and brain. The mammalian brain develops intimately within a maternal environment both prenatally *in utero* and postnatally during suckling. The human brain is further exceptional in that development continues long beyond weaning, and parts of the cortex undergo radical reorganization during the post-pubertal period [1]. Memory is a multicomponent process consisting of distinct phases that require distinct changes within the central nervous system. Unlike short-term memory, the formation and maintenance of long-term memory requires the synthesis of new protein within the brain [2], suggesting that it is mediated

M. Neidhart: DNA Methylation and Complex Human Disease.
DOI: http://dx.doi.org/10.1016/B978-0-12-420194-1.00016-6

by a mechanism or group of mechanisms that operate within the nucleus of a cell. Given that DNA methylation plays a major role in transcriptional regulation and is capable of perpetuation of potentially lifelong changes in gene expression, it is essential first to ask whether changes in DNA methylation are required for the formation and storage of behavioral memories. In the second part of this chapter we will present some evidence that DNA methylation in brain plays a role in alcoholism, drug addiction, and schizophrenia. Human studies with monozygotic twins have linked the environment to long-lasting epigenetic effects on the phenotype, in that psychiatric disease susceptibility is not concordant for genetically identical twins [3]. Moreover, young identical twins have similar amounts of epigenetically methylated DNA, but with aging they differ significantly in both the amounts and the patterning of DNA methylation, and hence gene expression [4].

16.2 CO-ADAPTATION OF IMPRINTED GENES

In genetic terms, as discussed in Chapter 14, "imprinting" refers to an intriguing pattern of inheritance wherein the mother's copy of a gene is differentially expressed relative to the father's copy of that same gene. In other words, the two alleles (the mother's and father's copies) of the same gene are somehow epigenetically tagged and subsequently handled differently during development and in the adult offspring of the mother and father. In traditional cases of genetic imprinting, one copy of the gene (for example, the copy inherited from the father) is fully silenced, leaving the mother's copy of the gene the exclusive source of cellular mRNA product of that gene. Epigenetic control of the expression of parent-specific alleles is a driving factor for regulating gene transcription broadly in the brain [5].

Interestingly, a subset of imprinted genes that are expressed in the developing hypothalamus are also expressed in the fetal placenta [6]. The functional outcome of the hypothalamus can be best viewed as a parent–infant co-adaptive process. These genes are all paternally expressed, but assuming their ancestral state to be biallelic expression then the regulatory process of silencing to enable only paternal expression primarily occurs in the matriline [7]. Considering the relative sparsity of imprinted genes discovered, their expression in certain tissues like the brain and placenta is greater than might be expected. Moreover, all of the genes so far identified that are expressed in both placenta and brain, especially the hypothalamus, are maternally imprinted. These distinct organ types (i.e., fetal placenta and maternal hypothalamus) function as one in the pregnant mother, although they are encoded by different genotypes. Early comparisons of the

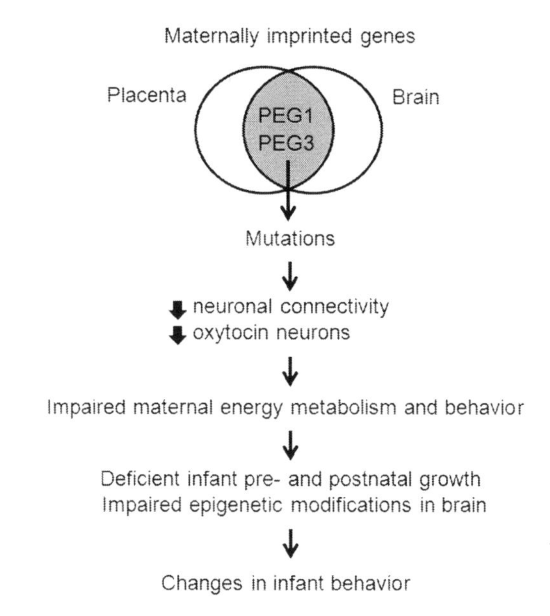

FIGURE 16.1 Influence of a mutation in imprinted genes on neuronal activity, energy metabolism, and behavior of mother and child.

phenotype of maternal hypothalamic versus fetal–placental expression regarding imprinted genes have shown the importance of paternal expressed gene proteins (PEG1/MEST and PEG3) [8–11]. Because imprinted genes are expressed according to parent-of-origin, a mother that carries a homozygous null mutation will, when mated with a wild-type father, produce offspring that are all wild-type normal, while a mutant father mated with a wild-type mother will produce mutant offspring that carry and express the mutant allele. In this way the effects of the mutation on the maternal phenotype and the infant phenotype can be investigated independently. The maternal consequences of expressing this targeted deletion have much in common with lesions of the maternal hypothalamus, namely reduced food intake, impaired maternal care, inability to maintain body temperature under a cold challenge, and a severe impairment in milk letdown [10–13]. Deficient PEG3 gene activity results in defective neuronal connectivity, as well as reduced oxytocin neurons in the hypothalamus [11]. As a consequence, maternal weight gain and fat reserves during pregnancy are impaired and pups suffer reduced pre- and postnatal growth, even though the litter size is smaller [12]. For the developing infant the mother provides the most significant environmental influence, shaping offspring brain development by producing long-term epigenetic modifications to neural and behavioral phenotypes (Figure 16.1). The mother's behavior is able to influence the development of brain regions that are important in the future regulation of maternal care in their daughters and boldness in their sons. Thus, non-heritable epigenetic modifications enable long-term stable changes in neural and behavioral phenotypes in response to environmental experiences. Conceptually these experiences have much in common with learning and memory, but differ in the timeframe whereby early life experiences may impact upon the behavioral phenotype at later periods in life [6].

16.3 MEMORY

Early studies tested the hypothesis that DNA methyltransferase (DNMT) activity might regulate the induction of hippocampal long-term potentiation (LTP), and this was indeed found to be the case [14]. Subsequent studies examined the role of DNA methylation in hippocampal-dependent forms of learning, such as contextual fear conditioning. These studies showed that dynamic changes in DNA methylation are critical in the consolidation of contextual fear-conditioned memories. Immediately following fear conditioning, mRNA levels of DNA-methyltransferase 3 (DNMT3A and DNMT3B) are upregulated in the cornu ammonis1 (CA1) of the hippocampus, and blocking DNMT activity prevents the normal memory consolidation [15]. Likewise, double knockout of DNMT1 and DNMT3A in adult neurons in mouse forebrain resulted in impaired escape latency and memory for platform location on the Morris water maze, as well as deficits in the 24-hour retention (but not immediate recall) of a contextual fear memory [16].

This consolidation is reliant upon the suppression of certain genes such as memory suppressors, for example protein phosphatase 1 (PP1), and the activation of other genes such as those involved in synaptic plasticity, for example REELIN, which is a large secreted extracellular matrix glycoprotein that helps regulate processes of neuronal migration and positioning in the developing brain by controlling cell–cell interactions. DNA methylation at the REELIN gene (which is associated with memory formation) is decreased following fear conditioning, whereas DNA methylation at the PP1 gene (which is viewed as a memory repressor) is enhanced following fear conditioning. Peak changes in the methylation status of these genes occur in CA1 1 hour following fear conditioning. Importantly, both of these changes returned to baseline levels 24 hours after conditioning, suggesting dynamic temporal regulation of both DNA methylation and demethylation in the hippocampus [15]. This showed that changes in DNA methylation may not be irreversible in post-mitotic neurons.

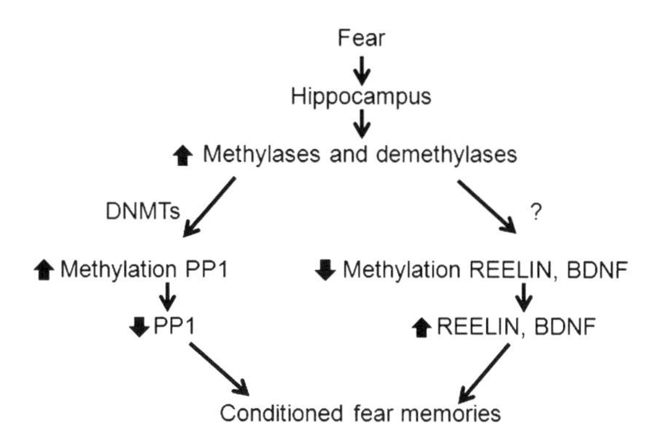

FIGURE 16.2 Promoter methylation of specific genes in the hippocampus and conditioned fear memories.

Furthermore, the gene encoding brain-derived neurotrophic factor (BDNF), which has been linked to the persistence of fear memories [17], also undergoes unique changes in DNA methylation as a result of fear conditioning [18]. Specifically, exon 4 of the BDNF gene undergoes significant suppression in DNA methylation in its promoter region following contextual fear conditioning, and this corresponds to a substantial increase in exon 4 mRNA that returns to baseline levels within 24 hours. This later change in DNA methylation was reversed by intrahippocampal infusions of the DNMT inhibitor, which impaired memory formation. In contrast, methylation at BDNF exons 1 and 6 is substantially decreased following context exposure alone, resulting in significantly increased mRNA for these exons (Figure 16.2).

Rat pups that receive low levels of stimulation develop increased stress responses, decreased response to reward, and decreased cognitive ability, and exhibit lower maternal care, in comparison to pups that receive high levels of tactile stimulation. The neuroendocrine system controls these behaviors, and the methylation of specific promoters play an essential role (Figure 16.3). The increased stress response has been associated with decreased expression of glucocorticoid receptors (GR) in the hippocampus due to a decrease in H3 acetylation and increase in DNA methylation of the 1–7 exons of the GR promoter in the hippocampus that contains a EGR-1 (also called NGFI-A) binding site [19]. The decreased levels of maternal care that they themselves exhibit as dams is associated with a decrease in the number of oxytocin receptors and estrogen receptor α (ERα) in the medial preoptic area that are related to increased DNA methylation of the ERα promoter in this brain region [20]. Thus, early life experiences can indeed have long-lasting profound effects upon the adult phenotype, and these may be mediated via epigenetic modifications of gene promoters in a brain

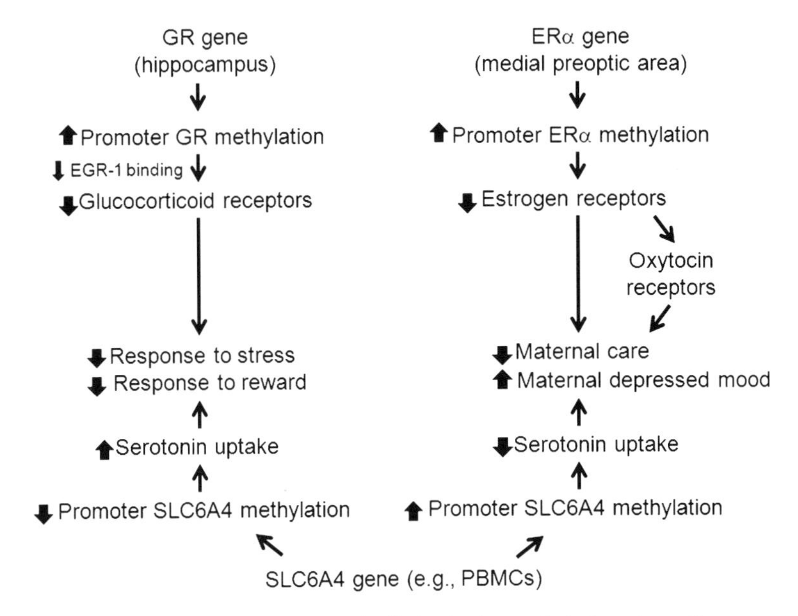

FIGURE 16.3 Neuroendocrine control of behavior by methylation of specific promoters.

region-specific manner. In rhesus macaques, higher methylation of the serotonin transporter SLC6A4 is associated with higher reactivity in adults that experienced early life stress as infants [21], whereas the DNA methylation level was not associated with rearing conditions. In humans, maternal depressed mood was associated with decreased DNA methylation of the promoter region of SLC6A4 in leukocytes of both maternal peripheral blood and neonatal cord blood [22]. Furthermore, the DNA methylation level of the CpG island was increased in subjects who had a history of childhood physical abuse [23]. In addition, higher levels of SLC6A4 promoter methylation occured in both T cells and monocytes in adult males with high childhood-limited aggression, who had lower *in vivo* serotonin synthesis in the orbitofrontal cortex as detected by positron emission tomography [24].

These findings suggest that changes in DNA methylation, even at the same gene within the same brain structure, are regulated in a highly complex fashion by different stimuli during learning, and that only some of these changes correspond to and are important for memory formation. These observations also reveal a general mechanism by which DNA methylation may contribute to neuronal function. However, if DNA methylation represents a molecular mechanism of memory storage, it is important to demonstrate a link between synaptic plasticity and changes in methylation. DNMTs, as the methylation status of REELIN and BDNF, play a role in both memory consolidation and neuronal plasticity [14–16,18]. Moreover, a follow-up to this report demonstrated that disruption of long-lasting LTP induced by DNMT inhibitors could be reversed by pretreatment with the histone deacetylase (HDAC) inhibitor trichostatin-A, suggesting a potential crosstalk between histone acetylation and DNA methylation [25].

Additionally, the prefrontal cortex receives inputs from a number of sensory and limbic systems and projects to areas implicated in motivation and action selection, making it the ideal storage site for specific memories. One possibility is that DNA methylation provides a regulatory mechanism prior to a learning event in order to generate cell-specific responses. For example, DNA methylation could modulate cell excitability both positively and negatively, thereby determining which neurons are eligible for participation in a given memory trace [26]. This mechanism would ensure that only a small fraction of neurons are capable of undergoing plastic changes at any given time. Such a function would be reminiscent of the role played by CREB during the formation of fear memories in the lateral amygdala [27]. Alternatively, the inherent connectivity of cortical circuits could engender selectivity in responsiveness, while DNA methylation serves to stamp in a specific distribution of synaptic weights after learning in order to preserve certain memories [26]. Finally, given that behavioral representations of a memory likely depend on multiple interacting circuits, another possibility is that DNA methylation may silence an entire node within a circuit to produce a given behavioral outcome. Critically, each of these hypotheses could potentially explain the remote memory deficits induced by DNMT inhibitors.

16.4 ALCOHOLISM

Chronic alcohol consumption causes well-documented vitamin B and folate deficiencies that negatively affect the methionine metabolism. This can result in excess levels of the S-adenosylmethionine (SAM) precursor homocysteine in the blood (i.e., homocysteinemia) and decreased SAM production [28–31]. In addition, the alcohol metabolite, acetaldehyde, may induce inhibition of DNMT1 [32]. Furthermore, alcohol-induced DNA damage and the resulting repair reactions can lead to demethylation of 5-methylcytosine nucleotides [33]. All these mechanisms can cause a global DNA hypomethylation (Figure 16.4). Alcohol-induced global DNA hypomethylation has been reported in several peripheral tissues of alcohol-related models (e.g., colon [34]) and may play a role in alcoholic liver disease [30], fetal alcohol syndrome [32], and colon cancer [29].

In the frontal cortex of chronic alcoholic patients, long terminal repeat (LTR)-containing retrotransposons, which usually are heavily methylated, are less methylated [35]; this is associated with increased

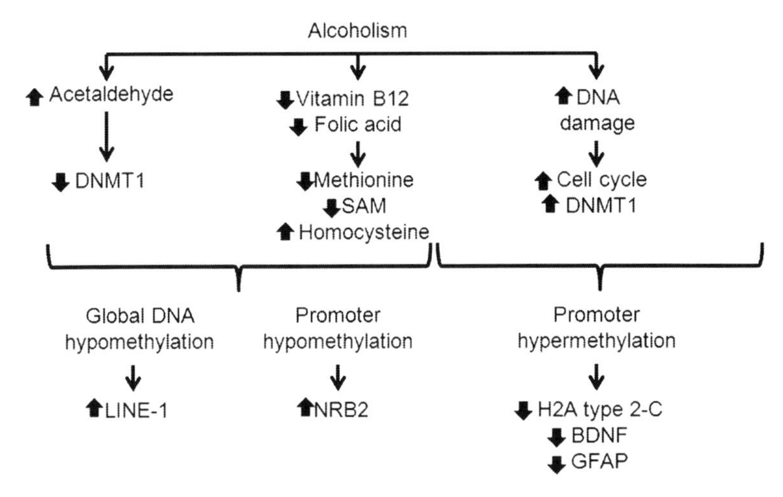

FIGURE 16.4 DNA methylation changes due to chronic alcohol consumption affecting peripheral tissues, as well as the central nervous system.

expression of these transcripts. This may be a sign of global DNA hypomethylation in the brain that is consistent with the majority of previous studies on alcohol-induced changes in DNA methylation. A microarray analysis revealed that about 20% of all promoters are differentially methylated between controls and alcoholics [36]; particularly, the promoter of HIST2H2AC encoding histone H2A type 2-C is more methylated in these patients.

Studies assessing epigenetic regulation of individual genes in the brain have shown that alcohol's effects on DNA methylation depend on a variety of factors, including the specific gene targets, developmental stage of exposure, and type of neuronal tissue affected [37]. Much of this work has focused on the central effects of prenatal alcohol exposure and on gene regulation in cell cultures. Prenatal exposure of rats to alcohol resulted in DNA hypermethylation and a reduced expression of BDNF in olfactory bulbs of rat pups, which was associated with loss of neurons in this brain region [38]. Similar results were obtained in a separate study where prenatal alcohol treatment of rats led to DNA hypermethylation and a decreased expression of a protein characteristically found in astrocytes (i.e., glial fibrillary acidic protein [GFAP]) in the brains of the pups [39]. In neural cell cultures, alcohol-induced downregulation of cell-cycle genes was paralleled by an increased DNMT activity and hypermethylation of the promoters of those genes [40]. Conversely, upregulation of the gene encoding a receptor subunit for the neurotransmitter glutamate (i.e., the NR2B glutamate receptor subunit) was associated with demethylation of CpG dinucleotides in the gene's promoter after chronic alcohol exposure [41].

16.5 DRUG ADDICTION

Reward processing is an integral neural event that ensures survival in an organism, as it reinforces positive behaviors and experiences. Perception of a reward is mediated by the brain reward system and uses dopamine as its principal neurotransmitter. The ventral tegmental area (VTA) is a midbrain region crucial to reward processing and represents a major dopaminergic output in the brain. Dopaminergic output from the VTA primarily projects to the nucleus accumbens (NAc), a principal center for reward processing, while the dorsal striatum, another region heavily implicated in addictive disorders, receives most of its dopaminergic input from midbrain neurons in the substantia nigra pars compacta.

Food reward has been shown to activate dopaminergic neurons in the VTA. Indeed, feeding behavior is heavily influenced by the expectation of pleasure and reward, and this proves to be a very powerful motivator of consumption. Associative learning for sucrose rewards increases methylation of learning-associated genes within dopaminergic neurons in the VTA, and inhibition of DNA methylation in this brain region, but not the NAc, prevents acquisition of the behavior [42].

Drug addiction involves long-term changes in cellular phenotype and gene expression, particularly in striatal brain areas, including the NAc [43,44]. Acute and chronic cocaine exposure promotes DNMT3A expression in the NAc [45,46] (Figure 16.5). Long after administration of an addictive substance, the drug and even other contextual cues can have powerful effects on behavior. Some of these long-term changes in gene expression and dependence are associated with epigenetic alterations (i.e., H3 acetylation of specific promoters [47]) as well as increased expression of DNA methyl binding proteins MeCP2 and MBD1 [48]. The binding of these proteins can be associated with both transcriptional activation and transcriptional repression, depending on its association with different transcription factors [49]. More specifically, chronic cocaine self-administration in rats increases striatal MeCP2 levels. Interestingly, when MeCP2 is locally knocked down in the striatum, rats decrease their cocaine intake levels [50]. Conversely, genetic ablation of MeCP2 in the NAc enhances amphetamine reward [51].

16.6 SCHIZOPHRENIA

In schizophrenia, even in genetically identical twins there is only 50% concordance [6]. Thus, determining the epigenetic contribution to such human disorders will be crucial though not necessarily easy, as the epigenetic regulation of gene expression will be both developmentally and

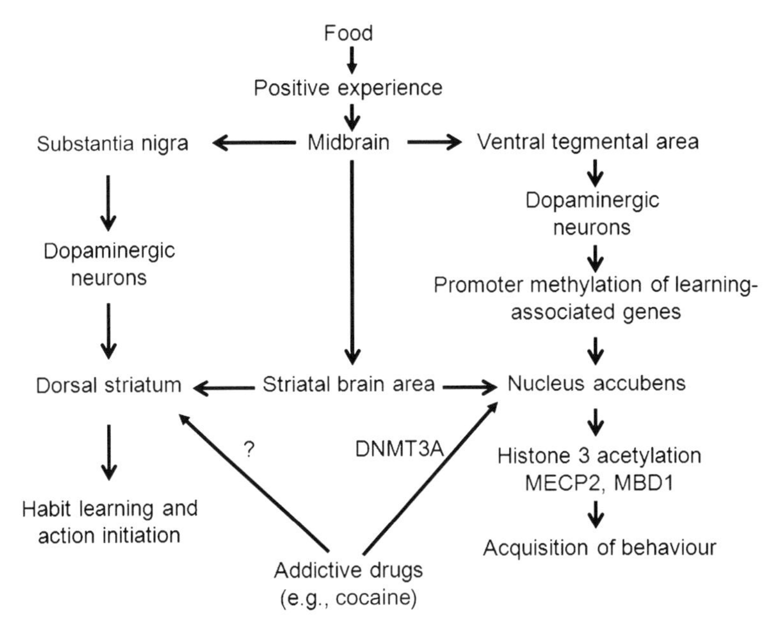

FIGURE 16.5 Mechanism of reward and drug addiction and influence of epigenetics on the activity of dopaminergic neurons in the ventral tegmental area, as well as on neurons of the nucleus accubens, which leads to the acquisition of a specific behavior.

regionally specific within the brain. A disbalanced expression of DNMT1 appears to be associated with psychiatric diseases (Figure 16.6).

Levels of REELIN and GAD67 are downregulated in the cortical and hippocampal tissue samples of schizophrenic patients. These proteins are used by GABAergic neurons, and abnormalities in their levels could result in some of the symptoms found in schizophrenics. These two genes have CpG islands in their promoters that are prone to being methylated. The lower expression of these genes is associated with an increased DNMT1 expression [52]. It has also been shown that a schizophrenic-type state can be induced in mice when they are chronically supplemented with L-methionine.

DNA methylation can also affect expression of BDNF. The BDNF protein is important for cognition, learning, and even vulnerability to early life trauma. Fear condition led to changes in DNA methylation levels in BDNF promoter regions in hippocampal neurons [17,18]. Inhibition of DNMT activity led to change in levels of BDNF in the hippocampus. Methylation of BDNF promoter has also been shown to be affected by postnatal social experiences, stressful environment, and social interaction deprivation. Furthermore, these stimuli have also been linked to increased anxiety and problems with cognition [53].

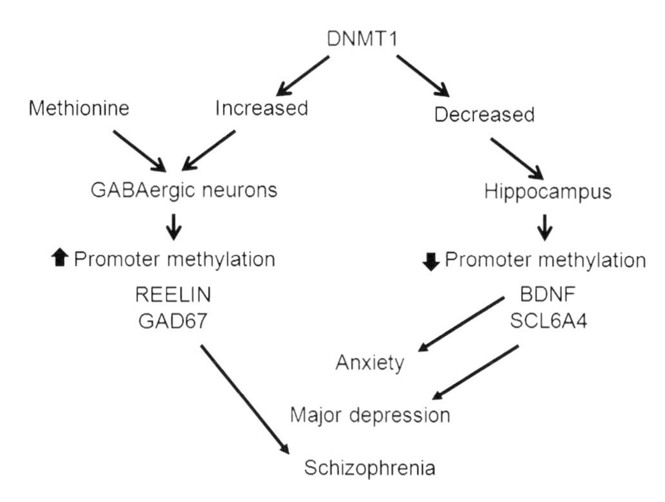

FIGURE 16.6 Role of a disbalanced expression of DNMT1 in psychiatric diseases.

Fluoxetine, a selective serotonin reuptake inhibitor used in the treatment of schizophrenia, has been shown to increase MBD1 and MeCP2 levels in areas of the adult rat brain with serotonergic receptors, such as the caudate putamen, frontal cortex, and dentate gyrus [48].

SLC6A4 encodes the serotonin transporter (SERT or 5-HTT). This protein is the target of many antidepressant medications. A repeat-length polymorphism in the promoter of this gene has been shown to affect the rate of serotonin uptake, and may play a role in sudden infant death syndrome, aggressive behavior in Alzheimer's disease patients, and post-traumatic stress disorder and depression susceptibility in people experiencing emotional trauma. A meta-analysis showed a significant association between serotonin transporter-linked promoter region (HTTLPR) and alcohol dependency [54].

A role of genome and environment interaction between SLC6A4 and stress has been reported in depression [55]. Thus, in lymphoblastoid cells, the DNA methylation level of SLC6A4 promoter tends to be higher in subjects with a lifetime history of major depression compared to those without a history of major depression [56]. DNA methylation status of the buccal cells, which are derived from ectoderm, might be more similar to that of neuronal cells compared with peripheral blood leukocytes (PBLs), which are derived from mesoderm [57]. Whereas there was no association between depressive symptoms and methylation level or HTTLPR genotype, depressive symptoms were more common among those with elevated methylation levels in S allele carriers. This result implies that an interaction of epigenetic and genetic factors involving SLC6A4 is related to depressive symptoms.

16.7 CONCLUSION

The study of epigenetics in neural systems is associated with a number of unique challenges [26]. One particular difficulty arises from the complexity of neuronal subtypes, even within the same brain region. It would not be surprising if the various types of interneurons and projection neurons were to exhibit drastically different epigenetic profiles, even at the same moment in time. Thus, an important challenge will be to unfold the many layers of epigenetic regulation that occur within separate cell types, and link each of these layers to overall neuronal and systems function in a meaningful way. A second and related challenge will be to understand how different neural circuits use epigenetic mechanisms in different ways for behavioral adaptation. This will be important not only for understanding the molecular mechanisms underlying the function of a single brain region, but also to understand how multiple regions integrate and store information at different speeds. Finally, it will be necessary to fully understand how the genes that undergo epigenetic modification following learning in turn influence synaptic and circuit plasticity, and understand how and why this may differ in discrete brain regions.

There are relatively few studies that focus on the role of DNA methylation and drug addiction, and to date there are no studies that look at the role of DNA demethylation. The process of DNA methylation has long been considered to be a stable, static process, but with our recent understanding of the molecular mechanisms of DNA demethylation catalyzed by TET dioxygenases and DNA glycosylase TDG, DNA methylation appears not to be as stable as previously thought. It will therefore be necessary to address the specific role that DNA demethylation machinery plays in the addiction process. Despite the challenges, regulation of DNA methylation states has the potential to serve as a molecular switch that can drive memory formation and shape vulnerability to drug abuse and psychiatric disorders.

References

[1] Sowell ER, Thompson PM, Toga AW. Mapping changes in the human cortex throughout the span of life. Neuroscientist 2004;10:372–92.

[2] Davis HP, Squire LR. Protein synthesis and memory: a review. Psychol Bull 1984;96:518–59.

[3] Caspi A, Moffitt TE. Gene-environment interactions in psychiatry: joining forces with neuroscience. Nat Rev Neurosci 2006;7:583–90.

[4] Fraga MF, Ballestar E, Paz MF, Ropero S, Setien F, Ballestar ML, et al. Epigenetic differences arise during the lifetime of monozygotic twins. Proc Natl Acad Sci USA 2005;102:10604–9.

[5] Gregg C, Zhang J, Weissbourd B, Luo S, Schroth GP, Haig D, et al. High-resolution analysis of parent-of-origin allelic expression in the mouse brain. Science 2010; 329:643–8.
[6] Keverne EB, Curley JP. Epigenetics, brain evolution and behavior. Front Neuroendocrinol 2008;29:398–412.
[7] Surani MA. Reprogramming of genome function through epigenetic inheritance. Nature 2001;414:122–8.
[8] Lefebvre L, Viville S, Barton SC, Ishino F, Keverne EB, Surani MA. Abnormal maternal behavior and growth retardation associated with loss of the imprinted gene Mest. Nat Genet 1998;20:163–9.
[9] Mayer W, Hemberger M, Frank HG, Grummer R, Winterhager E, Kaufmann P, et al. Expression of the imprinted genes MEST/Mest in human and murine placenta suggests a role in angiogenesis. Dev Dyn 2000;217:1–10.
[10] Hiby SE, Lough M, Keverne EB, Surani MA, Loke YW, King A. Paternal monoallelic expression of PEG3 in the human placenta. Hum Mol Genet 2001;10:1093–100.
[11] Li L, Keverne EB, Aparicio SA, Ishino F, Barton SC, Surani MA. Regulation of maternal behavior and offspring growth by paternally expressed Peg3. Science 1999;284:330–3.
[12] Curley JP, Pinnock SB, Dickson SL, Thresher R, Miyoshi N, Surani MA, et al. Increased body fat in mice with a targeted mutation of the paternally expressed imprinted gene Peg3. FASEB J 2005;19:1302–4.
[13] Curley JP, Barton S, Surani A, Keverne EB. Coadaptation in mother and infant regulated by a paternally expressed imprinted gene. Proc Biol Sci 2004;271:1303–9.
[14] Levenson JM, O'Riordan KJ, Brown KD, Trinh MA, Molfese DL, Sweatt JD. Regulation of histone acetylation during memory formation in the hippocampus. J Biol Chem 2004;279:40545–59.
[15] Miller CA, Sweatt JD. Covalent modification of DNA regulates memory formation. Neuron 2007;53:857–69.
[16] Feng J, Zhou Y, Campbell SL, Le T, Li E, Sweatt JD, et al. Dnmt1 and Dnmt3a maintain DNA methylation and regulate synaptic function in adult forebrain neurons. Nat Neurosci 2010;13:423–30.
[17] Bekinschtein P, Cammarota M, Katche C, Slipczuk L, Rossato JI, Goldin A, et al. BDNF is essential to promote persistence of long-term memory storage. Proc Natl Acad Sci USA 2008;105:2711–16.
[18] Lubin FD, Roth TL, Sweatt JD. Epigenetic regulation of BDNF gene transcription in the consolidation of fear memory. J Neurosci 2008;28:10576–86.
[19] Weaver IC. Epigenetic programming by maternal behavior and pharmacological intervention. Nature versus nurture: let's call the whole thing off. Epigenetics 2007;2:22–8.
[20] Champagne FA, Weaver IC, Diorio J, Dymov S, Szyf M, Meaney MJ. Maternal care associated with methylation of the estrogen receptor-alpha1b promoter and estrogen receptor-alpha expression in the medial preoptic area of female offspring. Endocrinology 2006;147:2909–15.
[21] Kinnally EL, Capitanio JP, Leibel R, Deng L, LeDuc C, Haghighi F, et al. Epigenetic regulation of serotonin transporter expression and behavior in infant rhesus macaques. Genes Brain Behav 2010;9:575–82.
[22] Devlin AM, Brain U, Austin J, Oberlander TF. Prenatal exposure to maternal depressed mood and the MTHFR C677T variant affect SLC6A4 methylation in infants at birth. PLoS One 2010;5:e12201.
[23] Beach SR, Brody GH, Todorov AA, Gunter TD, Philibert RA. Methylation at 5HTT mediates the impact of child sex abuse on women's antisocial behavior: an examination of the Iowa adoptee sample. Psychosom Med 2011;73:83–7.

[24] Wang D, Szyf M, Benkelfat C, Provencal N, Turecki G, Carmaschi D, et al. Peripheral SLC6A4 DNA methylation is associated with in vivo measures of human brain serotonin synthesis and childhood physical aggression. PLoS One 2012;7:e39501.
[25] Miller CA, Campbell SL, Sweatt JD. DNA methylation and histone acetylation work in concert to regulate memory formation and synaptic plasticity. Neurobiol Learn Mem 2008;89:599–603.
[26] Day JJ, Sweatt JD. Cognitive neuroepigenetics: a role for epigenetic mechanisms in learning and memory. Neurobiol Learn Mem 2011;96:2–12.
[27] Zhou Y, Won J, Karlsson MG, Shou M, Rogerson T, Balaji J, et al. CREB regulates excitability and the allocation of memory to subsets of neurons in the amygdala. Nat Neurosci 2009;12:1438–43.
[28] Blasco C, Caballeria J, Deulofeu R, Lligona A, Pares A, Lluis JM, et al. Prevalence and mechanisms of hyperhomocysteinemia in chronic alcoholics. Alcohol Clin Exp Res 2005;29:1044–8.
[29] Hamid A, Wani NA, Kaur J. New perspectives on folate transport in relation to alcoholism-induced folate malabsorption--association with epigenome stability and cancer development. FEBS J 2009;276:2175–91.
[30] Lu SC, Huang ZZ, Yang H, Mato JM, Avila MA, Tsukamoto H. Changes in methionine adenosyltransferase and S-adenosylmethionine homeostasis in alcoholic rat liver. Am J Physiol Gastrointest Liver Physiol 2000;279:G178–185.
[31] Shukla SD, Velazquez J, French SW, Lu SC, Ticku MK, Zakhari S. Emerging role of epigenetics in the actions of alcohol. Alcohol Clin Exp Res 2008;32:1525–34.
[32] Garro AJ, McBeth DL, Lima V, Lieber CS. Ethanol consumption inhibits fetal DNA methylation in mice: implications for the fetal alcohol syndrome. Alcohol Clin Exp Res 1991;15:395–8.
[33] Chen CH, Pan CH, Chen CC, Huang MC. Increased oxidative DNA damage in patients with alcohol dependence and its correlation with alcohol withdrawal severity. Alcohol Clin Exp Res 2011;35:338–44.
[34] Choi SW, Stickel F, Baik HW, Kim YI, Seitz HL, Mason JB. Chronic alcohol consumption induces genomic but not p53-specific DNA hypomethylation in rat colon. J Nutr 1999;129:1945–50.
[35] Ponomarev I, Wang S, Zhang L, Harris RA, Mayfield RD. Gene coexpression networks in human brain identify epigenetic modifications in alcohol dependence. J Neurosci 2012;32:1884–97.
[36] Manzardo AM, Henkhaus RS, Butler MG. Global DNA promoter methylation in frontal cortex of alcoholics and controls. Gene 2012;498:5–12.
[37] Ponomarev I. Epigenetic control of gene expression in the alcoholic brain. Alcohol Res 2013;35:69–76.
[38] Maier SE, Cramer JA, West JR, Sohrabji F. Alcohol exposure during the first two trimesters equivalent alters granule cell number and neurotrophin expression in the developing rat olfactory bulb. J Neurobiol 1999;41:414–23.
[39] Valles S, Pitarch J, Renau-Piqueras J, Guerri C. Ethanol exposure affects glial fibrillary acidic protein gene expression and transcription during rat brain development. J Neurochem 1997;69:2484–93.
[40] Hicks SD, Middleton FA, Miller MW. Ethanol-induced methylation of cell cycle genes in neural stem cells. J Neurochem 2010;114:1767–80.
[41] Marutha Ravindran CR, Ticku MK. Changes in methylation pattern of NMDA receptor NR2B gene in cortical neurons after chronic ethanol treatment in mice. Brain Res Mol Brain Res 2004;121:19–27.
[42] Day JJ, Childs D, Guzman-Karlsson MC, Kibe M, Moulden J, Song E, et al. DNA methylation regulates associative reward learning. Nat Neurosci 2013;16:1445–52.

[43] Tuesta LM, Zhang Y. Mechanisms of epigenetic memory and addiction. EMBO J 2014;33:1091–103.
[44] Hyman SE, Malenka RC, Nestler EJ. Neural mechanisms of addiction: the role of reward-related learning and memory. Annu Rev Neurosci 2006;29:565–98.
[45] Anier K, Malinovskaja K, Aonurm-Helm A, Zharkovsky A, Kalda A. DNA methylation regulates cocaine-induced behavioral sensitization in mice. Neuropsychopharmacology 2010;35:2450–61.
[46] LaPlant Q, Vialou V, Covington 3rd HE, Dumitriu D, Feng J, Warren BL, et al. Dnmt3a regulates emotional behavior and spine plasticity in the nucleus accumbens. Nat Neurosci 2010;13:1137–43.
[47] Kumar A, Choi KH, Renthal W, Tsankova NM, Theobald DE, Truong HT, et al. Chromatin remodeling is a key mechanism underlying cocaine-induced plasticity in striatum. Neuron 2005;48:303–14.
[48] Cassel S, Carouge D, Gensburger C, Anglard P, Burgun C, Dietrich JB, et al. Fluoxetine and cocaine induce the epigenetic factors MeCP2 and MBD1 in adult rat brain. Mol Pharmacol 2006;70:487–92.
[49] Chahrour M, Jung SY, Shaw C, Zhou X, Wong ST, Qin J, et al. MeCP2, a key contributor to neurological disease, activates and represses transcription. Science 2008;320:1224–9.
[50] Im HI, Hollander JA, Bali P, Kenny PJ. MeCP2 controls BDNF expression and cocaine intake through homeostatic interactions with microRNA-212. Nat Neurosci 2010;13:1120–7.
[51] Deng JV, Rodriguiz RM, Hutchinson AN, Kim IH, Wetsel WC, West AE. MeCP2 in the nucleus accumbens contributes to neural and behavioral responses to psychostimulants. Nat Neurosci 2010;13:1128–36.
[52] Gavin DP, Sharma RP. Histone modifications, DNA methylation, and schizophrenia. Neurosci Biobehav Rev 2010;34:882–8.
[53] Roth TL, Lubin FD, Sodhi M, Kleinman JE. Epigenetic mechanisms in schizophrenia. Biochim Biophys Acta 1790;869-877:2009.
[54] Arias A, Feinn R, Kranzler HR. Association of an Asn40Asp (A118G) polymorphism in the mu-opioid receptor gene with substance dependence: a meta-analysis. Drug Alcohol Depend 2006;83:262–8.
[55] Kendler KS, Kuhn JW, Vittum J, Prescott CA, Riley B. The interaction of stressful life events and a serotonin transporter polymorphism in the prediction of episodes of major depression: a replication. Arch Gen Psychiatry 2005;62:529–35.
[56] Philibert RA, Sandhu H, Hollenbeck N, Gunter T, Adams W, Madan A. The relationship of 5HTT (SLC6A4) methylation and genotype on mRNA expression and liability to major depression and alcohol dependence in subjects from the Iowa Adoption Studies. Am J Med Genet B, Neuropsychiatric Genet 2008;147B:543–9.
[57] Olsson CA, Foley DL, Parkinson-Bates M, Byrnes G, McKenzie M, Patton GC, et al. Prospects for epigenetic research within cohort studies of psychological disorder: a pilot investigation of a peripheral cell marker of epigenetic risk for depression. Biol Psychol 2010;83:159–65.

CHAPTER

17

DNA Methylation in Psychiatric Diseases

OUTLINE

M. Neidhart: *DNA Methylation and Complex Human Disease*.
DOI: http://dx.doi.org/10.1016/B978-0-12-420194-1.00017-8

17.1 INTRODUCTION

Psychiatric diseases place a tremendous burden on affected individuals, their caregivers, and the healthcare system. Although evidence exists for a strong inherited component to many of these conditions, dedicated efforts to identify DNA sequence-based causes have not been exceptionally productive, and very few pharmacologic treatment options are clinically available. Many features of psychiatric diseases are consistent with an epigenetic dysregulation, such as discordance of monozygotic twins, late age of onset, parent-of-origin and sex effects, and fluctuating disease course. In recent years, experimental technologies have advanced significantly, permitting in-depth studies of the epigenome and its role in maintenance of normal genomic functions, as well as disease etiopathogenesis [1]. Chapter 16 presented an overview regarding the role of DNA methylation in brain development and neurology. In the present chapter we will address more specific psychiatric diseases.

17.2 MAJOR PSYCHOSIS

Major psychosis is a classification that encompasses both schizophrenia and bipolar disorders. Schizophrenia is a multifactorial disease characterized by disordered thinking and concentration that results in psychotic thoughts (delusions and hallucinations), inappropriate emotional responses, erratic behavior, as well as social and occupational deterioration, while bipolar disorders represent a category of mood disorders in which affected individuals experience episodes of mania or hypomania interspersed with periods of depression, and may also suffer from delusions and hallucinations. Classically, psychosis research was aimed at defining genetic and environmental risk factors, but despite significant evidence of a heritable component derived from twin and adoption studies [2,3] many molecular genetic findings have not been replicated, and significant heterogeneity and small effect sizes are thought to plague genetic association studies [4].

17.2.1 Environmental Factors

The etiology of schizophrenia is now thought to be multifactorial, with multiple small-effect and fewer large-effect susceptibility genes interacting with several environmental factors. These factors may lead to developmentally mediated alterations in neuroplasticity, manifesting in a cascade of neurotransmitter and circuit dysfunctions and impaired connectivity with an onset around early adolescence [5]. Several environmental factors, such as antenatal maternal virus infections, obstetric complications entailing hypoxia as a common factor, or stress during neurodevelopment, have been identified to play a role in schizophrenia and bipolar disorder, possibly contributing to smaller hippocampal volumes [6].

17.2.2 Schizophrenia

An epigenomic study [7] of major psychosis revealed that, in the cortex, differences were discovered at loci involved in glutamatergic and ψ-aminobutyric acid (GABA)-ergic neurotransmission, brain development, mitochondrial function, stress response, and other disease-related functions, many of which correspond to psychosis-related changes in steady-state mRNA. In relation to the glutamatergic hypothesis, a lower degree of DNA methylation was observed in schizophrenia and combined male psychosis samples at two glutamate receptor genes, NR3B and the α-amino-3-hydroxy-5-methyl-4isoxazolepropionic acid receptor (AMPA) receptor-subunit gene GRIA2; the dysregulation of AMPA and N-methyl-D-aspartic acid (NMDA) receptors is an etiological component of major psychosis, and it has been shown that GRIA2 expression is altered in the prefrontal cortex and striatum of schizophrenia patients [8]. Hypomethylation was also detected at the vesicular glutamate transporter (VGLUT2) in schizophrenic women, and at secretogranin II (SCG2), which encodes a neuronal vesicle protein that stimulates glutamate release. On the other hand, a higher degree of methylation was observed in schizophrenic women at VGLUT1, a transporter protein that is downregulated in schizophrenic brains [9], and the glutaminase enzyme, GLS2, in schizophrenic males, which has previously been shown to exhibit altered expression in cases of schizophrenia [10]. In synergy with glutamatergic pathways, GABAergic pathways also show dysregulation in cases of major psychosis. Detected disruptions in such pathways included hypermethylation at the RNA-binding regulator of GABA(B) receptors, MARLIN-1, in psychotic females; the G protein-coupled inwardly rectifying potassium channel linked to GABA neurotransmission, KCNJ6, in psychotic males; as well as the HELT locus in psychotic females, which is known to determine GABAergic over glutamatergic neuronal fate in the mesencephalon. Several other intriguing loci were highlighted, such as hypermethylation at WNT1, a gene critical for neurodevelopment that is differentially expressed in schizophrenic brains [11], in females affected with major psychosis; and at AUTS2 in schizophrenic males, which spans a translocation breakpoint associated with autism and mental retardation. A highly significant hypermethylation was detected in both male and female samples at two loci: RPP21, which encodes a component of ribonuclease P, a complex that forms t-RNA molecules via 5′-end cleavage, and KEL, which encodes the Kell blood-group glycoprotein and causes McLeod syndrome when incorrectly expressed; SZ symptoms are manifested as part of McLeod syndrome. Network and gene ontology (GO) analyses were performed in order to determine relationships between the functionally linked pathways from the microarray dataset. The network analysis revealed a lower degree of modularity of DNA methylation "nodes" in the major psychosis samples, indicating that there is some degree of systemic epigenetic dysregulation involved in the disorder [11].

Both schizophrenia and bipolar diseases have also been examined using the candidate gene approach, as epigenetic downregulation of genes is emerging as a possible underlying mechanism of the GABAergic neuronal dysfunction in schizophrenia. Interesting candidate genes are (as mentioned in the previous chapter) the glutamate decarboxylase 67 kDa (GAD67/GAD1) and DNMT1. GAD67 catalyzes the conversion of glutamic acid to GABA. In cases of schizophrenia, the levels of this enzyme and several others involved in GABAergic neurotransmission, such as GAD65 and GABA plasma membrane transporter-1 (GAT-1), display decreased mRNA levels [12–14]. In addition to aberrant methylation at this locus, an analysis [15] showed that decreased GAD67 mRNA levels strongly correlated with upregulated HDAC1 in the prefrontal cortices of patients with schizophrenia. At the GAD67 promoter, these patients have been shown to display a biologically relevant deficit in repressive chromatin-associated DNA methylation [16]. Hypermethylation of GAD67 occurs through DNMT1 [17], which is upregulated in the GABAergic neurons and peripheral blood lymphocytes of schizophrenic patients, along with the *de novo* methyltransferase, DNMT3A [17,18]. Interestingly, nicotine has been shown to decrease DNMT1 mRNA expression in cortical and hippocampal GABAergic neurons in mice; this decrease results in GAD67 promoter demethylation, and is inversely related to an upregulation of cortical GAD67 protein [19]. This information is highly relevant, as schizophrenic patients tend to smoke tobacco at a rate that is two- to four-fold higher than in the general population [20], and are possibly drawn to the nicotine content for its effects on the aforementioned pathway (Figure 17.1).

17.2.3 Bipolar Diseases

In bipolar diseases, genomic imprinting has been suggested by statistical genetics but molecular approaches have not yielded the imprinted disease genes [21]. A study [22] applied methylation-sensitive representational difference analysis to lymphoblastoid cells derived from twins

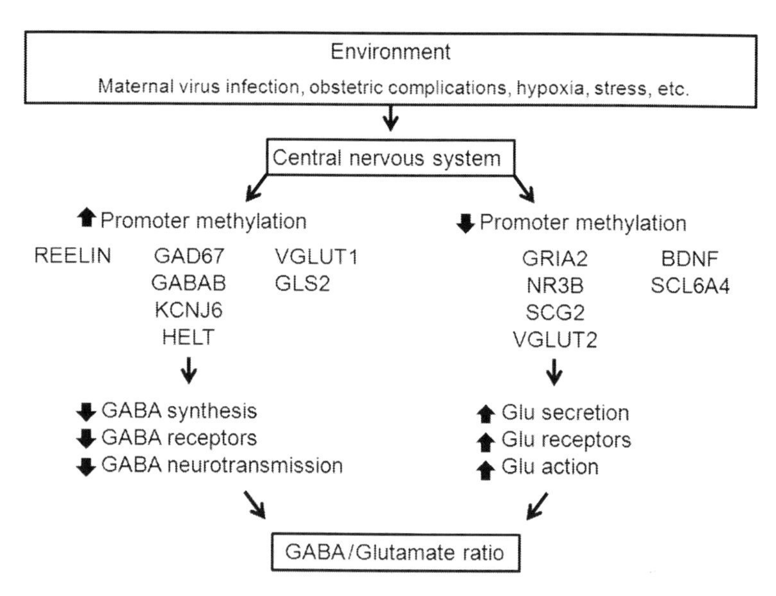

FIGURE 17.1 Changes in DNA methylation cause a GABA/glutamate neurotransmitter imbalance in the brain of patients with schizophrenia.

discordant for bipolar disorders. One detected gene, named peptidylprolyl isomerase Elike (PPIEL), was hypomethylated in bipolar disorder-affected twins, while a region of the spermine synthase (SMS) gene was hypermethylated versus unaffected twins; it has yet to be determined if either of these regions is biologically and functionally significant. In combined studies of epigenetics and DNA sequence, some interesting developments have been observed. As mentioned in Chapter 16, it has been shown that rare G variants of a G/A polymorphism in the potassium chloride co-transporter 3 gene (SLC12A6) may represent risk factors for bipolar disorders [23]. Variants containing the G allele are methylated at the adjacent cytosine, and this accompanied a decrease in gene expression in human lymphocytes [24]. In addition, as mentioned in Chapter 16, decreased serotonin transporter SLC6A4 methylation has been associated with antisocial behavior and aggressiveness [25–27].

17.3 ALZHEIMER'S DISEASE

Alzheimer's disease is a neurodegenerative disorder, and the most common form of dementia in the elderly; it is characterized by the accumulation of intracellular neurofibrillary tangles and extracellular amyloid plaques in the brain. Alzheimer's disease often presents with psychiatric symptoms, such as memory loss, mood swings, and irritability, that increase in severity as the disease progresses. Amyloid β (Aβ) denotes peptides of 36–43 amino acids that are crucially involved in Alzheimer's disease as the main component of the amyloid plaques found in the brains of these patients. The peptides result from the amyloid precursor protein, which is being cut by certain enzymes to yield Aβ. Aβ molecules can aggregate to form flexible soluble oligomers, which may exist in several forms and can be proteolytically degraded within the brain mainly by neprilysin (NEP) and insulin-degrading enzyme (IDE, insulysin). The endolytic degradation of Aβ peptides within microglia by NEP and related enzymes is significantly enhanced by apolipoprotein E (APOE). Similarly, Aβ extracellular degradation by IDE is facilitated by APOE. There are three major isoforms of APOE: APOE2, APOE3, and APOE4. Among them, APOE3 is the most common isoform in the population, while APOE4 has been shown to confer dramatically increased risk for late onset Alzheimer's disease. The majority of Alzheimer's disease research focuses on dysregulation of fibers and proteins, such as APOE4, but little ground has been gained with regard to determining the actual origins of their dysfunction. Consistent with the epigenetic hypothesis, abnormal levels of folate and homocysteine, signs of dysregulated methylation maintenance, have been detected in the brain of patients with Alzheimer's disease.

17.3.1 Environmental Factors

Numerous genetic and environmental risk factors are involved in the etiology and pathogenesis of AD, including alterations in the expression of thousands of genes, amyloid β-peptide (Aβ) deposition, tau hyperphosphorylation, inflammation, oxidative stress, energy metabolism, and aberrant re-entry into the cell cycle/apoptosis. It is worth noting that when Aβ-inducing mutations are absent, these molecular and genetic factors do not have absolute penetrance in causing the disorder [28]. Knowing that DNA methylation changes do interact with Alzheimer's disease and usually serve as causes, one may seek the origin of these changes in DNA methylation patterns. Such pursuit is worthwhile, since it may finally guide us to a new path for treating Alzheimer's disease. We divide possible causes of DNA methylation regulation in Alzheimer's disease into four aspects: aging, vitamin B deficiency, oxidative stress, and heavy metal exposure [29].

Aging is a widely accepted risk factor of Alzheimer's disease, and some reports have regarded methylation change related to Alzheimer's disease as an acceleration [30] or specialized case [31] of aging. These studies provide a chain of causation from aging to DNA methylation deregulation and then to this disease.

A deficiency of vitamin B may block the regular DNA methylation metabolic cycle. This indication is supported by experimental evidence in rats [32], and vitamin B deficiency is commonly used as a method to create DNA methylation metabolism disorder in current studies [33]. Moreover, homocysteine inhibits the dimerization of APOE3 and reduces APOE3-mediated high-density lipoprotein (HDL) generation [34]. It impairs APOE3 dimerization and APOE3's ability to generate HDL by binding to cysteine residues of APOE3. Therefore, hyperhomocysteinemia may promote the pathogenesis of Alzheimer's disease.

Causing the imbalance between DNA methylation and demethylation, oxidative stress is also known as an environmental factor interacting with DNA methylation. Study in SH-SY5Y neuroblastoma cells revealed that treatment with H_2O_2 may activate a DNMT inhibitor and result in the upregulation of amyloid precursor and β-secretase 1 (BACE1) through a transcription activator, leading to the upregulation of Aβ production [35]. Extracellular cleavage of amyloid precursor by BACE1 creates a soluble extracellular fragment and a cell membrane-bound fragment referred to as C99. In turn, cleavage of C99 within its transmembrane domain by γ-secretase releases the intracellular domain of amyloid precursor and produces Aβ.

Finally, infant exposure to lead (Pb) has been reported as another environmental factor contributing to Alzheimer's disease through DNA methylation regulation. Developmental exposure of rodents to the heavy metal lead has been shown to increase amyloid precursor and Aβ in the aging brain [36]. A study in aged monkeys showed that the group that was fed Pb in early life had lower DNMT activity. The downregulation of DNMT activity results in the hypomethylation of several genes involved in Aβ formation, such as amyloid precursor and BACE1, and causes upregulation of their protein expression; in turn, SP1 transcription factor increases, which finally results in Aβ formation and Alzheimer's disease [37]. These finding are confirmed by a genome-wide study on mice [38] and an *in vitro* study [39]. A latent early-life associated regulation model, which claims that environmental agents perturb gene regulation at early stage but do not have pathological results until later in life, was proposed to explain these phenomena [40].

17.3.2 Global DNA Hypomethylation

Most studies support the view that, on the whole-genome scale, the DNA methylation level in patients with Alzheimer's disease is lower in comparison with normal individuals. Direct evidence came from studies in monozygotic twins. A post-mortem study of rare sets of monozygotic twins discordant for Alzheimer's disease reported significantly reduced levels of DNA methylation in temporal neocortex neuronal nuclei of the Alzheimer's twin, which provided the potential for disease discordancy in spite of genetic similarities. With specific markers, decrement in DNA methylation was observed in Alzheimer's disease, reflected among neurons, reactive astrocytes, and microglia [28]. Research also recognized a dramatic decrease in patients with Alzheimer's disease of far-going epigenetic markers and regulators in neurons from control entorhinal cortex layer II, consistent with the high vulnerability of this brain region to AD pathology. The result indicates that global DNA and RNA methylation status are significantly diminished in AD in this region. This trend of global hypomethylation was further confirmed in a more recent and direct study, in which decrements of 5-methyl cytosine and 5-methyl hydroxycytosine were shown in Alzheimer's patients' hippocampus by a quantitative immunohistochemical method [41].

Another study [42] examined the methylation changes in selected genes from two cohorts from USA and Iceland, taking DNA samples at two time points from each subject, spaced either 11 or 16 years apart. In these two populations, time-dependent changes in global DNA methylation were observed within the same individual, with 8–10% of individuals showing changes that were greater than 20%; both gains and losses of methylation were detected. Similarly, another American study [43] measured DNA methylation in the blood of elderly subjects (55–92 years of age) over a span of 8 years. A progressive loss of DNA methylation in repetitive elements was found, particularly in ALU repeats, and this linear decline highly correlated with time since the first measurement [30,43]. Accordingly, the DNA methylation status of repetitive elements long interspersed element-1 (LINE-1) is increased in patients with Alzheimer's disease compared with healthy volunteers [44].

17.3.3 Changes in Methylation of Specific Genes

Using the Alzheimer's model human glioblastoma cell line H4-sw, which harbors a mutation allowing production of high levels of the toxic form Aβ, global methylation status was analyzed. Among the total 6296 differentially methylated CpG sites, 23% were shown to be hypermethylated while others were hypomethylated [45]. Results overall suggest that there is no correlation between amyloid precursor gene hypomethylation and Alzheimer's disease [46,47]. However, a study in SH-SY5Y neuroblastoma cells indicated that the promoter of amyloid precursor gene is capable of being hypomethylated [48].

A few genes decreased in methylation with age in control subjects, including some implicated in Alzheimer's disease and other psychiatric disorders: CITF [45], NXT2 [45], S100A2 [49], PSEN1 [33], BACE1 [50],

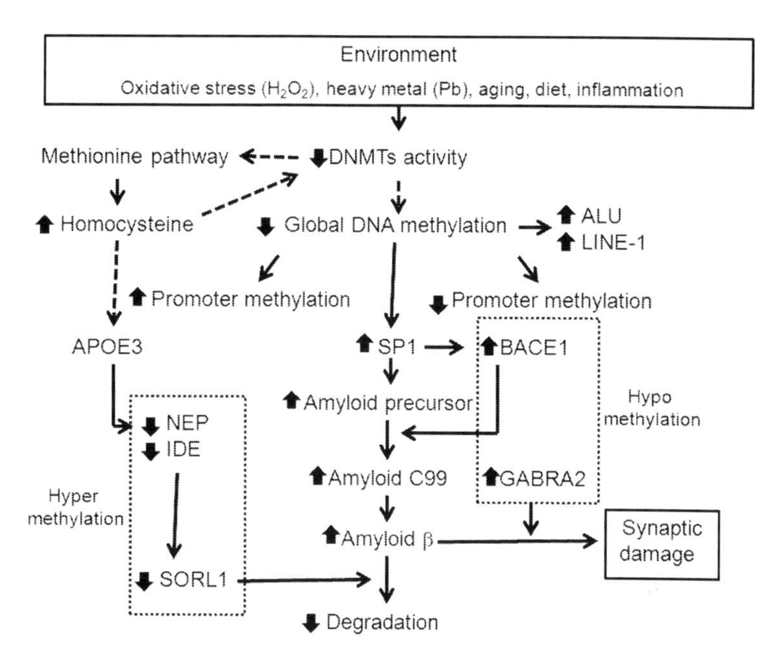

FIGURE 17.2 Changes in DNA methylation lead to an accumulation of amyloid β in in the brain of patients with Alzheimer's disease.

COX2 [51], BDNF [51], and NF-κB [51]. It is not clear whether this is associated with the pathogenesis or is a consequence of the disease (Figure 17.2).

S100A2 displays a complex chronology, but results in a slow, stochastic methylation decrease later in life [49]. It encodes a calcium binding protein from the S100 family. As part of normal brain aging, S100A2 protein accumulates in corpora amylacea, or polyglucosan bodies; subjects with neurodegenerative disorders experience a much greater accumulation of corpora amylacea [49], and this is consistent with the eventual decrease in S100A2 promoter methylation [30].

PSEN1 (the gene encoding PS1) promoter hypomethylation and the resulting overexpression of PSEN1 as factors may contribute to Alzheimer's disease. First, studies in neuroblastoma cell lines and mice [33] showed that PSEN1 can be overexpressed through DNA hypomethylation. Further investigations in mice ruled out the possibility that hypomethylation of PSEN1 promoter is the consequence of amyloid production. Similar hypomethylation was observed in a post-mortem study [31]. It has also been reported that BACE1 expression can be upregulated via demethylation of at least two CpG sites in the 5′-untranslated region in BV-2 microglial cells [50]. Whether this is the case in human samples of patients with Alzheimer's disease is not yet known.

In the Alzheimer's brain, cyclooxygenase-2 (COX-2, one of the arachidonic acid cascade markers) and BDNF promoter regions are hypomethylated, while the promoter region of cAMP response element-binding protein (CREB), which regulates the transcription of BDNF, was shown to be hypermethylated [51]. This study also found decreased methylation of the NF-κB promoter CpG region.

On the other hand, a large number of genes increased in methylation with age in control subjects, including several possibly implicated in Alzheimer's disease and schizophrenia: GAD1 [30], GABRA2 [30], CREB [51], DDR2 [45], SORBS3 [49], NEP [52], SORL1 [53], NGB [54], and SYP [51]. Hypermethylation of GAD1, GABRA2, NEP, and SORL1, at least, could be involved in development of the disease.

Increased methylation of GAD1, GABRA2, HOXA1 (homeobox protein Hox-A1), NEUROD1 (neurogenic differentiation 1), NEUROD2 (neurogenic differentiation 2), PGR (progesterone receptor), STK11 (serine/threonine kinase 11), and SYK (spleen tyrosine kinase) promoters was described in an early study [30]. Unfortunately, no confirmation followed. Most interestingly, GAD1 encodes one of several forms of glutamic acid decarboxylase, which is responsible for catalyzing the production of GABA (from L-glutamic acid). GABRA2 encodes a γ-aminobutyric acid receptor subunit (α2). It's important to note that glutamate receptors are centrally involved in synaptic targeting by Aβ [55].

It has been reported that Aβ causes NEP promoter region hypermethylation, which consequently suppresses NEP's expression in mRNA and protein levels, reduces the Aβ clearance, and probably elevates Aβ accumulation. This could be part of a vicious cycle which plays a role in the pathophysiology of Alzheimer's disease [52].

Sortilin-related receptor (SORL1, also known as SORLA, SORLA1, or LR11) is a neuronal APOE receptor. Its association with Alzheimer's disease is based on its reduction in Alzheimer's brains and its ability to lower Aβ levels [56]. The SORL1 gene showed differences in its expression among peripheral blood leukocytes, and it may act as a marker of aging in this tissue. It has been shown that SORL1 promoter DNA methylation could serve as one of the mechanisms in charge of the differences in expression observed between blood and brain for both healthy elders and Alzheimer's patients [53]. Thus, both decreased NEP and SORL1 could cause an accumulation of Aβ.

Finally, SORBS3 gained methylation over time and is more likely to be methylated in patients with Alzheimer's disease [49]. It encodes a neuronal/glial cell adhesion molecule. In Alzheimer's brain, a significant increase in DNA methylation at the promoter region of synaptophysin (SYP) has also been reported [51]. SYP is present in neuroendocrine cells and in virtually all neurons in the brain and spinal cord that participate in synaptic transmission. Elimination of synaptophysin in mice creates behavioral changes, such as increased exploratory behavior, impaired object novelty recognition, and reduced spatial learning [57].

17.4 AUTISM

Autism and related developmental disorders, such as Asperger's and Rett syndromes, fall under the broader class of autism spectrum disorders (ASDs). These disorders become apparent in young children and persist into adulthood, with deficits in social cognition regarded as the most characteristic feature of ASD, leading to restrictions in social communication. While autism itself is believed to have a particularly strong inherited basis relative to other developmental psychiatric syndromes, DNA sequence factors in the etiology of ASD are still largely unknown [58,59]. The prenatal factors associated with autism risk were provided by a meta-analysis [60]: advanced paternal and maternal age at birth, gestational diabetes, gestational bleeding, multiple birth, being first born compared to being third or after, and maternal birth abroad. In addition, parental immigration, especially maternal immigration but also paternal immigration, is a risk factor for ASD [61]. This association between migration and autism is more particularly observed in male children of immigrant parents living in urban areas compared to rural areas. Furthermore, evidence supports a contribution of imprinted genes in ASD, as well as paternal transmission, and perhaps the combination of this information and the lack of identified genetic markers will stimulate future epigenetic and epigenomic studies of ASD.

17.4.1 Environmental Factors

Pro-oxidant environmental stressors could modulate autism development through DNA methylation changes. A link between epigenetic regulation and antioxidant/detoxification capacity has been reported in many children with autism that showed genome-wide DNA hypomethylation and oxidative protein/DNA damage [62]. Deficits in antioxidant and methylation capacity could promote cellular damage and altered epigenetic gene expression.

17.4.2 Changes in Methylation of Specific Genes

Rett syndrome (RTT), a division of ASD, has been extensively studied and arises from loss of function mutations at the locus for methyl-CpG-binding protein 2 (MeCP2). Mouse models have been very useful

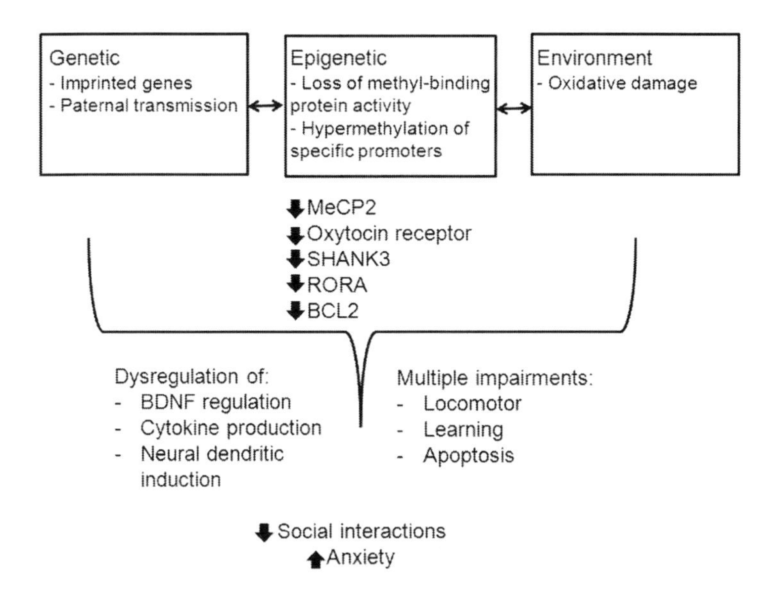

FIGURE 17.3 Together with genetic factors, the loss of methyl-binding proteins and specific promoter hypermethylation lead to neurological dysregulation and multiple impairments in patients with autism spectrum disorders.

in delineating the relationship between disturbances to MeCP2 and the disease [63]. In mice, deletion of MeCP2 mimics RTT syndrome, leading to locomotor impairments and reduction in brain size [64,65]. Mice with a truncated MeCP2 protein, similar to that of RTT patients, developed many features of RTT, such as tremors, motor impairments, hypoactivity, increased anxiety-related behavior, seizures, kyphosis, and stereotypic forelimb motions. In astrocytes cultured from a mouse model of RTT, MeCP2 deficiency causes significant abnormalities in BDNF regulation, cytokine production, and neuronal dendritic induction [66] (Figure 17.3).

The classic form of autism also appears to be connected to MeCP2 expression. Coding mutations affecting the protein are rarely detected in autism, but significantly increased MeCP2 promoter methylation has been found in autistic male frontal cortex compared with controls, and this inversely correlated with protein expression [67]; aberrant promoter methylation at MeCP2 has also been detected in female brain DNA [68]. Similarly, loss of methyl-CpG binding protein 1 (MBD1) leads to autism-like behavioral deficits in mice, namely reduced social interaction, learning deficits, anxiety, defective sensory motor gating, depression, and abnormal brain serotonin activity [69].

In ASD, DNA methylation has been also linked to reduced expression of the oxytocin receptor [70]. Oxytocin is a neuropeptide hormone correlated with social behaviors. In a mouse model, the oxytocin receptor gene expression is epigenetically regulated by DNA methylation of its promoter. Furthermore, methylation of specific sites in the gene promoter of the oxytocin receptor gene significantly inhibits its transcription in individuals with autism.

Genes that control synaptic molecules also show epigenetic regulation. Among them, the SHANK3 gene is subjected to a specific epigenetic control mechanism [71]. Indeed, DNA methylation regulates the tissue-specific expression of the SHANK3 gene [72]. The methylation of CpG-island 2 in the SHANK3 promoter seems to be involved in the tissue-regulated expression. SHANK3 mutant mice exhibit impaired social interaction and repetitive behaviors, like autism. In the central nervous system, the protein SHANK3 is mainly expressed in neurons, especially in their synapses, and is strictly associated with the cell adhesion proteins neuroligins (NLGN), acting as a scaffolding protein.

Hypermethylation of specific CpG sites in upstream promoter regions of BCL2 and retinoic acid receptor-related orphan receptor (RORA) genes were identified in autistic children compared to healthy developing twins [73]. The expression of these genes was found to be downregulated in the cerebellum of post-mortem brain tissues from subjects with ASDs. BCL2 is a known protein involved in the anti-apoptotic process. These findings suggest that a possible dysfunctional apoptotic event could lead to altered development of specific brain regions, resulting in decreased cognitive function. RORA belongs to the nuclear receptor-1 subfamily of nuclear hormone receptors, and is involved in the control of the neuronal oxidative stress. RORA is a novel candidate gene for ASD pathology, and could also be related to sex hormone involvement in autism [74].

17.5 PARKINSON'S DISEASE

Parkinson's disease is the second most common neurodegenerative disorder after Alzheimer's disease. More than 4 million people worldwide are affected with this disease. The prevalence of Parkinson's disease in industrialized countries is generally estimated at about 1–2% of the population over 60 years of age, and increases to 3–5% in people above 85 years old. This neurodegenerative disorder is characterized by progressive loss of substantia nigra dopaminergic neurons and striatal projections, causing the typical symptomatology: muscle rigidity, bradykinesia, tremor, and postural instability.

Although more than 90% of cases can be interpreted as sporadic PD, the greatest insights into PD etiology have come from the study of familial forms. Mutations in six genes have been identified as causes of PD [75]: SNCA (which encodes α-synuclein), PARK2 (parkin), PINK1

(PTEN-induced kinase protein 1), UCHL1 (ubiquitin carboxyl-terminal hydrolase isozyme L1), DJ1 (protein DJ-1), and LRRK2 (leucine-rich repeat serine/threonine-protein kinase 2). The hallmark neuropathological sign of Parkinson's disease is the presence of fibrillar aggregates of misfolded α-synuclein called Lewy bodies, which accumulate in the same sites where neuronal loss is found.

17.5.1 Environmental Factors

Environment and nutrition can significantly affect the risk and progression of neurodegenerative disorders, including Parkinson's disease.

Elevated plasma homocysteine levels have been documented in patients with Parkinson's disease [76]. A further increase in plasma homocysteine levels of blood cell samples was observed in individuals with both Parkinson's disease and the MTHFR C677T mutation [77]. Methylenetetrahydrofolate reductase (MTHFR) catalyzes the conversion of 5,10-methylenetetrahydrofolate to 5-methyltetrahydrofolate, a cosubstrate for homocysteine remethylation into methionine. Patients with Parkinson's disease undergoing regular treatment with L-Dopa had higher plasma homocysteine concentrations relative to healthy controls, indicating a possible methylated catabolism of the drug [78]. Dietary folate deficiency and elevated homocysteine levels have been found to be harmful to dopaminergic neurons in mouse models of Parkinson's disease [79]. A recent study found that hallmarks of neurodegeneration, such as amyloid precursor and α-synuclein, were related to markers of methylation, like S-adenosylmethionine (SAM) and its downstream byproduct S-adenosyl-L-homocysteine (SAH) in individuals with Parkinson's disease [80]. A higher SAM/SAH ratio, which indicates a higher methylation potential, was linked to better cognitive function.

Telomeric dysfunction has been discovered to be associated with development of age-related pathologies, and, similarly to AD, shortened telomeres were found present in patients with Parkinson's disease [81]. Telomere length is epigenetically regulated by DNA methylation, which in turn could be modulated by folate status. In humans, telomere length is associated with folate status [82,83].

Environmental exposure, including paraquat, is believed to be a risk for Parkinson's disease. Paraquat is a widely used herbicide. Interestingly, pretreatment of PC12 pheochromocytoma cells with a DNMT inhibitor sensitizes cells to paraquat exposition. Similar results were obtained using dopaminergic cells and treatments of MPP(+), 6-hydroxydopamine, and rotenone [84], suggesting that DNA methylation might modulate the effect of these toxins and perhaps play a role in susceptibility to Parkinson's disease [85]. These findings underlie a possible mechanism in which the environment influences the pathology of Parkinson's disease via DNA methylation modification. However, further analyses are necessary.

17.5.2 Changes in Methylation of Specific Genes

Considering that α-synuclein makes a major contribution to the formation of Lewy bodies and even to the entire pathogenesis of Parkinson's disease, a discussion was started regarding the impact DNA methylation has on α-synuclein [29]. An earlier study revealed that the DNA methylation pattern within the α-synuclein (SNCA) gene promoter region was altered in the blood samples of patients with alcoholism, which was significantly associated with their increased homocysteine levels [86] (see also Chapter 16, Section 16.4). This was the first study that showed a correlation between DNA methylation and α-synuclein in certain syndromes. Later, it was reported that the methylation of human SNCA intron 1 decreased gene expression while inhibition of DNA methylation activated its expression in the brains of patient with Parkinson's disease [87], which further strengthens the link. Although analysis of post-mortem brain did not reveal regional specific methylation differences in the putamen and anterior cingulate between Parkinson's disease and healthy individuals, methylation was found to be specifically and significantly reduced in the substantia nigra of such patients [88]. In addition, single CpG analysis reflected fluctuating methylation levels at different locations in various brain regions and stages of disease, even if the overall methylation levels in the promoter and intron 1 of the α-synuclein gene were reported to be similarly low in both patients and controls [89]. Furthermore, the interaction of DNMT and α-synuclein has also been tested. Reduction of nuclear DNMT1 levels was observed in post-mortem brain samples from patients with Parkinson's disease and in the brains of α-synuclein transgenic mice [90], underlying a mechanism in which DNMT1 might be excluded from the nucleus by α-synuclein, and the segregation of DNMT1 further resulted in hypomethylated CpG islands upstream of α-synuclein. Taken together, these results suggest a potential role of DNA methylation in α-synuclein neuropathogenesis.

Tumor necrosis factor α (TNF-α) is a critical inflammatory cytokine, and increased TNF-α is associated with dopaminergic cell death in Parkinson's disease. It has been suggested that a lesser extent of methylation of the TNF-α promoter in human substantia nigra cells could uphold the increased vulnerability of dopaminergic neurons to TNF-α regulated inflammatory reactions [91]. Because TNF-α overexpression induces apoptosis in neuronal cells and TNF-α levels were rather high in the cerebrospinal fluid of patients with Parkinson's disease [92], it can be speculated that DNA methylation is the reason for such overexpression of TNF-α.

As is well known, the PARKIN gene plays a relatively important part in the emergence and development of the disease. The methylation levels of the PARKIN gene promoters were analyzed in samples from patients heterozygous for PARKIN mutations, Parkinson's patients without PARKIN mutations, and normal controls; however, no significant difference was observed among the three groups, indicating that PARKIN promoter methylation alone is unlikely to impact the pathogenesis and development of the disease [93].

A large-scale sequencing analysis of post-mortem brain samples identified methylation and expression changes associated with Parkinson's disease risk variants in PARK16, GPNMB, and STX1B loci, suggesting that some other disease-related genes could also be epigenetically modified in Parkinson's brains [94]. PARK16 contains several genes, one or some of which (e.g., RAB7L1) increase the risk of Parkinson's disease. GPNMB encodes the transmembrane glycoprotein NMB, which could play a role in neurodegenerative disorders [95]. STX1B encodes syntaxin-1B, which is important for neuronal survival, possibly by regulating the secretion of neurotrophic factors, such as BDNF, from glial cells [96]. All these genes could be candidates for an epigenetic study. Clearly, there is the hypothesis that DNA methylation is associated with Parkinson's disease through a variety of genetic pathways (Figure 17.4), but the association is less clear than for Alzheimer's disease.

17.6 HUNTINGTON'S DISEASE

Huntington's disease, or Huntington's chorea, is the most common genetic cause of chorea in high-income countries, with a prevalence of about 1 : 10,000 people. This lethal neurodegenerative disease primarily affects the cerebral cortex and the striatum. Initial physical symptoms are chorea, rigidity, and dystonia, and these become more apparent as the disorder progresses. Cognitive abilities become gradually impaired, finally leading to dementia. This disease is caused by the expansion of CAG triplet repeats in the huntingtin (HTT) gene, which encodes an expanded polyglutamine (polyQ) stretch in the HTT protein. In addition, a highly polymorphic locus, D4S95, was identified and demonstrated to be tightly linked to the Huntington's disease gene [97]. Most studies aiming to find a clear correlation between Huntington's disease and DNA methylation focus on two specific subjects: the HTT gene and D4S95 locus. Later, evidences were presented involving a genetic imprinting mechanism in Huntington's disease [98] (Figure 17.5).

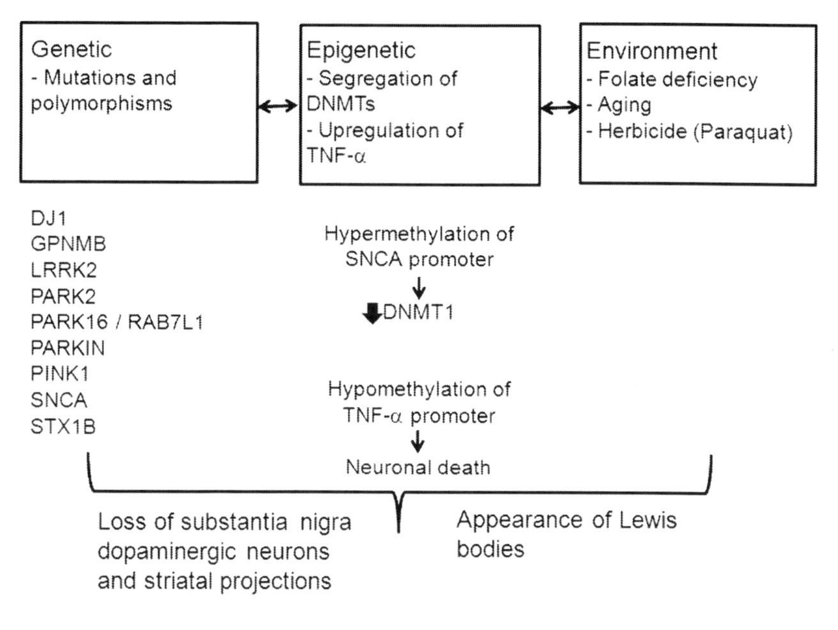

FIGURE 17.4 Together with genetic and environmental factors, the segregation of DNMT1 and upregulation of TNF-α expression by changes in DNA methylation contribute to the pathogenesis of Parkinson's disease.

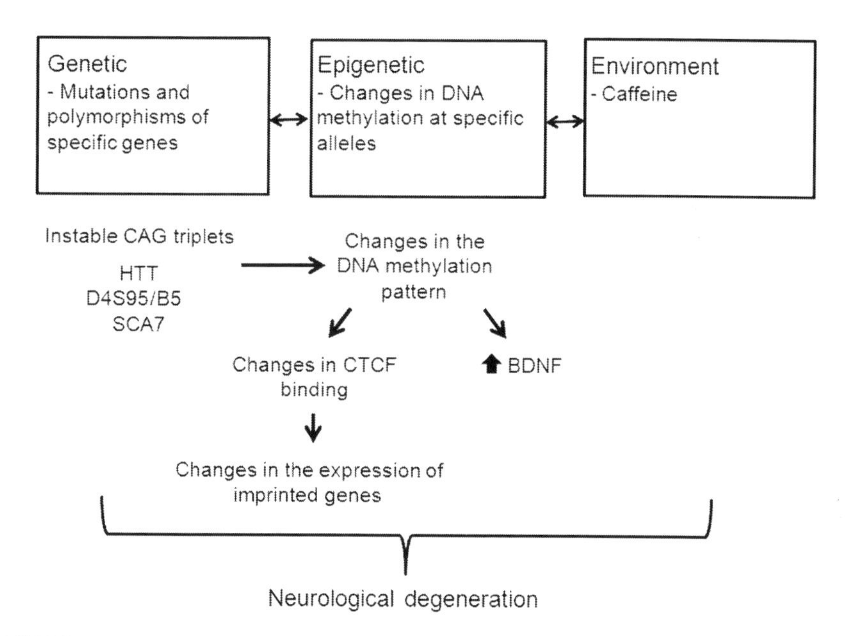

FIGURE 17.5 Modifications of the genetic sequence change DNA methylation, the binding of CTCF, and the expression of imprinted genes. Together with increased BDNF, this leads to neurological degeneration in Huntington's disease.

17.6.1 Environmental Factors

In addition to the genetic factors, environment also plays a vital role in the pathogenesis of Huntington's disease, but this has not been studied in this disease so far, with the exception that habitual caffeine intake is beneficially associated with the age of onset in patients with Huntington's disease [99].

17.6.2 Changes in Methylation of Specific Genes

Interestingly, methylation was found at the B5 allele of the D4S95 locus, which was not inherited in a Mendelian fashion, as its appearance depended on the methylation status of the human lymphoblastoid cells from which DNA samples were obtained [100]. Although a comparison between patients with Huntington's disease and normal controls showed no strong relevance between methylation and onset-age of the disease, a significant association of the patient's age with demethylation at D4S95 was found [101].

Another finding that genome-wide demethylation could accelerate instability of CTG/CAG trinucleotide repeats in mammalian cells implies that changes in methylation patterns during epigenetic reprograming may trigger the intergenerational repeat expansions, leading to instable HTT CAG triplets and neurological diseases like Huntington's disease [102]. Extensive changes in DNA methylation were reported to be linked to expression of mutant HTT gene, revealing the potential effects of DNA methylation alterations on neurogenesis and cognitive decline in patients with this disease [103].

It has been discovered that there is also a region of highly unstable CAG repeats at the human spinocerebellar ataxia type 7 (SCA7) locus, and this region contains binding sites for CTCF, a regulatory factor involved in genomic imprinting, chromatin remodeling, and DNA conformation change [104]. An investigation in transgenic mice found that CpG methylation of CTCF binding sites could further destabilize triplet repeat expansion [105], underpinning the role of DNA methylation in the regulation of neurological diseases.

Finally, a Huntington's disease-associated modification of BDNF gene expression was found in the hippocampus of female and male diseased mice, independent of methylation increases in the gene sequence, while there existed a pattern of sex-specific differences in the levels of methylation at individual CpG sites, suggesting that such differences might explain the differential regulation of BDNF expression in male and female brains [106]. Since it has been reported that the loss of BDNF gene transcription is likely a central factor in the progressive pathology of Huntington's disease, DNA methylation might be the explanation for such loss; this, however, needs further confirmation.

17.7 AMYOTROPHIC LATERAL SCLEROSIS

Amyotrophic lateral sclerosis (ALS) is an idiopathic, fatal neurodegenerative disease of the human motor system. The clinical hallmark of ALS is the presence of the lower motor neuron signs in brainstem and spinal cord, and the upper motor neuron signs in the motor cortex. Loss of these neurons leads to clinical phenotypes including muscle atrophy, weakness, fasciculation, spasticity, and cognitive dysfunction. Proposed pathogenic mechanisms for ALS include oxidative stress, glutamate excitotoxicity, impaired axonal transport, neurotrophic deprivation, neuroinflammation, apoptosis, altered protein turnover, etc. Furthermore, influences from astrocytes and microglia in the motor neuron microenvironment contribute to ALS pathogenesis [107].

ALS is traditionally classified into two categories: familial ALS (FALS) and sporadic ALS (SALS). FALS is predominantly hereditary, and almost always autosomal dominant, while X-linked or recessive FALS is rare. Several genes and their mutations have been found to be associated with ALS. Superoxide dismutase 1 (SOD1) mutations are the cause in about 20% of FALS cases. Moreover, TARDBP, which encodes TAR DNA-binding protein, and FUS, an RNA- binding protein fused in sarcoma, also contribute to FALS cases. SALS has been associated with another gene, ELP3, encoding the catalytic subunit of the histone acetyltransferase (HAT) complex elongator protein [75]. Additionally, ALS2, ATXN2, and some other genes [108] are also associated with ALS; these are discussed below.

17.7.1 Environmental Factors

Environmental exposure to heavy metals has been implicated in SALS, and functionally impaired detoxification of these metals may cause serious susceptibility to the disease. Metallothionein (MT) is a family of proteins that are involved in primary detoxification mechanism for heavy metals. As a matter of fact, no promoter methylation of human MT genes was evident in any SALS or control samples, implying the possibility that methylation at these gene promoters may not be a common cause of SALS [109].

17.7.2 Changes in Methylation of Specific Genes

Altered methylation of the ALSIN (ALS2) gene promoter was observed in hippocampal cells of individuals with a history of being abused in childhood [110]. Higher levels of promoter methylation were correlated with repression of ALS2 transcription, suggesting a role of DNA methylation in the regulation of this gene. In addition, methylation of the human glutamate transporter EAAT2 gene promoter has been reported to be associated with the silent state of the human EAAT2 gene. Since dysfunction of the EAAT2 transporter might contribute to the pathogenesis of ALS [111], it would be relevant to test regulation of the EAAT2 transporter via an epigenetic mechanism including DNA methylation in ALS models [112]. Studies also found that a group of genes, either unsuspected in SALS or in potential pathways of cell death, revealed changed methylation status in SALS brains [109].

In mouse models, the apoptosis process of motor neurons showed alterations in DNMT1, DNMT3a, and 5-methylcytosine, which is similar to those in human ALS, indicating that DNMT may mediate neuronal cell death through DNA methylation [113]. In addition, CpG methylation in human ATXN2 gene promoter is associated with pathogenic CAG expansions in spinocerebellar ataxia type 2 (SCA2) cases [114]. Since such aberrant expansions in ATXN2 were shown to contribute to ALS [115], we can expect that a link might exist between ATXN2 promoter methylation and pathogenesis of ALS.

17.8 CONCLUSION

In the fields of both severe psychosis and neurodegenerative diseases, although experimental evidence has revealed correlations between DNA methylation and these diseases, two major issues remain unclear. The first is the causal relationship between DNA methylation modifications and psychiatric diseases. In other words, do DNA methylation modifications precede the appearance of psychosis or neurodegenerative symptoms? Is there a proven mechanism demonstrating that DNA methylation modifications will finally lead to these diseases? Another question is: if DNA methylation modifications do cause psychosis or neurodegenerative diseases, what is the trigger for these methylation modifications? Environmental factors that favored the pathogenesis of psychiatric disorders are not well known yet, in particular during the perinatal period. This is also a crucial question since it may lead us to new paths for curing these diseases.

References

[1] Ptak C, Petronis A. Epigenetic approaches to psychiatric disorders. Dialogues Clin Neurosci 2010;12:25–35.

[2] Bertelsen A, Gottesman II. Schizoaffective psychoses: genetical clues to classification. Am J Med Genet 1995;60:7–11.

[3] Cardno AG, Gottesman II. Twin studies of schizophrenia: from bow-and-arrow concordances to star wars Mx and functional genomics. Am J Med Genet 2000;97:12–17.

[4] O'Donovan MC, Craddock NJ, Owen MJ. Genetics of psychosis; insights from views across the genome. Hum Genet 2009;126:3–12.

[5] Haller CS, Padmanabhan JL, Lizano P, Torous J, Keshavan M. Recent advances in understanding schizophrenia. F1000Prime Rep 2014;6:57.

[6] Schmitt A, Malchow B, Hasan A, Falkai P. The impact of environmental factors in severe psychiatric disorders. Front Neurosci 2014;8:19.

[7] Mill J, Tang T, Kaminsky Z, Khare T, Yazdanpanah S, Bouchard L, et al. Epigenomic profiling reveals DNA-methylation changes associated with major psychosis. Am J Hum Genet 2008;82:696–711.

[8] Gupta DS, McCullumsmith RE, Beneyto M, Haroutunian V, Davis KL, Meador-Woodruff JH. Metabotropic glutamate receptor protein expression in the prefrontal cortex and striatum in schizophrenia. Synapse 2005;57:123–31.

[9] Eastwood SL, Harrison PJ. Decreased expression of vesicular glutamate transporter 1 and complexin II mRNAs in schizophrenia: further evidence for a synaptic pathology affecting glutamate neurons. Schizophr Res 2005;73:159–72.

[10] Bruneau EG, McCullumsmith RE, Haroutunian V, Davis KL, Meador-Woodruff JH. Increased expression of glutaminase and glutamine synthetase mRNA in the thalamus in schizophrenia. Schizophr Res 2005;75:27–34.

[11] Miyaoka T, Seno H, Ishino H. Increased expression of Wnt-1 in schizophrenic brains. Schizophr Res 1999;38:1–6.

[12] Bullock WM, Cardon K, Bustillo J, Roberts RC, Perrone-Bizzozero NI. Altered expression of genes involved in GABAergic transmission and neuromodulation of granule cell activity in the cerebellum of schizophrenia patients. Am J Psychiatry 2008;165:1594–603.

[13] Hashimoto T, Volk DW, Eggan SM, Mimics K, Pierri JN, Sun Z. Gene expression deficits in a subclass of GABA neurons in the prefrontal cortex of subjects with schizophrenia. J Neurosci 2003;23:6315–26.

[14] Huang HS, Matevossian A, Whittle C, Kim SY, Schumacher A, Baker SP, et al. Prefrontal dysfunction in schizophrenia involves mixed-lineage leukemia 1-regulated histone methylation at GABAergic gene promoters. J Neurosci 2007;27:11254–62.

[15] Sharma RP, Grayson DR, Gavin DP. Histone deactylase 1 expression is increased in the prefrontal cortex of schizophrenia subjects: analysis of the National Brain Databank microarray collection. Schizophr Res 2008;98:111–17.

[16] Huang HS, Akbarian S. GAD1 mRNA expression and DNA methylation in prefrontal cortex of subjects with schizophrenia. PLoS One 2007;2:e809.

[17] Veldic M, Kadriu B, Maloku E, Agis-Balboa RC, Guidotti A, Davis JM, et al. Epigenetic mechanisms expressed in basal ganglia GABAergic neurons differentiate schizophrenia from bipolar disorder. Schizophr Res 2007;91:51–61.

[18] Zhubi A, Veldic M, Puri NV, Kadriu B, Caruncho H, Loza I, et al. An upregulation of DNA-methyltransferase 1 and 3a expressed in telencephalic GABAergic neurons of schizophrenia patients is also detected in peripheral blood lymphocytes. Schizophr Res 2009;111:115–22.

[19] Satta R, Maloku E, Zhubi A, Pibiri F, Hajos M, Costa E, et al. Nicotine decreases DNA methyltransferase 1 expression and glutamic acid decarboxylase 67 promoter methylation in GABAergic interneurons. Proc Natl Acad Sci USA 2008;105:16356–61.
[20] Lising-Enriquez K, George TP. Treatment of comorbid tobacco use in people with serious mental illness. J Psychiatry Neurosci 2009;34:E1–2.
[21] McGowan PO, Kato T. Epigenetics in mood disorders. Environ Health Prev Med 2008;13:16–24.
[22] Kuratomi G, Iwamoto K, Bundo M, Kusumi I, Kato N, Iwata N, et al. Aberrant DNA methylation associated with bipolar disorder identified from discordant monozygotic twins. Mol Psychiatry 2008;13:429–41.
[23] Meyer J, Johannssen K, Freitag CM, Schraut K, Teuber I, Hahner A, et al. Rare variants of the gene encoding the potassium chloride co-transporter 3 are associated with bipolar disorder. Int J Neuropsychopharmacol 2005;8:495–504.
[24] Moser D, Ekawardhani S, Kumsta R, Palmason H, Bock C, Athanassiadou Z, et al. Functional analysis of a potassium-chloride co-transporter 3 (SLC12A6) promoter polymorphism leading to an additional DNA methylation site. Neuropsychopharmacology 2009;34:458–67.
[25] Devlin AM, Brain U, Austin J, Oberlander TF. Prenatal exposure to maternal depressed mood and the MTHFR C677T variant affect SLC6A4 methylation in infants at birth. PLoS One 2010;5:e12201.
[26] Beach SR, Brody GH, Todorov AA, Gunter TD, Philibert RA. Methylation at SLC6A4 is linked to family history of child abuse: an examination of the Iowa Adoptee sample. Am J Med Genet B Neuropsychiatr Genet 2010;153B:710–13.
[27] Wang D, Szyf M, Benkelfat C, Provencal N, Turecki G, Caramaschi D, et al. Peripheral SLC6A4 DNA methylation is associated with in vivo measures of human brain serotonin synthesis and childhood physical aggression. PLoS One 2012;7:e39501.
[28] Mastroeni D, Grover A, Delvaux E, Whiteside C, Coleman PD, Rogers J. Epigenetic mechanisms in Alzheimer's disease. Neurobiol Aging 2011;32:1161–80.
[29] Lu H, Liu X, Deng Y, Qing H. DNA methylation, a hand behind neurodegenerative diseases. Front Aging Neurosci 2013;5:85.
[30] Siegmund KD, Connor CM, Campan M, Long TI, Weisenberger DJ, Biniszkiewicz D, et al. DNA methylation in the human cerebral cortex is dynamically regulated throughout the life span and involves differentiated neurons. PLoS One 2007;2:e895.
[31] Wang SC, Oelze B, Schumacher A. Age-specific epigenetic drift in late-onset Alzheimer's disease. PLoS One 2008;3:e2698.
[32] Miller CA, Campbell SL, Sweatt JD. DNA methylation and histone acetylation work in concert to regulate memory formation and synaptic plasticity. Neurobiol Learn Mem 2008;89:599–603.
[33] Fuso A, Cavallaro RA, Nicolia V, Scarpa S. PSEN1 promoter demethylation in hyperhomocysteinemic TgCRND8 mice is the culprit, not the consequence. Curr Alzheimer Res 2012;9:527–35.
[34] Minagawa H, Wantanabe A, Akatsu H, Adachi K, Ohtsuka C, Terayama Y, et al. Homocysteine, another risk factor for Alzheimer disease, impairs apolipoprotein E3 function. J Biol Chem 2010;285:38382–8.
[35] Gu X, Sun J, Li S, Wu X, Li L. Oxidative stress induces DNA demethylation and histone acetylation in SH-SY5Y cells: potential epigenetic mechanisms in gene transcription in Abeta production. Neurobiol Aging 2013;34:1069–79.
[36] Basha MR, Wei W, Bakheet SA, Benitez N, Siddigi HK, Ge YW, et al. The fetal basis of amyloidogenesis: exposure to lead and latent overexpression of amyloid precursor protein and beta-amyloid in the aging brain. J Neurosci 2005;25:823–9.
[37] Wu J, Basha MR, Brock B, Cox DP, Cardozo-Pelaez F, McPherson CA, et al. Alzheimer's disease (AD)-like pathology in aged monkeys after infantile exposure to

environmental metal lead (Pb): evidence for a developmental origin and environmental link for AD. J Neurosci 2008;28:3–9.
[38] Bihaqi SW, Huang H, Wu J, Zawia NH. Infant exposure to lead (Pb) and epigenetic modifications in the aging primate brain: implications for Alzheimer's disease. J Alzheimers Dis 2011;27:819–33.
[39] Bihaqi SW, Zawia NH. Alzheimer's disease biomarkers and epigenetic intermediates following exposure to Pb *in vitro*. Curr Alzheimer Res 2012;9:555–62.
[40] Lahiri DK, Maloney B. The "LEARn" (latent early-life associated regulation) model: an epigenetic pathway linking metabolic and cognitive disorders. J Alzheimers Dis 2012;30(Suppl 2):S15–30.
[41] Chouliaras L, Mastroeni D, Delvaux E, Grover A, Kenis G, Hof PR, et al. Consistent decrease in global DNA methylation and hydroxymethylation in the hippocampus of Alzheimer's disease patients. Neurobiol Aging 2013;34:2091–9.
[42] Bjornsson HT, Sigurdsson MI, Fallin MD, Irizarry RA, Aspelund T, Cui H, et al. Intra-individual change over time in DNA methylation with familial clustering. JAMA 2008;299:2877–83.
[43] Bollati V, Schwartz J, Wright R, Litonjua A, Tarantini L, Suh H, et al. Decline in genomic DNA methylation through aging in a cohort of elderly subjects. Mech Ageing Dev 2009;130:234–9.
[44] Bollati V, Galimberti D, Pergoli L, Dalla Valle E, Barretta F, Cortini F, et al. DNA methylation in repetitive elements and Alzheimer disease. Brain Behav Immun 2011;25:1078–83.
[45] Sung HY, Choi EN, Ahn Jo, S, Oh S, Ahn JH. Amyloid protein-mediated differential DNA methylation status regulates gene expression in Alzheimer's disease model cell line. Biochem Biophys Res Commun 2011;414:700–5.
[46] Barrachina M, Ferrer I. DNA methylation of Alzheimer disease and tauopathy-related genes in postmortem brain. J Neuropathol Exp Neurol 2009;68:880–91.
[47] Brohede J, Rinde M, Winblad B, Graff C. A DNA methylation study of the amyloid precursor protein gene in several brain regions from patients with familial Alzheimer disease. J Neurogenet 2010;24:179–81.
[48] Guo X, Wu X, Ren L, Liu G, Li L. Epigenetic mechanisms of amyloid-beta production in anisomycin-treated SH-SY5Y cells. Neuroscience 2011;194:272–81.
[49] Hoyaux D, et al. S100 proteins in Corpora amylacea from normal human brain. Brain Res 2000;867:280–8.
[50] Byun CJ, Seo J, Jo SA, Park YJ, Klug M, Rehli M, et al. DNA methylation of the 5′-untranslated region at +298 and +351 represses BACE1 expression in mouse BV-2 microglial cells. Biochem Biophys Res Commun 2012;417:387–92.
[51] Rao JS, Keleshian VL, Klein S, Rapoport SI. Epigenetic modifications in frontal cortex from Alzheimer's disease and bipolar disorder patients. Transl Psychiatry 2012;2:e132.
[52] Chen KL, Wang SS, Yang YY, Yuan RY, Chen RM, Hu CJ. The epigenetic effects of amyloid-beta(1-40) on global DNA and neprilysin genes in murine cerebral endothelial cells. Biochem Biophys Res Commun 2009;378:57–61.
[53] Furuya TK, da Silva PM, Payao SL, Rasmussen LT, de Labio RW, Bertolucci PH, et al. SORL1 and SIRT1 mRNA expression and promoter methylation levels in aging and Alzheimer's disease. Neurochem Int 2012;61:973–5.
[54] Zhang W, Tian Z, Sha S, Cheng LY, Philipsen S, Tan-Un KC. Functional and sequence analysis of human neuroglobin gene promoter region. Biochim Biophys Acta 2011;1809:236–44.
[55] Paula-Lima AC, Brito-Moreira J, Ferreira ST. Deregulation of excitatory neurotransmission underlying synapse failure in Alzheimer's disease. J Neurochem 2013;126:191–202.
[56] Offe K, Dodson SE, Shoemaker JT, Fritz JJ, Gearing M, Levy AI, et al. The lipoprotein receptor LR11 regulates amyloid beta production and amyloid precursor protein trafic in endosomal compartments. J Neurosci 2006;26:1596–603.

[57] Schmitt U, Tanimoto N, Seeliger M, Schaeffel F, Leube RE. Detection of behavioral alterations and learning deficits in mice lacking synaptophysin. Neuroscience 2009;162:234–43.
[58] Hallmayer J, Glasson EJ, Bower C, Petterson B, Croen L, Grether J, et al. On the twin risk in autism. Am J Hum Genet 2002;71:941–6.
[59] Piggot J, Shirinyan D, Shemmassian S, Vazirian S, Alarcon M. Neural systems approaches to the neurogenetics of autism spectrum disorders. Neuroscience 2009;164:247–56.
[60] Gardener H, Spiegelman D, Buka SL. Prenatal risk factors for autism: comprehensive meta-analysis. Br J Psychiatry 2009;195:7–14.
[61] Lord C. Fetal and sociocultural environments and autism. Am J Psychiatry 2013;170:355–8.
[62] Melnyk S, Fuchs GJ, Schulz E, Lopez M, Kahler SG, Fussell JJ, et al. Metabolic imbalance associated with methylation dysregulation and oxidative damage in children with autism. J Autism Dev Disord 2012;42:367–77.
[63] Kriaucionis S, Bird A. DNA methylation and Rett syndrome. Hum Mol Genet 2003;12 Spec No 2:R221–7.
[64] Guy J, Hendrich B, Holmes M, Martin JE, Bird A. A mouse Mecp2-null mutation causes neurological symptoms that mimic Rett syndrome. Nat Genet 2001;27:322–6.
[65] Collins AL, Levenson JM, Vilaythong AP, Richman R, Armstrong DL, Noebels JL, et al. Mild overexpression of MeCP2 causes a progressive neurological disorder in mice. Hum Mol Genet 2004;13:2679–89.
[66] Maezawa I, Swanberg S, Harvey D, LaSalle JM, Jin LW. Rett syndrome astrocytes are abnormal and spread MeCP2 deficiency through gap junctions. J Neurosci 2009;29:5051–61.
[67] Nagarajan RP, Hogart AR, Gwye Y, Martin MR, LaSalle JM. Reduced MeCP2 expression is frequent in autism frontal cortex and correlates with aberrant MECP2 promoter methylation. Epigenetics 2006;1:e1–11.
[68] Nagarajan RP, Patzel KA, Martin M, Yasui DH, Swanberg SE, Hertz-Picciotto I, et al. MECP2 promoter methylation and X chromosome inactivation in autism. Autism Res 2008;1:169–78.
[69] Allan AM, Liang X, Luo Y, Pak C, Li X, Szulwach KE, et al. The loss of methyl-CpG binding protein 1 leads to autism-like behavioral deficits. Hum Mol Genet 2008;17:2047–57.
[70] Mamrut S, Harony H, Sood R, Shahar-Gold H, Gainer H, Shi YJ, et al. DNA methylation of specific CpG sites in the promoter region regulates the transcription of the mouse oxytocin receptor. PLoS One 2013;8:e56869.
[71] Beri S, Tonna N, Menozzi G, Bonaglia MC, Sala C, Giorda R. DNA methylation regulates tissue-specific expression of Shank3. J Neurochem 2007;101:1380–91.
[72] Uchino S, Waga C. SHANK3 as an autism spectrum disorder-associated gene. Brain Dev 2013;35:106–10.
[73] Nguyen A, Rauch TA, Pfeifer GP, Hu VW. Global methylation profiling of lymphoblastoid cell lines reveals epigenetic contributions to autism spectrum disorders and a novel autism candidate gene, RORA, whose protein product is reduced in autistic brain. FASEB J 2010;24:3036–305.
[74] Hu VW. Is retinoic acid-related orphan receptor-alpha (RORA) a target for gene–environment interactions contributing to autism? Neurotoxicology 2012;33:1434–5.
[75] Urdinguio RG, Sanchez-Mut JV, Esteller M. Epigenetic mechanisms in neurological diseases: genes, syndromes, and therapies. Lancet. Neurol 2009;8:1056–72.
[76] O'Suilleabhain PE, Sung V, Hernandez C, Lacritz L, Dewey Jr. RB, Bottiglieri T, et al. Elevated plasma homocysteine level in patients with Parkinson disease: motor, affective, and cognitive associations. Arch Neurol 2004;61:865–8.

[77] Brattstrom L. Plasma homocysteine and MTHFR C677T genotype in levodopa-treated patients with PD. Neurology 2001;56:281.

[78] Blandini F, Fancellu R, Martignoni E, Mangiagalli A, Pacchetti C, Samuele A, et al. Plasma homocysteine and l-dopa metabolism in patients with Parkinson disease. Clin Chem 2001;47:1102–4.

[79] Duan W, Ladenheim B, Cutler RG, Kruman II, Cadet JL, Mattson MP. Dietary folate deficiency and elevated homocysteine levels endanger dopaminergic neurons in models of Parkinson's disease. J Neurochem 2002;80:101–10.

[80] Obeid R, Schadt A, Dillmann U, Kostopoulos P, Fassbender K, Herrmann W. Methylation status and neurodegenerative markers in Parkinson disease. Clin Chem 2009;55:1852–60.

[81] Guan JZ, Maeda T, Sugano M, Oyama J, Higuchi Y, Suzuki T, et al. A percentage analysis of the telomere length in Parkinson's disease patients. J Gerontol A Biol Sci Med Sci 2008;63:467–73.

[82] Paul L, Cattaneo M, D'Angelo A, Sampietro F, Fermo I, Razzari C, et al. Telomere length in peripheral blood mononuclear cells is associated with folate status in men. J Nutr 2009;139:1273–8.

[83] Paul L. Diet, nutrition and telomere length. J Nutr Biochem 2011;22:895–901.

[84] Wang Y, Wang X, Li R, Yang ZF, Wang YZ, Gong XL, et al. A DNA methyltransferase inhibitor, 5-aza-2'-deoxycytidine, exacerbates neurotoxicity and upregulates Parkinson's disease-related genes in dopaminergic neurons. CNS Neurosci Ther 2013;19:183–90.

[85] Kong M, Ba M, Liang H, Ma L, Yu Q, Yu T, et al. 5'-Aza-dC sensitizes paraquat toxic effects on PC12 cell. Neurosci Lett 2012;524:35–9.

[86] Bonsch D, Lenz B, Kornhuber J, Bleich S. DNA hypermethylation of the alpha synuclein promoter in patients with alcoholism. Neuroreport 2005;16:167–70.

[87] Jowaed A, Schmitt I, Kaut O, Wullner U. Methylation regulates alpha-synuclein expression and is decreased in Parkinson's disease patients' brains. J Neurosci 2010;30:6355–9.

[88] Matsumoto L, Takuma H, Tamaoka A, Kurisaki H, Date H, Tsuji S, et al. CpG demethylation enhances alpha-synuclein expression and affects the pathogenesis of Parkinson's disease. PLoS One 2010;5:e15522.

[89] de Boni L, Tierling S, Roeber S, Walter J, Giese A, Kretzschmar HA. Next-generation sequencing reveals regional differences of the alpha-synuclein methylation state independent of Lewy body disease. Neuromolecular Med 2011;13:310–20.

[90] Desplats P, Spencer B, Coffee E, Patel P, Michael S, Patrick C, et al. Alpha-synuclein sequesters Dnmt1 from the nucleus: a novel mechanism for epigenetic alterations in Lewy body diseases. J Biol Chem 2011;286:9031–7.

[91] Pieper HC, Evert BO, Kaut O, Riederer PF, Waha A, Wullner U. Different methylation of the TNF-alpha promoter in cortex and substantia nigra: Implications for selective neuronal vulnerability. Neurobiol Dis 2008;32:521–7.

[92] Mogi M, Harada M, Narabayashi H, Inagaki H, Minami M, Nagatsu T. Interleukin (IL)-1 beta, IL-2, IL-4, IL-6 and transforming growth factor-alpha levels are elevated in ventricular cerebrospinal fluid in juvenile parkinsonism and Parkinson's disease. Neurosci Lett 1996;211:13–16.

[93] Cai M, Tian J, Zhao GH, Luo W, Zhang BR. Study of methylation levels of parkin gene promoter in Parkinson's disease patients. Int J Neurosci 2011;121:497–502.

[94] Plagnol V, Howson JM, Smyth DJ, Walker N, Hafler JP, Wallace C, et al. Genome-wide association analysis of autoantibody positivity in type 1 diabetes cases. PLoS Genet 2011;7:e1002216.

[95] Tanaka H, Shimazawa M, Kimura M, Takata M, Tsuruma K, Yamada M, et al. The potential of GPNMB as novel neuroprotective factor in amyotrophic lateral sclerosis. Sci Rep 2012;2:573.

[96] Kofuji T, Fujiwara T, Sanada M, Mishima T, Akagawa K. HPC-1/syntaxin 1A and syntaxin 1B play distinct roles in neuronal survival. J Neurochem 2014;130:514–25.
[97] Wasmuth JJ, Hewitt J, Smith B, Allard D, Haines JL, Skarecky D, et al. A highly polymorphic locus very tightly linked to the Huntington's disease gene. Nature 1988;332:734–6.
[98] Farrer LA, Cupples LA, Kiely DK, Conneally PM, Myers RH. Inverse relationship between age at onset of Huntington disease and paternal age suggests involvement of genetic imprinting. Am J Hum Genet 1992;50:528–35.
[99] Simonin C, Duru C, Salleron J, Hincker P, Charles P, Delval A, et al. Association between caffeine intake and age at onset in Huntington's disease. Neurobiol Dis 2013;58:179–82.
[100] Pritchard CA, Cox DR, Myers RM. Methylation at the Huntington disease-linked D4S95 locus. Am J Hum Genet 1989;45:335–6.
[101] Reik W, Dean W, Walter J. Epigenetic reprogramming in mammalian development. Science 2001;293:1089–93.
[102] Gorbunova V, Seluanov A, Mittelman D, Wilson JH. Genome-wide demethylation destabilizes CTG.CAG trinucleotide repeats in mammalian cells. Hum Mol Genet 2004;13:2979–89.
[103] Ng CW, Yildirim F, Yap YS, Dalin S, Matthews BJ, Velez PJ, et al. Extensive changes in DNA methylation are associated with expression of mutant huntingtin. Proc Natl Acad Sci USA 2013;110:2354–9.
[104] Filippova GN, Thienes CP, Penn BH, Cho DH, Hu YJ, Moore JM, et al. CTCF-binding sites flank CTG/CAG repeats and form a methylation-sensitive insulator at the DM1 locus. Nat Genet 2001;28:335–43.
[105] Libby RT, Hagerman KA, Pineda VV, Lau R, Cho DH, Baccam SL, et al. CTCF cis-regulates trinucleotide repeat instability in an epigenetic manner: a novel basis for mutational hot spot determination. PLoS Genet 2008;4:e1000257.
[106] Zajac MS, Pang TY, Wong N, Weinrich B, Leang LS, Craig JM, et al. Wheel running and environmental enrichment differentially modify exon-specific BDNF expression in the hippocampus of wild-type and pre-motor symptomatic male and female Huntington's disease mice. Hippocampus 2010;20:621–36.
[107] de Carvalho M, Swash M. Amyotrophic lateral sclerosis: an update. Curr Opin Neurol 2011;24:497–503.
[108] Ferraiuolo L, Kirby J, Grierson AJ, Sendtner M, Shaw PJ. Molecular pathways of motor neuron injury in amyotrophic lateral sclerosis. Nat Rev Neurol 2011;7:616–30.
[109] Morahan JM, Yu B, Trent RJ, Pamphlett R. A genome-wide analysis of brain DNA methylation identifies new candidate genes for sporadic amyotrophic lateral sclerosis. Amyotroph Lateral Scler 2009;10:418–29.
[110] Labonte B, Suderman M, Maussion G, Navaro L, Yerko V, Mahar I, et al. Genome-wide epigenetic regulation by early-life trauma. Arch Gen Psychiatry 2012;69:722–31.
[111] Rothstein JD, Van Kammen M, Levey AI, Martin LJ, Kuncl RW. Selective loss of glial glutamate transporter GLT-1 in amyotrophic lateral sclerosis. Ann Neurol 1995;38:73–84.
[112] Yang Y, Gozen O, Vidensky S, Robinson MB, Rothstein JD. Epigenetic regulation of neuron-dependent induction of astroglial synaptic protein GLT1. Glia 2010;58:277–86.
[113] Chestnut BA, Chang Q, Price A, Lesuisse C, Wong M, Martin LJ. Epigenetic regulation of motor neuron cell death through DNA methylation. J Neurosci 2011;31:16619–36.
[114] Laffita-Mesa JM, Bauer PO, Kouri V, Pena Serrano L, Roskams J, Almaguer Gotay D, et al. Epigenetics DNA methylation in the core ataxin-2 gene promoter: novel physiological and pathological implications. Hum Genet 2012;131:625–38.
[115] Lahut S, Omur O, Uyan O, Agim ZS, Ozoguz A, Parman Y, et al. ATXN2 and its neighbouring gene SH2B3 are associated with increased ALS risk in the Turkish population. PLoS One 2012;7:e4295.

CHAPTER

18

DNA Methylation in Cellular Mechanisms of Neurodegeneration

OUTLINE

18.1 INTRODUCTION

In Chapter 17, we presented state of the art knowledge regarding Alzheimer's, Parkinson's, and Huntington's diseases, and amyotrophic lateral sclerosis, mainly regarding human diseases. The present chapter

M. Neidhart: DNA Methylation and Complex Human Disease.
DOI: http://dx.doi.org/10.1016/B978-0-12-420194-1.00018-X

focuses on the cellular mechanisms responsible for neurodegeneration, including the knowledge acquired from animal models. The designation "neurodegeneration" is an umbrella term for the progressive loss of structure or function of neurons, including death of neurons. As research progresses, many similarities appear that relate these different diseases to one another on a subcellular level. Thus, there are many parallels between different neurodegenerative disorders, including atypical protein assemblies as well as induced cell death [1,2]. In a rare set of monozygotic twins discordant for Alzheimer's disease, significantly reduced levels of DNA methylation were observed in temporal neocortex neuronal nuclei of the Alzheimer's twin [3]. These findings are consistent with the hypothesis that epigenetic mechanisms may mediate at the molecular level the effects of life events on risk for neurodegenerative diseases, and provide a potential explanation for discordance in such diseases in spite of genetic similarities. Epigenetic therapy in neurodegenerative diseases includes supplements with compounds related to the methionine and transmethylation pathway. Folate, other B group vitamins (B2, B6, and B12), and homocysteine participate in one-carbon metabolism, the metabolic pathway required for S-adenosylmethionine (SAM) production. In this chapter we will present evidence of an impaired one-carbon metabolism and reduced DNA methylation in various neurodegenerative diseases [4].

18.2 PROTEIN MISFOLDING

Several neurodegenerative diseases are classified as proteopathies, as they are associated with the aggregation of misfolded proteins. The three most important are α-synuclein, β-amyloid, and hyperphosphorylated tau protein.

18.2.1 α-Synuclein

In cultured cells, a CpG-rich region of the SNCA gene that encodes α-synuclein has been identified in which the methylation status can be altered along with changes in protein expression [5]. Post-mortem brain analysis revealed regional non-specific DNA methylation differences in this CpG region in the anterior cingulate and putamen among controls and patients with Parkinson's disease; however, in the substantia nigra of these patients, DNA methylation of SNCA was significantly decreased. This CpG region may function as an intronic regulatory element for the SNCA gene. In peripheral blood leukocytes of patients with Parkinson's disease, controversial reports have been presented regarding methylation of the SNCA promoter, probably depending on the sequences that are analyzed [6,7]. The possible role of α-synuclein in membrane damage and dopamine neuron cell death is presented below.

18.2.2 Amyloid Precursor Protein

The presence of extracellular amyloid plaques is a neuropathological hallmark of Alzheimer's disease. An early study [8] compared the methylation status of a CpG island in the amyloid precursor protein gene (APP) in DNA extracted from the more plaque-vulnerable cortex regions with DNA from the more plaque-resistant cerebellum, using material from six familial Alzheimer's disease cases. Bisulfite sequencing of a 188-bp fragment in the APP associated CPG island showed no methylation in any brain region. Human neuroblastoma cells and transgenic amyotrophic lateral sclerosis mice maintained under conditions of vitamin B deficiency showed presenilin1 (PSEN1) promoter demethylation, with subsequent increased production of presenilin1, β-secretase1 (BACE1) and APP proteins, and β-amyloid deposition in the animal's brains. By contrast, SAM supplementation induced an opposite tendency, restored PSEN1 methylation levels, and reduced progression of the sclerosis-like features induced by B vitamin deficiency in mice [9–11].

The Swedish mutation of amyloid precursor protein (APP-sw) has been reported to dramatically increase β-amyloid production through aberrant cleavage at the β-secretase site, causing early-onset Alzheimer's disease. DNA methylation has been reported to be associated with the pathogenesis [3], but, as mentioned above, APP in general is not transcriptionally regulated by methylation [8]. A more recent study [12] investigated whether this is also the case for the targets of the APP-sw protein, by analyzing genome-wide interplay between promoter CpG DNA methylation and gene expression in an APP-sw-expressing Alzheimer's model cell line. To identify genes whose expression is regulated by DNA methylation status, the authors performed integrated analysis of CpG methylation and mRNA expression profiles, and identified three target genes of the APP-sw mutant; hypomethylated CBP80/CBP20-dependent translation initiation factor (CTIF) and nuclear exporting factor 2 (NXT2), as well as hypermethylated discoidin domain receptor 2 (DDR2). Treatment with the demethylating agent 5-aza-2′-deoxycytidine restored mRNA expression of these three genes, implying methylation-dependent transcriptional regulation. Thus, the APP-sw mutation causes an alteration in DNA methylation of specific genes and changes gene expression. This may contribute to the dramatic onset of the disease.

Tauopathies are a class of neurodegenerative diseases associated with the pathological aggregation of tau protein in the human brain. An early study [13] found no difference in the methylation of specific genes (MAPT, APP, and PSEN1) in the brain of such patients. However, this study used total brain region homogenates and does not differentiate among cell types, or pathological and healthy regions. In addition, many other genes have to be investigated.

18.3 PROTEIN DEGRADATION

Neurodegenerative diseases can be associated with the accumulation of intracellular toxic proteins. Diseases caused by the aggregation of proteins are known as proteinopathies, and they are primarily caused by aggregates in the following structures: cytosol, nucleus, endoplasmic reticulum, or extracellular released proteins. There are two mechanisms by which cells remove troublesome proteins or organelles: the ubiquitin–proteasome and autophagy–lysosome pathways.

FK506 binding protein 51 kDa (FKBP51, encoded by the FKBP5 gene) forms a mature chaperone complex with HSP90 that prevents tau degradation. In human brains, FKBP51 levels increased relative to age and Alzheimer's disease, corresponding with DNA demethylation of the regulatory regions of the FKBP5 gene [14]. Evidence also exists that the regulation of autophagy-related genes by DNA methylation is altered in frontotemporal dementia [15]. Further investigations, however, are needed.

18.4 MEMBRANE DAMAGE

Damage to the membranes of organelles by monomeric or oligomeric proteins could also contribute to these diseases. α-Synuclein can damage membranes by inducing membrane curvature, and extensive tubulation and vesiculation were observed when these proteins were incubated with artificial phospholipid vesicles. We mentioned above that SNCA encoding α-synuclein can be hypermethylated in defined regions of the central nervous system in neurodegeneration. An antagonistic effect of the synuclein proteins on the secretory functions of the endoplasmic reticulum and the Golgi apparatus appears to simultaneously influence trafficking of the dopamine transporter and other membrane proteins [16] (Figure 18.1). In turn, there is a large body of evidence from morphological, molecular biological, and toxicological studies indicating that dopamine transporters are responsible for the selectivity of dopamine neuron cell death in Parkinson's disease [17]. This link has to be further investigated.

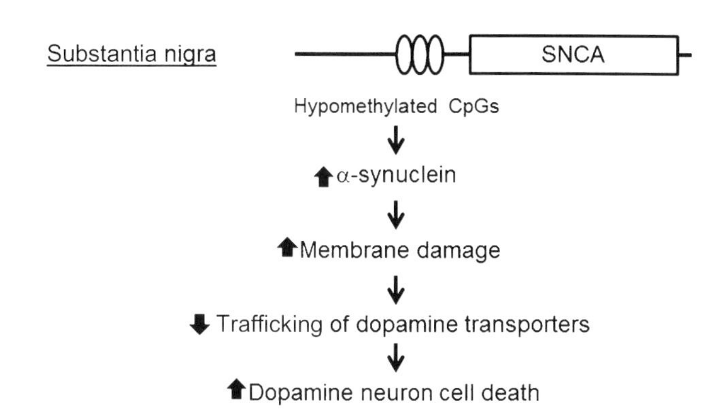

FIGURE 18.1 Mechanisms by which α-synuclein induces dopamine neuron cell death in Parkinson's disease.

18.5 MITOCHONDRIAL DYSFUNCTION

Mitochondrial DNA methylation sites have not yet been analyzed in neurodegenerative diseases or chromosomal aneuploidy, with the exception of the finding that it is hypomethylated in Down's syndrome [18], in spite of the increase of mitochondrial DNA content. Recently, mitochondrial DNMT3A and 5-methylcytosine levels were evaluated in human superoxide dismutase-1 (SOD1) transgenic mouse models of amyotrophic lateral sclerosis [19]. Superoxide dismutases are responsible for destroying free superoxide radicals (O_2^-) in the body. DNMT3A and 5-methylcytosine levels are reduced significantly in the skeletal muscle and spinal cord of these sclerotic mice. DNA pyrosequencing revealed significant abnormalities in 16S rRNA gene methylation. Most importantly, immunofluorescence showed that 5-methylcytosine immunoreactivity is sequestered into autophagosomes. These are the key structures in macroautophagy – i.e., the intracellular degradation system for cytoplasmic contents (e.g., abnormal intracellular proteins, excess or damaged organelles). Although mitochondrial DNA methylation has not yet been investigated in other neurodegenerative diseases, the suggested central role for integrity and mitochondrial function in the etiology of Alzheimer's disease, Parkinson's disease, and dementia indicates that it could play a role in these disorders [20].

The most common form of cell death in neurodegeneration is through the intrinsic mitochondrial apoptotic pathway (Figure 18.2). This pathway controls the activation of caspase-9 by regulating the release of cytochrome c from the mitochondrial intermembrane space. An early study [21] examined how homocysteine, S-adenosylhomocysteine (SAH), and SAM may synergistically induce neuronal apoptosis of BV-2 microglial cells.

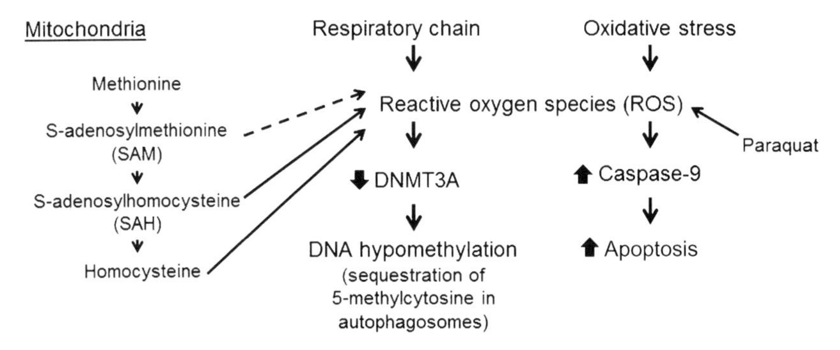

FIGURE 18.2 Mitochondrial dysfunction due to accumulation of ROS resulting in DNA hypomethylation and apoptosis.

Incubation of BV-2 cells with these products of the methionine pathway led to marked apoptosis of these cells. The combination of all three products markedly induced intracellular levels of reactive oxygen species (ROS) and significantly decreased the mitochondrial potential. The combination significantly elevated caspase-9 and caspase-3 activities. The combination also induced mitochondrial hypomethylation, as indicated by significantly decreased 5-methylcytosine levels and SAM/SAH ratios. These products induced apoptosis, possibly by generation of ROS and induction of mitochondrial hypomethylation. ROS are normal by-products of mitochondrial respiratory chain activity. ROS concentration is mediated by mitochondrial antioxidants such as manganese superoxide dismutase and glutathione peroxidase.

Overproduction of ROS is a central feature of all neurodegenerative disorders. Dietary SAM supplementation reduced oxidative stress [22] and delayed β-amyloid and tau pathology in transgenic sclerotic mice [23], suggesting a possible role of SAM as a neuroprotective dietary supplement in amyotrophic lateral sclerosis. In addition to the generation of ROS, mitochondria are also involved with life-sustaining functions, including calcium homeostasis, programmed cell death, mitochondrial fission and fusion, lipid concentration of the mitochondrial membranes, and the mitochondrial permeability transition. Mitochondrial disease leading to neurodegeneration is likely, at least on some level, to involve all of these functions.

Furthermore, as presented in Chapter 17, environmental exposure, including paraquat, is considered a risk for Parkinson's disease. In a study [24], PC12 pheochromocytoma cells were pretreated with methyltransferase inhibitor 5′-aza-2-deoxycytidine for 24 hours and then exposed to paraquat for 12 hours. The cell activity remarkably decreased and apoptotic cells increased after combined treatment. Moreover, compared with paraquat treatment alone, after being exposed to the combination, the level of ROS also increased. These findings suggest that DNA hypomethylation could sensitize paraquat toxic effects by oxidative stress increment and mitochondrial deficit.

18.6 AXONAL TRANSPORT

Axonal swelling and spheroids have been observed in many different neurodegenerative diseases. This suggests not only that defective axons are present in diseased neurons, but also that they may cause certain pathological insult due to accumulation of organelles. Axonal transport can be disrupted by a variety of mechanisms, including damage to kinesin and cytoplasmic dynein, microtubules, cargoes, and mitochondria. When axonal transport is severely disrupted a degenerative pathway known as Wallerian-like degeneration is often triggered.

KIF1A encoding kinesin is directly involved in the microtubule-based transport of dense-core vesicles in mammalian neurons [25]. Methylation of KIF1A is known to be frequent and to show higher levels in various cancer [26]. However, no information is available regarding neurodegenerative diseases.

S-adenosylmethionine synthetase, also known as methionine adenosyltransferase (MAT), refers to an enzyme that catalyzes the formation of SAM by joining methionine and ATP. Cycloleucine, an inhibitor of MAT, has been used to produce an experimental model of subacute combined degeneration of the spinal cord. Brain concentrations of SAM are reduced and levels of methionine become greatly elevated [27]. The morphological effects of cycloleucine are considered to be the result of SAM deficiency impairing transmethylation processes known to be important in the formation and stabilization of myelin through the methylation of myelin basic protein.

18.7 PROGRAMMED CELL DEATH

Programmed cell death is death of a cell in any form, mediated by an intracellular program. There are, however, situations in which these mediated pathways are artificially stimulated due to injury or disease. Three forms can be distinguished: apoptosis, autophagy, and cytoplasmic cell death.

Apoptosis is a form of programmed cell death in multicellular organisms. It involves a series of biochemical events leading to a characteristic cell morphology and death. Caspases (cysteine-aspartic acid proteases) cleave at very specific amino acid residues. There are two types of caspases: initiators and effectors. Initiator caspases cleave inactive forms of effector caspases. This activates the effectors, which in turn cleave other proteins, resulting in apoptotic initiation. We have already discussed the role of caspase-9 in mitochondria in response to ROS.

Autophagy was also mentioned above, and is essentially a form of intracellular phagocytosis in which a cell actively consumes damaged organelles or misfolded proteins by encapsulating them into an autophagosome, which fuses with a lysosome to destroy the contents of the autophagosome. Many neurodegenerative diseases show unusual protein aggregates. This could potentially be a result of an underlying autophagic defect common to multiple neurodegenerative diseases. Further investigations are needed.

In mitochondria, ROS induced a deficiency in DNMT3A and a caspase-dependent apoptosis. Reduced DNMT3, however, didn't induce apoptosis. Conversely, it has been shown that truncation mutation of the DNMT3A catalytic domain and DNMT3A interfering RNA blocked apoptosis of cultured neurons [28]. Inhibition of DNMT catalytic activity with small molecules RG108 and procainamide protects motor neurons from excessive DNA methylation and apoptosis in cell culture and in a mouse model of amyotrophic lateral sclerosis. Thus, it has been concluded that at least motor neurons can engage epigenetic mechanisms to cause their degeneration, involving DNMTs and increased DNA methylation.

Finally, homocysteine is a neurotoxic amino acid that accumulates in several neurological disorders. In cortical neurons, homocysteine increased cytosolic calcium, ROS, and phospho-tau immunoreactivity, and externalized phosphatidyl-serine (indicative of apoptosis) [29]. Apoptosis after treatment with homocysteine is reduced by co-treatment with 3-aminobenazmidine, an inhibitor of poly-ADP-ribosome polymerase (PARP), according to the concept that ATP depletion by PARP-mediated repair of DNA strand breakage mediates homocysteine-induced apoptosis. Homocysteine accumulation reduces cellular levels of SAM, while co-treatment with SAM reduces apoptosis, suggesting that inhibition of critical methylation reactions may mediate homocysteine-induced apoptosis. Thus, homocysteine compromises neuronal homeostasis by multiple divergent routes. Overall, these results suggest that natural compounds found in the diet, including folate, vitamins, polyphenols, flavonoids, methionine, and betaine, alter the availability of methyl groups and influence the activity of DNMTs, thereby representing potential "epigenetic" preventive factors for neurodegeneration.

18.8 EARLY LIFE EXPERIENCES

Correlation between maternal diet and fetal neurodegeneration has been reported mainly in animal studies. Performance in Morris maze experiments is affected in pups born to mice fed on a high fat diet during the gestational and lactation periods, and these results were attributed to decreased cell proliferation [30]. Similarly, studies have shown that maternal folate depletion results in oxidative stress and epigenetic changes in the offspring [31], which ultimately lead to neurodegeneration. Further elevated levels of homocysteine in the mother were shown to increase oxidative stress in the pups' brains, leading to apoptosis, as marked by DNA fragmentation [32]. A high dose of iron at the neonatal stage has similarly been shown to result in neurodegeneration of the midbrain at a later age. Pups with a higher iron dose reduced dopaminergic neurons at the age of 24 months as compared to that of 2-month-old pups. This indicates that there are long-term effects of neonatal iron exposure that are associated with degenerative changes [33]. Conversely, an omega-3 fatty acid-rich maternal diet is neuroprotective. This was shown by a study where omega-3 fatty acid supplementation to the mother resulted in neonate protection from LPS-induced brain injury [34]. Therefore, a balanced diet during pregnancy has been suggested to protect offspring from neurodegenerative diseases.

As presented in Chapters 13 and 14, the fetus becomes adapted to a new environment depending on the environmental stimuli in the uterus by means of physiological and hormonal alterations, and prepares itself for the upcoming conditions in postnatal life – a phenomenon called fetal programming [35]. It takes cues from the maternal health status and shows adaptive responses to survive in the maternal environment. Adaptive responses may be in the form of metabolic changes, hormonal release, or sensitivity of the target organs to hormones, which in turn affects the development of target organs, leading to physiologic and metabolic disturbances. Thus, reduced growth or body size can be considered as a fetal adaptive response toward the small uterus size of the mother with no immediate consequences in the newborn, but which may lead to physiologic changes that can cause diseases in later life.

Reversal of induced changes may be possible if associated epigenetic modifications and physiologic gene expression changes can be switched back to "normal." Cognitive impairment because of an imbalanced maternal diet has been tested by leptin treatment, as leptin receptors are present in brain regions and are known to regulate neuronal excitability and long-term potentiation [36]. Peroxisome proliferator activator receptor α (PGC1α) regulates the expression of genes involved in bioenergetics. PGC1α expression in the offspring of under-fed female rats was returned to normal by an exogenous supply of leptin [37].

18.9 CONCLUSION

Available treatments for neurodegenerative diseases such as Alzheimer's disease, Parkinson's disease, amyotrophic lateral sclerosis, and Huntington's disease do not arrest disease progression but mainly help in keeping patients from deteriorating, for a limited period of time. Increasing evidence suggests that epigenetic mechanisms such as DNA methylation and histone tail modifications are dynamically regulated in neurons, and play a fundamental role in learning and memory processes. In addition, both global and gene-specific epigenetic changes and deregulated expression of the writer and eraser proteins of epigenetic marks are believed to contribute to the onset and progression of neurodegeneration. Studies in animal models of neurodegenerative diseases have highlighted the potential role of epigenetic drugs, including methyl donor compounds, in ameliorating the cognitive symptoms and preventing or delaying the motor symptoms of the disease, thereby opening the way for a potential application in human pathology.

References

[1] Rubinsztein DC. The roles of intracellular protein-degradation pathways in neurodegeneration. Nature 2006;443:780–6.

[2] Bredesen DE, Rao RV, Mehlen P. Cell death in the nervous system. Nature 2006;443:796–802.

[3] Mastroeni D, McKee A, Grover A, Rogers J, Coleman PD. Epigenetic differences in cortical neurons from a pair of monozygotic twins discordant for Alzheimer's disease. PLoS One 2009;4:e6617.

[4] Coppede F. One-carbon metabolism and Alzheimer's disease: focus on epigenetics. Curr Genomics 2010;11:246–60.

[5] Iwata A. Neurodegeneration and epigenetics [in Japanese]. Nihon Shinkei Seishin Yakurigaku Zasshi 2012;32:269–73.

[6] Song Y, Ding H, Yang J, Lin Q, Xue J, Zhang Y, et al. Pyrosequencing analysis of SNCA methylation levels in leukocytes from Parkinson's disease patients. Neurosci Lett 2014;569:85–8.

[7] Tan YY, Wu L, Zhao ZB, Wang Y, Xiao Q, Liu J, et al. Methylation of alpha-synuclein and leucine-rich repeat kinase 2 in leukocyte DNA of Parkinson's disease patients. Parkinsonism Relat Disord 2014;20:308–13.

[8] Brohede J, Rinde M, Winblad B, Graff C. A DNA methylation study of the amyloid precursor protein gene in several brain regions from patients with familial Alzheimer disease. J Neurogenet 2010;24:179–81.

[9] Fuso A, Seminara L, Cavallaro RA, D'Anselmi F, Scarpa S. S-adenosylmethionine/homocysteine cycle alterations modify DNA methylation status with consequent deregulation of PS1 and BACE and beta-amyloid production. Mol Cell Neurosci 2005;28:195–204.

[10] Fuso A, Nicolia V, Cavallaro RA, Ricceri L, D'Anselmi F, Coluccia P, et al. B-vitamin deprivation induces hyperhomocysteinemia and brain S-adenosylhomocysteine, depletes brain S-adenosylmethionine, and enhances PS1 and BACE expression and amyloid-beta deposition in mice. Mol Cell Neurosci 2008;37:731–46.

[11] Fuso A, Nicolia V, Ricceri L, Cavallaro RA, Isopi E, Mangia F, et al. S-adenosylmethionine reduces the progress of the Alzheimer-like features induced by B-vitamin deficiency in mice. Neurobiol Aging 2012;33(1482):e1–16.

[12] Sung HY, Choi EN, Ahn Jo, S, Oh S, Ahn JH. Amyloid protein-mediated differential DNA methylation status regulates gene expression in Alzheimer's disease model cell line. Biochem Biophys Res Commun 2011;414.
[13] Barrachina M, Ferrer I. DNA methylation of Alzheimer disease and tauopathy-related genes in postmortem brain. J Neuropathol Exp Neurol 2009;68:880–91.
[14] Blair LJ, Nordhues BA, Hill SE, Scaglione KM, O'Leary III JC, Fontaine SN, et al. Accelerated neurodegeneration through chaperone-mediated oligomerization of tau. J Clin Invest 2013;123.
[15] Ferrari R, Hernandez DG, Nalls MA, Rohrer JD, Ramasamy A, Kwok JB, et al. Frontotemporal dementia and its subtypes: a genome-wide association study. Lancet Neurol 2014;13:686–99.
[16] Oaks AW, Sidhu A. Parallel mechanisms for direct and indirect membrane protein trafficking by synucleins. Commun Integr Biol 2013;6:e26794.
[17] Storch A, Ludolph AC, Schwarz J. Dopamine transporter: involvement in selective dopaminergic neurotoxicity and degeneration. J Neural Transm 2004;111:1267–86.
[18] Infantino V, Castegna A, Iacobazzi F, Spera I, Scala I, Andria G, et al. Impairment of methyl cycle affects mitochondrial methyl availability and glutathione level in Down's syndrome. Mol Genet Metab 2011;102:378–82.
[19] Wong M, Gertz B, Chestnut BA, Martin LJ. Mitochondrial DNMT3A and DNA methylation in skeletal muscle and CNS of transgenic mouse models of ALS. Front Cell Neurosci 2013;7:279.
[20] Coskun P, Wyrembak J, Schriner SE, Chen HW, Marciniack C, Laferla F, et al. A mitochondrial etiology of Alzheimer and Parkinson disease. Biochim Biophys Acta 1820;553-564:2012.
[21] Lin HC, Yang CM, Liu CL, Hu ML. Synergistic effects of homocysteine, S-adenosylhomocysteine and adenosine on apoptosis in BV-2 murine microglial cells. Biofactors 2008;34:81–95.
[22] Tchantchou F, Graves M, Falcone D, Shea TB. S-adenosylmethionine mediates glutathione efficacy by increasing glutathione S-transferase activity: implications for S-adenosylmethionine as a neuroprotective dietary supplement. J Alzheimers Dis 2008;14:323–8.
[23] Lee S, Lemere CA, Frost JL, Shea TB. Dietary supplementation with S-adenosyl methionine delayed amyloid-beta and tau pathology in 3 × Tg-AD mice. J Alzheimers Dis 2012;28:423–31.
[24] Kong M, Ba M, Liang H, Ma L, Yu Q, Yu T, et al. 5′-Aza-dC sensitizes paraquat toxic effects on PC12 cell. Neurosci Lett 2012;524:35–9.
[25] Lo KY, Kuzmin A, Unger SM, Petersen JD, Silverman MA. KIF1A is the primary anterograde motor protein required for the axonal transport of dense-core vesicles in cultured hippocampal neurons. Neurosci Lett 2011;491:168–73.
[26] Guerrero-Preston R, Hadar T, Ostrow KL, Soudry E, Echenique M, Ili-Gangas C, et al. Differential promoter methylation of kinesin family member 1a in plasma is associated with breast cancer and DNA repair capacity. Oncol Rep 2014;32: 505–12.
[27] Lee CC, Surtees R, Duchen LW. Distal motor axonopathy and central nervous system myelin vacuolation caused by cycloleucine, an inhibitor of methionine adenosyltransferase. Brain 1992;115(Pt 3):935–55.
[28] Martin LJ, Wong M. Aberrant regulation of DNA methylation in amyotrophic lateral sclerosis: a new target of disease mechanisms. Neurotherapeutics 2013;10:722–33.
[29] Ho PI, Ortiz D, Rogers E, Shea TB. Multiple aspects of homocysteine neurotoxicity: glutamate excitotoxicity, kinase hyperactivation and DNA damage. J Neurosci Res 2002;70:694–702.

[30] White CL, Pistell PJ, Purpera MN, Gupta S, Fernandez-Kim SO, Hise TL, et al. Effects of high fat diet on Morris maze performance, oxidative stress, and inflammation in rats: contributions of maternal diet. Neurobiol Dis 2009;35:3–13.
[31] Langie SA, Achterfeldt S, Gorniak JP, Halley-Hogg KJ, Oxley D, van Schooten FJ, et al. Maternal folate depletion and high-fat feeding from weaning affects DNA methylation and DNA repair in brain of adult offspring. FASEB J 2013;27:3323–34.
[32] Koz ST, Gouwy NT, Demir N, Nedzvetsky VS, Etem E, Baydas G. Effects of maternal hyperhomocysteinemia induced by methionine intake on oxidative stress and apoptosis in pup rat brain. Int J Dev Neurosci 2010;28:325–9.
[33] Kaur D, Peng J, Chinta SJ, Rajagopalan S, Di Monte DA, Cherny RA, et al. Increased murine neonatal iron intake results in Parkinson-like neurodegeneration with age. Neurobiol Aging 2007;28:907–13.
[34] Tuzun F, Kumral A, Dilek M, Ozbal S, Ergur B, Yesilirmak DC, et al. Maternal omega-3 fatty acid supplementation protects against lipopolysaccharide-induced white matter injury in the neonatal rat brain. J Matern Fetal Neonatal Med 2012;25:849–54.
[35] Gluckman PD, Hanson MA. Maternal constraint of fetal growth and its consequences. Semin Fetal Neonatal Med 2004;9:419–25.
[36] Signore AP, Zhang F, Weng Z, Gao Y, Chen J. Leptin neuroprotection in the CNS: mechanisms and therapeutic potentials. J Neurochem 2008;106:1977–90.
[37] Kakuma T, Wang ZW, Pan W, Unger RH, Zhou YT. Role of leptin in peroxisome proliferator-activated receptor gamma coactivator-1 expression. Endocrinology 2000;141:4576–82.

CHAPTER

19

DNA Methylation and Autoimmunity

OUTLINE

19.1 INTRODUCTION

Autoimmunity is the failure of an organism in recognizing its own constituent parts as *self*, thus leading to an immune response against its own cells and tissues. Any disease that results from such an aberrant immune response is termed an autoimmune disease. Several mechanisms are thought to be operative in the pathogenesis of autoimmune diseases, against a backdrop of genetic predisposition and environmental modulation. Certain individuals are genetically susceptible to developing autoimmune diseases. This susceptibility is associated with multiple genes plus other risk factors. However, genetically predisposed individuals do not always develop autoimmune diseases. In addition, it is important to note that not in all diseases that are labeled "autoimmune" have breakdown of tolerance, clonal expansion of autoreactive T cells, or pathological autoantibodies been documented. Often,

M. Neidhart: DNA Methylation and Complex Human Disease.
DOI: http://dx.doi.org/10.1016/B978-0-12-420194-1.00019-1

it has been sufficient to call a disease "autoimmune" when reactions or dysfunctions of the immune cells can be observed [1].

First, clear autoimmune diseases are diabetes mellitus type 1, systemic lupus erythematosus, Hashimoto's thyroiditis, and Graves' disease. Diabetes mellitus type 1 is a form of diabetes mellitus that results from autoimmune destruction of the insulin-producing β cells in the pancreas [2]. The appearance of diabetes-related autoantibodies has been shown to be able to predict the appearance of diabetes type 1 before any hyperglycemia arises, the main ones being islet cell autoantibodies, insulin autoantibodies, autoantibodies targeting the 65-kDa isoform of glutamic acid decarboxylase, autoantibodies targeting the phosphatase-related IA-2 molecule, and zinc transporter autoantibodies (ZnT8) [3]. Systemic lupus erythematosus is a systemic autoimmune connective tissue disease that mainly affects young women and can target any part of the body [4]. The immune system attacks the body's cells and tissues, resulting in inflammation and tissue damage. Autoantibodies to ribonucleoproteins (nRNP A and nRNP C) initially targeted restricted, proline-rich motifs. Antibody binding subsequently spread to other epitopes. The similarity and cross-reactivity between the initial targets of nRNP and Smith (Sm) autoantibodies identifies a likely commonality in cause and a focal point for intermolecular epitope spreading. SLE does run in families, but no single causal gene has been identified. Instead, multiple genes appear to influence a person's chance of developing lupus when triggered by environmental factors. Clearly, the onset of lupus and its flares are triggered by various environmental factors in genetically susceptible individuals. Hashimoto's thyroiditis or chronic lymphocytic thyroiditis is an autoimmune disease in which the thyroid gland is attacked by a variety of cell- and antibody-mediated immune processes. Diagnosis is made by detecting elevated levels of antithyroid peroxidase antibodies in the serum. Here again, both genetic predisposition and environmental factors serve as triggers of the disease [5]. Graves' disease (or Flajani-Basedow-Graves' disease) of the thyroid gland frequently causes it to enlarge to twice its size or more (goiter) and become overactive, with related hyperthyroid symptoms. The immune system produces antibodies to the receptor for thyroid-stimulating hormone. These antibodies cause hyperthyroidism because they bind to the TSH receptor and chronically stimulate it [6].

In the second case, the autoimmune reaction is triggered by an infectious agent, such as Sjögren's syndrome. This is a chronic autoimmune disease in which the leukocytes destroy the exocrine glands, specifically the salivary and lacrimal glands, which produce saliva and tears, respectively. Viral proteins, engulfed molecules, or degraded self-structures may initiate autoimmunity by molecular mimicry and increase the chances of Sjögren's syndrome development. The disease is diagnosed based on the detection of circulating antinuclear autoantibodies, such as anti-Sjögren's syndrome A (anti-SSA/Ro) or B (anti-SSB/La), and/or the presence of Chisholm grade-3 or -4 nodular inflammatory lymphocyte infiltrates in a minor salivary gland biopsy. As we will see below, many autoimmune diseases are believed to be triggered at origin by an infectious agent, but many of have been found only to be modulated by them.

A third group comprises diseases that are called "autoimmune" but have many components of an allergic reaction; these include celiac disease and Churg-Strauss syndrome. Celiac disease is caused by a reaction to gliadin, a prolamin (gluten protein) found in wheat, and similar proteins found in the crops of the tribe Triticeae. Upon exposure to gliadin, and specifically to three peptides found in prolamins, the enzyme tissue transglutaminase modifies the protein, and the immune system cross-reacts with the small-bowel tissue, causing an inflammatory reaction. The patient can develop multiple autoimmune antibodies, such as IgA antiendomysial, antitransglutaminase, antireticulin, and antigliadin antibodies. Churg-Strauss syndrome (also known as eosinophilic granulomatosis with polyangiitis, or allergic granulomatosis) is an autoimmune condition that causes inflammation of small and medium-sized blood vessels (vasculitis) in persons with a history of airway allergic hypersensitivity (atopy). Diagnostic markers include eosinophil granulocytes and granulomas in affected tissue, and antineutrophil cytoplasmic antibodies against neutrophil granulocytes.

A fourth group includes diseases that are called "autoimmune," but in fact have important inflammatory components. These include sarcoidosis, rheumatoid arthritis, ankylosing spondylitis, and systemic sclerosis. Sarcoidosis is a disease involving abnormal collections of inflammatory cells (granulomas) that can form as nodules in multiple organs. The granulomas are most often located in the lungs or its associated lymph nodes, but any organ can be affected. Several infectious agents appear to be significantly associated with sarcoidosis, but none of the known associations is specific enough to suggest a direct causative role. The only argument in favor of an autoimmune disease is the prevalence of Th1 $CD4^+$ helper cells. Rheumatoid arthritis is considered to be an autoimmune disease that results in a chronic, systemic inflammatory disorder that may affect many tissues and organs, but principally attacks flexible (synovial) joints. It can be a disabling and painful condition, which can lead to substantial loss of functioning and mobility if not adequately treated. When RA is clinically suspected, serological analyses are required, such as testing for the presence of rheumatoid factor (RF, non-specific antibodies that bind the Fc part of IgGs). A negative RF does not rule out RA; rather, the arthritis is then called seronegative. Because of this low specificity, new serological tests have been developed that test for the presence of anticitrullinated protein antibodies. The most common tests for ACPAs are the anti-CCP (cyclic citrullinated peptide) test and the anti-MCV assay (antibodies against mutated citrullinated vimentin). The effector cells of joint destruction are synovial fibroblasts, not leukocytes. Ankylosing spondylitis (Bechterew's disease) is a chronic inflammatory disease of the axial skeleton, with variable involvement of peripheral joints and nonarticular structures. Approximately 90% of AS patients express the HLA-B27 genotype, meaning there is a strong genetic association. Systemic sclerosis is a connective tissue disease. It is characterized by thickening of the skin caused by accumulation of collagen, and by injuries to the smallest arteries. Sytemic sclerosis is associated with the production of anti-topoisomerase antibodies (i.e. Scl-70). Early physiopathogenic events appear to be endothelial injury and imbalance in vascular repair with the activation of endothelial cells, the immune system and platelets, with the release of multiple mediators such as Th2 proinflammatory cytokines and growth factors, triggering a sequence of simultaneous or cascading events that involve several intracellular signaling pathways. The most important result of these events is the hyperactivation of fibroblasts, the main effector cells of fibrosis, which will then produce large amounts of extracellular matrix constituents and secrete multiple growth factors and cytokines that perpetuate the process [7].

Finally, in some diseases, the evidences for "classical" autoimmunity are even weaker, for example in Addison's disease and polymyositis/ dermatomyositis. Addison's disease is a rare, chronic endocrine system disorder in which the adrenal glands do not produce sufficient steroid hormones (glucocorticoids and often mineralocorticoids) [8]. In some patients, autoantibodies against steroidogenic enzymes have been reported. Polymyositis ("inflammation of many muscles") is a type of chronic inflammation of the muscles (inflammatory myopathy) [9]. Its cause is unknown, and may involve viruses and autoimmune factors. Dermatomyositis is a connective-tissue disease related to polymyositis. This disease most frequently affects the skin and muscles, but it is also a systemic disorder that may also affect the joints, esophagus, lungs and, less commonly, heart. Mixed B- and T-lymphocyte perivascular inflammatory infiltrates occur.

Thus, it is very important to differentiate the underlying pathological mechanism in the so-called "autoimmune" diseases in human. Many of them came by this label because the corresponding animal models are highly inflammatory and are triggered by antigen mimicry. The "classical" concept of autoimmunity, however, requires the identification of autoantigens/neoepitopes that are presented by, for example, macrophages, which in turn trigger the clonal expansion of T cells and production of autoantibodies by B cells. An increased production of pro-inflammatory cytokines, or a switch from a functional T-cell population to another, is not enough.

19.2 ENVIRONMENTAL FACTORS

A "classical" autoimmune disease can be induced or triggered by infectious agents, which can also determine its clinical manifestations. Most infectious agents, such as viruses, bacteria, and parasites, can induce autoimmunity via different mechanisms. In many cases, it is not a single infection but rather the burden of infections from childhood that is responsible for the induction of autoimmunity. The development of an autoimmune disease after infection tends to occur in genetically susceptible individuals. By contrast, some infections can protect individuals from specific autoimmune diseases [10]. Thus, in fact, an interesting inverse relationship exists between infectious diseases and autoimmune diseases. In areas where multiple infectious diseases are endemic, autoimmune diseases are quite rarely seen. The reverse, to some extent, seems to hold true. The "hygiene hypothesis" attributes these correlations to the immune-manipulating strategies of pathogens. Thus, parasite infection often is associated with reduced activity of autoimmune disease [11]. The putative mechanism is that the parasite attenuates the host immune response in order to protect itself. This may provide a serendipitous benefit to a host that also suffers from immune dysfunctions. The details of parasite immune modulation are not yet known, but may include secretion of anti-inflammatory agents or interference with the host immune signaling.

In diabetes mellitus type 1, genetic predisposition, although clearly important, cannot completely explain the onset and occurrence of the disease; hence it has been proposed that changes in the environment and/or changes in how we respond to our environment must contribute to a rising incidence [2]. An interesting observation was the strong association of certain microbial organisms with autoimmune diseases – for example, *Klebsiella pneumoniae* and coxsackievirus B have been strongly correlated with ankylosing spondylitis [12] and diabetes mellitus type 1 [2,13]. In rheumatoid arthritis, clinical and animal model studies have suggested that infections by many microorganisms, such as *Porphyromonas gingivalis*, *Proteus mirabilis*, Epstein-Barr virus, and mycoplasma, contribute to the pathogenesis [14]. This has been explained by the tendency of the infecting organism to produce superantigens that are capable of polyclonal activation of B lymphocytes, and production of large amounts of antibodies of varying specificities, some of which may be self-reactive. However, not only pathogenic microorganisms but also non-pathogenic commensal microorganisms induce pro-inflammatory or regulatory immune responses within the host. Commensal microbiota might play a critical role in the development of diabetes type 1 [15], Hashimoto's disease [5], celiac disease [16], and rheumatoid arthritis [17]. In Sjögren's syndrome, infectious agents may play both causative and protective roles [18].

Certain chemical agents and drugs can also be associated with the genesis of autoimmune conditions, or conditions that simulate autoimmune diseases. The most striking of these is drug-induced lupus erythematosus [4,19]. Usually, withdrawal of the offending drug cures the symptoms in a patient. In lupus, other environmental factors include cigarette smoke, alcohol, occupationally and non-occupationally related chemicals, ultraviolet light, infections, sex hormones, and certain medications and vaccines [4]. In Churg-Strauss syndrome, exposure to crystalline silica stands out as the most convincing environmental risk factor [20]. Such an association has also been made for systemic lupus erythematosus, rheumatoid arthritis, systemic sclerosis, and antineutrophil cytoplasmic antibody (ANCA)-associated vasculitides [21]. Cigarette smoking is now established as a major risk factor for both incidence and severity of rheumatoid arthritis [22]. This may relate to abnormal citrullination of proteins, the occurrence of smoking correlating with the presence of antibodies to citrullinated peptides (anti-CCP and anti-MCV mentioned in the introduction) [23]. Regarding ankylosing spondylitis, drinking and cooking oil may be risk exposure factors combined with predisposing genes [24]. Interestingly, high dietary intake of omega-3 fatty acids could give rise to anti-inflammatory effects in both rheumatoid arthritis [25] and ankylosing spondylitis [26].

The autoimmune thyroid diseases are prototypical organ-specific autoimmune diseases, but the mechanisms that trigger the autoimmune response to the thyroid are still unclear. Epidemiological data point to an interaction between genetic susceptibility and environmental triggers as the key factor leading to the breakdown of tolerance and the development of disease [27]. Of the known environmental factors, infection, diet, iodine, and smoking appear most important [28].

Because sarcoidosis most commonly involves the lungs, eyes, and skin, the search for environmental causes has centered on exposure to airborne antigens. Some of the earliest studies of sarcoidosis reported associations with exposure to irritants found in rural settings, such as emissions from wood-burning stoves, and tree pollen [29]. Associations between sarcoidosis and exposure to inorganic particles [30], insecticides, and moldy environments [31] have been reported. Occupational studies have shown positive associations with metalworking [32], firefighting [33,34], and the handling of building supplies [35].

Finally, the relative incidence of dermatomyositis may follow a latitudinal gradient in the northern hemisphere that may be explained by the immunomodulatory action of ultraviolet radiation [9].

19.3 CHANGES IN DNA METHYLATION

In recent years, accumulating evidence has demonstrated that, in addition to genetics, other complementary mechanisms are involved in the pathogenesis of autoimmunity – in particular, epigenetics [36]. Epigenetic marks can be affected by age and other environmental triggers, providing a plausible link between environmental factors and the onset and development of various human diseases.

Diabetes mellitus type 1 is one of the most heritable common diseases, and among autoimmune diseases it has the largest range of concordance rates in monozygotic twins [37]. This fact, coupled with evidence of various epigenetic modifications of gene expression, provides convincing proof of the complex interplay between genetic and environmental factors. DNA methylation patterns of seven CpGs proximal to the transcription start site in the insulin (INS) gene promoter revealed that patients with diabetes mellitus type 1 have reduced methylation of CpG-19, -135, and -234, and increased methylation of CpG-180, compared with controls [38]. A novel method, the highly sensitive methylation-specific quantitative polymerase chain reaction assay, was successfully applied in the detection of circulating β-cell DNA in peripheral blood from diabetic mice, and holds great promise for future monitoring of β-cell death and the prediction of diabetes onset [39]. Another study [40], an epigenome-wide association study involving 27,458 different CpG sites within 14,475 promoters in the monocytes of 15 diabetes mellitus type 1-discordant monozygotic twin pairs, identified 132 different CpG sites where differences significantly correlated with diabetic state. This study included identification of diabetes mellitus type 1-specific methylation variable positions in patients before the appearance of specific autoantibodies, thus demonstrating that disease-specific methylation variable positions antedate clinical disease, including epigenetic changes in the HLA class II gene HLA-DQB1 or in GAD2, which encodes GAD65, a major diabetes mellitus type 1 autoantigen involved in the pathogenesis [41].

Similarly to diabetes mellitus type 1, for Graves' disease and Hashimoto's thyroiditis, the comparison of concordance rates between monozygotic and dizygotic twins provides irrefutable evidence of a genetic component [42], regarding both etiology and production of specific autoantibodies. However, again, the lack of complete phenotypic concordance in monozygotic twin pairs indicates that environmental and/or epigenetic factors are also of importance. Genetic data point to the involvement of both shared and unique genes. Among the shared susceptibility genes, HLA-DRβ1-Arg74 (human leukocyte antigen DR containing an arginine at position β74) confers the strongest risk. Recent genome-wide analyses have revealed new putative candidate genes [28].

A DNMT1 polymorphism (+32204GG) was found more frequently in patients with intractable Graves' disease than in patients with disease in remission [43]. Genomic DNA showed significantly lower levels of global methylation in individuals with this polymorphism. Interestingly, the same study reported that the methionine synthase reductase (MTRR) gene showed a frequent polymorphism (+66AA) in patients with severe Hashimoto's thyroiditis compared to those with mild disease. MTRR regulates the synthesis of L-methionine, the precursor of S-adenosylmethionine, the methyl donor involved in DNA methylation.

The disease concordance rate of systemic lupus erythematosus in monozygotic twins is lower than in diabetes mellitus type 1 or autoimmune thyroiditis; this clearly suggests an epigenetic contribution to the disease development [44]. A role of DNA hypomethylation in systemic lupus erythematosus has been demonstrated in studies showing demethylating drug-induced T-cell autoreactivity and lupus-like symptoms in mice, and impaired DNA methylation in $CD4^+$ T cells from active lupus patients [45,46]. Many methylation-sensitive genes are overexpressed in SLE $CD4^+$ T cells, functionally contributing to lupus development; these include CD11A causing autoreactivity, perforin increasing apoptosis, and CD70 and CD40L enhancing B–T cell interaction for autoantibody production [47]. Genome-wide DNA methylation analysis has shown severe hypomethylation of type I IFN-regulated genes in lupus $CD4^+$ T cells, $CD19^+$ B cells, and $CD14^+$ monocytes, suggesting epigenetically mediated type I IFN hyper-responsiveness in SLE patients [48]. Interestingly, healthy individuals carrying the lupus-risk IRAK1–MECP2 haplotype with increased levels of a specific MECP2 transcript isoform in stimulated T cells showed a similar hypomethylation of IFN-regulated genes, providing evidence for genetic–epigenetic interaction associated with SLE-risk variants [49]. Located on the X chromosome, the S196F variant of IRAK1 captures the association signals of an SLE risk haplotype shared by multiple ancestries [50]. The risk 196F allele confers increased NF-κB activity and is associated with decreased mRNA levels of MECP2 but not IRAK1, suggesting contributions of both IRAK1 and MECP2 to lupus susceptibility.

In Sjögren's syndrome several epigenetic mechanisms are defective, including DNA demethylation that predominates in epithelial cells, and abnormal chromatin positioning associated with autoantibody production [51]. Global DNA methylation is reduced in labial salivary glands when comparing biopsy sections from patients with Sjögren's syndrome to controls [52]. The global DNA demethylation is associated with a decrease in DNMT1 and an increase in the cell cycle inhibitor GADD45A. DNA methylation levels were inversely correlated with SS severity and B-cell infiltration. A genome-wide analysis of DNA methylation in naïve $CD4^+$ $CD45RA^+$ T cells was recently performed in patients with Sjögren's syndrome [52]. Among 485,000 CpG sites tested within the entire genome, 753 CpG motifs were differentially methylated, with the majority (311 genes, i.e. 75%) being demethylated. Demethylated genes in patients with Sjögren's syndrome include lymphotoxin-α (previously known as tumor necrosis factor-β) involved in T-cell activation, genes implicated in the type I interferon pathway, and genes encoding for membrane water channel proteins. In addition, analysis of gene promoter DNA methylation status in $CD4^+$ T cells has revealed demethylation and overexpression of CD70 (TNSF7), and an absence of epigenetic regulation for interferon regulatory factor IRF5 [53,54].

The familial tendency in celiac disease led to the study of genetic markers (e.g., HLA-B8). The concordance of celiac disease appeared in a great majority of monozygotic twins. Nevertheless, additional, nongenetic factors seem likely [55]. In Churg-Strauss syndrome and patients with reactive eosinophilias, changes in X-chromosome inactivation patterns occurred [56]. However, no more information is available. Sarcoidosis also has a strong familial heritability. This might be the reason why no study has investigated epigenetics changes in this disease, bar one that reported early changes in telomere methylation [57]. Nevertheless, this was not an argument for diabetes mellitus type 1, for example. Clearly, more investigations in epigenetics are needed for these diseases.

Ankylosing spondylitis also has a very strong genetic predisposition, but not all HLA-B27+ individuals developed the disease. Interestingly, a recent study [58] has shown an association between methylation of SOCS1 and cytokine production in this disease. SOCS1 is a cytokine-inducible negative regulator of cytokine signaling. Methylation of SOCS1 is significantly associated with the severity of patient's spondylopathy, sacroiliitis, and C-reactive protein. Patients with ankylosing spondylitis exhibited higher serum IL-6 and TNF-α levels. Importantly, patients with high serum IL-6 or TNF-α levels demonstrated significantly higher SOCS1 methylation. SOCS1 methylation can be detected in HLA-B27+ patients with ankylosing spondylitis, but not in HLA-B27+ healthy controls. This deserves further investigation.

A special case is rheumatoid arthritis, in which twin concordance is surprisingly low [59], suggesting an important contribution of the environment to the pathogenesis [60]. More importantly, genome-wide association studies have thus far reported genomic variants that only account for a minority of cases [61]. Rheumatoid arthritis is a complex disease where predetermined and stochastic factors conspire to confer disease susceptibility. In light of the diverse responses to targeted therapies, rheumatoid arthritis might represent a final common clinical phenotype that reflects many pathogenic pathways. Therefore, it might be appropriate to begin thinking about rheumatoid arthritis as a syndrome rather than a disease [62]. Although there is controversy regarding the involvement of epigenetic and genetic factors in rheumatoid arthritis etiology, it is becoming obvious that the two systems interact with each other and are ultimately responsible for the development of this syndrome [63]. Early studies found widespread DNA hypomethylation in rheumatoid arthritis synovial fibroblasts [64], including hypomethylation of the LINE1 retrotransposons [65] and of the CXCL12 gene promoter [66]. In this case, loss of the repressive DNA methylation signal results in increased expression of multiple genes [64]. A recent genome-wide study on rheumatoid arthritis synovial fibroblasts revealed a number of differentially (hypo- and hyper-) methylated genomic regions [67]. Most of the affected genes appear to be involved in inflammation, matrix remodeling, leukocyte recruitment, and immune responses. DNA hypomethylation appears to be due to an intrinsically activated polyamine recycling pathway that consumes S-adenosylmethionine [68].

In systemic sclerosis, concordance for the disease among identical twins is low; however, concordance for autoantibodies associated with systemic sclerosis and for fibroblast gene expression profiles is higher [69]. Orchestrated by environmental factors, epigenetic modifications can drive genetically predisposed individuals to develop autoimmunity, and are thought to represent the crossroads between the environment and genetics in systemic sclerosis [70]. Potential mechanisms for environmentally induced systemic autoimmunity include interference with immune tolerance, activation of the immune system, induction of genetic alterations, and dysregulation of epigenetic patterns [71]. The observation that fibroblasts taken from patients with systemic sclerosis keep their altered behavior for multiple passages when cultured outside the pathological context is one of the strongest indicators to date for the fundamental role of epigenetic regulation in this disease.

In systemic sclerosis [72] and Addison's disease [73], hypomethylation of $CD4^+$ T cells has been reported. This resembles the autoreactive T cells described in systemic lupus erythematosus [74]. Thus, demethylation of specific regulatory elements contributes to perforin overexpression in $CD4^+$ lupus T cells [74] and to CD40L overexpression in $CD4^+$ systemic sclerosis T cells [72].

No such information is available regarding polymyositis/dermatomyositis.

19.4 CONCLUSION

Immune-mediated pulmonary diseases are a group of diseases that result from immune imbalance initiated by allergens or of unknown causes. Inflammatory responses without restrictions cause tissue damage and remodeling, which leads to cellular hyperactivity, destruction of tissue architecture, and a resultant loss of function. Recent studies have identified that epigenetic changes regulate molecular pathways in immune-related diseases. Aberrant DNA methylation status, dysregulation of histone modifications, and altered microRNA expression could change the transcription activity of genes involved in the development of these diseases, which contributes to skewed differentiation of T cells and proliferation and activation of other cells, such as endothelial cells and fibroblasts; this is accompanied by overproduction of inflammatory cytokines and excessive accumulation or destruction of extracellular matrix, depending on the disease. Aside from this, epigenetics also explains how environmental exposure influences gene transcription without genetic changes and/or with a defined genetic backgrounds. It mostly acts as a mediator of the interaction between environmental factors and genetic factors. Identification of the abnormal epigenetic markers in diseases provides novel biomarkers for prediction and diagnosis, and affords novel therapeutic targets for these difficult clinical problems.

References

[1] Selmi C. Autoimmunity in 2012. Clin Rev Allergy Immunol 2013;45:290–301.

[2] Richardson SJ, Morgan NG, Foulis AK. Pancreatic pathology in type 1 diabetes mellitus. Endocr Pathol 2014;25:80–92.

[3] Xie Z, Chang C, Zhou Z. Molecular mechanisms in autoimmune type 1 diabetes: a critical review. Clin Rev Allergy Immunol 2014;47:174–92.

[4] Mak A, Tay SH. Environmental factors, toxicants and systemic lupus erythematosus. Int J Mol Sci 2014;15:16043–56.

[5] Mori K, Nakagawa Y, Ozaki H. Does the gut microbiota trigger Hashimoto's thyroiditis? Discov Med 2012;14:321–6.

[6] Mansourian AR. The immune system which adversely alter thyroid functions: a review on the concept of autoimmunity. Pak J Biol Sci 2010;13:765–74.

[7] Zimmermann AF, Pizzichini MM. Update on etiopathogenesis of systemic sclerosis. Rev Bras Reumatol 2013;53:516–24.

[8] Napier C, Pearce SH. Autoimmune Addison's disease. Presse Med 2012;41:e626–635.

[9] Meyer A, Meyer N, Schaeffer M, Gottenberg JE, Geny B, Sibilia J. Incidence and prevalence of inflammatory myopathies: a systematic review. Rheumatology (Oxford) 2014.

[10] Kivity S, Agmon-Levin N, Blank M, Shoenfeld Y. Infections and autoimmunity – friends or foes? Trends Immunol 2009;30:409–14.

[11] Maizels RM, McSorley HJ, Smyth DJ. Helminths in the hygiene hypothesis: sooner or later? Clin Exp Immunol 2014;177:38–46.

[12] Rashid T, Wilson C, Ebringer A. The link between ankylosing spondylitis, Crohn's disease, Klebsiella, and starch consumption. Clin Dev Immunol 2013;2013:872632.

[13] Cekin Y, Ozkaya E, Gulkesen H, Akcurin S, Colak D. Investigation of enterovirus infections, autoimmune factors and HLA genotypes in patients with T1DM. Minerva Endocrinol 2014;39:67–74.

[14] Li S, Yu Y, Yue Y, Zhang Z, Su K. Microbial infection and rheumatoid arthritis. J Clin Cell Immunol 2013;4.

[15] Nielsen DS, Krych L, Buschard K, Hansen CH, Hansen AK. Beyond genetics. Influence of dietary factors and gut microbiota on type 1 diabetes. FEBS Lett 2014.
[16] Bustos Fernandez LM, Lasa JS, Man F. Intestinal microbiota: its role in digestive diseases. J Clin Gastroenterol 2014;48:657–66.
[17] Taneja V. Arthritis susceptibility and the gut microbiome. FEBS Lett 2014;588 (22):4244–9.
[18] Kivity S, Arango MT, Ehrenfeld M, Tehori O, Shoenfeld Y, Anaya JM, et al. Infection and autoimmunity in Sjögren's syndrome: a clinical study and comprehensive review. J Autoimmun 2014;51:17–22.
[19] Ahronowitz I, Fox L. Severe drug-induced dermatoses. Semin Cutan Med Surg 2014;33:49–58.
[20] Gibelin A, Maldini C, Mahr A. Epidemiology and etiology of wegener granulomatosis, microscopic polyangiitis, churg-strauss syndrome and goodpasture syndrome: vasculitides with frequent lung involvement. Semin Respir Crit Care Med 2011;32:264–73.
[21] Gomez-Puerta JA, Gedmintas L, Costenbader KH. The association between silica exposure and development of ANCA-associated vasculitis: systematic review and meta-analysis. Autoimmun Rev 2013;12:1129–35.
[22] Sundstrom B, Johansson I, Rantapaa-Dahlqvist S. Interaction between dietary sodium and smoking increases the risk for rheumatoid arthritis: results from a nested case–control study. Rheumatology (Oxford) 2014;54(3):487–93.
[23] Krol A, Garred P, Heegaard N, Christensen A, Hetland M, Stengaard-Pedersen K, et al. Interactions between smoking, increased serum levels of anti-CCP antibodies, rheumatoid factors, and erosive joint disease in patients with early, untreated rheumatoid arthritis. Scand J Rheumatol 2015;44(1):8–12.
[24] Ding N, Hu Y, Zeng Z, Liu S, Liu L, Yang T, et al. Case-only designs for exploring the interaction between FCRL4 gene and suspected environmental factors in patients with ankylosing spondylitis. Inflammation 2015;38(2):632–6.
[25] Cleland LG, James MJ, Proudman SM. The role of fish oils in the treatment of rheumatoid arthritis. Drugs 2003;63:845–53.
[26] Sundstrom B, Stalnacke K, Hagfors L, Johansson G. Supplementation of omega-3 fatty acids in patients with ankylosing spondylitis. Scand J Rheumatol 2006;35:359–62.
[27] Huber A, Menconi F, Corathers S, Jacobson EM, Tomer Y. Joint genetic susceptibility to type 1 diabetes and autoimmune thyroiditis: from epidemiology to mechanisms. Endocr Rev 2008;29:697–725.
[28] Tomer Y, Huber A. The etiology of autoimmune thyroid disease: a story of genes and environment. J Autoimmun 2009;32:231–9.
[29] Iannuzzi MC, Rybicki BA, Teirstein AS. Sarcoidosis. N Engl J Med 2007;357:2153–65.
[30] Rybicki BA, Amend KL, Maliarik MJ, Iannuzzi MC. Photocopier exposure and risk of sarcoidosis in African-American sibs. Sarcoidosis Vasc Diffuse Lung Dis 2004;21:49–55.
[31] Newman LS, Rose CS, Bresnitz EA, Rossman MD, Barnard J, Frederick M, et al. A case–control etiologic study of sarcoidosis: environmental and occupational risk factors. Am J Respir Crit Care Med 2004;170:1324–30.
[32] Kucera GP, Rybicki BA, Kirkey KL, Coon SW, Major ML, Maliarik MJ, et al. Occupational risk factors for sarcoidosis in African-American siblings. Chest 2003;123:1527–35.
[33] Prezant DJ, Dhala A, Goldstein A, Janus D, Ortiz F, Aldrich TK, et al. The incidence, prevalence, and severity of sarcoidosis in New York City firefighters. Chest 1999;116:1183–93.
[34] Izbicki G, Chavko R, Banauch GI, Weiden MD, Berger KI, Aldrich TK, et al. World Trade Center "sarcoid-like" granulomatous pulmonary disease in New York City Fire Department rescue workers. Chest 2007;131:1414–23.

[35] Barnard J, Rose C, Newman L, Canner M, Martyny J, McCammon C, et al. Job and industry classifications associated with sarcoidosis in A Case–Control Etiologic Study of Sarcoidosis (ACCESS). J Occup Environ Med 2005;47:226–34.
[36] Lu Q. The critical importance of epigenetics in autoimmunity. J Autoimmun 2013;41:1–5.
[37] Stankov K, Benc D, Draskovic D. Genetic and epigenetic factors in etiology of diabetes mellitus type 1. Pediatrics 2013;132:1112–22.
[38] Fradin D, Le Fur S, Mille C, Naoui N, Groves C, Zelenika D, et al. Association of the CpG methylation pattern of the proximal insulin gene promoter with type 1 diabetes. PLoS One 2012;7:e36278.
[39] Husseiny MI, Kuroda A, Kaye AN, Nair I, Kandeel F, Ferreri K. Development of a quantitative methylation-specific polymerase chain reaction method for monitoring beta cell death in type 1 diabetes. PLoS One 2012;7:e47942.
[40] Rakyan VK, Beyan H, Down TA, Hawa MI, Maslau S, Aden D, et al. Identification of type 1 diabetes-associated DNA methylation variable positions that precede disease diagnosis. PLoS Genet 2011;7:e1002300.
[41] Herold KC, Vignali DA, Cooke A, Bluestone JA. Type 1 diabetes: translating mechanistic observations into effective clinical outcomes. Nat Rev Immunol 2013;13:243–56.
[42] Brix TH, Hegedus L. Twins as a tool for evaluating the influence of genetic susceptibility in thyroid autoimmunity. Ann Endocrinol 2011;72:103–7.
[43] Arakawa Y, Wantanabe M, Inoue N, Sarumaru M, Hidaka Y, Iwatani Y. Association of polymorphisms in DNMT1, DNMT3A, DNMT3B, MTHFR and MTRR genes with global DNA methylation levels and prognosis of autoimmune thyroid disease. Clin Exp Immunol 2012;170:194–201.
[44] Deng Y, Tsao BP. Advances in lupus genetics and epigenetics. Curr Opin Rheumatol 2014;26:482–92.
[45] Hedrich CM, Tsokos GC. Epigenetic mechanisms in systemic lupus erythematosus and other autoimmune diseases. Trends Mol Med 2011;17:714–24.
[46] Zhang Y, Zhao M, Sawalha AH, Richardson B, Lu Q. Impaired DNA methylation and its mechanisms in CD4(+)T cells of systemic lupus erythematosus. J Autoimmun 2013;41:92–9.
[47] Javierre BM, Richardson B. A new epigenetic challenge: systemic lupus erythematosus. Adv Exp Med Biol 2011;711:117–36.
[48] Absher DM, Li X, Waite LL, Gibson A, Roberts K, Edberg J, et al. Genome-wide DNA methylation analysis of systemic lupus erythematosus reveals persistent hypomethylation of interferon genes and compositional changes to CD4 + T-cell populations. PLoS Genet 2013;9:e1003678.
[49] Koelsch KA, Webb R, Jeffries M, Dozmorov MG, Frank MB, Guthridge JM, et al. Functional characterization of the MECP2/IRAK1 lupus risk haplotype in human T cells and a human MECP2 transgenic mouse. J Autoimmun 2013;41:168–74.
[50] Kaufman KM, Zhao J, Kelly JA, Hughes T, Adler A, Sanchez E, et al. Fine mapping of Xq28: both MECP2 and IRAK1 contribute to risk for systemic lupus erythematosus in multiple ancestral groups. Ann Rheum Dis 2013;72:437–44.
[51] Konsta OD, Thabet Y, Le Dantec C, Brooks WH, Tzioufas AG, Pers JO, et al. The contribution of epigenetics in Sjögren's Syndrome. Front Genet 2014;5:71.
[52] Thabet Y, Le Dantec C, Ghedira I, Devauchelle V, Cornec D, Pers JO, et al. Epigenetic dysregulation in salivary glands from patients with primary Sjögren's syndrome may be ascribed to infiltrating B cells. J Autoimmun 2013;41:175–81.
[53] Yin H, Zhao M, Wu X, Gao F, Luo Y, Ma L, et al. Hypomethylation and overexpression of CD70 (TNFSF7) in $CD4^+$ T cells of patients with primary Sjögren's syndrome. J Dermatol Sci 2010;59:198–203.

[54] Gestermann N, Koutero M, Belkhir R, Tost J, Mariette X, Miceli-Richard C. Methylation profile of the promoter region of IRF5 in primary Sjögren's syndrome. Eur Cytokine Netw 2012;23:166–72.
[55] Losowsky MS. A history of coeliac disease. Dig Dis 2008;26:112–20.
[56] Chang HW, Leong KH, Koh DR, Lee SH. Clonality of isolated eosinophils in the hypereosinophilic syndrome. Blood 1999;93:1651–7.
[57] Maeda T, Guan JZ, Higuchi Y, Oyama J, Makino N. Aging-related alterations of subtelomeric methylation in sarcoidosis patients. J Gerontol A Biol Sci Med Sci 2009;64:752–60.
[58] Lai NS, Chou JL, Chen GC, Liu SQ, Lu MC, Chan MW. Association between cytokines and methylation of SOCS-1 in serum of patients with ankylosing spondylitis. Mol Biol Rep 2014;41:3773–80.
[59] Oliver JE, Silman AJ. Risk factors for the development of rheumatoid arthritis. Scand J Rheumatol 2006;35:169–74.
[60] Kobayashi S, Momohara S, Kamatani N, Okamoto H. Molecular aspects of rheumatoid arthritis: role of environmental factors. FEBS J 2008;275:4456–62.
[61] Bogdanos DP, Smyk DS, Rogopoulou EI, Mytilinaiou MG, Heneghan MA, Selmi C, et al. Twin studies in autoimmune disease: genetics, gender and environment. J Autoimmun 2012;38:J156–169.
[62] Firestein GS. The disease formerly known as rheumatoid arthritis. Arthritis Res Ther 2014;16:114.
[63] Glant TT, Mikecz K, Rauch TA. Epigenetics in the pathogenesis of rheumatoid arthritis. BMC Med 2014;12:35.
[64] Karouzakis E, Gay RE, Michel BA, Gay S, Neidhart M. DNA hypomethylation in rheumatoid arthritis synovial fibroblasts. Arthritis Rheum 2009;60:3613–22.
[65] Neidhart M, Rethage J, Kuchen S, Kunzler P, Crowl RM, Billingham ME, et al. Retrotransposable L1 elements expressed in rheumatoid arthritis synovial tissue: association with genomic DNA hypomethylation and influence on gene expression. Arthritis Rheum 2000;43:2634–47.
[66] Karouzakis E, Rengel Y, Jungel A, Kolling C, Gay RE, Michel BA, et al. DNA methylation regulates the expression of CXCL12 in rheumatoid arthritis synovial fibroblasts. Genes Immun 2011;12:643–52.
[67] Nakano K, Whitaker JW, Boyle DL, Wang W, Firestein GS. DNA methylome signature in rheumatoid arthritis. Ann Rheum Dis 2013;72:110–17.
[68] Neidhart M, Karouzakis E, Jungel A, Gay RE, Gay S. Inhibition of spermidine/spermine N1-acetyltransferase activity: a new therapeutic concept in rheumatoid arthritis. Arthritis Rheumatol 2014;66:1723–33.
[69] Allanore Y, Wipff J, Kahan A, Boileau C. Genetic basis for systemic sclerosis. Joint Bone Spine 2007;74:577–83.
[70] Broen JC, Radstake TR, Rossato M. The role of genetics and epigenetics in the pathogenesis of systemic sclerosis. Nat Rev Rheumatol 2014;10:671–81.
[71] Mora GF. Systemic sclerosis: environmental factors. J Rheumatol 2009;36:2383–96.
[72] Lian X, et al. DNA demethylation of CD40l in $CD4^+$ T cells from women with systemic sclerosis: a possible explanation for female susceptibility. Arthritis Rheum 2012;64:2338–45.
[73] Bjanesoy TE, Andreassen BK, Bratland E, Reiner A, Islam S, Husebye ES, et al. Altered DNA methylation profile in Norwegian patients with Autoimmune Addison's Disease. Mol Immunol 2014;59:208–16.
[74] Kaplan MJ, Lu Q, Wu A, Attwood J, Richardson B. Demethylation of promoter regulatory elements contributes to perforin overexpression in $CD4^+$ lupus T cells. J Immunol 2004;172:3652–61.

CHAPTER

20

DNA Methylation in Lymphocyte Development

OUTLINE

M. Neidhart: *DNA Methylation and Complex Human Disease.*
DOI: http://dx.doi.org/10.1016/B978-0-12-420194-1.00020-8

20.1 INTRODUCTION

The main goal of any immune response is to clear or control infection, and this is certainly the function of a primary T-cell response. However, T-cell responses serve another function, which is to generate memory T cells that can remember the pathogens they encounter and mount swift protective responses against reinfection. It is now well established that $CD8^+$ T cells undergo changes with the development of memory, such that they respond to antigen stimulation more quickly, mobilize effector functions more rapidly, and possess the stem-cell-like quality of self-renewal. During their development from progenitors, lymphocytes make a series of cell-fate decisions. These decisions reflect and require changes in overall programs of gene expression. To maintain cellular identity, programs of gene expression must be iterated through mitosis in a heritable manner by epigenetic processes, which include DNA methylation, methyl-CpG-binding proteins, histone modifications, transcription factors, and higher-order chromatin structure. DNA methylation acts in concert with other epigenetic processes to limit the probability of aberrant gene expression and to stabilize, rather than initiate, cell-fate decisions. In particular, DNA methylation appears to be a non-redundant repressor of CD8 expression in TCR-$\gamma\delta$ T cells and Th2 cytokine expression in Th1 and $CD8^+$ T cells [1]. This chapter focuses on the role of DNA methylation in T- and B-cell development.

20.2 HEMATOPOIETIC CELLS

Each stage of hematopoiesis has been analyzed for genome-wide DNA methylation patterns using microarray technology [2]. These analyses have revealed a program of epigenetic modifications associated with lymphocyte development. Many genes that are initially methylated in hematopoietic stem cells undergo selective demethylation in a tissue- or lineage-specific manner. These genes include lymphocyte-specific protein tyrosine kinase (LCK), which undergoes demethylation in T cells and encodes a SRC family kinase that is responsible for initiating signaling downstream of the T cell receptor, as well as POU domain class 2-associating factor 1 (POU2AF1), which encodes a B cell-specific coactivator and undergoes demethylation in B cells [3] (see Figure 20.1). As opposed to these demethylation events, another set of genes is actively subject to *de novo* methylation during lineage-specific development. These genes include, for example, dachshund homolog 1 (DACH1), which is packaged in an open chromatin conformation in

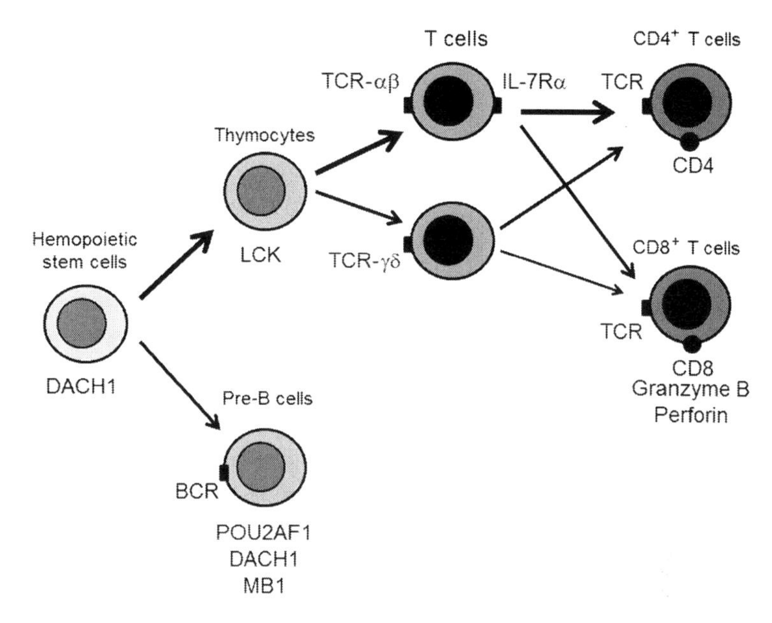

FIGURE 20.1 Genes controlled by DNA methylation and involved in T- and B-cell development.

stem cells, and undergoes silencing in double negative ($CD4^-$, $CD8^-$) thymocytes. Although the precise mechanism for gene-specific *de novo* methylation is not known, many of these events occur at Polycomb target sites, probably through the recruitment of DNA methyltransferase 3A (DNMT3A) and DNMT3B [4–6]. Recent evidence shows that DNA methylation also has a direct role in regulating hemopoietic stem cells' self-renewal and commitment to lymphoid versus myeloid cell fates [5,6]. The same principles that govern the differentiation of hemotopoeitic stem cells into lymphoid or myeloid progenitor cells also appear to direct the process of B- or T-cell commitment and the terminal differentiation to antigen receptor-producing cells through gene rearrangement.

20.3 T-CELL SUBPOPULATIONS

20.3.1 T-cell Development

DNMT1 is required for proper maturation of thymic progenitors, as shown from studies in T cell-specific conditional DNMT1 knockout mice [7]. Conditional deletion of DNMT1 in early double-negative ($CD4^-$, $CD8^-$) thymocytes is followed by cellular and DNA replication and DNA demethylation. This is associated with a profound reduction in cells of the TCR-αβ lineage. This results from apoptosis of cells as levels of DNA methylation fall below a critical threshold. In addition to the attrition of TCR-αβ lineage cells, the numbers of TCR-γδ cells are increased in these mice, which appears to reflect in part their more limited intrathymic replication and, as a consequence, less profound DNA demethylation. A similar, though less severe, deficiency of TCR-αβ cells is seen in thymocytes lacking the lymphocyte-specific helicase (LSH), in association with markedly reduced and aberrant DNA methylation [8–10].

20.3.2 TCR-αβ Versus TCR-γδ

The DNA of TCR loci is methylated in double-negative ($CD4^-$, $CD8^-$) thymocytes. Activation of the TCR-β gene proceeds through a pathway involving IL-7Rα signaling and demethylation of CpG sites [11–13] (Figure 20.1). These events open the locus to the V(D)J recombination machinery, and allow expression of the TCR-β chain as part of a pre-TCR complex. Pre-TCR signaling triggers ATP-dependent BAF chromatin remodeling complexes that affect expression of multiple survival and differentiation genes, as well as excluding expression of the other TCR-β allele. Failure of TCR-γ rearrangement in IL-7Rα knockout mice is associated with persistent DNA methylation at the TCR-γ enhancer, suggesting that DNA demethylation either is necessary for rearrangement or occurs as a consequence of other processes that are essential [12]. However, DNMT1-deficient thymocytes exhibit normal V(D)J recombination and allelic exclusion [7]. This does not exclude the possibility that demethylation is necessary for recombination to occur, but does suggest that DNA demethylation is not a limiting factor in V(D)J recombination and neither are proper levels of DNA methylation required to enforce allelic exclusion. V(D)J recombination at the TCR-α, -β, and -γ and immunoglobulin heavy chain loci is also not impaired in the absence of lymphocyte-specific helicase LSH [9]. The irreversible silencing of terminal transferase expression in mature thymocytes is also relevant to T-cell development.

20.3.3 CD4 Versus CD8

The CD4 gene is a target of reversible repression in double-negative thymocytes and then permanent silencing as double-positive cells differentiate toward the single-positive $CD8^+$ state. This silencing is an epigenetically regulated event (Figure 20.1), but DNA methylation does not appear to be necessary. Instead, there are specific repressive pathways in $CD8^+$ T cells mediated by Runx3 interacting in *cis* with a defined CD4 silencer sequence [14]. Demethylation of the CD8α and CD8β genes occurs during thymocyte maturation from the double-negative to the $CD4^+CD8^+$ stage. The retention of demethylated CD8 genes in single-positive $CD4^+$ cells provided early evidence for the notion that these cells are derived from previously $CD8^+$ precursors [15], which was later proved by lineage tracing studies. A role for DNA methylation in the regulation of CD8 expression was also seen in conditional $DNMT1^{-/-}$ mice [7], in which aberrant CD8α and CD8β expression occurred in a large fraction of TCR-γδ lineage cells. *De novo* remethylation of the CD8α gene has been associated with the death of misselected thymic emigrants [16]. However, other mechanisms must be able to silence the CD8 gene after commitment to the CD4 lineage, since double-positive $CD4^+CD8^+$ T cells are not found in the periphery of conditional $DNMT1^{-/-}$ mice [7]. Demethylation of specific promoters (granzyme B and interferon-γ) is associated with the acquisition of differential gene expression profiles unique to effector $CD8^+$ T cells [17]. On the other hand, homeostasis genes, such as transcription factor 7 (TCF7), expressed highly in naïve and memory cells, have reduced expression and increased promoter methylation in effector relative to naive $CD8^+$ T cells.

20.4 T-CELL FUNCTIONS

20.4.1 Interleukin-2

A role for DNA methylation in the regulation of T-cell function was suggested by early studies in which the expression of interleukin-2 (IL-2) was shown to be enhanced by agents that inhibit DNA methylation [18]. Subsequent studies have confirmed that this gene is quiescent, the IL-2 promoter DNA is methylated, and the locus is contained in inactive chromatin in resting naïve T cells [19,20]. Early after TCR-dependent activation, one or both alleles become activated, depending on the strength of stimulus. Although IL-2 expression occurs in the absence of cell division, cell division increases the probability and frequency of these events. In naïve CD4 and CD8 T cells from conditional $DNMT1^{-/-}$ mice, IL-2 expression is increased about five-fold following activation *in vitro* [21]. This relatively modest increase may reflect the fact that the IL-2 gene normally undergoes rapid DNA demethylation in wild-type cells in response to activation. Remarkably, demethylation in the IL-2 promoter appears to commence within hours after T-cell stimulation *in vitro*, within 20 minutes of T-cell stimulation *in vivo*, and well before entry of cells into S phase. This suggests the occurrence of active catalytic DNA demethylation targeted to a specific subset of CpG within ~800 bp of the IL-2 transcription initiation site [19]. These studies also showed that promoter methylation represses IL-2 transcription, supporting the functional importance of demethylation for expression of IL-2.

20.4.2 Interferon-γ

Interferon-γ (IFN-γ) was one of the first cytokines shown to be regulated epigenetically. A role for DNA methylation in the regulation of IFN-γ in T cells is supported by the fact that its expression is increased after treatment with DNA methylation inhibitors [22,23]. In addition, in mice, DNA methylation in the promoter correlates inversely with differences in expression of IFN-γ between neonatal and adult or "naïve" and "memory" T cells [24–26], as well as between Th1 and Th2 cells [22,27]. Accordingly, *de novo* methylation in T-cell populations correlates with reduced expression [28]. The first of these features is reproduced in DNMT1$^{-/-}$ T cells, in which expression of IFN-γ is increased about 5- to 10-fold following activation [21]. Increased IFN-γ expression appears to be primarily, if not solely, due to DNA demethylation in the IFN-γ gene [7,29]. Several more distal regulatory elements flanking the INFG gene have been identified [30]. Consistent with findings in Dnmt1$^{-/-}$ T cells, naïve T cells lacking the methyl-CpG binding protein MBD2 produce increased amounts of IFN-γ in response to activation and before entering the cell cycle in the absence of cell division [31].

20.4.3 Interleukin-4/Interleukin-13 Locus

There are multiple components of epigenetic machinery that target these regulatory sequence elements in the IL-4/IL-13 locus. DNA methylation was identified as an important level of control of IL-4 gene expression, as for IFN-γ, through the mapping of CpG site methylation changes across the gene during T-cell differentiation and the analysis of the effects of inhibiting DNMT1 [7,21,32]. IL-4 silencing also occurs as a naïve T cell develops into a Th1 cell, and this silencing is associated with DNA methylation/histone deacetylation of the GATA-3 and IL-4 genes [33]. As Th2 cells develop, increased expression of GATA-3 is implicated in the changes that occur within the IL-4 gene locus, including DNA demethylation [34]. Human Th2 cells polarized from adult CD45RA$^+$ cells changed DNA methylation in the IL-4/IL-13 locus, less in the IL-4 promoter, but a demethylation around at the IL-13 promoter and first exon occurs [35].

Knockout of a methyl-DNA binding protein, MBD2, shows a similar phenotype to that of DNMT1. MBD2 knockout mice did not show any significant developmental arrest during T-lymphocyte development [31]. However $MDB2^{-/-}$ $CD4^{+}$ T cells produced higher levels of IFN-γ (as mentioned above) and IL-4 in both Th1- and Th2-skewing conditions compared with wild-type cells. MBD2 regulates IL-4 by directly competing with GATA3 protein at the key regulatory regions in the IL4 gene.

20.4.4 Interleukin-17

Epigenetic regulation has also proven essential for the more recently discovered T helper cell subtypes: the Th17 lineage and regulatory T cells [30]. Even though interleukin-17 (IL-17) production from T helper cells was observed many years ago, it was not until recently that Th17 cells were accepted as a separate lineage of development from naïve cells, apart from Th1 and Th2. Whether this differentiation is regulated by DNA methylation is as yet unknown.

20.4.5 Regulatory T cells

The most widely used marker for the T regulatory (Treg) population today is the transcription factor FOXP3. Its expression is required for the development of Tregs, and is closely linked to a suppressive phenotype [30]. Mouse studies of the epigenetic regulation of *Foxp3* have focused on a conserved element in intron 1 of the FOXP3 gene. The Treg-specific demethylated region (TSDR) shows transcriptional activity and differential methylation between $CD4^{+}CD25^{-}$ and T regulatory cells [36]. This region binds the cyclic AMP response element binding protein (CREB) in a methylation dependent manner, and it also binds STAT5 in Tregs [37]. Furthermore, demethylation occurs at intron 1 in $CD25^{low}$ cells in the presence of transforming growth factor-β (TGF-β) [38]. However, the TGF-β induced demethylation was not as pronounced as in the freshly isolated $CD25^{high}$ population [36]. Interestingly, 5-azacytidine treatment induces expression of FOXP3 in murine $CD25^{low}$ cells, similar to the presence of TGF-β [38]. Such studies in mice [38,39] suggested that although TGF-β is able to induce partial demethylation, additional chromatin modulating agents apart from the ones downstream of TGF-β signaling are also needed for the complete conversion of conventional T cells to committed Tregs. Similarly, in humans, complete demethylation of the FOXP3 gene is only seen in freshly isolated $CD4^{+}CD25^{high}$ cells, whereas stimulated $CD25^{low}$ cells, that transiently express FOXP3, remain partially methylated [40].

Importantly, one epigenetic event differs between the mouse and the human setting; TGF-β induces demethylation in the previously mentioned intronic region and the promoter in murine but not in human CD25^{low} cells [40,41]. Both DNMT1 and DNMT3B are associated with the FOXP3 locus in non-Treg CD4^{+} T cells [42].

20.5 B-CELL DEVELOPMENT

Once commitment to the lymphoid lineage has been achieved, B cells are generated through further differentiation steps leading to the formation of the early B cell precursors, pro-B and pre-B cells, the immature B cells, and the terminally differentiated plasma cells and germinal-center B cells [43] (Figure 20.1). The MB1 gene encodes the Igα subunit (CD79a) of the pre-B cell receptor (BCR) and BCR, and is a target gene for immunoglobulin enhancer-binding factors E2A, transcription factor COE1 (encoded by the EBF gene), and paired box protein Pax5. These transcription factors cooperate and mediate epigenetic events to regulate MB1 expression [44]. The MB1 gene is methylated at CpG dinucleotides in hematopoietic stem cells and gradually demethylated during B-cell commitment in accordance with its pattern of expression. COE1 and E2A contribute to the CpG demethylation and nucleosomal remodeling of the MB1 promoter, which is necessary for its transcriptional activation by Pax5. Interestingly, in embryonic stem cells Pax5 is silenced by DNA methylation and becomes activated in multipotent hematopoietic progenitors. The enhancer contains binding sites for the transcription factors PU.1, IRF4, IRF8, and NF-κB, suggesting that these regulators play a role in the sequential enhancer activation in hematopoietic progenitors and during B-cell development [45].

20.6 INFECTION AND INFLAMMATION

Genome-wide studies to profile transcriptional and epigenetic changes during infection have revealed that dynamic changes in DNA methylation patterns and histone modifications accompany transcriptional signatures that define and regulate cytotoxic CD8^{+} T-cell differentiation states [46]. Thus, it is becoming increasingly clear that what an activated CD8^{+} T cell and its daughter cells will become, or have the potential to become, is influenced or programmed early during the primary response to infection [47]. We mentioned above that epigenetic changes, such as methylation of GAT3A, IL4, IL13, and FOXP3, accompany the differentiation of naïve CD4^{+} T cells into Th1 effectors cells [1] (Figure 20.2). Epigenetic modifications, including DNA demethylation,

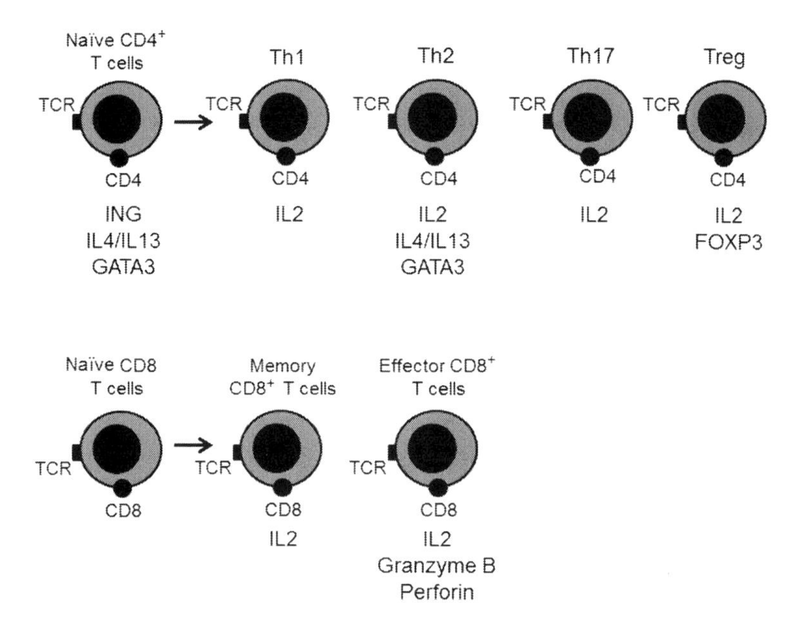

FIGURE 20.2 Genes controlled by DNA methylation and involved in T-cell activation.

also occur in $CD8^+$ T cells early after infection at the effector stage and are maintained during the development of memory [17,47]. These processes lead to the stable epigenetic modification of an entire program of genes involved in T-cell survival, metabolism, proliferation, and effector function, which serve as a molecular basis for cellular memory during T-cell responses. Perforin is a cytolytic protein found in the granules of cytotoxic $CD4^+$ T lymphocytes and NK cells. Based on studies in which T cells were treated with 5-azacytidine [48], it has been suggested that DNA methylation represses expression of its gene in human $CD8^+$ T and NK cells.

Inflammatory factors signal strongly to the developing $CD8^+$ T cell, increasing its potential to become a potent effector while decreasing its potential to develop into a long-lived memory cell. Epigenetic modification is an integral part of memory $CD8^+$ T-cell differentiation. It is possible that the varying external signals may influence epigenetic programming in an activated $CD8^+$ T cell that determines its potential to differentiate into a potent effector and a long-lived memory T cell [47].

20.7 TUMOR CLEARANCE

Failure of the immune system to detect formation of malignant cells may lead to tumor development. Tumors have several strategies to overcome the intrinsic immune defense, including downregulation of MHC molecules to avoid recognition, and secretion of immunosuppressive molecules such as TGF-β and indoleamine-pyrrole 2,3-dioxygenase (IDO). Interestingly, the presence of tumor-infiltrating lymphocytes correlates positively with survival in many cancers. However, tumor-infiltrating lymphocytes are poor producers of IFN-γ, and have a low proliferative rate compared to lymphocytes from the sentinel node [30]. The sentinel node is the first lymph node to receive lymphatic drainage from the tumor, and is therefore also the location for tumor recognition and the initial clonal expansion of tumor-reactive T cells. The INFG promoter in $CD4^+$ tumor-infiltrating lymphocytes is hypermethylated compared to $CD4^+$ cells from the sentinel node, as expected for a non-naïve T cell. In addition, 5-azacytidine treatment restores IFN-γ production from tumor-infiltrating lymphocytes [49]. These observations suggest an altered cytokine profile of tumor-infiltrating lymphocytes, induced by the tumor and regulated by methylation of cytokine loci as an additional tumor immune escape mechanism.

20.8 ENVIRONMENTAL FACTORS AND AUTOIMMUNITY

One example of how the environment uses epigenetic mechanisms to establish a gene expression pattern in cells is through diet. A key amino acid needed for the *de novo* methylation process is L-methionine [50]. The aging process itself appears to be another way in which epigenetic changes occur to affect disease development and behavioral changes, without a change in the DNA sequence itself [51]. Thus, environmental factors may exert an effect on a naïve T cell as it is developing into a Th1 or Th2 cell in a way that will change the effector cell phenotype. Also, environmental factors may affect the differentiated T cell itself, changing which cytokines are produced and passing this change along as the cell divides. A failure to maintain epigenetic homeostasis in the immune response due to factors including environmental influences leads to aberrant gene expression, contributing to immune dysfunction and, in some cases, the development of autoimmunity in genetically predisposed individuals [52,53]. This is exemplified by systemic lupus erythematosus [52], where environmentally induced epigenetic changes contribute to disease pathogenesis in those genetically predisposed. Similar interactions between genetically determined susceptibility and environmental factors are implicated in other systemic autoimmune diseases, such as rheumatoid arthritis and scleroderma, as well as in organ-specific autoimmunity. The skin is exposed to a wide variety of environmental agents, including UV radiation, and is prone to the development of autoimmune conditions such as atopic dermatitis, psoriasis, and some forms of vitiligo, depending on environmental and genetic influences. In addition, dysregulated DNA methylation of CD4$^+$ cells has been implicated in the pathogenesis of autoimmune disorders [52]. CD4$^+$ T cells lose their ability to distinguish MHC-self peptides from non-self when treated with the demethylating agent 5-azacytidine. The reason for this involves increased expression of LFA-1, an adhesion molecule that stabilizes the MHC–TCR interaction at the immunological synapse [30]. Demethylated CD4$^+$ T cells kill autologous macrophages in an MHC restricted manner. The cytotoxicity is mediated by heightened levels of LFA-1 and perforin (otherwise only used by CD8$^+$ cytotoxic T cells and NK cells, as mentioned above). This results in the release and accumulation of apoptotic material, which triggers the production of anti-DNA antibodies from B cells [54]. The B cells are activated due to overexpression of CD70 and CD40 ligand, as well as hyper-production of IFN-γ and IL-4 by the CD4$^+$ T cells [55,56]. Hypomethylation of the CD70 promoter regulatory elements is also reported in Sjögren's syndrome [57]; it is mainly involved in the interaction with CD27 and alternative activation of naïve T cells (Figure 20.3).

Demethylation of the promoter regions of adhesion molecules, co-stimulatory molecules, and cytokines in CD4$^+$ T cells results in autoreactivity that triggers release of apoptotic material. Failure of clearance

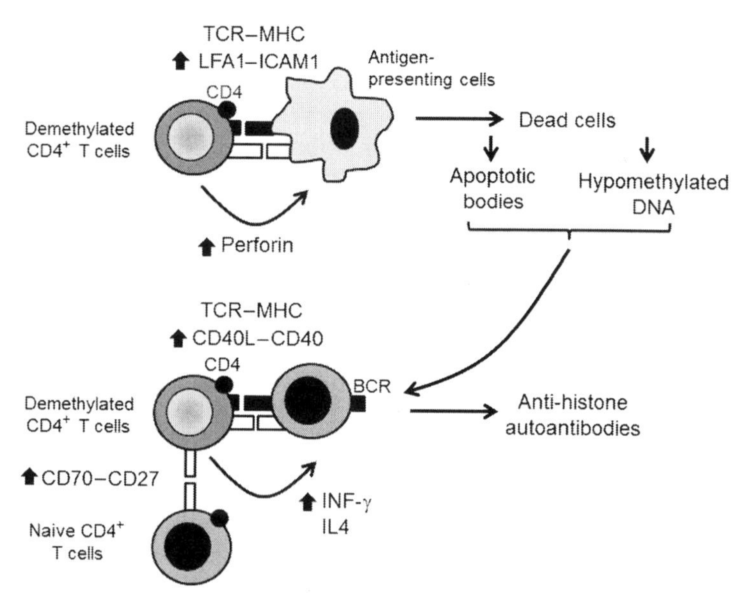

FIGURE 20.3 Genes expressed in demethylated CD4$^+$ T cells and involved in the pathogenesis of systemic lupus erythematosus.

due to the absence of macrophages induces autoantibody production that is further enforced by the elevated levels of co-stimulatory molecules and cytokine release from T cells. Interestingly, T cells from lupus patients display genome-wide demethylation, elevated expression of LFA-1 and perforin, and B-cell activating properties, suggesting that similar mechanisms are involved as in the murine model of this disease. This is further supported by the observation that patients with active lupus show deficiency in ERK signaling, known to regulate DNMT1 [52]. Hypomethylation of gene promoters has been described, which permits transcriptional activation and therefore functional changes in the cells, and also hypomethylation of the ribosomal RNA gene cluster [58]. In addition to genes involved in cellular interactions (CD70 and CD40 ligand) and cytotoxicity (perforin), mentioned above, many other genes undergo demethylation in systemic lupus erythematosus, including those playing a role in autoreactivity (ITGAL/CD11a) [59], antigen presentation (CSF3R) [58], inflammation (MMP14) [58], and cytokine pathways (CSF3R, IL-6, and IFNGR2) [58,60].

Finally, it is important to note that $CD5^+$ B lymphocytes predominate in systemic lupus erythematosus. Normally, exon 1B of CD5 is silenced by methylation and its product degraded in the proteasomes, but it is not the case in this disease [61,62]. Blockade of the interleukin-6 autocrine loop in lupus B cells restores DNA methylation status, thus opening new perspectives for therapy [62,63]. Autoreactive B cells are characterized by their inability to induce DNA methylation that prolongs their survival [64]. In addition, there is evidence that hypomethylated DNA is immunogenic, that antihistone autoantibodies in patients with systemic lupus erythematosus bind epigenetic-sensitive hot spots, and that epigenetically-modified ribonucleoprotein-derived peptides can modulate the disease [65]. Certain microRNAs, RFX1, defective ERK pathway signaling, GADD45α, and DNA hydroxymethylation have been proposed as potential mechanisms leading to DNA hypomethylation [66].

20.9 CONCLUSION

Epigenetics regulates important inherited events in the T-cell compartment of the immune system. These range from some of the earliest events in developing thymocytes to possibly some of the latest events in memory T cells. Since alterations in TCR repertoires, IFN-γ, expression and IL-4 expression are strongly implicated in autoimmune and allergic diseases, it has been proposed [1] that some diseases may possess an underlying "epigenetic immunopathology." Environmentally induced epigenetic changes may be either harmful or beneficial to the survival of the host [67]. It is probable that environmentally induced epigenetic changes might explain how certain Th1/Th2-associated diseases occur.

References

[1] Wilson CB, Makar KW, Shnyreva M, Fitzpatrick DR. DNA methylation and the expanding epigenetics of T cell lineage commitment. Semin Immunol 2005;17:105–19.

[2] Ji H, Ehrlich LI, Seita J, Murakami P, Doi A, Lindau P, et al. Comprehensive methylome map of lineage commitment from haematopoietic progenitors. Nature 2010;467:338–42.

[3] Cedar H, Bergman Y. Epigenetics of haematopoietic cell development. Nat Rev Immunol 2011;11:478–88.

[4] Vire E, Brenner C, Deplus R, Blanchon L, Fraga M, Didelot C, et al. The Polycomb group protein EZH2 directly controls DNA methylation. Nature 2006;439:871–4.

[5] Broske AM, et al. DNA methylation protects hematopoietic stem cell multipotency from myeloerythroid restriction. Nat Genet 2009;41:1207–15.

[6] Trowbridge JJ, Snow JW, Kim J, Orkin SH. DNA methyltransferase 1 is essential for and uniquely regulates hematopoietic stem and progenitor cells. Cell stem cell 2009;5:442–9.

[7] Lee PP, Fitzpatrick DR, Beard C, Jessup HK, Lehar S, Makar KW, et al. A critical role for Dnmt1 and DNA methylation in T cell development, function, and survival. Immunity 2001;15:763–74.

[8] Dennis K, Fan T, Geiman T, Yan Q, Muegge K. Lsh, a member of the SNF2 family, is required for genome-wide methylation. Genes Dev 2001;15:2940–4.

[9] Geiman TM, Muegge K. Lsh, an SNF2/helicase family member, is required for proliferation of mature T lymphocytes. Proc Natl Acad Sci USA 2000;97:4772–7.

[10] Yan Q, Cho E, Lockett S, Muegge K. Association of Lsh, a regulator of DNA methylation, with pericentromeric heterochromatin is dependent on intact heterochromatin. Mol Cell Biol 2003;23:8416–28.

[11] Chattopadhyay S, Whitehurst CE, Schwenk F, Chen J. Biochemical and functional analyses of chromatin changes at the TCR-beta gene locus during CD4-CD8- to $CD4^+CD8^+$ thymocyte differentiation. J Immunol 1998;160:1256–67.

[12] Durum SK, Candeias S, Nakajima H, Leonard WJ, Baird AM, Berg LJ, et al. Interleukin 7 receptor control of T cell receptor gamma gene rearrangement: role of receptor-associated chains and locus accessibility. J Exp Med 1998;188:2233–41.

[13] Sakamoto S, Ortaldo JR, Young HA. Methylation patterns of the T cell receptor beta-chain gene in T cells, large granular lymphocytes, B cells, and monocytes. J Immunol 1988;140:654–60.

[14] Taniuchi I, Osato M, Egawa T, Sunshine MJ, Bae SC, Komori T, et al. Differential requirements for Runx proteins in CD4 repression and epigenetic silencing during T lymphocyte development. Cell 2002;111:621–33.

[15] Carbone AM, Marrack P, Kappler JW. Demethylated CD8 gene in $CD4^+$ T cells suggests that $CD4^+$ cells develop from $CD8^+$ precursors. Science 1988;242:1174–6.

[16] Pestano GA, Zhou Y, Trimble LA, Daley J, Weber GF, Cantor H. Inactivation of misselected CD8 T cells by CD8 gene methylation and cell death. Science 1999;284:1187–91.

[17] Scharer CD, Barwick BG, Youngblood BA, Ahmed R, Boss JM. Global DNA methylation remodeling accompanies CD8 T cell effector function. J Immunol 2013;191:3419–29.

[18] Ballas ZK. The use of 5-azacytidine to establish constitutive interleukin 2-producing clones of the EL4 thymoma. J Immunol 1984;133:7–9.

[19] Bruniquel D, Schwartz RH. Selective, stable demethylation of the interleukin-2 gene enhances transcription by an active process. Nat Immunol 2003;4:235–40.

[20] Ward SB, Hernandez-Hoyos G, Chen F, Waterman M, Reeves R, Rothenberg EV. Chromatin remodeling of the interleukin-2 gene: distinct alterations in the proximal versus distal enhancer regions. Nucleic Acids Res 1998;26:2923–34.

[21] Makar KW, Wilson CB. DNA methylation is a nonredundant repressor of the Th2 effector program. J Immunol 2004;173:4402–6.

[22] Young HA, Ghosh P, Ye J, Lederer J, Lichtman A, Gerard JR, et al. Differentiation of the T helper phenotypes by analysis of the methylation state of the IFN-gamma gene. J Immunol 1994;153:3603–10.

[23] Farrar WL, Ruscetti FW, Young HA. 5-Azacytidine treatment of a murine cytotoxic T cell line alters interferon-gamma gene induction by interleukin 2. J Immunol 1985;135:1551–4.

[24] Melvin AJ, McGurn ME, Bort SJ, Gibson C, Lewis DB. Hypomethylation of the interferon-gamma gene correlates with its expression by primary T-lineage cells. Eur J Immunol 1995;25:426–30.

[25] Fitzpatrick DR, Shirley KM, McDonalrd LE, Bielefeldt-Ohmann H, Kay GF, Kelso A. Distinct methylation of the interferon gamma (IFN-gamma) and interleukin 3 (IL-3) genes in newly activated primary $CD8^+$ T lymphocytes: regional IFN-gamma promoter demethylation and mRNA expression are heritable in CD44(high)$CD8^+$ T cells. J Exp Med 1998;188:103–17.

[26] White GP, Watt PM, Holt BJ, Holt PG. Differential patterns of methylation of the IFN-gamma promoter at CpG and non-CpG sites underlie differences in IFN-gamma gene expression between human neonatal and adult CD45RO − T cells. J Immunol 2002;168:2820–7.

[27] Katamura K, Fukui T, Kiyomasu T, Iio J, Tai G, Uneo H, et al. IL-4 and prostaglandin E2 inhibit hypomethylation of the 5′ regulatory region of IFN-gamma gene during differentiation of naive $CD4^+$ T cells. Mol Immunol 1998;35:39–45.

[28] Mikovits JA, Young HA, Vertino P, Issa JP, Pitha PM, Turcoski-Corrales S, et al. Infection with human immunodeficiency virus type 1 upregulates DNA methyltransferase, resulting in *de novo* methylation of the gamma interferon (IFN-gamma) promoter and subsequent downregulation of IFN-gamma production. Mol Cell Biol 1998;18:5166–77.

[29] Makar KW, Perez-Melgosa M, Shnyreva M, Weaver WM, Fitzpatrick DR, Wilson CB. Active recruitment of DNA methyltransferases regulates interleukin 4 in thymocytes and T cells. Nat Immunol 2003;4:1183–90.

[30] Janson PC, Winerdal ME, Winqvist O. At the crossroads of T helper lineage commitment-Epigenetics points the way. Biochim Biophys Acta 1790;906–919:2009.

[31] Hutchins AS, Mullen AC, Lee HW, Sykes KJ, High FA, Hendrich BD, et al. Gene silencing quantitatively controls the function of a developmental trans-activator. Mol Cell 2002;10:81–91.

[32] Lee DU, Agarwal S, Rao A. Th2 lineage commitment and efficient IL-4 production involves extended demethylation of the IL-4 gene. Immunity 2002;16:649–60.

[33] Grogan JL, Mohrs M, Harmon B, Lacy DA, Sedat JW, Locksley RM. Early transcription and silencing of cytokine genes underlie polarization of T helper cell subsets. Immunity 2001;14:205–15.

[34] Bird JJ, Brown DR, Mullen AC, Moskowitz NH, Mahowald MA, Sider JR, et al. Helper T cell differentiation is controlled by the cell cycle. Immunity 1998;9:229–37.

[35] Santangelo S, Cousins DJ, Winkelmann NE, Staynov DZ. DNA methylation changes at human Th2 cytokine genes coincide with DNase I hypersensitive site formation during CD4(+) T cell differentiation. J Immunol 2002;169:1893–903.

[36] Floess S, Freyer J, Siewert C, Baron U, Olek S, Polansky J, et al. Epigenetic control of the foxp3 locus in regulatory T cells. PLoS Biol 2007;5:e38.
[37] Yao Z, Kanno Y, Kerenyi M, Stephens G, Durant L, Watford WT, et al. Nonredundant roles for Stat5a/b in directly regulating Foxp3. Blood 2007;109:4368–75.
[38] Kim HP, Leonard WJ. CREB/ATF-dependent T cell receptor-induced FoxP3 gene expression: a role for DNA methylation. J Exp Med 2007;204:1543–51.
[39] Polansky JK, Kretschmer K, Freyer J, Floess S, Garbe A, Baron U, et al. DNA methylation controls Foxp3 gene expression. Eur J Immunol 2008;38:1654–63.
[40] Janson PC, Winerdal ME, Martis P, Thorn M, Ohlsson R, Winqvist O. FOXP3 promoter demethylation reveals the committed Treg population in humans. PloS one 2008;3:e1612.
[41] Baron U, et al. DNA demethylation in the human FOXP3 locus discriminates regulatory T cells from activated FOXP3(+) conventional T cells. Eur J Immunol 2007;37:2378–89.
[42] Lal G, Bromberg JS. Epigenetic mechanisms of regulation of Foxp3 expression. Blood 2009;114:3727–35.
[43] Parra M. Epigenetic events during B lymphocyte development. Epigenetics 2009;4:462–8.
[44] Maier H, Ostraat R, Gao H, Fields S, Shinton SA, Medina KL, et al. Early B cell factor cooperates with Runx1 and mediates epigenetic changes associated with mb-1 transcription. Nat Immunol 2004;5:1069–77.
[45] Decker T, Pasca di Magliano M, McManus S, Sun Q, Bonifer C, Tagoh H, et al. Stepwise activation of enhancer and promoter regions of the B cell commitment gene Pax5 in early lymphopoiesis. Immunity 2009;30:508–20.
[46] Gray SM, Kaech SM, Staron MM. The interface between transcriptional and epigenetic control of effector and memory CD8(+) T-cell differentiation. Immunol Rev 2014;261:157–68.
[47] Pearce EL, Shen H. Making sense of inflammation, epigenetics, and memory $CD8^+$ T-cell differentiation in the context of infection. Immunol Rev 2006;211:197–202.
[48] Lu Q, Wu A, Ray D, Deng C, Attwood J, Hanash S, et al. DNA methylation and chromatin structure regulate T cell perforin gene expression. J Immunol 2003;170:5124–32.
[49] Janson PC, Marits P, Thorn M, Ohlsson R, Winqvist O. CpG methylation of the IFNG gene as a mechanism to induce immunosuppression in tumor-infiltrating lymphocytes. J Immunol 2008;181:2878–86.
[50] Jiang YH, Bressler J, Beaudet AL. Epigenetics and human disease. Annu Rev Genomics Hum Genet 2004;5:479–510.
[51] Martin GM. Epigenetic drift in aging identical twins. Proc Natl Acad Sci USA 2005;102:10413–14.
[52] Strickland FM, Richardson BC. Epigenetics in human autoimmunity. Epigenetics in autoimmunity - DNA methylation in systemic lupus erythematosus and beyond. Autoimmunity 2008;41:278–86.
[53] Javierre BM, Esteller M, Ballestar E. Epigenetic connections between autoimmune disorders and haematological malignancies. Trends Immunol 2008;29:616–23.
[54] Yung R, Powers D, Johnson K, Amento E, Carr D, Laing T, et al. Mechanisms of drug-induced lupus. II. T cells overexpressing lymphocyte function-associated antigen 1 become autoreactive and cause a lupuslike disease in syngeneic mice. J Clin Invest 1996;97:2866–71.
[55] Oelke K, Lu Q, Richardson D, Wu A, Deng C, Hanash S, et al. Overexpression of CD70 and overstimulation of IgG synthesis by lupus T cells and T cells treated with DNA methylation inhibitors. Arthritis Rheum 2004;50:1850–60.

[56] Lu Q, Wu A, Tesmer L, Ray D, Yousif N, Richardson B. Demethylation of CD40LG on the inactive X in T cells from women with lupus. J Immunol 2007;179:6352–8.

[57] Yin H, Zhao M, Wu X, Gao F, Luo Y, Ma L, et al. Hypomethylation and overexpression of CD70 (TNFSF7) in CD4^{+} T cells of patients with primary Sjogren's syndrome. J Dermatol Sci 2010;59:198–203.

[58] Javierre BM, Richardson B. A new epigenetic challenge: systemic lupus erythematosus. Adv Exp Med Biol 2011;711:117–36.

[59] Lu Q, Kaplan M, Ray D, Ray D, Zacharek S, Gutsch D, et al. Demethylation of ITGAL (CD11a) regulatory sequences in systemic lupus erythematosus. Arthritis Rheum 2002;46:1282–91.

[60] Mi XB, Zeng FQ. Hypomethylation of interleukin-4 and -6 promoters in T cells from systemic lupus erythematosus patients. Acta Pharmacol Sin 2008;29:105–12.

[61] Youinou P, Renaudineau Y. CD5 expression in B cells from patients with systemic lupus erythematosus. Crit Rev Immunol 2011;31:31–42.

[62] Garaud S, Le Dantec C, Jousse-Joulin S, Hanrotel-Saliou C, Saraux A, Mageed RA, et al. IL-6 modulates CD5 expression in B cells from patients with lupus by regulating DNA methylation. J Immunol 2009;182:5623–32.

[63] Renaudineau Y, Youinou P. Epigenetics and autoimmunity, with special emphasis on methylation. Keio J Med 2011;60:10–16.

[64] Garaud S, Youinou P, Renaudineau Y. DNA methylation and B-cell autoreactivity. Adv Exp Med Biol 2011;711:50–60.

[65] Thabet Y, Canas F, Ghedira I, Youinou P, Mageed RA, Renaudineau Y. Altered patterns of epigenetic changes in systemic lupus erythematosus and auto-antibody production: is there a link?. J Autoimmun 2012;39:154–60.

[66] Zhang Y, Zhao M, Sawalha AH, Richardson B, Lu Q. Impaired DNA methylation and its mechanisms in CD4(+)T cells of systemic lupus erythematosus. J Autoimmun 2013;41.

[67] Sanders VM. Epigenetic regulation of Th1 and Th2 cell development. Brain Behav Immun 2006;20:317–24.

CHAPTER

21

DNA Methylation in Stem Cell Diseases

OUTLINE

21.1 INTRODUCTION

Stem cells are undifferentiated cells that can differentiate into specialized cells and can divide to produce more stem cells. In mammals, there are two broad types of stem cells: embryonic stem cells, which are isolated from the inner cell mass of blastocysts, and adult stem cells, which are found in various tissues. In adult organisms, stem cells and progenitor cells act as a repair system for the body, replenishing adult tissues. In a developing embryo, stem cells are pluripotent and can differentiate into all the specialized cells (ectoderm, endoderm, and

M. Neidhart: DNA Methylation and Complex Human Disease.
DOI: http://dx.doi.org/10.1016/B978-0-12-420194-1.00021-X

mesoderm), but also maintain the normal turnover of regenerative organs, such as blood, skin, or intestinal tissues. The ability to culture stem cells and direct their differentiation into specific cell types *in vitro* provides a valuable experimental system for modeling pluripotency, development, and cellular differentiation. High-throughput profiling of the transcriptomes and epigenomes of pluripotent stem cells and their differentiated derivatives has led to identification of patterns characteristic of each cell type, discovery of new regulatory features in the epigenome, and early insights into the complexity of dynamic interactions among regulatory elements [1]. This chapter focuses on DNA methylation within the context of pluripotency and differentiation.

21.2 STEM CELL DIFFERENTIATION

Totipotent stem cells can differentiate into embryonic and extraembryonic cell types. These cells are produced from the fusion of an egg and sperm cell, and can construct a complete, viable organism. Cells produced by the first few divisions of the fertilized egg are also totipotent. In mouse development, after fertilization and before pronuclear fusion, zygotic genome activation, and first-cell division, the paternal genome is demethylated by enzymatic modification of 5-methylcytosine, whereas the maternal genome is passively demethylated during subsequent rounds of replication via nuclear exclusion of an oocyte-specific DNMT1 isoform [2]. Thus, the totipotent zygote is essentially devoid of DNA methylation except at imprinted regions. The genome is gradually remethylated during subsequent cleavage divisions that generate the morula and early blastocyst. The genomes of pluripotent stem cells derived from the inner cell mass/epiblast are highly methylated. Erasure of gametic methylation patterns and genomic remethylation in the developing zygote is necessary for specification of initial lineage commitment.

Embryonic stem cells are derived from the inner cell mass of a blastocyst, an early-stage embryo. These cells are pluripotent and give rise during development to all derivatives of the three primary germ layers: ectoderm, endoderm, and mesoderm. They can develop into each of the hundreds of cell types of the adult body under specific conditions. An embryonic stem cell is also defined by the expression of several transcription factors and cell surface proteins. The transcription factors OCT4, NANOG, and SOX2 form the core regulatory network that ensures the suppression of genes that lead to differentiation and the maintenance of pluripotency. Although embryonic stem cells express more mRNA species than differentiated cells, the global level of 5-methylcytosine at CpG sites is similar in embryonic stem cells and differentiated cells [3]. However, 5-methylcytosine at non-CpG sites is uniquely abundant in embryonic stem cells [4,5]. The non-CpG 5-methylcytosine is enriched in gene bodies and disappears on differentiation of embryonic stem cells; however, the functions of non-CpG 5-methylcytosine, including their connection to open chromatin, are largely unknown. New 5-methylcytosine is established by the *de novo* DNA methyltransferases DNMT3A and DNMT3B. However, DNA methylation is not required for pluripotency because the genomes of DNMT3A, DNMT3B, and DNMT1 triple-knockout mouse embryonal stem cells are hypomethylated, yet the cells retain self-renewal and pluripotency,but exhibit a host of differentiation defects [6]. This suggests that establishment of zygotic methylation patterns is crucial for germ layer specification and lineage commitment. Similarly, to evaluate their role in hematopoiesis, mice with conditional knockouts of DNMTs have been generated and demonstrate the importance of DNA methylation in the hematopoietic stem cell compartment. Specifically, loss of DNMT1 in hematopoietic stem cells leads to dysregulation of lineage output, with a skewing toward myelopoiesis, and defects in self-renewal [7,8], while a conditional knockout of DNMT3A alone drives a loss in differentiation potential after serial transplant [9], and loss of both DNMT3A and DNMT3B leads to an even more severe arrest of hematopoietic stem cell differentiation [10].

Adult (somatic) stem cells maintain and repair the tissue in which they are localized. Pluripotent somatic stem cells are rare and generally small in number, but they can be found in umbilical cord blood (mesenchymal stem cells) and other tissues, such as the bone marrow (hematopoietic stem cells), endothelium, adipose tissue, dental pulp, etc. Most adult stem cells are multipotent (i.e., lineage-restricted). The DNA methylation machinery works in conjunction with other modes of epigenetic regulation to regulate gene expression via local chromatin structure and higher order genomic topology. Genes regulated by methylation usually contain a low density of promoter CpG sites. Most low CpG-density promoters are methylated in embryonal stem cells and subsequently demethylated in somatic stem cells and expressed in a lineage or cell-type specific manner during differentiation [11–13]. Areas termed CpG islands, which are regions of high CpG density found within or near proximal promoters or transcription start sites, are typically devoid of DNA methylation. Often, there is a progressive loss of methylation as differentiation proceeds [5,13]. Promoter methylation and silencing of the pluripotency regulators OCT4 and NANOG occurs during lineage specification, and specifid HOX gene clusters become progressively methylated [14]. Inhibition of DNA methylation has been shown to result in strong induction of cardiomyocyte differentiation [15]. In addition, a

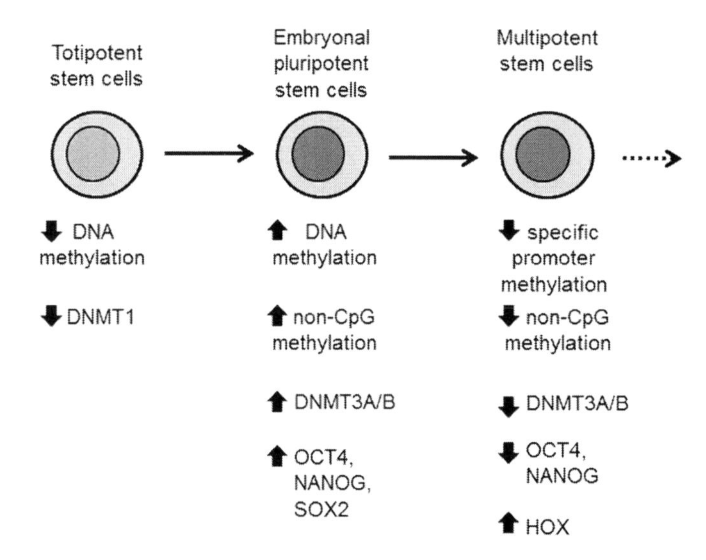

FIGURE 21.1 Role of DNA methylation in the development of stem cells.

subset of genes involved in cardiac structure undergoes cardiomyocyte-specific demethylation during development [16,17].

Figure 21.1 illustrates the role of DNA methylation in the development of stem cells.

21.3 INDUCED PLURIPOTENT STEM CELLS

Somatic cells can be reprogrammed into induced pluripotent stem cells (iPSCs) by the transient expression of reprogramming factors (i.e., OCT3/4, SOX2, c-MYC, and KLF4). During the reprogramming process, somatic cells acquire the ability to undergo unlimited proliferation, which is also an important characteristic of cancer cells, while their underlying DNA sequence remains unchanged [18]. The fact that the DNA methylation status at imprinted loci is associated with the quality of iPSCs supports the notion that proper remodeling of epigenetic modifications is essential to achieve successful reprogramming [19].

Genomic imprints are established in the male and female germline when the parental alleles can be independently marked. As mentioned above, this establishment occurs after imprint erasure and is part of the widespread epigenetic reprogramming and genome-wide demethylation that is essential for totipotency [2,20,21]. However, it is now known that at least some of the germline-specific reprogramming events can be bypassed during reprogramming when iPSCs are derived from differentiated somatic cells [22].

An initial report describing the reprogramming of mouse embryonic fibroblasts to a pluripotent state showed that several imprinted genes (H19, PEG1, PEG3, and SNRPN) maintained proper allele-specific DNA methylation after reprogramming [23]. Subsequent studies characterizing imprinting during the induction of human iPSCs showed that loss of imprinting is an exceedingly rare but observable event that is evident at early stages in the reprogramming process, is highly cell-line specific, and is maintained through multiple passages [24,25]. Interestingly, maintenance of the state of imprinting is also evident in iPSCs generated from fibroblasts of patients with Angelman syndrome or Prader-Willi syndrome, with pathological errors in imprinting and expression being retained through reprogramming and subsequent culture [26]. Thus, it appears that imprints present in the somatic cell of origin are, for the most part, faithfully retained in iPSCs after reprogramming [1].

A significant and functionally crucial exception to these trends involves errors during iPSC reprogramming at the DIK1–DIO3 imprinted cluster. Hypermethylation across this cluster causes aberrant repression and is dependent on the inappropriate recruitment of DNMT3A [27]. Recent follow-up work [28] has reported that treatment of iPSCs with ascorbic acid (vitamin C) during passage and reprogramming ensures the maintenance of euchromatic marks across this cluster; ascorbic acid treatment inhibits the recruitment of DNMT3A by a highly specific but unknown mechanism, and increases iPSC pluripotency and reprogramming efficiency.

21.4 ENVIRONMENTAL FACTORS AND AGING

Certain epigenetic factors are thought to play a role in mediating aging. Changes in DNA methylation and histone modifications were investigated as markers of aging. Four hypermethylated CpG sites (associated with genes NPTX2, TRIM58, GRIA2, and KCNQ1DN) have been identified, and an additional hypomethylated CpG site (BIRC4BP) was found to be an epigenetic aging signature across different tissues [29]. To address whether aged hematopoietic stem cells have altered methylation patterns that contribute to changes in their functional potential, recent works have looked at DNA methylation profiles in fetal, young, and old murine hematopoietic stem cells [30–32], as well as in human progenitor cells [33]. These studies have shown locus-specific differences in DNA methylation profiles associated with aging of the hematopoietic stem cells compartment, with some regions gaining methylation whereas other regions become more hypomethylated.

Hypermethylated regions are enriched for targets of the Polycomb Repressive Complex 2 (PRC2), which establish the repressive H3K27me3 histone modification, suggesting an interplay between these two epigenetic marks. In these cells, the age-associated diminution of expression of PRC2 core components histone-lysine N-methyltransferase (EZH2), and Polycomb proteins EED and SUZ12, correspond with increased DNA methylation at targets of the complex, suggesting that loss of repression by PRC2 binding at selected targets allows these regions to become accessible to *de novo* DNA methylation [34].

21.5 DISRUPTION OF DIFFERENTIATION AND DEVELOPMENT

21.5.1 Epithelial to Mesenchymal Transition

Endothelial stem cells are one of three types of multipotent stem cells found in bone marrow. They give rise to progenitor cells, which are intermediate stem cells that lose potency, and eventually produce endothelial cells, which create the thin-walled endothelium that lines the inner surface of blood vessels and lymphatic vessels. The epithelial–mesenchymal transition (EMT) is a process by which epithelial cells lose their cell polarity and cell–cell adhesion and gain migratory and invasive properties to become mesenchymal stem cells; these are multipotent stromal cells that can differentiate into a variety of cell types. EMT is essential for numerous developmental processes, including mesoderm, neural tube formation, and wound healing. A growing body of evidence suggests that EMT plays a central role during tumor metastasis and frequently imparts a stem cell-like phenotype and therapeutic resistance to tumor cells. The induction of EMT is accompanied by a dynamic reprogramming of the epigenome involving changes in DNA methylation and several post-translational histone modifications [35]. These changes in turn promote the expression of mesenchymal genes or repress those associated with an epithelial phenotype. EMT of cancer cells is partly under reversible epigenetic control [36]. Differences between cell types are guided by the expression of tissue-specific transcription factors and consolidation of associated epigenetic states. Therefore, the epigenome of a cancer cell is determined in part by the cell of origin for that cancer and includes passenger hypermethylation events at genes not required in that particular lineage [37]. In order for metastatic colonization and the formation of macrometastases to occur, tumor cells frequently undergo a reversal of EMT referred to as the mesenchymal-to-epithelial transition (MET). Thus, a high degree of epigenetic plasticity is required in order to induce and reverse EMT during tumor progression [35].

21.5.2 Hematopoietic Stem Cell Compartment

Hematopoietic stem cells are also derived from the mesoderm, located in the bone marrow, and give rise to all blood cells: the myeloid (including monocytes, granulocytes, erythrocytes, and platelets) and lymphoid lineages (T, B, and NK cells). Age-associated phenotypes of the hematopoietic system include increased incidence of many hematological malignancies encompassing leukemias (AML, CML, CLL, CMML), myelodysplastic syndromes (MDS), myeloproliferative disorders (CMD), myeloma, and lymphomas (Hodgkin's and non-Hodgkin's). Interestingly, many of the mutations found in the hematopoietic stem cell compartment leading to malignant transformation are genes involved in epigenetic regulation, such as DNMT3A and EZH2, further implicating the importance of maintaining faithful epigenetic control in hematopoietic stem cells [34,38]. Thus, the DNMT3A gene is commonly mutated in human cases of AML [39,40], suggesting that epigenetic perturbation can lead to differentiation block and subsequent malignant transformation. Furthermore, aberrant DNA methylation patterns play an important role in the emergence and progression of MDS, a prototypic disease of the elderly. The aberrant methylation profiles of hematopoietic stem cells isolated from MDS patients [41], together with the many studies also showing that key epigenetic regulators are mutated in those patients [42], implicate methyl-silencing as a dominant mechanism in MDS. Interestingly, while in MDS the loss of epigenetic fidelity is typically manifest as hypermethylation, the progression to AML is typically associated with global hypomethylation [43]. One potential mechanism behind this loss of methylation that is leading to leukemia might be attributed to dysregulation of DNMT1. However, loss of DNMT1 in a mouse model of MLL-AF9-induced AML led to a reduction in leukemic progression and appears to have little effect on normal hematopoiesis [44].

It has long been debated whether cancer cells arise by dedifferentiation or instead originate from stem cells or early progenitors by a differentiation block. Polycomb repressors mark genes in stem cells encoding master regulators of differentiation and development, poised either to be turned on to coordinate differentiation of a lineage or to be fully repressed if not needed in that particular lineage [45]. These genes occupied by Polycomb repressors in stem cells are particularly prone to acquiring CpG island hypermethylation during cell proliferation, aging, and, particularly, malignant transformation [46]. Although the genes affected by this process are primarily those not required or expressed in that particular cell lineage, cancer cells do also show evidence of silencing of genes essential for differentiation of their cell of origin [46–48]. This predisposition of Polycomb target genes to aberrant permanent epigenetic silencing is consistent with a model in which stem cells slowly acquire irreversible silencing of poised master regulators required for successful differentiation. As a consequence, some stem cells lose their ability to properly differentiate while retaining their self-renewal capabilities and become attractive candidates for malignant transformation by subsequent genetic and epigenetic events. One provocative implication of this model is that the first steps of oncogenesis may in some cases be an epigenetic defect affecting the differentiation capabilities of stem cells, as opposed to a gatekeeper mutation.

Global DNA hypomethylation is not the only aberrant methylation profile associated with cancers, as various cancers also demonstrate DNA hypermethylation at specific promoter-associated regions [49]. These regions of increased DNA methylation associated with cancer are typically over-represented for targets of the Polycomb group [50], similar to what is seen in aged hematopoietic stem cells. These Polycomb targets are largely key regulators of development, and are repressed to maintain plasticity in embryonic stem cells [51]. Dysregulation or permanent silencing of typically plastic regions that regulate stem cells' choice between differentiation and self-renewal due to aberrant accumulation of DNA methylation may inhibit differentiation and lead to a self-renewing population that could serve as a pool for mutation accrual [50].

21.5.3 Mesenchymal Stem Cell Compartment

Mesenchymal stem cells are multipotent stromal cells that can differentiate into a variety of cell types, including fibroblasts, osteoblasts, chondrocytes, and adipocytes. Both DNA methylation and histone modification appear to be important factors for tissue- and cell-specific differentiation, specifically osteoblastic [52], chondrogenic [53,54], and adipocytic [55] differentiation. The differentiation of osteoblasts and osteoclasts is accompanied by profound changes in gene expression. It has been shown that DNA methylation negatively regulates the expression of several genes associated with different stages of osteoblast differentiation, such as the RANKL/OPG system [52]. Importantly, epigenetic mechanisms during cartilage development and onset of joint diseases have potential value in the treatment of degenerative joint diseases [56,57]. The mechanical environment plays a definitive role in regulating chondrogenesis of mensenchymal stem cells [58]. Mechanical stimulation can alter the epigenetic state by reducing DNA methylation, thereby resulting in the upregulation of osteogenic gene expression [59]. As mentioned above, OCT4, NANOG, and SOX2 are three transcription factors essential for the maintenance of stem cell pluripotency. The expression of NANOG and SOX2 is inversely correlated with the DNA methylation pattern in the promoter region of equine bone marrow-derived mesenchymal stem cells [60]. Compared with embryonal stem cells, mensenchymal stem cells had more methylations on NANOG and OCT4 promoters, suggesting that pluripotency was restricted in the later cells [54]. Cell type-specific DNA methylation patterns are thought to be established prior to the terminal differentiation of adult progenitor cells. Demethylation of the COL10A1 promoter correlated with the induction of type X collagen during mesenchymal stem cell-derived chondrogenesis [61]. The removal of DNA methylation by 5-azacytidine treatment altered the myogenic lineage commitment of C2C12 myoblasts and induced spontaneous osteogenic and adipogenic differentiation [62]. Similarly, another DNA demethylating agent also stimulated osteogenic differentiation in human bone marrow-derived mesenchymal stem cells [63]. SOX genes (i.e., SOX5, SOX6, and SOX9) play an essential role in chondrogenic differentiation. During chondrogenic induction of human synovium-derived stem cells, DNA methylation levels of CpG-rich promoters related to chondrocyte phenotypes are largely hypomethylated [64]. Correspondingly, increased methylation in promoter regions of SOX5 and SOX9 genes explains the low expression of these respective genes in a surgically induced rat osteoarthritis model [65]. In patients with osteoarthritis, demethylation of the MMP13 promoter in chondrocytes was found to be responsible for the increase in expression of this matrix-degrading enzyme [66]. Control of this enzyme may have analogous effects in deterring the progression of osteoarthritis. Glucosamine is a commonly prescribed drug for alleviating OA, although the mechanism of action behind its therapeutic efficacy is not yet clear. However, it was demonstrated that glucosamine and an NF-κB inhibitor prevent cytokine-induced demethylation of a specific CpG site in the IL-1β promoter, resulting in decreased expression of this pro-inflammatory cytokine [67]. In rheumatoid arthritis, multipotent synovial fluid fibroblast-like cells are found that have the capacity to destroy the joint [68,69].

21.6 CONCLUSION AND OUTLOOK

There is an expanding base of knowledge shedding light on the role of epigenetics and stem cell regulation; however, the complicated nature of the interactions between these mechanisms (DNA methylation, histone modifications, and non-coding RNA) and the implications of dysregulation of these interactions is just beginning to be understood [34]. Although there may be some universal facts regarding the interplay between these epigenetic regulators, there appear to be many unique interactions that may ultimately be cell-type specific. Developments in technology will allow for smaller numbers of cells to be investigated for these epigenetic marks, and provide more information as to steady-state epigenetic regulations of cell populations with limited numbers, such as somatic stem cells. Ultimately, epigenetic modifications are not permanent, raising the prospect that re-writing the epigenetic marks of aging could be used to alter the potential of somatic stem cells from the elderly. This idea was tested in a recent study [70] showing that aged hematopoietic progenitors reprogrammed to induced pluripotent (iPS) cells and then re-differentiated to blood cells no longer exhibited the hallmark functional features of aged stem cells. Recently, it has become obvious that joint diseases such as osteoarthritis and rheumatoid arthritis are at least in part caused by deregulation of the development and function of mesenchymal stem cells [71–73]. Examples of changes in DNA methylation and histone modification occurring in synovial fibroblasts during the disease process are presented in Chapter 22.

References

[1] Thiagarajan RD, Morey R, Laurent LC. The epigenome in pluripotency and differentiation. Epigenomics 2014;6:121–37.
[2] Li E. Chromatin modification and epigenetic reprogramming in mammalian development. Nat Rev Genet 2002;3:662–73.
[3] Meissner A, Mikkelsen TS, Gu H, Wernig M, Hanna J, Sivachenko A, et al. Genome-scale DNA methylation maps of pluripotent and differentiated cells. Nature 2008;454:766–70.
[4] Lister R, Pelizzola M, Dowen RH, Hawkins RD, Hon G, Tonti-Filippini J, et al. Human DNA methylomes at base resolution show widespread epigenomic differences. Nature 2009;462:315–22.
[5] Laurent L, Wong E, Li G, Huynh T, Tsirigos A, Ong CT, et al. Dynamic changes in the human methylome during differentiation. Genome Res 2010;20:320–31.
[6] Tsumura A, Hayakawa T, Kumaki Y, Takebayashi S, Sakaue M, Matsuoka C, et al. Maintenance of self-renewal ability of mouse embryonic stem cells in the absence of DNA methyltransferases Dnmt1, Dnmt3a and Dnmt3b. Genes Cells 2006;11:805–14.
[7] Broske AM, Vockentanz L, Kharazi S, Huska MR, Mancini E, Scheller M, et al. DNA methylation protects hematopoietic stem cell multipotency from myeloerythroid restriction. Nat Genet 2009;41:1207–15.
[8] Trowbridge JJ, Snow JW, Kim J, Orkin SH. DNA methyltransferase 1 is essential for and uniquely regulates hematopoietic stem and progenitor cells. Cell Stem Cell 2009;5:442–9.
[9] Challen GA, Sun D, Jeong M, Luo M, Jelinek J, Berg JS, et al. Dnmt3a is essential for hematopoietic stem cell differentiation. Nat Genet 2012;44:23–31.

[10] Challen GA, Sun D, Mayle A, Jeong M, Luo M, Rodriguez B, et al. Dnmt3a and dnmt3b have overlapping and distinct functions in hematopoietic stem cells. Cell Stem Cell 2014;15:350–64.
[11] Fouse SD, Shen Y, Pellegrini M, Cole S, Meissner A, Van Neste L, et al. Promoter CpG methylation contributes to ES cell gene regulation in parallel with Oct4/Nanog, PcG complex, and histone H3 K4/K27 trimethylation. Cell Stem Cell 2008;2:160–9.
[12] Mikkelsen TS, Ku M, Jaffe DB, Issac B, Lieberman E, Giannoukos G, et al. Genome-wide maps of chromatin state in pluripotent and lineage-committed cells. Nature 2007;448:553–60.
[13] Nazor KL, Altun G, Lynch C, Tran H, Harness JV, Slavin I, et al. Recurrent variations in DNA methylation in human pluripotent stem cells and their differentiated derivatives. Cell Stem Cell 2012;10:620–34.
[14] Boland MJ, Nazor KL, Loring JF. Epigenetic regulation of pluripotency and differentiation. Circ Res 2014;115:311–24.
[15] Abbey D, Seshagiri PB. Aza-induced cardiomyocyte differentiation of P19 EC-cells by epigenetic co-regulation and ERK signaling. Gene 2013;526:364–73.
[16] Gu Y, Liu GH, Plongthongkum N, Benner C, Yi F, Qu J, et al. Global DNA methylation and transcriptional analyses of human ESC-derived cardiomyocytes. Protein Cell 2013.
[17] Chaturvedi P, Tyagi SC. Epigenetic mechanisms underlying cardiac degeneration and regeneration. Int J Cardiol 2014;173:1–11.
[18] Ohnishi K, Semi K, Yamada Y. Epigenetic regulation leading to induced pluripotency drives cancer development in vivo. Biochem Biophys Res Commun 2014.
[19] Stadtfeld M, Apostolou E, Akutsu H, Fukuda A, Follett P, Natesan S, et al. Aberrant silencing of imprinted genes on chromosome 12qF1 in mouse induced pluripotent stem cells. Nature 2010;465:175–81.
[20] Surani MA, Hayashi K, Hajkova P. Genetic and epigenetic regulators of pluripotency. Cell 2007;128:747–62.
[21] Hajkova P, Ancelin K, Waldmann T, Lacoste N, Lange UC, Cesari F, et al. Chromatin dynamics during epigenetic reprogramming in the mouse germ line. Nature 2008;452:877–81.
[22] Plasschaert RN, Bartolomei MS. Genomic imprinting in development, growth, behavior and stem cells. Development 2014;141:1805–13.
[23] Wernig M, Meissner A, Foreman R, Brambrink T, Ku M, Hochedlinger K, et al. In vitro reprogramming of fibroblasts into a pluripotent ES-cell-like state. Nature 2007;448:318–24.
[24] Pick M, Stelzer Y, Bar-Nur O, Mayshar Y, Eden A, Benvenisty N. Clone- and gene-specific aberrations of parental imprinting in human induced pluripotent stem cells. Stem Cells 2009;27:2686–90.
[25] Hiura H, Toyoda M, Okae H, Sakurai M, Miyauchi N, Sato A, et al. Stability of genomic imprinting in human induced pluripotent stem cells. BMC Genet 2013;14:32.
[26] Chamberlain SJ, Chen PF, Ng KY, Bourgois-Rocha F, Lemtiri-Chlieh F, Levine ES, et al. Induced pluripotent stem cell models of the genomic imprinting disorders Angelman and Prader-Willi syndromes. Proc Natl Acad Sci USA 2010;107:17668–73.
[27] Carey BW, Markoulaki S, Hanna JH, Faddah DA, Buganim Y, Kim J, et al. Reprogramming factor stoichiometry influences the epigenetic state and biological properties of induced pluripotent stem cells. Cell Stem Cell 2011;9:588–98.
[28] Stadtfeld M, Apostolou E, Ferrari F, Choi J, Walsh RM, Chen T, et al. Ascorbic acid prevents loss of Dlk1–Dio3 imprinting and facilitates generation of all-iPS cell mice from terminally differentiated B cells. Nat Genet 2012;44: 398–405, S391–392
[29] Koch CM, Wagner W. Epigenetic-aging signature to determine age in different tissues. Aging 2011;3:1018–27.

[30] Beerman I, Bock C, Garrison BS, Smith ZD, Gu H, Meissner A, et al. Proliferation-dependent alterations of the DNA methylation landscape underlie hematopoietic stem cell aging. Cell Stem Cell 2013;12:413–25.
[31] Sun D, Luo M, Jeong M, Rodriguez B, Xia Z, Hannah R, et al. Epigenomic profiling of young and aged HSCs reveals concerted changes during aging that reinforce self-renewal. Cell Stem Cell 2014;14:673–88.
[32] Taiwo O, Wilson GA, Emmett W, Morris T, Bonnet D, Schuster E, et al. DNA methylation analysis of murine hematopoietic side population cells during aging. Epigenetics 2013;8:1114–22.
[33] Bocker MT, Hellwig I, Breiling A, Eckstein V, Ho AD, Lyko F. Genome-wide promoter DNA methylation dynamics of human hematopoietic progenitor cells during differentiation and aging. Blood 2011;117:e182–9.
[34] Beerman I, Rossi DJ. Epigenetic regulation of hematopoietic stem cell aging. Exp Cell Res 2014;329:192–9.
[35] Bedi U, Mishra VK, Wasilewski D, Scheel C, Johnsen SA. Epigenetic plasticity: a central regulator of epithelial-to-mesenchymal transition in cancer. Oncotarget 2014;5:2016–29.
[36] De Craene B, Berx G. Regulatory networks defining EMT during cancer initiation and progression. Nat Rev Cancer 2013;13:97–110.
[37] Sproul D, Kitchen RR, Nestor CE, Dixon JM, Sims AH, Harrison DJ, et al. Tissue of origin determines cancer-associated CpG island promoter hypermethylation patterns. Genome Biol 2012;13:R84.
[38] Shih AH, Abdel-Wahab O, Patel JP, Levine RL. The role of mutations in epigenetic regulators in myeloid malignancies. Nat Rev Cancer 2012;12:599–612.
[39] Ley TJ, Ding L, Walter MJ, McLellan MD, Lamprecht T, Larson DE, et al. DNMT3A mutations in acute myeloid leukemia. N Engl J Med 2010;363:2424–33.
[40] Yan XJ, Xu J, Gu ZH, Pan CM, Lu G, Shen Y, et al. Exome sequencing identifies somatic mutations of DNA methyltransferase gene DNMT3A in acute monocytic leukemia. Nat Genet 2011;43:309–15.
[41] Will B, Zhou L, Vogler TO, Ben-Neriah S, Schinke C, Tamari R, et al. Stem and progenitor cells in myelodysplastic syndromes show aberrant stage-specific expansion and harbor genetic and epigenetic alterations. Blood 2012;120:2076–86.
[42] Itzykson R, Fenaux P. Epigenetics of myelodysplastic syndromes. Leukemia 2014; 28:497–506.
[43] Feinberg AP, Vogelstein B. Hypomethylation distinguishes genes of some human cancers from their normal counterparts. Nature 1983;301:89–92.
[44] Trowbridge JJ, Sinha AU, Zhu N, Li M, Armstrong SA, Orkin SH. Haploinsufficiency of Dnmt1 impairs leukemia stem cell function through derepression of bivalent chromatin domains. Genes Dev 2012;26:344–9.
[45] Bernstein BE, Mikkelsen TS, Xie X, Kamal M, Huebert DJ, Cuff J, et al. A bivalent chromatin structure marks key developmental genes in embryonic stem cells. Cell 2006;125:315–26.
[46] Teschendorff AE, Menon U, Gentry-Maharaj A, Ramus SJ, Weisenberger DJ, Shen H, et al. Age-dependent DNA methylation of genes that are suppressed in stem cells is a hallmark of cancer. Genome Res 2010;20:440–6.
[47] Berman BP, Weisenberger DJ, Aman JF, Hinoue T, Ramjan Z, Liu Y, et al. Regions of focal DNA hypermethylation and long-range hypomethylation in colorectal cancer coincide with nuclear lamina-associated domains. Nat Genet 2012;44:40–6.
[48] Easwaran H, Johnstone SE, Van Neste L, Ohm J, Mosbruger T, Wang Q, et al. A DNA hypermethylation module for the stem/progenitor cell signature of cancer. Genome Res 2012;22:837–49.
[49] Esteller M. Relevance of DNA methylation in the management of cancer. Lancet Oncol 2003;4:351–8.

[50] Widschwendter M, Fiegl H, Egle D, Mueller-Holzner E, Spizzo G, Marth C, et al. Epigenetic stem cell signature in cancer. Nat Genet 2007;39:157–8.

[51] Boyer LA, Plath K, Zeitlinger J, Brambrink T, Medeiros LA, Lee TI, et al. Polycomb complexes repress developmental regulators in murine embryonic stem cells. Nature 2006;441:349–53.

[52] Delgado-Calle J, Sanudo C, Fernandez AF, Garcia-Renedo R, Fraga MF, Riancho JA. Role of DNA methylation in the regulation of the RANKL–OPG system in human bone. Epigenetics 2012;7:83–91.

[53] Herlofsen SR, Byrne JC, Hoiby T, Wang L, Issner R, Zhang X, et al. Genome-wide map of quantified epigenetic changes during in vitro chondrogenic differentiation of primary human mesenchymal stem cells. BMC Genomics 2013;14:105.

[54] Yannarelli G, Pacienza N, Cuniberti L, Medin J, Davies J, Keating A. Brief report: The potential role of epigenetics on multipotent cell differentiation capacity of mesenchymal stromal cells. Stem Cells 2013;31:215–20.

[55] Pinnick KE, Karpe F. DNA methylation of genes in adipose tissue. Proc Nutr Soc 2011;70:57–63.

[56] Barter MJ, Bui C, Young DA. Epigenetic mechanisms in cartilage and osteoarthritis: DNA methylation, histone modifications and microRNAs. Osteoarthritis Cartilage 2012;20:339–49.

[57] Karouzakis E, Gay RE, Gay S, Neidhart M. Epigenetic deregulation in rheumatoid arthritis. Adv Exp Med Biol 2011;711:137–49.

[58] Kelly DJ, Jacobs CR. The role of mechanical signals in regulating chondrogenesis and osteogenesis of mesenchymal stem cells. Birth Defects Res C Embryo Today 2010;90:75–85.

[59] Arnsdorf EJ, Tummala P, Castillo AB, Zhang F, Jacobs CR. The epigenetic mechanism of mechanically induced osteogenic differentiation. J Biomech 2010;43:2881–6.

[60] Hackett CH, Greve L, Novakofski KD, Fortier LA. Comparison of gene-specific DNA methylation patterns in equine induced pluripotent stem cell lines with cells derived from equine adult and fetal tissues. Stem Cells Dev 2012;21:1803–11.

[61] Zimmermann P, Boeuf S, Dickhut A, Boehmer S, Olek S, Richter W. Correlation of COL10A1 induction during chondrogenesis of mesenchymal stem cells with demethylation of two CpG sites in the COL10A1 promoter. Arthritis Rheum 2008;58:2743–53.

[62] Hupkes M, van Someren EP, Middelkamp SH, Piek E, van Zoelen EJ, Dechering KJ. DNA methylation restricts spontaneous multi-lineage differentiation of mesenchymal progenitor cells, but is stable during growth factor-induced terminal differentiation. Biochim Biophys Acta 2011;1813:839–49.

[63] El-Serafi AT, Oreffo RO, Roach HI. Epigenetic modifiers influence lineage commitment of human bone marrow stromal cells: Differential effects of 5-aza-deoxycytidine and trichostatin A. Differentiation 2011;81:35–41.

[64] Ezura Y, Sekiya I, Koga H, Muneta T, Noda M. Methylation status of CpG islands in the promoter regions of signature genes during chondrogenesis of human synovium-derived mesenchymal stem cells. Arthritis Rheum 2009;60:1416–26.

[65] Kim SY, Im GI. The expressions of the SOX trio, PTHrP (parathyroid hormone-related peptide)/IHH (Indian hedgehog protein) in surgically induced osteoarthritis of the rat. Cell Biol Int 2011;35:529–35.

[66] Bui C, Barter MJ, Scott JL, Xu Y, Galler M, Reynard LN, et al. cAMP response element-binding (CREB) recruitment following a specific CpG demethylation leads to the elevated expression of the matrix metalloproteinase 13 in human articular chondrocytes and osteoarthritis. FASEB J 2012;26:3000–11.

[67] Imagawa K, de Andres MC, Hashimoto K, Pitt D, Itoi E, Goldring MB, et al. The epigenetic effect of glucosamine and a nuclear factor-kappa B (NF-κB) inhibitor on

primary human chondrocytes – implications for osteoarthritis. Biochem Biophys Res Commun 2011;405:362–7.

[68] Neidhart M, Seemayer CA, Hummel KM, Michel BA, Gay RE, Gay S. Functional characterization of adherent synovial fluid cells in rheumatoid arthritis: destructive potential *in vitro* and *in vivo*. Arthritis Rheum 2003;48:1873–80.

[69] Jones EA, English A, Henshaw K, Kinsey SE, Markham AF, Emery P, et al. Enumeration and phenotypic characterization of synovial fluid multipotential mesenchymal progenitor cells in inflammatory and degenerative arthritis. Arthritis Rheum 2004;50:817–27.

[70] Wahlestedt M, Norddahl GL, Sten G, Ugale A, Frisk MA, Mattsson R, et al. An epigenetic component of hematopoietic stem cell aging amenable to reprogramming into a young state. Blood 2013;121:4257–64.

[71] El-Jawhari JJ, El-Sherbiny YM, Jones EA, McGonagle D. Mesenchymal stem cells, autoimmunity and rheumatoid arthritis. QJM 2014;107:505–14.

[72] Jones E, Churchman SM, English A, Buch MH, Horner EA, Burgoyne CH, et al. Mesenchymal stem cells in rheumatoid synovium: enumeration and functional assessment in relation to synovial inflammation level. Ann Rheum Dis 2010;69:450–7.

[73] Mohanty ST, Kottam L, Gambardella A, Nicklin MJ, Coulton L, Hughes D, et al. Alterations in the self-renewal and differentiation ability of bone marrow mesenchymal stem cells in a mouse model of rheumatoid arthritis. Arthritis Res Ther 2010;12:R149.

CHAPTER 22

DNA Methylation and Rheumatology

OUTLINE

22.1 INTRODUCTION

Over the past few years, the dominant role of T cells in autoimmunity and systemic lupus erythematosus has been established. As discussed in Chapter 20, the field of epigenetics research has provided us with new insights into the global DNA hypomethylation in lupus T cells and its consequences [1,2]. In Chapter 21 we mentioned that, in osteoarthritis, alterations in the development of chondrocytes from mesenchymal stem cells can occur, especially regarding the control

M. Neidhart: DNA Methylation and Complex Human Disease.
DOI: http://dx.doi.org/10.1016/B978-0-12-420194-1.00022-1

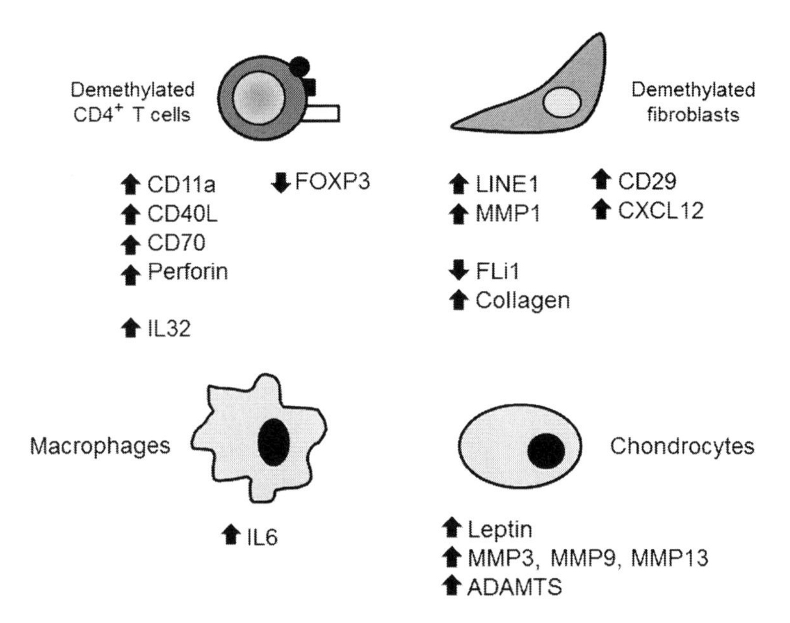

FIGURE 22.1 Several cell types are affected by epigenetic alterations in rheumatic diseases, and multiple genes are affected by changes in DNA methylation.

of specific collagen genes by promoter DNA methylation [3]. In rheumatoid arthritis, synovial fibroblasts that are also derived from mesenchymal stem cells can give rise to progeny termed "transformed-like" synoviocytes, which are key players in the perpetuation of joint inflammation and destruction [4,5]. This is in great part due to a global DNA hypomethylation [6]. In systemic sclerosis, both T cells and dermal fibroblasts show epigenetic changes [7]. The fact that fibroblasts taken from the joints of patients with rheumatoid arthritis or from the skin of patients with systemic sclerosis kept their pathological characteristics over several passages *in vitro* suggests the involvement of epigenetic modifications. In the present chapter, we present an overview of the consequences of a changed DNA methylation in rheumatic diseases (Figure 22.1).

22.2 SYSTEMIC LUPUS ERYTHEMATOSUS

Systemic lupus erythematosus is an autoimmune connective tissue disease that can affect any part of the body. The disease does run in families, but no single causal gene has been identified. Instead, multiple genes appear to influence a person's chance of developing lupus when triggered by environmental factors. Thus, alterations in the epigenome have been implicated in both idiopathic and drug-induced lupus. Lower levels of genomic DNA methylation have been reported in peripheral T cells from lupus patients compared with healthy controls [2,8], and adoptive transfer of T cells that have been treated with the DNMT1 inhibitor 5-aza-2′-deoxycytidine induces a lupus-like condition in syngeneic mice. Incubation of human T cells with this agent results in alterations in gene expression similar to those found in idiopathic lupus, including upregulation of LFA-1/CD11a. It is important to note that drugs associated with the development of iatrogenic lupus, procainamide and hydralazine, have been shown to be functional inhibitors of DNMT1, potentially resulting in DNA hypomethylation [9,10] (for further details, see Chapter 20).

22.3 SYSTEMIC SCLEROSIS

Systemic sclerosis (scleroderma) is another autoimmune connective tissue disease. It is characterized by thickening of the skin caused by accumulation of collagen, and by injuries to the smallest arteries. Similarly to the findings in systemic lupus erythematosus, $CD4^+$ T cells from patients with systemic sclerosis show a general reduction in the levels of DNA methylation associated with a concurrent decreased expression of methylation-regulating genes such as DNMT1, MBD3, and MBD4 [11]. The promoter of CD70 is substantially hypomethylated and probably contributes to the overexpression of CD70 by $CD4^+$ T cells in patients [12]. In addition, $CD4^+$ T cells from patients with systemic sclerosis have high expression of CD40L; this gene is located on the X chromosome and is thought to be important in the pathogenesis of systemic sclerosis, given that its blockade attenuates skin fibrosis and autoimmunity in the tight-skin mouse model [13]. CD40L upregulation specifically occurred in female patients with systemic sclerosis and correlated with hypomethylation of its promoter, suggesting that the altered methylation pattern of this gene promoter contributes to the increased susceptibility of women to the disease [14]. Supporting this hypothesis, the DNA methylation profile of peripheral blood mononuclear cells from monozygotic twins discordant for systemic sclerosis showed that only X chromosome sites were consistently either hypermethylated or hypomethylated [15]. Another report indicated that increased DNA methylation of regulatory sequences in FOXP3 from $CD4^+$ T cells from patients with systemic sclerosis affected the expression of this key transcription factor, which is required for the generation of regulatory T cells.

A central feature of systemic sclerosis is tissue fibrosis mediated by interstitial fibroblasts. These cells have an altered phenotype characterized by excessive deposition of extracellular matrix proteins, including collagens. The maintenance of this phenotype *in vitro* has been correlated with higher levels of DNMT1 [16,17]. In contrast to $CD4^+$ T cells, altered DNA methylation has been more robustly demonstrated in fibroblasts from patients with systemic sclerosis. Skin biopsy samples from patients with systemic sclerosis and derivative fibroblasts showed considerably higher levels of the methylation-regulating genes DNMT1, MBD1, and MECP2 than samples from healthy controls, indicative of global DNA hypermethylation in fibroblasts from those patients [16]. Treatment of patient dermal fibroblasts with with 5-aza-2′-deoxycytidine results in reduced expression of collagen, but has no effect on fibroblasts from healthy controls. A major suppressor of collagen transcription is FLi1, which is downregulated in SSc dermal fibroblasts [16,18]. This alteration is associated with higher methylation of a CpG island in the FLi1 promoter. Thus, higher specific DNA methylation is present in patient dermal fibroblasts associated with an overproduction of collagen. Hypermethylation in these fibroblasts also affects Wnt signaling, one of the central profibrotic pathways in systemic sclerosis, by silencing expression of the Wnt antagonists DKK1 and SFRP1 [19]. Treatment with 5-aza-2′-deoxycytidine restored the expression of both genes, thereby blocking activation of the Wnt pathway and reducing bleomycin-induced fibrosis.

22.4 OSTEOARTHRITIS

The term "osteoarthritis" designates a group of mechanical abnormalities involving degradation of joints, including articular cartilage and subchondral bone. Thus, in this disease, studies of the role of the epigenome have concentrated on chondrocytes. Genomic DNA methylation levels were found to be similar in chondrocytes from osteoarthrtic and healthy joints [20]. However, the levels of methylation of the leptin promoter were lower in chondrocytes isolated from severely involved cartilage compared with minimally involved or normal cartilage, and were associated with greater expression of this catabolic cytokine and its downstream target matrix metalloprotein-3 (MMP-3) [21]; similar findings have been reported in MMP-9, MMP-13, and ADAMTS [22]. Of particular note is the finding of lower ADAMTS-4 promoter methylation and higher expression in lesional compared with non-lesional chondrocytes [23]. Nitric oxide (NO), a key signaling molecule, is produced at high levels by activated chondrocytes and mediates IL-1β-induced suppression of cartilage proteoglycan synthesis. Lesional chondrocytes express high levels of inducible NO synthesis (iNOS) and have reduced methylation of a nuclear factor κB (NF-κB) enhancer element 5.8 kb upstream of the iNOS transcriptional start site [24].

22.5 RHEUMATOID ARTHRITIS

Rheumatoid arthritis is a complex autoimmune disease that results in a chronic, systemic inflammatory disorder and can affect many tissues and organs, but principally attacks flexible joints. Global DNA hypomethylation has also been reported in peripheral blood mononuclear cells [25] and in $CD3^+$ T cells [8] of patients with rheumatoid arthritis, compared with controls. However, the number of cells affected is much less than in systemic lupus erythematosus. A single CpG motif in the IL-6 promoter, ~1 kb upstream of the transcriptional start site, was significantly less methylated in peripheral blood mononuclear cells from patients with rheumatoid arthritis compared with controls [26]; it correlated with higher lipopolysaccharide-induced IL-6 mRNA levels by monocyte-derived macrophages. As in systemic lupus erythematosus and systemic sclerosis, higher expression of CD40L and lower promoter methylation is found in rheumatoid arthritis $CD4^+$ T cells [27]. It is important to note that rheumatoid arthritis is an infiltrative disease, and it is always questionable whether peripheral blood leukocytes are representative for the whole cell population. In addition, even a purified cell population such as $CD4^+$ or $CD8^+$ T cells includes a heterogeneous mixture of cell types. Nevertheless, the trend toward DNA hypomethylation and increased CD40L appears the same as in the two other autoimmune diseases mentioned previously.

The rheumatoid arthritis synovial fibroblasts are the effectors of tissue destruction via the production of a range of disease-related molecules, including chemokines, adhesion molecules, and matrix-degrading enzymes [28]. Engraftment of normal human cartilage and rheumatoid arthritis synovial fibroblasts into the severe combined immunodeficiency mouse revealed this aggressive phenotype to be maintained for up to 60 days and to be independent of cells of the immune system [29]. Rheumatoid arthritis synovial fibroblasts expressed LINE-1 transcripts and proteins, and showed a global DNA hypomethylation [6,30,31]. Methylation array studies [32,33] reported differential methylation patterns compared to non-aggressive osteoarthritis synovial fibroblasts, including genes regulating cell adhesion, transendothelial migration, and extracellular matrix interactions. Furthermore, treating osteoarthritis synovial fibroblasts with the DNA demethylating agent 5-aza-deoxycytidine resulted in conversion to an rheumatoid arthritis fibroblast-like phenotype [6]. Rheumatoid arthritis synovial fibroblasts produce more CXCL12 than normal cells due to promoter methylation changes; in turn, stimulation with CXCL12 activates the transcription of matrix-degrading enzymes through the CXCR7 receptor [34]. Chapter 23 will focus on rheumatoid arthritis synovial fibroblasts in more detail.

22.6 JUVENILE RHEUMATOID ARTHRITIS

Juvenile rheumatoid arthritis is the most common form of arthritis in children and adolescents (below 16 years). It is an autoimmune, non-infective, inflammatory joint disease. It differs significantly from arthritis commonly seen in adults, having less marked cartilage erosion. Increased production of pro-inflammatory cytokines and impaired balance of T cells are considered to be responsible for the disease. Interestingly, a significant effect of methotrexate was reported in a genome-scale assessment of DNA methylation in $CD4^+$ T cells from patients with juvenile rheumatoid arthritis, where the removal of methotrexate-exposed cases from the analysis considerably reduced the number of differently methylated genes, reduced methylation of the IL32 gene being the most important [35]. This cytokine induces the production of TNF-α from macrophages. A recent study [36] has shown an association between the disease-related changes in the environment surrounding synovial fibroblasts in juvenile rheumatoid arthritis and the impaired regenerative potential and proliferative capacities of these cells.

22.7 DERMATOMYOSITIS

An early study [11] reported that the DNA of $CD4^+$ T cells from patients with systemic lupus erythematosus and systemic sclerosis are hypomethylated, but this is not the case in dermatomyositis. In inflamed muscles from children with juvenile dermatomyositis, another study [37] showed that the WT1 gene is hypomethylated. Accordingly, the WT1 transcription factor is present in juvenile dermatomyositis muscle, but is undetectable in normal muscle. This work suggested that affected muscles of those children have the capacity to be repaired, and that homeobox and WT1 genes are epigenetically marked to facilitate this repair process.

22.8 HYPERMOBILITY SYNDROMES

Hypermobility, also termed ligamentous laxity, may present in different parts of the body at different times throughout childhood and adolescence. Hypermobility may be associated with collagen disorders that affect vital organ systems. The major collagen disorders are inherited in an autosomal dominant fashion – for example, FBN1 (fibrillin-1) or TGF-β receptor mutations. Other hypermobility syndromes can be caused by mutations in collagen genes directly: COL1A1, COL1A2,

COL5A1, and COL5A2 are the most prominent. Epigenetic modifications have not been investigated as yet. However, it is interesting to note that FBN1 hypermethylation occurs in colorectal cancer [38], and TGF-β receptor (I and II) hypermethylation is reported in gastric cardia dysplasia [39]. Lower TGF-β signaling decreased the expression of DNMT1 and DNMT3A, as well as increasing methylation of the COL1A1 gene [40]. Thus, the methylation status of these genes has to be determined.

22.9 CONCLUSION

There is increasing evidence implicating the epigenome with the development of inflammatory and age-related rheumatic diseases [41]. The complexity of the epigenetic signature and its dynamic nature, the differences between cell types and tissues, and the potential effects of inflammation on the epigenome complicate studies in rheumatic diseases. Initial attempts have concentrated on candidate genes in specific cell types that are known to be implicated in diseases, such as synovial fibroblasts in rheumatoid arthritis or dermal fibroblasts in systemic sclerosis, $CD4^+$ T cells in systemic lupus erythematosus, and chondrocytes in osteoarthritis. It is important to expand these investigations to other rheumatic diseases, such as dermatomyositis/polymyositis and hypermobility syndromes. Additional issues to be considered are that epigenetic differences may arise secondary to disease or therapies.

References

[1] Konya C, Paz Z, Tsokos GC. The role of T cells in systemic lupus erythematosus: an update. Curr Opin Rheumatol 2014;26:493–501.

[2] Javierre BM, Richardson B. A new epigenetic challenge: systemic lupus erythematosus. Adv Exp Med Biol 2011;711:117–36.

[3] Ezura Y, Sekiya I, Koga H, Muneta T, Noda M. Methylation status of CpG islands in the promoter regions of signature genes during chondrogenesis of human synovium-derived mesenchymal stem cells. Arthritis Rheum 2009;60:1416–26.

[4] Okamoto T. The epigenetic alteration of synovial cell gene expression in rheumatoid arthritis and the roles of nuclear factor kappaB and Notch signaling pathways. Mod Rheumatol 2005;15:79–86.

[5] El-Jawhari JJ, El-Sherbiny YM, Jones EA, McGonagle D. Mesenchymal stem cells, autoimmunity and rheumatoid arthritis. QJM 2014;107:505–14.

[6] Karouzakis E, Gay RE, Michel BA, Gay S, Neidhart M. DNA hypomethylation in rheumatoid arthritis synovial fibroblasts. Arthritis Rheum 2009;60:3613–22.

[7] Broen JC, Radstake TR, Rossato M. The role of genetics and epigenetics in the pathogenesis of systemic sclerosis. Nat Rev Rheumatol 2014;10:671–81.

[8] Richardson B, Scheinbart L, Strahler J, Gross L, Hanash S, Johnson M. Evidence for impaired T cell DNA methylation in systemic lupus erythematosus and rheumatoid arthritis. Arthritis Rheum 1990;33:1665–73.

[9] Cornacchia E, Golbus J, Maybaum J, Strahler J, Hanash S, Richardson B. Hydralazine and procainamide inhibit T cell DNA methylation and induce autoreactivity. J Immunol 1988;140:2197–200.
[10] Scheinbart LS, Johnson MA, Gross LA, Edelstein SR, Richardson BC. Procainamide inhibits DNA methyltransferase in a human T cell line. J Rheumatol 1991;18:530–4.
[11] Lei W, Luo Y, Lei W, Luo Y, Yan K, Zhao S, et al. Abnormal DNA methylation in $CD4^+$ T cells from patients with systemic lupus erythematosus, systemic sclerosis, and dermatomyositis. Scand J Rheumatol 2009;38:369–74.
[12] Jiang H, Xiao R, Lian X, Kanekura T, Luo Y, Yin Y, et al. Demethylation of TNFSF7 contributes to CD70 overexpression in $CD4^+$ T cells from patients with systemic sclerosis. Clin Immunol 2012;143:39–44.
[13] Komura K, Fujimoto M, Yanaba K, Matsushita T, Matsushita Y, Horikawa M, et al. Blockade of CD40/CD40 ligand interactions attenuates skin fibrosis and autoimmunity in the tight-skin mouse. Ann Rheum Dis 2008;67:867–72.
[14] Lian X, Xiao R, Hu X, Kanekura T, Jiang H, Li Y, et al. DNA demethylation of CD40l in $CD4^+$ T cells from women with systemic sclerosis: a possible explanation for female susceptibility. Arthritis Rheum 2012;64:2338.
[15] Selmi C, Feghali-Bostwick CA, Lleo A, Lombardi SA, De Santis M, Cavaciocchi F, et al. X chromosome gene methylation in peripheral lymphocytes from monozygotic twins discordant for scleroderma. Clin Exp Immunol 2012;169:253–62.
[16] Wang Y, Fan PS, Kahaleh B. Association between enhanced type I collagen expression and epigenetic repression of the FLI1 gene in scleroderma fibroblasts. Arthritis Rheum 2006;54:2271–9.
[17] Qi Q, Guo Q, Tan G, Mao Y, Tang H, Zhou C, et al. Predictors of the scleroderma phenotype in fibroblasts from systemic sclerosis patients. J Eur Acad Dermatol Venereol 2009;23:160–8.
[18] Kubo M, Czuwara-Ladykowska J, Moussa O, Markiewicz M, Smith E, et al. Persistent down-regulation of Fli1, a suppressor of collagen transcription, in fibrotic scleroderma skin. Am J Pathol 2003;163:571–81.
[19] Dees C, Schlottmann I, Funke R, Distler A, Palumbo-Zerr K, Zerr P, et al. The Wnt antagonists DKK1 and SFRP1 are downregulated by promoter hypermethylation in systemic sclerosis. Ann Rheum Dis 2014;73:1232–9.
[20] Sesselmann S, Söder S, Voigt R, Haag J, Grogan SP, Aigner T. DNA methylation is not responsible for p21WAF1/CIP1 down-regulation in osteoarthritic chondrocytes. Osteoarthritis Cartilage 2009;17:507–12.
[21] Iliopoulos D, Malizos KN, Tsezou A. Epigenetic regulation of leptin affects MMP-13 expression in osteoarthritic chondrocytes: possible molecular target for osteoarthritis therapeutic intervention. Ann Rheum Dis 2007;66:1616–21.
[22] Roach HI, Yamada N, Cheung KS, Tilley S, Clarke NM, Oreffo RO, et al. Association between the abnormal expression of matrix-degrading enzymes by human osteoarthritic chondrocytes and demethylation of specific CpG sites in the promoter regions. Arthritis Rheum 2005;52:3110–24.
[23] Cheung KS, Hashimoto K, Yamada N, Roach HI. Expression of ADAMTS-4 by chondrocytes in the surface zone of human osteoarthritic cartilage is regulated by epigenetic DNA de-methylation. Rheumatol Int 2009;29:525–34.
[24] de Andres MC, Imagawa K, Hashimoto K, Gonzalez A, Roach HI, Goldring MB, et al. Loss of methylation in CpG sites in the NF-kappaB enhancer elements of inducible nitric oxide synthase is responsible for gene induction in human articular chondrocytes. Arthritis Rheum 2013;65:732–42.
[25] Liu CC, Fang TJ, Ou TT, Wu CC, Li RN, Lin YC, et al. Global DNA methylation, DNMT1, and MBD2 in patients with rheumatoid arthritis. Immunol Lett 2011;135:96–9.

[26] Nile CJ, Read RC, Akil M, Duff GW, Wilson AG. Methylation status of a single CpG site in the IL6 promoter is related to IL6 messenger RNA levels and rheumatoid arthritis. Arthritis Rheum 2008;58:2686–93.
[27] Liao J, Liang G, Xie S, Zhao H, Zuo X, Li F, et al. CD40L demethylation in CD4(+) T cells from women with rheumatoid arthritis. Clin Immunol 2012;145:13–18.
[28] Muller-Ladner U, Ospelt C, Gay S, Distler O, Pap T. Cells of the synovium in rheumatoid arthritis. Synovial fibroblasts. Arthritis Res Ther 2007;9:223.
[29] Muller-Ladner U, Kriegsmann J, Franklin BN, Matsumoto S, Geiler T, Gay RE, et al. Synovial fibroblasts of patients with rheumatoid arthritis attach to and invade normal human cartilage when engrafted into SCID mice. Am J Pathol 1996;149:1607–15.
[30] Neidhart M, Rethage J, Kuchen S, Künzler P, Crowl RM, Billingham ME, et al. Retrotransposable L1 elements expressed in rheumatoid arthritis synovial tissue: association with genomic DNA hypomethylation and influence on gene expression. Arthritis Rheum 2000;43:2634–47.
[31] Kuchen S, Seemayer CA, Rethage J, von Knoch R, Kuenzler P, Michel BA, et al. The L1 retroelement-related p40 protein induces p38delta MAP kinase. Autoimmunity 2004;37:57–65.
[32] Wilkinson J. Study reveals a DNA methylome signature in rheumatoid arthritis. Epigenomics 2012;4:481.
[33] Nakano K, Whitaker JW, Boyle DL, Wang W, Firestein GS. DNA methylome signature in rheumatoid arthritis. Ann Rheum Dis 2013;72:110–17.
[34] Karouzakis E, Rengel Y, Jüngel A, Kolling C, Gay RE, Michel BA, et al. DNA methylation regulates the expression of CXCL12 in rheumatoid arthritis synovial fibroblasts. Genes Immun 2011;12:643–52.
[35] Ellis JA, Munro JE, Chavez RA, Gordon L, Joo JE, Akikusa JD, et al. Genome-scale case-control analysis of $CD4^+$ T-cell DNA methylation in juvenile idiopathic arthritis reveals potential targets involved in disease. Clin Epigenetics 2012;4:20.
[36] Lazic E, Jelusic M, Grcevic D, Marusic A, Kovacic N. Osteoblastogenesis from synovial fluid-derived cells is related to the type and severity of juvenile idiopathic arthritis. Arthritis Res Ther 2012;14:R139.
[37] Wang M, Xie H, Shrestha S, Sredni S, Morgan GA, Pachman LM. Methylation alterations of WT1 and homeobox genes in inflamed muscle biopsy samples from patients with untreated juvenile dermatomyositis suggest self-renewal capacity. Arthritis Rheum 2012;64:3478–85.
[38] Guo Q, Song Y, Zhang H, Wu X, Xia P, Dang C. Detection of hypermethylated fibrillin-1 in the stool samples of colorectal cancer patients. Med Oncol 2013;30:695.
[39] Guo W, Dong Z, Guo Y, Kuang G, Yang Z, Shan B. Concordant repression and aberrant methylation of transforming growth factor-beta signaling pathway genes occurs early in gastric cardia adenocarcinoma. Mol Biol Rep 2012;39:9453–62.
[40] Pan X, Chen Z, Huang R, Yao Y, Ma G. Transforming growth factor beta1 induces the expression of collagen type I by DNA methylation in cardiac fibroblasts. PloS One 2013;8:e60335.
[41] Gay S, Wilson AG. The emerging role of epigenetics in rheumatic diseases. Rheumatology 2014;53:406–14.

CHAPTER

23

DNA Methylation in Synovial Fibroblasts

OUTLINE

23.1 INTRODUCTION

Rheumatoid arthritis is a complex disease where predetermined and stochastic factors conspire to confer disease susceptibility [1]. The susceptibility to develop the disease is determined in part by inherited risk factors. The single nucleotide polymorphisms associated with rheumatoid arthritis are dispersed widely across the genome, with notable concentration in genes that participate in adaptive and innate immune responses. Multiple genome-wide association studies have

M. Neidhart: DNA Methylation and Complex Human Disease.
DOI: http://dx.doi.org/10.1016/B978-0-12-420194-1.00023-3

identified scores of disease-associated polymorphisms. By far the greatest genetic risk is conferred by the class II major histocompatibility gene HLA-DR, which participates in antigen presentation to T lymphocytes [2]. The critical regions of the encoded protein have been well characterized and are located in and around the antigen-binding groove. However, the observation that identical twins only have perhaps a 15% concordance rate for rheumatoid arthritis indicates that inherited DNA sequences account for a minority of risk and might not be as important as other influences [3]. In other words, whole genome sequencing of patients ignores over 80% of disease risk. Furthermore, most animal models of arthritis have a strong inflammatory component, but have also taught us that there are different ways to get arthritis. Thus, even the disease in humans represents a final common clinical phenotype that reflects many pathogenic pathways. Therefore, it might be appropriate to begin thinking about rheumatoid arthritis as a syndrome rather than a disease [1]. Another question regards the role of the adaptive and innate immune system in this disease. Epstein-Barr virus [4], parvoviruses (B19) [5], retroviruses (HTLV-1) [6], and certain bacteria (*Borrelia burgdorferi*) [7] are among the candidates as triggers for various forms of inflammatory arthropathies that mimic rheumatoid arthritis. Bacteria and their products have been conclusively linked to many forms of inflammatory reactive arthritis [8]. For instance, rheumatoid arthritis-like diseases can be induced in certain inbred strains of rats with bacterial fragments, such as streptococcal cell wall proteoglycan [9]. The factors that allow an abnormal immune response, once initiated, to become permanent and chronic are not clearly understood. As mentioned above, a genetic predisposition exists involving the HLA-DRB1 molecule (so-called "shared epitope") [10] and specific gene polymorphisms (e.g., PTPN22 and TRAF1-C5) [11]. These genetic factors may interact with environmental risk factors. The only clearly identified risk factor for rheumatoid arthritis is cigarette smoking [12], although non-smokers can also develop the disease. Other environmental factors likely exist, modulating the risk of acquiring RA. Hormonal factors in the individual may explain some features of the disease (e.g., prolactin and cortisol) [13–15], such as the higher occurrence in women [16], the not infrequent onset after childbirth [17], and the modulation of disease risk by hormonal medications [18]. How altered regulatory thresholds allow the triggering of a specific autoimmune response is uncertain. Smoking promotes protein citrullination and production of anticitrullinated protein/peptide autoantibodies [19]. This may take place several years before any symptoms occur. B lymphocytes produce rheumatoid factors (mostly IgM that binds the Fc part of other immunoglobulins) and anticitrullinated protein autoantibodies in large quantities. They activate macrophages through Fc receptor and complement binding, which in turn start an intense inflammatory response. High and chronic production of TNF-α appears sufficient in transgenic mice to develop an inflammatory polyarthritis [20]. This contributes to inflammation of the synovial membrane, in terms of edema, vasodilation, and infiltration by activated T cells [21]. The T-cell repertoire appears specifically restricted in patients with synovitis and positive serology for anticitrullinated protein autoantibodies [22]; however, the proof for clonal expansion of T cells in response to an autoantigen in rheumatoid arthritis has not yet been provided. The disease progresses with hyperplasia of the synovial membrane, neoangiogenesis, production of matrix-degrading enzymes, and tissue damage [23]. Synovial fibroblasts have been recognized as the major effector cells of joint destruction [24]. Interestingly, it has been reported that progressive joint destruction in rheumatoid arthritis can occur in the absence of $CD4^+$ T cells [25]. Rheumatoid arthritis synovial fibroblasts maintained their aggressive behavior in the absence of immune cells, as shown in a SCID mice model in which synovial fibroblasts are implanted with a piece of human cartilage under the kidney capsule [26]. This model demonstrated, for example, that chondrocytes contribute significantly to the degradation of cartilage by releasing factors that stimulate rheumatoid arthritis synovial fibroblasts. Among those, IL-1β-mediated mechanisms might be of particular importance [27]. The intrinsically activated phenotype of rheumatoid arthritis synovial fibroblasts was associated with increased expression of oncogenes, contributing to the designation "transformed-like" fibroblasts [28]. The search for other similarities with cancer cells, however, resulted in a few more obvious observations, such as the overproduction of matrix-degrading enzymes (e.g., MMP-1) [29]. The fact that oncogenic transformation, including aggressive behavior, can be obtained in fibroblasts upon infection with retroviruses prompted the first research in this direction [6,30]. In this chapter, we review the research in this field that led to the discovery that it is essential to consider epigenetic modifications in rheumatoid arthritis in order to understand the cellular mechanism and develop new therapeutic strategies.

23.2 ENDOGENOUS RETROVIRUSES

As mentioned above, although rheumatoid arthritis has been widely suspected to have an infectious etiology, it has remained difficult to prove this hypothesis. An early study [31] reported virus-like particles with retroviral C-type morphology in rheumatoid arthritis synovial fluid. However, the involvement of an exogenous retrovirus has not yet been confirmed [32]. During the search for exogenous retroviruses in rheumatoid arthritis synovial fibroblasts another phenomena became obvious, namely, the increased expression of human endogenous retroviruses (HERVs) that should normally be silenced by DNA methylation [33–35]. HERVs are fossil viruses that began to be integrated into the human genome some 30–40 million years ago and now make up 8% of the genome. In rheumatoid arthritis synovial fibroblasts, roles for full-length LINE-1 [36,37] and HERV-K (HML-2) [38,39] have been proposed. LINE-1 and HERV-K are also expressed in many other diseases, such as cancer and systemic lupus erythematosus, in which they are associated with global DNA hypomethylation [38,40–42]. This revealed the major problems of these cells and explained why they show an overproduction of multiple proteins in part involved in the aggressive behavior [33,43].

23.3 DNA HYPOMETHYLATION

An early study [33] reported that, after incubation of synovial fibroblasts with the demethylating drug 5-aza-cytidine, LINE-1 transcripts appeared in a time- and dose-dependent manner. Compared with osteoarthritis synovial fibroblasts (which served as normal controls), rheumatoid arthritis synovial fibroblasts were found to be more sensitive to the drug. This difference in sensitivity was the first indication that the DNA of rheumatoid arthritis synovial fibroblasts is globally hypomethylated. More recently, it has been confirmed by bisulfite modification and sequencing that rheumatoid arthritis synovial fibroblasts show a global DNA hypomethylation *in situ* and *in vitro* [43]. This is associated with a 15% demethylation of the LINE-1 gene [18], similar to the findings in many cancer cell lines [40]. Mostly the LINE-1 ORF1 p40 protein is expressed in rheumatoid arthritis synovial fibroblasts [36], which may influence gene expression [33,37]. The next step is to search for the reason for this global DNA hypomethylation. DNA hypomethylation is not restricted to fibroblasts, but also affects peripheral blood leukocytes [44] and $CD4^+$ T cells [45].

23.4 DNA METHYLTRANSFERASE 1

The levels of DNMT1 transcripts in rheumatoid arthritis synovial fibroblasts are mostly normal [46], but the amount of DNMT1 protein can be reduced [43]. This is particularly the case during stimulation with growth factors or pro-inflammatory cytokines [43,46]. Thus, the cells do not produce enough DNMT1 during replication, which causes a progressive DNA demethylation in daughter cells. In contrast, peripheral blood leukocytes of patients with rheumatoid arthritis have higher expressions of DNMT1 and methyl-binding protein (MBD2) transcripts compared to healthy controls. In rheumatoid arthritis synovial fibroblasts, a deficiency in substrate (i.e., SAM) may favor degradation of the enzyme (i.e., DNMT1) [47]. Alternatively, polymorphisms in DNMTs have been identified as risk factors for disease, including systemic lupus erythematosus [48] and autoimmune thyroid diseases [49]. However, DNMT polymorphisms in rheumatoid arthritis have not been investigated so far. Finally, it has been hypothesized [44,50] that a metabolic pathway is intrinsically activated in rheumatoid arthritis synovial fibroblasts, competing for the availability of SAM.

23.5 S-ADENOSYLMETHIONINE AND ADENOSYLHOMOCYSTEINE

SAM is an essential precursor molecule found in all living organisms; it is derived from L-methionine and is the major substrate used in transmethylation, as well as in aminopropylation and transsulfuration. SAM is critical not only for DNA methylation, but also for other metabolic reactions such as nucleic acid synthesis, histone methylation, and polyamine synthesis. The production of SAM as a methyl donor for DNA methylation involves a complex interplay between metabolic substrate, enzymatic activity, and co-factor availability. S-adenosylhomocysteine (SAH) is produced during DNA methylation; it is an inhibitor of DNMTs and its removal is a prerequisite for correct DNA methylation. Homocysteine produced by the hydrolysis of SAH can be recycled into L-methionine or exported into the extracellular fluid. Impairment of this pathway may lead to elevated plasma homocysteine levels. In rheumatoid arthritis, hyperhomocysteinemia is associated with increased risk for cardiovascular diseases [51,52]. Methotrexate therapy is considered the most common causative factor [53]. However, rheumatoid arthritis synovial fibroblasts "spontaneously" release more homocysteine than normal cells [47,54]; in turn, homocysteine increases IL-6 production in a paracrine manner [54]. The rate of passage through the one-carbon cycle is influenced by genetic polymorphisms in genes encoding the enzymes involved in this pathway. Polymorphisms of the methylene tetrahydrofolate reductase (MTHFR) have been widely studied in this regard. MTHFR catalyzes the conversion of 5,10-methylenetetrahydrofolate to 5-methyltetrahydrofolate, a co-substrate for homocysteine remethylation to L-methionine. In rheumatoid arthritis, the A1298C [55] and/or CT677 [56–58] mutations of MTHFR have been associated with increased homocysteine levels, higher cytokine levels, and susceptibility to methotrexate toxicity and cardiovascular diseases. A genome-wide association study confirmed that variations in the MTHFR gene are a major determinant of serum homocysteine levels [59].

Taken together, these observations clearly indicate that a dysfunction of the one-carbon metabolism is involved in the systemic clinical picture of rheumatoid arthritis.

23.6 RECYCLING PATHWAY OF POLYAMINES

Biosynthesis of polyamines is tightly regulated by enzymes, including ornithine decarboxylase (ODC), which regulates the conversion of ornithine into putrescine; and spermidine synthase (SpS) and spermine synthase (SmS), which regulate the biosynthesis of spermidine and spermine, respectively (Figure 23.1). In this regard, SpS and SmS need decarboxy-SAM that is provided by SAM after enzymatic conversion mediated by S-adenosylmethionine decarboxylase (AMD). SAM is synthesized from the essential amino acid L-methionine. Putrescine holds the AMD domains in a more active state, inducing realignment of charged residues to the active site and thus enhancing the conversion of SAM. In addition, a recycling pathway that reconverts polyamines into putrescine is under the control of spermidine/spermine N1-acetyltransferase (SSAT1). Moreover, expression of the SSAT1 gene is regulated by polyamine-modulated factor 1-binding protein 1 (PMFBP1)

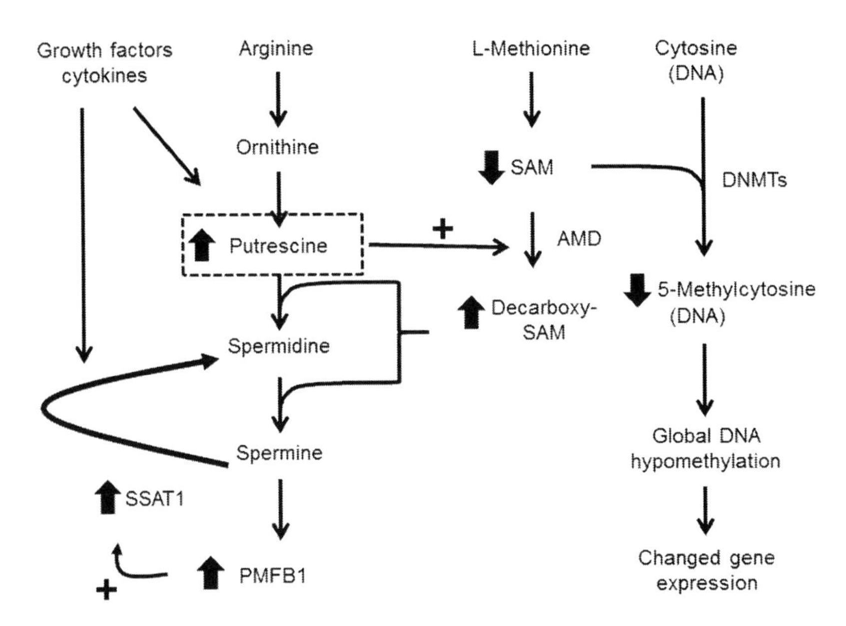

FIGURE 23.1 Model integrating the polyamine metabolism and the transmethylation pathway.

and nuclear factor E2-related factor 2. In rheumatoid arthritis synovial fibroblasts, expression of SSAT1, AMD, and PMFBP1 is significantly increased compared with osteoarthritis synovial fibroblasts [47]. Various stimuli, including growth factors, hormones, and cytokines, activate both ODC and SSAT1, dramatically increasing the need for SAM, which, under such conditions, is predominantly used in the biosynthesis and recycling of polyamines. Polyamines have an important role in cell proliferation, migration, and transcription, as well as in stabilization of nucleic acids and proteins. Pro-inflammatory cytokines increase ODC and SSAT1 activities in synovial fibroblasts, triggering the synthesis of polyamines and their recycling [60]. However, the level of putrescine in rheumatoid arthritis synovial fibroblasts is also elevated in the absence of stimulation. Dysfunction in the regulation of ODC does not cause elevated levels of putrescine, which instead could result from enhanced recycling of polyamines, a process that is controlled by SSAT1. During recycling of polyamines, intermediaries are produced in the form of diacetylpolyamines, which can be transported outside of cells through an exporter system involving solute carrier family 3 member 2 (SLC3A2)/CD98. Urinary diacetylspermine (DASp) is a marker for cancer and liver diseases. The levels of DASp in cell culture supernatants of rheumatoid arthritis synovial fibroblasts, as well as the expression of SLC3A2/CD98 on their cell surface are significantly elevated [47]. Accordingly, the concentrations of urinary DASp are also increased in patients with rheumatoid arthritis compared with patients with osteoarthritis or healthy control subjects [61,62]. Increased SSAT1 activity produces more putrescine and diacetylpolyamines [63]. High putrescine levels in turn enhance the stability of AMD, which is responsible for the conversion of SAM into decarboxy-SAM. Most importantly, in rheumatoid arthritis synovial fibroblasts the parameters of polyamine metabolism negatively correlated with the intracellular levels of SAM, the expression of DNMT1, and the 5-methylcytosine content of DNA [47]. Thus, indeed, overexpression of AMD and/or SSAT1 may result in a decrease in SAM. Transfection of interfering RNA targeting SSAT1 increases the 5-methylcytosine content of DNA within 21 days [50]. Similarly, berenil (diminazene aceturate, an inhibitor of SSAT1) partially restores DNA methylation. In addition, berenil increases the levels of DNMT1, decreases the levels of AMD, putrescine, activation markers, and MMP1, and alters the adhesion of rheumatoid arthritis synovial fibroblasts. As expected, berenil was more efficient in cells with higher levels of SSAT1. Most promising is that the combination of berenil and SAM reduced the invasiveness of rheumatoid arthritis synovial fibroblasts in the SCID mice model of cartilage destruction by 70%. This would be the first therapy that specifically targets the effector cells of joint destruction.

23.7 DIFFERENTIAL METHYLATION OF SPECIFIC PROMOTERS

By treating normal fibroblasts with the DNMT inhibitor 5-methylcytosine, a phenotype that mimics rheumatoid arthritis synovial fibroblasts can be obtained, with increased adhesion molecules, cytokine receptors, matrix-degrading enzymes, and changes in the transcription factor pattern (e.g., Wnt signaling) [43]. Similar findings are reported in rat fibroblasts [64]. Another study [65] showed that basal expression of CXCL12 is higher in rheumatoid arthritis synovial fibroblasts than in osteoarthritis synovial fibroblasts, due to the demethylation of its promoter. By high-throughput methylome analysis, a recent work [66–69] reported numerous hypo- and hypermethylated genes in rheumatoid arthritis synovial fibroblasts: CAPN8, CASP1, DPP4, CHI3L1, IL6R, STAT3, MAP3K5, MEFV, and WISP3 were hypomethylated, while EBF3, IRX1, TGFBR2, and FOXO1 appeared to be hypermethylated. The functions of these genes in the context of rheumatoid arthritis synovial fibroblasts have to be confirmed and clarified. Recently, the promoter of the transcription factor TBX5 has been found to be demethylated in rheumatoid arthritis synovial fibroblasts [70]. It increases the production of IL8, CXCL2, and CCL20, which might be responsible for the attraction of leukocytes into the synovial membrane.

23.8 MicroRNAs INTERFERING WITH DNA METHYLATION

MicroRNAs (miRNAs) are important regulators of a variety of fundamental biologic processes. Rheumatoid arthritis synovial fibroblasts expressed a specific profile of microRNAs – for example, miR-155, which limits the upregulation of MMP expression [71]. In the context of DNA methylation, interesting miRNAs are those in which the promoter is controlled by methylation and/or those that target epigenetic modulators [68]. For both, this is the case for miR-203, which can be silenced by DNA methylation [72] and may target DNMTs [73]. In rheumatoid arthritis synovial fibroblasts, miR-203 is hypermethylated [74]. It is interesting that DNMT transcripts are decreased after treatment of rheumatoid arthritis synovial fibroblasts with a supplement of L-methionine (N. Gaur, unpublished). Simultaneously, specific microRNAs targeting DNMTs, such as miR-29c and miR-203, are upregulated, thereby limiting the restoring effect of L-methionine on DNA methylation. Thus, the use of antagomirs against these miRNAs may increase the effectiveness of methyl donors. It appears that the methylation of miR-203 is a mechanism that should favor the restoration of DNA methylation in rheumatoid arthritis synovial fibroblasts.

23.9 CONCLUSION

Great efforts have been made to understand the intrinsically activated phenotype of rheumatoid arthritis synovial fibroblasts that leads to aggressive behavior and joint destruction. Global DNA hypomethylation probably mainly affects repetitive sequences. The effectiveness of methyl donors in restoring DNA methylation is limited by at least two factors: an upregulation of the polyamine recycling pathway in some patients and a counter-regulatory mechanism involving microRNAs that target DNMTs. Thus, the future will be to try combinations of methyl donors such as SAM, L-methionine, or betaine with specific inhibitors of SSAT1 and/or antagomirs against specific microRNAs. This work also makes clear how important it is to develop models that integrate epigenetics and metabolic pathways.

References

[1] Firestein GS. The disease formerly known as rheumatoid arthritis. Arthritis Res Ther 2014;16:114.

[2] Wellcome Trust Case Control Consortium. Genome-wide association study of 14,000 cases of seven common diseases and 3,000 shared controls. Nature 2007;447:661–78.

[3] Silman AJ, MacGregor AJ, Thomson W, Holligan S, Carthy D, Farhan A, et al. Twin concordance rates for rheumatoid arthritis: results from a nationwide study. Br J Rheumatol 1993;32:903–7.

[4] Draborg AH, Duus K, Houen G. Epstein-Barr virus in systemic autoimmune diseases. Clin Dev Immunol 2013;2013:535738.

[5] Colmegna I, Alberts-Grill N. Parvovirus B19: its role in chronic arthritis. Rheum Dis Clin North Am 2009;35:95–110.

[6] Habu K, Nakayama-Yamada J, Asano M, Saijo S, Itagaki K, Horai R, et al. The human T cell leukemia virus type I-tax gene is responsible for the development of both inflammatory polyarthropathy resembling rheumatoid arthritis and noninflammatory ankylotic arthropathy in transgenic mice. J Immunol 1999;162:2956–63.

[7] Nardelli DT, Callister SM, Schell RF. Lyme arthritis: current concepts and a change in paradigm. Clin Vaccine Immunol 2008;15:21–34.

[8] Girschick HJ, Guilherme L, Inman RD, Latsch K, Rihl M, Sherer Y, et al. Bacterial triggers and autoimmune rheumatic diseases. Clin Exp Rheumatol 2008;26:S12–17.

[9] Wilder RL. Streptococcal cell wall arthritis. Curr protoc Immunol 2001. Chapter 15, Unit 15.10.

[10] Gregersen PK, Silver J, Winchester RJ. The shared epitope hypothesis. An approach to understanding the molecular genetics of susceptibility to rheumatoid arthritis. Arthritis Rheum 1987;30:1205–13.

[11] Plenge RM, Seielstad M, Padyukov L, Lee AT, Remmers EF, Ding B, et al. TRAF1-C5 as a risk locus for rheumatoid arthritis – a genomewide study. N Engl J Med 2007;357:1199–209.

[12] Padyukov L, Silva C, Stolt P, Alfredsson L, Klareskog L. A gene–environment interaction between smoking and shared epitope genes in HLA-DR provides a high risk of seropositive rheumatoid arthritis. Arthritis Rheum 2004;50:3085–92.

[13] Adan N, Ledesma-Colunga MG, Reyes-Lopez AL, Martinez de la Escalera G, Clapp C. Arthritis and prolactin: a phylogenetic viewpoint. Gen Comp Endocrinol 2014;203: 132–6.

[14] Masi AT, Rehman AA, Cutolo M, Aldag JC. Do women with premenopausal-onset rheumatoid arthritis have relative insufficiency or imbalance of adrenocortical steroids? Ann N Y Acad Sci 2014;1317:7–16.

[15] Neidhart M. Prolactin in autoimmune diseases. Proc Soc Exp Biol Med 1998;217: 408–19.

[16] Hughes GC, Choubey D. Modulation of autoimmune rheumatic diseases by oestrogen and progesterone. Nat Rev Rheumatol 2014.

[17] Brennan P, Silman A. Breast-feeding and the onset of rheumatoid arthritis. Arthritis Rheum 1994;37:808–13.

[18] Oliver JE, Silman AJ. Risk factors for the development of rheumatoid arthritis. Scand J Rheumatol 2006;35:169–74.

[19] Sakkas LI, Bogdanos DP, Katsiari C, Platsoucas CD. Anti-citrullinated peptides as autoantigens in rheumatoid arthritis-relevance to treatment. Autoimmun Rev 2014;13:1114–20.

[20] Keffer J, Probert L, Cazlaris H, Georgopoulos S, Kaslaris E, Kioussis D, et al. Transgenic mice expressing human tumour necrosis factor: a predictive genetic model of arthritis. EMBO J 1991;10:4025–31.

[21] Cope AP, Schulze-Koops H, Aringer M. The central role of T cells in rheumatoid arthritis. Clin Exp Rheumatol 2007;25:S4–11.

[22] Cantaert T, Brouard S, Thurlings RM, Pallier A, Salinas GF, Braud C, et al. Alterations of the synovial T cell repertoire in anti-citrullinated protein antibody-positive rheumatoid arthritis. Arthritis Rheum 2009;60:1944–56.

[23] Muller-Ladner U, Gay RE, Gay S. Molecular biology of cartilage and bone destruction. Curr Opin Rheumatol 1998;10:212–19.

[24] Pap T, Muller-Ladner U, Gay RE, Gay S. Fibroblast biology. Role of synovial fibroblasts in the pathogenesis of rheumatoid arthritis. Arthritis Res 2000;2:361–7.

[25] Muller-Ladner U, Kriegsmann J, Gay RE, Koopman WJ, Gay S, Chatham WW. Progressive joint destruction in a human immunodeficiency virus-infected patient with rheumatoid arthritis. Arthritis Rheum 1995;38:1328–32.

[26] Geiler T, Kriegsmann J, Keyszer GM, Gay RE, Gay S. A new model for rheumatoid arthritis generated by engraftment of rheumatoid synovial tissue and normal human cartilage into SCID mice. Arthritis Rheum 1994;37:1664–71.

[27] Pap T, van der Laan WH, Aupperle KR, Gay RE, Verheijen JH, Firestein GS, et al. Modulation of fibroblast-mediated cartilage degradation by articular chondrocytes in rheumatoid arthritis. Arthritis Rheum 2000;43:2531–6.

[28] Trabandt A, Gay RE, Gay S. Oncogene activation in rheumatoid synovium. APMIS 1992;100:861–75.

[29] Eck SM, Blackburn JS, Schmucker AC, Burrage PS, Brinckerhoff CE. Matrix metalloproteinase and G protein coupled receptors: co-conspirators in the pathogenesis of autoimmune disease and cancer. J Autoimmun 2009;33:214–21.

[30] Grassmann R, Aboud M, Jeang KT. Molecular mechanisms of cellular transformation by HTLV-1 Tax. Oncogene 2005;24:5976–85.

[31] Stransky G, Vernon J, Aicher WK, Moreland LW, Gay RE, Gay S. Virus-like particles in synovial fluids from patients with rheumatoid arthritis. Br J Rheumatol 1993;32:1044–8.

[32] Seemayer CA, Kolb SA, Neidhart M, Ohshima S, Gay RE, Michel BA, et al. Absence of inducible retroviruses from synovial fibroblasts and synovial fluid cells of patients with rheumatoid arthritis. Arthritis Rheum 2002;46:2811–13.

[33] Neidhart M, Rethage J, Kuchen S, Künzler P, Crowl RM, Billingham ME, et al. Retrotransposable L1 elements expressed in rheumatoid arthritis synovial tissue: association with genomic DNA hypomethylation and influence on gene expression. Arthritis Rheum 2000;43:2634–47.

[34] Ogasawara H, Okada M, Kaneko H, Hishikawa T, Sekigawa I, Iida N, et al. Quantitative comparison of human endogenous retrovirus mRNA between SLE and rheumatoid arthritis. Lupus 2001;10:517–18.

[35] Ehlhardt S, Seifert M, Schneider J, Ojak A, Zang KD, Mehraein Y. Human endogenous retrovirus HERV-K(HML-2) Rec expression and transcriptional activities in normal and rheumatoid arthritis synovia. Journal Rheumatol 2006;33:16–23.

[36] Kuchen S, Seemayer CA, Rethage J, von Knoch R, Kuenzler P, Michel BA, et al. The L1 retroelement-related p40 protein induces p38delta MAP kinase. Autoimmunity 2004;37:57–65.

[37] Neidhart M, Karouzakis E, Schumann GG, Gay RE, Gay S. Trex-1 deficiency in rheumatoid arthritis synovial fibroblasts. Arthritis Rheum 2010;62:2673–9.

[38] Subramanian RP, Wildschutte JH, Russo C, Coffin JM. Identification, characterization, and comparative genomic distribution of the HERV-K (HML-2) group of human endogenous retroviruses. Retrovirology 2011;8:90.

[39] Balada E, Ordi-Ros J, Vilardell-Tarres M. Molecular mechanisms mediated by human endogenous retroviruses (HERVs) in autoimmunity. Rev Med Virol 2009;19:273–86.

[40] Carreira PE, Richardson SR, Faulkner GJ. L1 retrotransposons, cancer stem cells and oncogenesis. FEBS J 2014;281:63–73.

[41] Nakkuntod J, Avihingsanon Y, Mutirangura A, Hirankarn N. Hypomethylation of LINE-1 but not Alu in lymphocyte subsets of systemic lupus erythematosus patients. Clin Chim Acta 2011;412:1457–61.

[42] Nakkuntod J, Sukkapan P, Avihingsanon Y, Mutirangura A, Hirankarn N. DNA methylation of human endogenous retrovirus in systemic lupus erythematosus. J Hum Genet 2013;58:241–9.

[43] Karouzakis E, Gay RE, Michel BA, Gay S, Neidhart M. DNA hypomethylation in rheumatoid arthritis synovial fibroblasts. Arthritis Rheum 2009;60:3613–22.

[44] Liu CC, Fang TJ, Ou TT, Wu CC, Li RN, Lin YC, et al. Global DNA methylation, DNMT1, and MBD2 in patients with rheumatoid arthritis. Immunol Lett 2011;135:96–9.

[45] Liu Y, Chen Y, Richardson B. Decreased DNA methyltransferase levels contribute to abnormal gene expression in "senescent" CD4(+)CD28(−) T cells. Clin Immunol 2009;132:257–65.

[46] Nakano K, Boyle DL, Firestein GS. Regulation of DNA methylation in rheumatoid arthritis synoviocytes. J Immunol 2013;190:1297–303.

[47] Karouzakis E, Gay RE, Gay S, Neidhart M. Increased recycling of polyamines is associated with global DNA hypomethylation in rheumatoid arthritis synovial fibroblasts. Arthritis Rheum 2012;64:1809–17.

[48] Park BL, Kim LH, Shin HD, Park YW, Uhm WS, Bae SC. Association analyses of DNA methyltransferase-1 (DNMT1) polymorphisms with systemic lupus erythematosus. J Hum Genet 2004;49:642–6.

[49] Arakawa Y, Watanabe M, Inoue N, Sarumaru M, Hidaka Y, Iwatani Y. Association of polymorphisms in DNMT1, DNMT3A, DNMT3B, MTHFR and MTRR genes with global DNA methylation levels and prognosis of autoimmune thyroid disease. Clin Exp Immunol 2012;170:194–201.

[50] Neidhart M, Karouzakis E, Jungel A, Gay RE, Gay S. Inhibition of spermidine/spermine N1-acetyltransferase activity: a new therapeutic concept in rheumatoid arthritis. Arthritis Rheumatol 2014;66:1723–33.

[51] Roubenoff R, Dellaripa P, Nadeau MR, Abad LW, Muldoon BA, Selhub J, et al. Abnormal homocysteine metabolism in rheumatoid arthritis. Arthritis Rheum 1997;40:718–22.

[52] De Bree A, Verschuren WM, Kromhout D, Kluijtmans LA, Blom HJ. Homocysteine determinants and the evidence to what extent homocysteine determines the risk of coronary heart disease. Pharmacol Rev 2002;54:599–618.

[53] El Bouchti I, Sordet C, Kuntz JL, Sibilia J. Severe atherosclerosis in rheumatoid arthritis and hyperhomocysteinemia: is there a link? Joint Bone Spine 2008;75:499–501.

[54] Lazzerini PE, Selvi E, Lorenzini S, Capecchi PL, Ghittoni R, Bisogno S, et al. Homocysteine enhances cytokine production in cultured synoviocytes from rheumatoid arthritis patients. Clin Exp Rheumatol 2006;24:387–93.

[55] Berkun Y, Levartovsky D, Rubinow A, Orbach H, Aamar S, Grenader T, et al. Methotrexate related adverse effects in patients with rheumatoid arthritis are associated with the A1298C polymorphism of the MTHFR gene. Ann Rheum Dis 2004;63:1227–31.

[56] van Ede AE, Laan RF, Blom HJ, Huizinga TW, Haagsma CJ, Giesendorf BA, et al. The C677T mutation in the methylenetetrahydrofolate reductase gene: a genetic risk factor for methotrexate-related elevation of liver enzymes in rheumatoid arthritis patients. Arthritis Rheum 2001;44:2525–30.

[57] Hammons AL, Summers CM, Woodside JV, McNulty H, Strain JJ, Young IS, et al. Folate/homocysteine phenotypes and MTHFR 677C > T genotypes are associated with serum levels of monocyte chemoattractant protein-1. Clin Immunol 2009;133:132–7.

[58] Fujimaki C, Hayashi H, Tsuboi S, Matsuyama T, Kosuge K, Yamada H, et al. Plasma total homocysteine level and methylenetetrahydrofolate reductase 677C > T genetic polymorphism in Japanese patients with rheumatoid arthritis. Biomarkers 2009;14:49–54.

[59] Tanaka T, Scheet P, Giusti B, Bandinelli S, Piras MG, Usala G, et al. Genome-wide association study of vitamin B6, vitamin B12, folate, and homocysteine blood concentrations. Am J Hum Genet 2009;84:477–82.

[60] Furumitsu Y, Yukioka K, Yukioka M, Ochi T, Morishima Y, Matsui-Yuasa I, et al. Interleukin-1beta induces elevation of spermidine/spermine N1-acetyltransferase activity and an increase in the amount of putrescine in synovial adherent cells from patients with rheumatoid arthritis. J Rheumatol 2000;27:1352–7.

[61] Furumitsu Y, Yukioka K, Kojima A, Yukioka M, Shichikawa K, Ochi T, et al. Levels of urinary polyamines in patients with rheumatoid arthritis. J Rheumatol 1993;20:1661–5.

[62] Yukioka K, Wakitani S, Yukioka M, Furumitsu Y, Shichikawa K, Ochi T, et al. Polyamine levels in synovial tissues and synovial fluids of patients with rheumatoid arthritis. J Rheumatol 1992;19:689–92.

[63] Persson L. Polyamine homoeostasis. Essays Biochem 2009;46:11–24.

[64] Miao CG, Huang C, Huang Y, Yang YY, He X, Zhang L, et al. MeCP2 modulates the canonical Wnt pathway activation by targeting SFRP4 in rheumatoid arthritis fibroblast-like synoviocytes in rats. Cell Signal 2013;25:598–608.

[65] Karouzakis E, Rengel Y, Jüngel A, Kolling C, Gay RE, Michel BA, et al. DNA methylation regulates the expression of CXCL12 in rheumatoid arthritis synovial fibroblasts. Genes Immun 2011;12:643–52.

[66] Wilkinson J. Study reveals a DNA methylome signature in rheumatoid arthritis. Epigenomics 2012;4:481.

[67] Nakano K, Whitaker JW, Boyle DL, Wang W, Firestein GS. DNA methylome signature in rheumatoid arthritis. Ann Rheum Dis 2013;72:110–17.

[68] de la Rica L, Urquiza JM, Gómez-Cabrero D, Islam AB, López-Bigas N, Tegnér J, et al. Identification of novel markers in rheumatoid arthritis through integrated analysis of DNA methylation and microRNA expression. J Autoimmun 2013;41:6–16.

[69] Park SH, Kim SK, Choe JY, Moon Y, An S, Park MJ, et al. Hypermethylation of EBF3 and IRX1 genes in synovial fibroblasts of patients with rheumatoid arthritis. Mol Cells 2013;35:298–304.
[70] Stanczyk J, Pedrioli DM, Brentano F, Sanchez-Pernaute O, Kolling C, Gay RE, et al. Altered expression of MicroRNA in synovial fibroblasts and synovial tissue in rheumatoid arthritis. Arthritis Rheum 2008;58:1001–9.
[71] Karouzakis E, Trenkmann M, Gay RE, Michel BA, Gay S, Neidhart M. Epigenome analysis reveals TBX5 as a novel transcription factor involved in the activation of rheumatoid arthritis synovial fibroblasts. J Immunol 2014;193:4945–51.
[72] Huang YW, Kuo CT, Chen JH, Goodfellow PJ, Huang TH, Rader JS, et al. Hypermethylation of miR-203 in endometrial carcinomas. Gynecol Oncol 2014;133: 340–5.
[73] Sandhu R, Rivenbark AG, Mackler RM, Livasy CA, Coleman WB. Dysregulation of microRNA expression drives aberrant DNA hypermethylation in basal-like breast cancer. Int J Oncol 2014;44:563–72.
[74] Stanczyk J, Ospelt C, Karouzakis E, Filer A, Raza K, Kolling C, et al. Altered expression of microRNA-203 in rheumatoid arthritis synovial fibroblasts and its role in fibroblast activation. Arthritis Rheum 2011;63:373–81.

CHAPTER

24

DNA Methylation in Osteoporosis

OUTLINE

24.1 INTRODUCTION

Bone is a complex connective tissue characterized by a calcified extracellular matrix. This mineralized matrix is constantly being formed and resorbed throughout life, allowing the bone to adapt to daily mechanical loads and maintain skeletal properties and composition. The imbalance between bone formation and bone resorption leads to changes in bone mass. Osteoporosis is a progressive bone disease that is characterized by a decrease in bone mass and density, which can lead to an increased risk of fracture. The DNA polymorphisms identified so far in genome-wide studies explain less than 10% of the genetic risk, suggesting that other factors, and specifically environmental factors and epigenetic mechanisms, are involved in the pathogenesis of this disorder [1]. The form of osteoporosis most common in women after menopause is referred to as postmenopausal osteoporosis [2].

M. Neidhart: DNA Methylation and Complex Human Disease.
DOI: http://dx.doi.org/10.1016/B978-0-12-420194-1.00024-5

Senile osteoporosis occurs after age 75, and is seen in both females and males (ratio 2 : 1) [3]. Secondary osteoporosis may arise from chronic predisposing medical problems or disease, or prolonged use of medications such as glucocorticoids. The risk of osteoporosis fractures can be associated with lifestyle, including diet, alcohol, exercise, drug abuse, and cigarette smoking [4]. The underlying mechanism in all cases of osteoporosis is an imbalance between bone resorption and bone formation. After menopause, lack of estrogen increases bone resorption [2]. In addition, deficiency of calcium and vitamin D leads to impaired bone deposition [5]. The parathyroid glands react to low calcium levels by secreting parathyroid hormone, which increases bone resorption to ensure sufficient calcium in the circulation. The activation of osteoclasts is regulated by various molecular signals, such as nuclear factor κB ligand (RANKL) [6]. This molecule is produced by osteoblasts and other cells and stimulates RANK (receptor activator of nuclear factor κB). Osteoprotegerin (OPG) binds RANKL before it binds to RANK, thereby suppressing bone resorption. A decrease in genomic methylation commonly occurs in aging cells. There are positive correlations between global DNA hypomethylation (ALU) in blood cells and several age-related phenotypes in bone and body fat [7]. On the other hand, the risk to develop osteoporosis could even originate in prenatal life [4]. In this chapter, we will review the evidence that DNA methylation contributes to the pathogenesis of osteoporosis.

24.2 EARLY DEVELOPMENT

The pathology of bone development during intrauterine life could be a factor for osteoporosis. Moreover, the placental transfer of nutrients plays an important role in the building of bones of fetuses [4]. Various environmental factors, including nutrition state and maternal stress, may affect the epigenetic state of a number of genes during fetal development of bones. Gene–environment interactions are involved in fetal bone development, and the changes may cause osteoporosis later in life. A specific genotype may result in a phenotype only in certain environments. For example, interactions between the genome and fetal environment might establish basal levels of circulating growth hormone and contribute to accelerated bone loss. Many of these interactions occur primarily during early life, in fetuses and neonates. Feeding pregnant rats with a low-protein diet resulted in a reduced bone mass in the offspring that persisted up to 75 weeks of age and was accompanied by impaired proliferation and differentiation of bone marrow stromal cells, a population including osteoblast precursors [8]. Several environmental factors have been shown to directly influence methylation patterns and expression levels of some genes. Moreover, in some cases epigenetic alterations are transmissible beyond a single generation [9]. Genome-wide methylation profiling of bone samples has revealed differentially methylated regions in osteoarthritic and osteoporotic hip fractures [10]. Interestingly, these regions were enriched in genes associated with cell differentiation and skeletal embryogenesis, such as those in the homeobox superfamily, suggesting the existence of a developmental component in the predisposition to these disorders. It has been shown that maternal dietary restriction in rats induces changes in the methylation status of the genes coding for the glucocorticoid receptor and the peroxisomal proliferator-activated receptor alpha (PPARα) which persist after weaning and are transmitted to future generations [11].

After adjustment for age and gender, there are two major determinants of hip fracture risk: tall maternal height, and low rate of childhood growth. The hip fracture risk is elevated among babies born short. Fracture subjects are shorter at birth but of average height by the age of 7 years. Various studies suggest that epigenetic modulation of the hypothalamo-pituitary–adrenocortical axis represents a second mechanism by which poor maternal nutrition as well as environment may cause epigenetic changes in bones of fetuses and neonates. Protein restriction during mid- and late pregnancy was found to be associated with reduced methylation of important CpG-rich islands in the promoter region of the gene for the glucocorticoid receptor (GR) [12]. Prenatal glucocorticosteroids and neonatal behavioral manipulation (especially stress) promoted changes in histone acetylation and DNA methylation in a transcriptional factor binding site of GR [12]. These data suggest that the epigenetic mechanisms that respond to *in utero* or early-life exposures play a critical role in the growth/development of bones.

The role of the Wnt signaling pathway in the pathogenesis of osteoporosis is recognized [13]. Interestingly, wingless-type MMTV integration site family member 2 (WNT2) promoter methylation in the human placenta was found to be associated with low birth weight percentile in the neonate [14]. Normally, the WNT2 gene is highly expressed in the human placenta. High WNT2 promoter methylation is found only in placental tissue of low birth-weight neonates and not in the cord blood. The results showed that WNT2 expression can be downregulated in the placenta by DNA methylation of its promoter, and that high methylation is an epigenetic variant associated with reduced fetal growth potential. This observation supported the hypothesis that various epigenetic changes in the placenta responsible for nutrients transfer to the fetus play roles in abnormal fetal bone development and contribute to osteoporosis in adults. Moreover, the results support the view that low birth weight as well as low body mass in general may result from maternal, fetal, placental, and environmental factors.

24.3 OSTEOBLAST DIFFERENTIATION

Osteoblasts are the major cellular component of bone. They arise from mesenchymal stem cells and are responsible for deposition of a type I collagen matrix and mineralization. During osteogenic differentiation, DNA methylation remains relatively constant [15], with only transient changes in the expression of endogenous retroelements [16]. Nevertheless, it has been shown that DNA demethylation induced by 5-azacytidine facilitates osteogenic gene expression and differentiation [17]. Likewise, it has been reported that reduced DNA methylation of other CpG islands in the promoter regions of osteocalcin (BGLAP) and osteopontin (OPN) genes is associated with osteogenic differentiation [18,19]. Osteocalcin plays a role in bone mineralization and calcium ion homeostasis, while osteopontin is an important factor in bone remodeling. Thus, DNA methylation takes part in regulating the function of bone tissue cells [20]. The focal point of all skeletal pathologies is the deregulation of bone remodeling, mediated by bone-forming osteoblasts and bone-resorbing osteoclasts. In order to keep both processes in balance, the activity, differentiation, and apoptosis of both cell types have to be tightly regulated. In particular, the differentiation of osteoblasts and osteoclasts is accompanied by profound changes in gene expression. It has been shown that histone deacetylation and DNA methylation negatively regulate the expression of several genes associated with different stages of osteoblast differentiation.

The change in cell shape between bone-forming osteoblasts and osteocytes is accompanied by different gene expression profiles. It is known that alkaline phosphatase activity, an enzyme critical for bone mineralization, is high in osteoblasts, the unique bone-forming cell, whereas it is reduced in osteocytes, which do not produce bone [21]. Many studies [17,22–24] have shown that promoter methylation of the gene encoding a special form of alkaline phosphatase (ALPL, normally found in kidney, liver, and bone) is a major controller of its expression, since DNA demethylating agents repeatedly increased alkaline phosphatase expression and activity in osteoblasts under osteogenic and non-osteogenic conditions. More specifically, the degree of methylation in the CpG island located in the proximal region of the ALPL gene has been shown to be inversely associated with alkaline phosphatase levels. Accordingly, alkaline phosphatase-expressing osteoblasts displayed the least methylated ALPL gene, while it was hypermethylated in osteocytes, which do not express alkaline phosphatase. Bone lining cells, which represent a differentiation stage somewhere between osteoblasts and osteocytes, exhibited an intermediate methylation status; this indicates that ALPL methylation may increase progressively during osteoblastic differentiation, thus contributing notably to the phenotypic change associated with this process [24]. A proposed function of this form of the enzyme (so-called "tissue non-specific isoenzyme") is matrix mineralization.

Feedback from physical activity maintains bone mass, and feedback from osteocytes limits the size of the bone-forming unit. A number of mechanisms regulate bone density, including stress on the bone. An important additional mechanism is secretion by osteocytes, buried in the matrix, of sclerostin, a protein that interferes with a pathway that maintains osteoblast activity. Thus, when the osteon reaches a limiting size, it self-inactivates the bone synthesis pathway. Interestingly, the transition from osteoblasts toward osteocytes was accompanied by a progressive decrease in levels of the sclerostin gene (SOST) promoter methylation, enabling sclerostin to be expressed in an osteocyte-specific manner [25]. The SOST gene promoter region showed increased CpG methylation in patients treated for osteoporosis [26]. This observation suggest that increased SOST promoter methylation is a compensatory counteracting mechanism, which lowers serum sclerostin concentrations in an attempt to promote bone formation.

Many genes influencing osteogenesis, such as osterix (transcription factor SP7) [27], the osteogenic protein DLX-5 [27], bone morphogenic protein 7 (BMP7, OP-1) [28], aromatase [29], and estrogen receptors [30], are also regulated by DNA methylation. Aromatase is a specific component of the cytochrome P450 enzyme system that is responsible for the transformation of C19 androgen precursors into C18 estrogenic compounds. With aging, individual differences in aromatase activity and thus in estrogen levels may significantly affect bone loss and fracture risk [31]. In osteoblasts, increased intracellular homocysteine could increase methylation of the promoter of estrogen receptor alpha (ERα), decreasing its transcription and reducing the effect of estrogen, thereby favoring the development of osteoporosis [32]. Homocysteine is known to affect bone remodeling through several known mechanisms, such as increase in osteoclast activity, decrease in osteoblast activity, and direct action on bone matrix [33]. A tissue-specific accumulation of homocysteine in bone may be a promising mechanism explaining the adverse effects of hyperhomocysteinemia on bone [34].

As mentioned above, another critical factor in osteoclastogenesis is RANKL, which initiates the cascade of events for osteoclast maturation. RANKL is produced by many cell types, such as immune, vascular, and stromal cells. However, it is well accepted that osteoblasts, and probably osteocytes, are the major sources of this cytokine in bone tissue. Interestingly, DNA methylation at the proximal promoter of the RANKL gene inhibits its expression in a murine system, which results in impaired osteoclastogenesis [35]. Indeed, the methylation/demethylation of a CpG island located in the vicinity of the RANKL promoter in human bone cells regulates their activity [21]. The RANKL/OPG balance at the bone microenvironment is considered a major determinant of bone mass. DNA methylation and histone modifications cooperate to regulate OPG expression in nasopharyngeal carcinomas [36]. The methylation of CpG-rich regions in the OPG gene may also regulate OPG levels in human osteoblastic cells and non-neoplastic bone tissue [21,37].

24.4 VARIATIONS IN THE RESPONSE TO VITAMIN D

Epigenetic changes may also be associated with changes in DNA methylation of genes within or located near to 25-hydroxyvitamin D (vitamin D) response elements. Modified expression of the genes encoding placental calcium transporters by epigenetic regulation may indicate that maternal vitamin D status influences bone mineral homeostasis in the neonate. Thus, a study [38] regarding methylation and vitamin D receptors and placental calcium transporters suggests that epigenetic regulation might explain how maternal vitamin D levels affect bone mineralization in the neonate. Vitamin D and calcium availability during pregnancy have been shown to influence both fetal skeletal development and childhood bone mass [39]. It has been hypothesized that osteoporosis results from an alteration in the manner in which serum vitamin D is assimilated by the cells. A study [40] investigated DNA methylation levels of cytochrome P450 (CYP) enzymes (CYP2R1, CYP24A1, CYP27A1, and CYP27B1) as potential biomarkers predicting vitamin D response variation. Compared to non-responders, responders had significantly lower baseline DNA methylation levels in the promoter regions of CYP2R1 and CYP24A1. Baseline CYP2R1 and CYP24A1 methylation negatively correlated with the serum level of vitamin D. Further investigations are needed.

Figure 24.1 illustrates factors influencing the epigenome that affect bone formation and may contribute to osteoporosis.

24.5 CONCLUSION

Common skeletal disorders, such as osteoporosis and osteoarthritis, are the result of a complex interplay of genetic and acquired factors. However, the genetic factors so far identified explain less than 10% of the genetic risk. This suggests that mechanisms not related to the DNA sequence may be involved in the development of these diseases. This could be the case for DNA methylation and its mediators. DNA

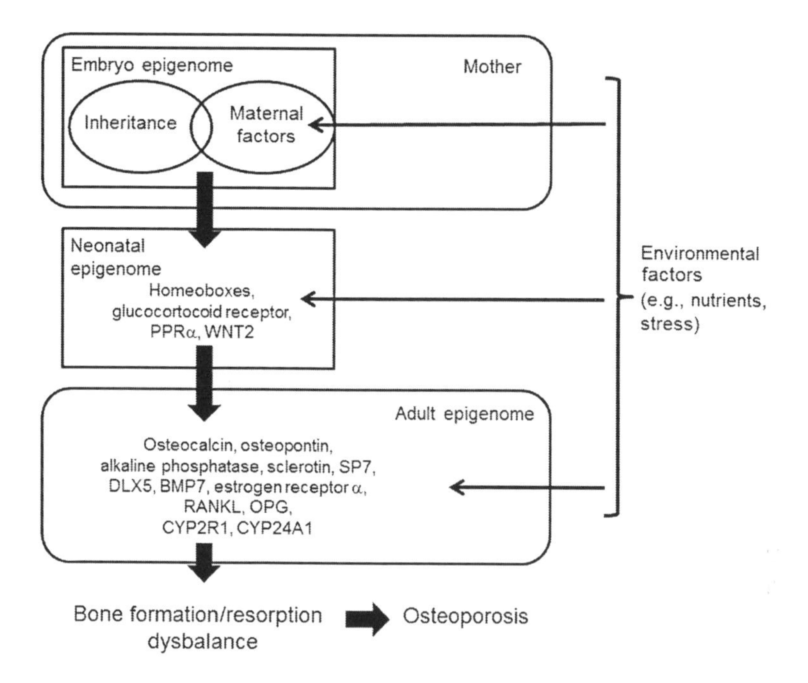

FIGURE 24.1 Factors influencing the epigenome, from early life to adult, affecting bone formation and resorption, may contribute to the pathogenesis of osteoporosis.

methylation marks are heritable at least through cell divisions, control gene response to environment, change with aging, and underlie cell commitment and the spatiotemporal control of gene expression. DNA methylation plays an important role in the differentiation of cells of the osteoblastic and osteoclastic lineages. Therefore, it is tempting to speculate that the aberrant phenotypes observed in bone diseases might be the consequence of a combination of intrinsic and environmental factors, including gene sequence variations and epigenetic signatures. These changes in the epigenome may occur in early life and in adults (Figure 24.1) and lead to diseases like osteoarthritis and osteoporosis. However, it is important to note that DNA methylation is only one of the mechanisms underlying gene expression. Thus, the integration of knowledge from both epigenenomics and genomics, together with other "omics" (i.e., transcriptomics, proteomics), will be essential for full understanding of the underlying mechanisms that govern the initiation and progression of bone diseases. Although there is still a long way to go, further studies in bone epigenetics may open a new door for drug development combining genetic and epigenetic strategies.

References

[1] Delgado-Calle J, Riancho JA. The role of DNA methylation in common skeletal disorders. Biology 2012;1:698–713.
[2] Diab DL, Watts NB. Postmenopausal osteoporosis. Curr Opin Endocrinol Diabetes Obes 2013;20:501–9.
[3] Khosla S. Pathogenesis of age-related bone loss in humans. J Gerontol A Biol Sci Med Sci 2013;68:1226–35.
[4] Bocheva G, Boyadjieva N. Epigenetic regulation of fetal bone development and placental transfer of nutrients: progress for osteoporosis. Interdiscip Toxicol 2011; 4:167–72.
[5] Aggarwal S, Nityanand. Calcium and vitamin D in post menopausal women. Indian J Endocrinol Metab 2013;17:S618–20.
[6] You L, Chen L, Pan L, Chen JY. New insights into the gene function of osteoporosis. Front Biosci 2013;18:1088–97.
[7] Jintaridth P, Tungtrongchitr R, Preutthipan S, Mutirangura A. Hypomethylation of Alu elements in post-menopausal women with osteoporosis. PloS One 2013;8:e70386.
[8] Oreffo RO, Lashbrooke B, Roach HI, Clarke NM, Cooper C. Maternal protein deficiency affects mesenchymal stem cell activity in the developing offspring. Bone 2003;33:100–7.
[9] Jiang YH, Bressler J, Beaudet AL. Epigenetics and human disease. Annu Rev Genomics Hum Genet 2004;5:479–510.
[10] Delgado-Calle J, Fernández AF, Sainz J, Zarrabeitia MT, Sañudo C, García-Renedo R, et al. Genome-wide profiling of bone reveals differentially methylated regions in osteoporosis and osteoarthritis. Arthritis Rheum 2013;65:197–205.
[11] Lillycrop KA, Phillips ES, Torrens C, Hanson MA, Jackson AA, Burdge GC. Feeding pregnant rats a protein-restricted diet persistently alters the methylation of specific cytosines in the hepatic PPAR alpha promoter of the offspring. Br J Nutr 2008; 100:278–82.
[12] Weaver IC, Cervoni N, Champagne FA, D'Alessio AC, Sharma S, Seckl JR, et al. Epigenetic programming by maternal behavior. Nat Neurosci 2004;7:847–54.
[13] Garcia-Ibarbia C, Delgado-Calle J, Casafont I, Velasco J, Arozamena J, Pérez-Núñez MI, et al. Contribution of genetic and epigenetic mechanisms to Wnt pathway activity in prevalent skeletal disorders. Gene 2013;532:165–72.
[14] Ferreira JC, Choufani S, Grafodatskaya D, Butcher DT, Zhao C, Chitayat D, et al. WNT2 promoter methylation in human placenta is associated with low birthweight percentile in the neonate. Epigenetics 2011;6:440–9.
[15] Hakelien AM, Bryne JC, Harstad KG, Lorenz S, Paulsen J, Sun J, et al. The regulatory landscape of osteogenic differentiation. Stem Cells 2014;32:2780–93.
[16] Kang MI, Kim HS, Jung YC, Kim YH, Hong SJ, Kim MK, et al. Transitional CpG methylation between promoters and retroelements of tissue-specific genes during human mesenchymal cell differentiation. J Cell Biochem 2007;102:224–39.
[17] Locklin RM, Oreffo RO, Triffitt JT. Modulation of osteogenic differentiation in human skeletal cells *in vitro* by 5-azacytidine. Cell Biol Int 1998;22:207–15.
[18] Arnsdorf EJ, Tummala P, Castillo AB, Zhang F, Jacobs CR. The epigenetic mechanism of mechanically induced osteogenic differentiation. J Biomech 2010;43:2881–6.
[19] Villagra A, Gutiérrez J, Paredes R, Sierra J, Puchi M, Imschenetzky M, et al. Reduced CpG methylation is associated with transcriptional activation of the bone-specific rat osteocalcin gene in osteoblasts. J Cell Biochem 2002;85:112–22.
[20] Vrtacnik P, Marc J, Ostanek B. Epigenetic mechanisms in bone. Clin Chem Lab Med 2014;52:589–608.

[21] Delgado-Calle J, Garmilla P, Riancho JA. Do epigenetic marks govern bone mass and homeostasis? Curr Genomics 2012;13:252–63.
[22] Vaes BL, Lute C, van der Woning SP, Piek E, Vermeer J, Blom HJ, et al. Inhibition of methylation decreases osteoblast differentiation via a non-DNA-dependent methylation mechanism. Bone 2010;46:514–23.
[23] El-Serafi AT, Oreffo RO, Roach HI. Epigenetic modifiers influence lineage commitment of human bone marrow stromal cells: Differential effects of 5-aza-deoxycytidine and trichostatin A. Differentiation 2011;81:35–41.
[24] Delgado-Calle J, Sañudo C, Sánchez-Verde L, García-Renedo RJ, Arozamena J, Riancho JA. Epigenetic regulation of alkaline phosphatase in human cells of the osteoblastic lineage. Bone 2011;49:830–8.
[25] Delgado-Calle J, Sañudo C, Bolado A, Fernández AF, Arozamena J, Pascual-Carra MA, et al. DNA methylation contributes to the regulation of sclerostin expression in human osteocytes. J Bone Miner Res 2012;27:926–37.
[26] Reppe S, Noer A, Grimholt RM, Halldórsson BV, Medina-Gomez C, Gautvik VT, et al. Methylation of bone SOST, its mRNA, and serum sclerostin levels correlate strongly with fracture risk in postmenopausal women. J Bone Miner Res 2015; 30:249–56.
[27] Lee JY, Lee YM, Kim MJ, Choi JY, Park EK, Kim SY, et al. Methylation of the mouse DIx5 and Osx gene promoters regulates cell type-specific gene expression. Mol Cells 2006;22:182–8.
[28] Loeser RF, Im HJ, Richardson B, Lu Q, Chubinskaya S. Methylation of the OP-1 promoter: potential role in the age-related decline in OP-1 expression in cartilage. Osteoarthritis Cartilage 2009;17:513–17.
[29] Demura M, Bulun SE. CpG dinucleotide methylation of the CYP19 I.3/II promoter modulates cAMP-stimulated aromatase activity. Mol Cell Endocrinol 2008; 283:127–32.
[30] Penolazzi L, Lambertini E, Giordano S, Sollazzo V, Traina G, del Senno L, et al. Methylation analysis of the promoter F of estrogen receptor alpha gene: effects on the level of transcription on human osteoblastic cells. J Steroid Biochem Mol Biol 2004;91:1–9.
[31] Gennari L, Merlotti D, Nuti R. Aromatase activity and bone loss. Adv Clin Chem 2011;54:129–64.
[32] Lv H, Ma X, Che T, Chen Y. Methylation of the promoter A of estrogen receptor alpha gene in hBMSC and osteoblasts and its correlation with homocysteine. Mol Cell Biochem 2011;355:35–45.
[33] Vacek TP, Kalani A, Voor MJ, Tyagi SC, Tyagi N. The role of homocysteine in bone remodeling. Clin Chem Lab Med 2013;51:579–90.
[34] Herrmann M, et al. Hyperhomocysteinemia induces a tissue specific accumulation of homocysteine in bone by collagen binding and adversely affects bone. Bone 2009;44:467–75.
[35] Kitazawa R, Kitazawa S. Methylation status of a single CpG locus 3 bases upstream of TATA-box of receptor activator of nuclear factor-kappaB ligand (RANKL) gene promoter modulates cell- and tissue-specific RANKL expression and osteoclastogenesis. Mol Endocrinol 2007;21:148–58.
[36] Lu TY, Kao CF, Lin CT, Huang DY, Chiu CY, Huang YS, et al. DNA methylation and histone modification regulate silencing of OPG during tumor progression. J Cell Biochem 2009;108:315–25.
[37] Delgado-Calle J, Sañudo C, Fernández AF, García-Renedo R, Fraga MF, Riancho JA. Role of DNA methylation in the regulation of the RANKL-OPG system in human bone. Epigenetics 2012;7:83–91.

[38] Wood CL, Wood AM, Harker C, Embleton ND. Bone mineral density and osteoporosis after preterm birth: the role of early life factors and nutrition. Int J Endocrinol 2013;2013:902513.
[39] Mahon P, Harvey N, Crozier S, Inskip H, Robinson S, Arden N, et al. Low maternal vitamin D status and fetal bone development: cohort study. J Bone Miner Res 2010;25:14–19.
[40] Zhou Y, Zhao LJ, Xu X, Ye A, Travers-Gustafson D, Zhou B, et al. DNA methylation levels of CYP2R1 and CYP24A1 predict vitamin D response variation. J Steroid Biochem Mol Biol 2014;144 Pt A:207–14.

CHAPTER

25

Epigenetic Therapies

OUTLINE

25.1 INTRODUCTION

Abnormal DNA methylation events occur early during carcinogenesis, resulting in premalignant states [1,2]. These altered methylation patterns include global DNA hypomethylation and promoter-specific hypermethylation. Tumor suppressor gene p16INK4A/CDKN2A provides one such example of increasing frequency of promoter-specific gain of DNA

M. Neidhart: DNA Methylation and Complex Human Disease.
DOI: http://dx.doi.org/10.1016/B978-0-12-420194-1.00025-7

hypermethylation during lung carcinogenesis, starting from 17% of promoter methylation in lung airway basal cell hyperplasia and rising to 24% methylation in squamous metaplasia and 50% methylation in carcinoma *in situ* [3]. Inhibition of DNMTs correlates with reduction in tumorigenicity and increased expression of tumor suppressor genes [4,5]. Hence, DNMTs are considered valuable targets for the design of specific anticancer strategies [5,6]. A similar approach has been used to reverse the pathological development of cancer stem cells. These cells, which are completely different from each other in their differentiated forms, still retain the complete complement of the genes in their genome which they have inherited from their ancestral embryonic stem cell. Also, they still have the potential to become totipotent under specific circumstances. This process is accompanied by the erasure of different epigenetic marks such as CpG methylation and chromatin remodeling through different types of covalent histone modifications, termed epigenetic reprogramming. The generation of cancer stem cells in the process of carcinogenesis may also involve similar kind of epigenetic reprogramming where, in contrast, it leads to the loss of expression of genes specific to the differentiated state and regaining of stem cell-specific characteristics [7]. In these cases, epigenetic therapy might be more complicated than simply re-expressing some "tumor suppressor genes" by demethylating agents. In recent years, "nutriepigenetics," which focuses on the influence of dietary agents on epigenetic mechanism(s), has emerged as an exciting novel area in epigenetics research [8]. Targeting of aberrant epigenetic modifications has gained considerable attention in cancer chemoprevention research because, unlike genetic changes, epigenetic alterations are reversible and occur during early carcinogenesis. Methyl donor supplements can reverse the DNA hypomethylation that occurs in many cancers and other diseases. Interestingly, a combination therapy of demethylating agents and methyl donors has been proposed in, for example, breast cancer. This chapter will focus on recent advances regarding therapies targeting the DNA methylation machinery.

25.2 DEMETHYLATING AGENTS

25.2.1 Pharmaceutical Compounds

In cancer, it is important to distinguish between the causes – either mutation or epimutation of "tumor suppressor genes." In the latter case, most of the hypermethylated genes are regulators of proliferation and differentiation [9]. A strategy is to disrupt the methylation machinery by inhibiting DNA methyltransferases (DNMTs). Broadly, DNMT inhibitors can be divided into nucleoside-derived and non-nucleoside inhibitors. The former class includes the two parenterally administered DNMT inhibitors 5-azacitidine and decitabine. 5-Azacitidine was initially evaluated in patients with acute myelogenous leukemia [10]. Although promising, the clinical toxicities observed were greatly limiting and the drug failed to progress due to the lack of a sufficiently wide therapeutic window. Following identification of the hypomethylating mechanism of action of the drug, several clinical studies [6] using low-dose 5-azacitidine demonstrated clinical effectiveness. Both 5-azacitidine and decitabine act as irreversible covalent inhibitors of all catalytically active DNMT isoforms following incorporation into DNA [11]. Although structurally similar, 5-azacitidine may be somewhat less selective and have more pleiotropic effects based on its ability to affect various RNA processing systems, including general ribosomal transcription [12]. New drugs, such as SGI-110 and zebularine, are in development and appear even more promising [11,13,14]. Zebularine has been reported to have better stability and less toxicity compared to the older agents, although definitive clinical benefits have not yet been established [15]. The disruption of DNMT function by different means has been shown to induce re-expression of many silenced tumor suppressor genes in different types of cancers, as well as silencing of tumor promoter genes [16–18]. Demethylating agents may have applications not only in cancer but also in neurological diseases [19]. For example, in mice, treatment with DNMT inhibitors can confer stroke protection after mild ischemia; furthermore, haploinsufficiency for DNMT1 in mice is associated with smaller infarction volumes after acute ischemia and stroke [20,21]. When administered directly into the brain tissue of mice and rats, DNMT inhibitors disrupt synaptic plasticity and hippocampal learning and memory, and they are powerful modulators of reward and addiction behaviors [22–25].

25.2.2 Natural Compounds

Not only pharmaceutical drugs but also natural compounds, such as black raspberry extracts, can affect the methylation machinery. Interleukin-10 knockout mice fed with 5% black raspberries had significantly less colonic ulceration compared with knockout mice that consumed a control diet [16]; this is because black raspberry-derived anthocyanins are associated with decreased expression of DNMT1 and DNMT3B [4,26]. Similarly, oral administration of black raspberry powder decreased promoter methylation of tumor suppressor genes in tumors from patients with colorectal cancer [26]. The anthocyanins in black raspberries are responsible, at least in part, for their cancer-inhibitory effects. Annurca apple is an apple variety from southern Italy that is rich in polyphenols. Annurca apple polyphenol extracts inhibited the expression of DNMT1 and DNMT3B in colon cancer cells [27]. Acetyl-keto-β-boswellic acid, derived from the plant *Boswellia serrata*, an Indian frankincense, also inhibits DNMT activity in colorectal cancer cell lines [28]. Furthermore, the flavonolignan silibinin, which is the main pharmacologically active component of the milk thistle plant (*Silybum marianum*), with anticancer properties, is able to significantly inhibit DNMT activity in colon cancer cells [29]. Similarly, genistein, present in soybean, modulates DNMTs and reactivates tumor suppressor genes in esophageal cancer cells [30,31]. Epigallocatechin-3-gallate, a major component of green tea polyphenols, decreases DNMT1 expression in cancer cells [18,32] and can restore estrogen receptor expression in breast cancer [18]. Two common catechol-containing coffee polyphenols, caffeic acid and chlorogenic acid, inhibit DNMT1 levels in breast ancer cells [33]. *Thymus serpyllum* (wild thyme) is an aromatic medicinal plant possessing several biological properties, including anticancer activity, and extracts of this plant were found to inhibit DNMTs in breast cancer cells [34]. Kazinol Q, a natural product from formosan plants, was found to act as an inhibitor of DNMTs [35]. In addition, a more recent study [36] confirmed that natural compounds such as epigallocatechin-3-gallate, genistein, withaferin A, curcumin, resveratrol, and guggulsterone all inhibit DNMT expression in breast cancer cells.

25.3 CANCER STEM CELLS – TARGETED THERAPIES

Cancer stem cells ensure the renewal of tumor cells; therefore, targeting these tumor-initiating cells is a logical strategy to eliminate the cause of disease. The idea emerged from a study investigating acute myeloid leukemia, in which the pattern of methylation of specific CpGs is correlated with tumor prognosis [37]. Specifically, promoter methylation-mediated silencing of ASCL2 and LGR5 leads to this poor prognosis, while their re-expression is associated with reduced tumor growth [38]. Interestingly, ASCL2 and LGR5 are cancer stem cell markers associated with Wnt signaling target proteins [7]. During the process of somatic cell differentiation, gradual alterations in the patterns of DNA methylation occurred compared to their stem cells [7,39]. Cancer stem cells undergo nuclear reprogramming, including an active DNA demethylation to regain stem-cell state and pluripotency [40,41]. Simultaneously, the cellular differentiation markers are progressively silenced by hypermethylation [42]. Tumor-initiating cells must undergo a similar series of methylation changes and chromatin remodeling to finally give rise to potential cancer stem cells. Thus, during the process of carcinogenesis the epigenetic modulators may function via two mechanisms; either these factors may facilitate the binding of the overexpressed transcription factors by exposing the target oncogenic DNA sites, functioning as the passenger events of carcinogenesis; or they might even initiate the transcription factor overexpression, thus playing the key functional role of driving events [7]. The process can be divided into two phases: DNA demethylation, and reprogramming. Epigenetic interventions can be tried in both.

25.3.1 DNA Demethylation

Global DNA hypomethylation is a hallmark of many cancers. Erasing the epigenetic marks in a somatic cell will allow it to be reprogrammed. Interestingly, CD133 (prominin 1), a universal cancer stem cell marker, is directly regulated by epigenetic modifications in ovarian cancer. The $CD133^+$ population is distinguishable from the $CD133^-$ population by retaining a promoter hypomethylated state [43]. $CD133^+$ brain tumor cells show the cancer stem cell characteristics, such as initiation of neurospheres exhibiting self-renewal, differentiation, and proliferation. CD133 expression is regulated through an epigenetic mechanism in human gliomas, where hypomethylation of promoter regions of higher activity induced lower CD133 expression [44]. Similar DNA hypomethylation was reported to regulate CD133 expression in colorectal and glioblastoma tumors. In these types of tumors, CD133-negative cells show higher methylation of promoter CpGs, while cells with higher CD133 expression lack such methylation [45]. Another study in hepatocellular carcinoma reported that TGF-β is involved in the epigenetic regulation of CD133 expression by inhibiting the expression of DNMT1 and DNMT3B, which causes demethylation of the transcriptionally active region of the CD133 promoter [46]. We will discuss the possibility of using methyl donors as remethylating agents, restoring global DNA methylation not only in inflammatory diseases but also in many types of cancer, even in combination with a DNMT1 inhibitor [47].

25.3.2 Reprogramming

Active DNA demethylation and remethylation are considered requirements for cells to regain the stem-cell state during induction of pluripotency [40,41] and cancer transformation [48]. Thus, *de novo* methylation has been shown to play a very important role in the regulation of stem cell characteristics [49]. A recent glioblastoma iPSC reprogramming study [50] showed that some of the tumors can be reprogrammed to lose their malignant behavior, by methylation-induced silencing of cancer-promoting pathways depending on their lineage identity. In

DNMT3A-null mice, each passage resulted in lowered differentiation capacity of hematopoietic stem cells. DNMT3A is indispensable for the proper differentiation of these cells [51]. Higher enrichment of DNMT3B at the CpG islands of hypermethylated genes suggests that this enzyme functions more importantly in furnishing the aberrant methylation pattern observed in the case of cancer stem cells, which helps them to maintain an undifferentiated state [52]. In lung cancer stem cell populations, knockdown of DNMT1 reduced the stem cell properties, suggesting the possibility that DNMT1 inhibition might prove to be an important therapeutic strategy to eliminate tumor cells [53]. Prolonged exposure of stem/progenitor cells to higher doses of DNMT inhibitors has proven to induce differentiation in the progenitor cells [54]. Administration of DNMT inhibitor at a low dose was reported to be capable of reducing the stem cell-like properties of $ALDH^+$ ovarian cancer stem cells by reprogramming these cells into a more differentiated state [55]. Similarly, DNMT inhibitors can be utilized to reset the cancer stem cells toward a differentiated phenotype, thus rendering different types of cancers more susceptible to available therapeutic options.

25.4 METHYL DONORS

Methyl donors such as L-methionine, S-adenosylmethionine (SAM), or betaine supplements are involved in "nutriepigenetics." The DNMTs catalyze the transfer of a methyl group from the methyl donor (i.e., SAM) to the C-5 of cytosine in DNA. Although *in vitro* and animal evidence linking methyl donors and DNA methylation is fairly extensive, epidemiological evidence is less comprehensive [56].

25.4.1 Neurological Disorders

Methyl donors, particularly SAM, have been reported to be beneficial in several neurological disorders, such as depression [57–59] and neurodegeneration [60]. In mice models of neurodegeneration, for example, vitamin B6 deficiency leads to PSEN1 promoter demethylation and amyloid β deposition in the brains. By contrast, SAM supplementation induces an opposite tendency, restoring PSEN1 methylation levels and reducing the progression of Alzheimer disease-like features [61]. Thus, in patients, methyl donors are being investigated for possible therapeutic effects to rescue the memory and cognitive decline found in Alzheimer's disease [60]. There is also a possibility that methyl donors help to rescue in drug addiction. In mice, SAM modulated cocaine-induced DNA methylation by inhibiting both promoter-associated

CpG-island hyper- and hypomethylation in the nucleus accumbens, but not in the reference tissue cerebellum [62]. The modulating effect of SAM is in part due to decreased methyltransferase activity via downregulation of the DNMT3A transcript.

25.4.2 Cancer Metastasis, Oncogene Expression, and Neoangiogenesis

As mentioned above, demethylating agents such as 5-azacytidine and other newer compounds are currently being tested on various cancers, including solid tumors, with the hope that so-called "tumor suppressor genes" are re-expressed. However, such drugs could also induce methylated prometastatic genes by DNA demethylation and thus induce cancer cell invasiveness [47]. In colorectal cancer, SAM reverses the hypomethylation of TIMP2, an inhibitor of matrix metalloproteinases (including MMP2) and reduces the invasive potential of these tumor cells [63]. An interesting combination of 5-azacytidine and SAM has been tested in breast cancer cell lines. SAM is able to reverse global- and gene-specific demethylation induced by 5-azacytidine; it prevents the activation of prometastatic genes uPA and MMP2, resulting in inhibition of cell invasiveness while augmenting the growth inhibitory effects of 5-azacytidine and its effects on tumor suppressor genes. Thus, a combination of drugs acting on the DNA methylation machinery at different levels could represent a new option for epigenetic therapy of cancer.

In precancerous cells, a global DNA hypomethylation can activate oncogene transcription, thus promoting carcinogenesis and tumor development. For example, in gastric cancer cells and colon cancer cells, C-MYC and H-RAS promoters are hypomethylated early in the disease. SAM treatment results in a heavy methylation of these promoters, which consequently downregulated transcripts and protein levels [64]. In these cancer cells, DNA hypomethylation correlates also with vascular endothelial growth factor-C (VEGFC) expression, which triggers the neovascularization essential for solid tumor survival. In gastric cancer cells, the VEGFC promoter is unmethylated [65]. After treatment with SAM, the VEGFC promoter becomes highly methylated and VEGFC expression is downregulated. SAM also significantly inhibits tumor growth *in vitro* and *in vivo*.

25.4.3 Inflammation, Degeneration, Chronic Pain, and Autoimmunity

Methyl donors, particularly SAM, were found or suggested to be safe and beneficial in several other non-neoplastic disorders, such as psychiatric disorders (mentioned above), liver disease [66], osteoarthritis [67–69], chronic pain [70], chronic fatigue syndrome, and fibromyalgia [71]. In macrophages, SAM is able to lower lipopolysaccharide-induced expression of the pro-inflammatory cytokine TNF-α and increase the expression of the anti-inflammatory cytokine IL-10 [72]. This is associated with changes in specific gene promoter methylation. In humans, preoperative administration of SAM has been found useful and safe in reducing hepatic ischemia–reperfusion injury in partial hepatectomy, especially for hepatocellular carcinoma patients whose disease is associated with chronic hepatitis B virus infection and cirrhosis [73]; this is because, in SAM-treated patients, plasma levels of TNF-α and IL-6 are reduced. Similarly, SAM reduces inflammation-induced colon cancer and inhibits several pathways important in colon carcinogenesis, in particular the IL-6-dependent signaling pathway [74].

The effect of SAM on immune reactions is beginning to be evaluated. Incubation of T and B lymphocytes with SAM inhibits both TCR-mediated T-cell proliferation and BCR (anti-IgM)-triggered B-cell proliferation in a dose-dependent manner [75]. However, in autoimmunity, methyl donors can both exacerbate or ameliorate disease, as shown in a transgenic murine lupus model [76]. In rheumatoid arthritis, methyl donors have not been evaluated in patients so far. However, *in vitro* treatment of rheumatoid synovial fibroblasts with SAM decreased their aggressiveness [77]. The efficiency of SAM can be optimized by inhibiting the polyamine recycling pathway that is intrinsically upregulated in those cells and is possibly responsible for an increased consumption of SAM.

25.5 POLYAMINE RECYCLING INHIBITORS

The biosynthesis and recycling pathways of polyamines were presented in Chapter 23. Brooks [78–80] proposed a hypothesis that some autoimmune diseases occur due to a loss of dose compensation of X-linked polyamine genes at Xp22.1, which impacts intracellular methylation [25]. In this respect, the two genes of interest are spermine synthase (SmS) and spermine/spermidine N1-acethyltransferase (SSAT1). Overexpression of SSAT1 increases recycling of polyamines. The resulting decrease in SAM would hamper methylation in the cell, and this could account for the aberrant DNA methylation observed in cancer cells, autoimmune T cells, and rheumatoid arthritis synovial fibroblasts. Confirming this hypothesis, SSAT1 was found to be increased in rheumatoid arthritis synovial fibroblasts. A study [81] showed that an intrinsically activated PMFBP1/SSAT1-dependent pathway is an important factor leading to global DNA hypomethylation in these cells. Recently, Berenil® (diminazene aceturate), an inhibitor of SSAT1 activity, has been shown to increase the 5-methylcytosine content of DNA in rheumatoid arthritis synovial fibroblasts [77]. In addition, Berenil increases the levels of DNMT1, decreases the level of activation marker and matrix metalloproteinase-1 (MMP1), and alters the adhesion capability of these cells. As expected, Berenil is more efficient in cells with higher levels of SSAT1. Most interestingly, the combination of Berenil and SAM reduced the invasiveness of rheumatoid arthritis synovial fibroblasts into cartilage by 70%. Epigenetic changes induced by increased levels of SSAT1 also play a role in myeloid leukemia [82]. In glioblastoma multiforme, increased SSAT1 is associated with resistance to ionizing radiation, histone 3 hyperacethylation, and poor prognosis [83]. Reduced SSAT1 expression in the brain, however, can cause psychiatric disorders [84]. Clearly, further investigations and more specific pharmaceutical inhibitors are needed.

25.6 CONCLUSION

New demethylating drugs are in development, many natural compounds can be tested, and a combination of demethylating drug and methyl donors appears promising in the treatment of epimutations in cancer. Missing is information regarding the differences between diverse methyl donors, i.e., SAM, L-methionine, and betaine, not only in their effectiveness in various pathological states, but also in off-target effects. Regarding cancer stem cells, deeper understanding of the role of different DNMTs in setting up the aberrant CpG methylation patterns will help the researchers to appropriately utilize available DNMT inhibitors for reversing the hypermethylation. Finally, a promising research area appears to be inhibition of the polyamine recycling pathway, but this is delayed because of the absence of an appropriate and specific inhibitor of SSAT1.

References

[1] Esteller M. Epigenetic lesions causing genetic lesions in human cancer: promoter hypermethylation of DNA repair genes. Eur J Cancer 2000;36:2294–300.
[2] Baylin SB, Jones PA. A decade of exploring the cancer epigenome - biological and translational implications. Nat Rev Cancer 2011;11:726–34.
[3] Belinsky SA. Role of the cytosine DNA-methyltransferase and p16INK4a genes in the development of mouse lung tumors. Exp Lung Res 1998;24:463–79.
[4] Subramaniam D, Thombre R, Dhar A, Anant S. DNA methyltransferases: a novel target for prevention and therapy. Front Oncol 2014;4:80.
[5] Khan H, Vale C, Bhagat T, Verma A. Role of DNA methylation in the pathogenesis and treatment of myelodysplastic syndromes. Semin Hematol 2013;50:16–37.

[6] Khan C, Pathe N, Fazal S, Lister J, Rossetti JM. Azacitidine in the management of patients with myelodysplastic syndromes. Ther Adv Hematol 2012;3:355–73.
[7] Shukla S, Meeran SM. Epigenetics of cancer stem cells: Pathways and therapeutics. Biochim Biophys Acta 2014;1840:3494–502.
[8] Thakur VS, Deb G, Babcook MA, Gupta S. Plant phytochemicals as epigenetic modulators: role in cancer chemoprevention. AAPS J 2014;16:151–63.
[9] Easwaran H, Johnstone SE, Van Neste L, Ohm J, Mosbruger T, Wang Q, et al. A DNA hypermethylation module for the stem/progenitor cell signature of cancer. Genome Res 2012;22:837–49.
[10] Griffiths EA, Gore SD. Epigenetic therapies in MDS and AML. Adv Exp Med Biol 2013;754:253–83.
[11] Dhanak D, Jackson P. Development and classes of epigenetic drugs for cancer. Biochem Biophys Res Commun 2014. [in press].
[12] Hollenbach PW, Nguyen AN, Brady H, Williams M, Ning Y, Richard N, et al. A comparison of azacitidine and decitabine activities in acute myeloid leukemia cell lines. PloS One 2010;5:e9001.
[13] Coral S, Parisi G, Nicolay HJ, Colizzi F, Danielli R, Fratta E, et al. Immunomodulatory activity of SGI-110, a 5-aza-2′-deoxycytidine-containing demethylating dinucleotide. Cancer Immunol Immunother 2013;62:605–14.
[14] Foulks JM, Parnell KM, Nix RN, Chau S, Swierczek K, Saunders M, et al. Epigenetic drug discovery: targeting DNA methyltransferases. J Biomol Screen 2012;17:2–17.
[15] Song SH, Han SW, Bang YJ. Epigenetic-based therapies in cancer: progress to date. Drugs 2011;71:2391–403.
[16] Wang LS, Kuo CT, Huang TH, Yearsley M, Oshima K, Stoner GD, et al. Black raspberries protectively regulate methylation of Wnt pathway genes in precancerous colon tissue. Cancer Prev Res 2013;6:1317–27.
[17] Meeran SM, Patel SN, Chan TH, Tollefsbol TO. A novel prodrug of epigallocatechin-3-gallate: differential epigenetic hTERT repression in human breast cancer cells. Cancer Prev Res 2011;4:1243–54.
[18] Meeran SM, Patel SN, Li Y, Shukla S, Tollefsbol TO. Bioactive dietary supplements reactivate ER expression in ER-negative breast cancer cells by active chromatin modifications. PloS One 2012;7:e37748.
[19] Jakovcevski M, Akbarian S. Epigenetic mechanisms in neurological disease. Nat Med 2012;18:1194–204.
[20] Endres M, Fan G, Meisel A, Dirnagl U, Jaenisch R. Effects of cerebral ischemia in mice lacking DNA methyltransferase 1 in post-mitotic neurons. Neuroreport 2001; 12:3763–6.
[21] Endres M, et al. DNA methyltransferase contributes to delayed ischemic brain injury. J Neurosci 2000;20:3175–81.
[22] Levenson JM, Roth TL, Lubin FD, Miller CA, Huang IC, Desai P, et al. Evidence that DNA (cytosine-5) methyltransferase regulates synaptic plasticity in the hippocampus. J Biol Chem 2006;281:15763–73.
[23] Han J, Li Y, Wang D, Wei C, Yang X, Sui N. Effect of 5-aza-2-deoxycytidine microinjecting into hippocampus and prelimbic cortex on acquisition and retrieval of cocaine-induced place preference in C57BL/6 mice. Eur J Pharmacol 2010;642:93–8.
[24] Miller CA, Sweatt JD. Covalent modification of DNA regulates memory formation. Neuron 2007;53:857–69.
[25] LaPlant Q, Vialou V, Covington III HE, Dumitriu D, Feng J, et al. Dnmt3a regulates emotional behavior and spine plasticity in the nucleus accumbens. Nat Neurosci 2010;13:1137–43.

[26] Wang LS, Kuo CT, Cho SJ, Seguin C, Siddiqui J, Stoner K, et al. Black raspberry-derived anthocyanins demethylate tumor suppressor genes through the inhibition of DNMT1 and DNMT3B in colon cancer cells. Nutr Cancer 2013;65:118–25.
[27] Fini L, Selgrad M, Fogliano V, Graziani G, Romano M, Hotchkiss E, et al. Annurca apple polyphenols have potent demethylating activity and can reactivate silenced tumor suppressor genes in colorectal cancer cells. J Nutr 2007;137:2622–8.
[28] Shen Y, Takahashi M, Byun HM, Link A, Sharma N, Balaguer F, et al. Boswellic acid induces epigenetic alterations by modulating DNA methylation in colorectal cancer cells. Cancer Biol Ther 2012;13:542–52.
[29] Kauntz H, Bousserouel S, Gosse F, Raul F. Epigenetic effects of the natural flavonolignan silibinin on colon adenocarcinoma cells and their derived metastatic cells. Oncol Lett 2013;5:1273–7.
[30] Fang MZ, Jin Z, Wang Y, Liao J, Yang GY, Wang LD, et al. Promoter hypermethylation and inactivation of O(6)-methylguanine-DNA methyltransferase in esophageal squamous cell carcinomas and its reactivation in cell lines. Int J Oncol 2005;26:615–22.
[31] Huang YW, Kuo CT, Stoner K, Huang TH, Wang LS. An overview of epigenetics and chemoprevention. FEBS Lett 2011;585:2129–36.
[32] Gilbert ER, Liu D. Flavonoids influence epigenetic-modifying enzyme activity: structure - function relationships and the therapeutic potential for cancer. Curr Med Chem 2010;17:1756–68.
[33] Lee WJ, Shim JY, Zhu BT. Mechanisms for the inhibition of DNA methyltransferases by tea catechins and bioflavonoids. Mol Pharmacol 2005;68:1018–30.
[34] Bozkurt E, Atmaca H, Kisim A, Uzunoglu S, Uslu R, Karaca B. Effects of Thymus serpyllum extract on cell proliferation, apoptosis and epigenetic events in human breast cancer cells. Nutr Cancer 2012;64:1245–50.
[35] Weng JR, Lai IL, Yang HC, Lin CN, Bai LY. Identification of kazinol Q, a natural product from Formosan plants, as an inhibitor of DNA methyltransferase. Phytother Res 2014;28:49–54.
[36] Mirza S, Sharma G, Parshad R, Gupta SD, Pandya P, Ralhan R. Expression of DNA methyltransferases in breast cancer patients and to analyze the effect of natural compounds on DNA methyltransferases and associated proteins. J Breast Cancer 2013;16:23–31.
[37] Deneberg S, Guardiola P, Lennartsson A, Qu Y, Gaidzik V, Blanchet O, et al. Prognostic DNA methylation patterns in cytogenetically normal acute myeloid leukemia are predefined by stem cell chromatin marks. Blood 2011;118:5573–82.
[38] de Sousa EMF, Colak S, Buikhuisen J, Koster J, Cameron K, de Jong JH, et al. Methylation of cancer-stem-cell-associated Wnt target genes predicts poor prognosis in colorectal cancer patients. Cell Stem Cell 2011;9:476–85.
[39] Meissner A, Mikkelsen TS, Gu H, Wernig M, Hanna J, Sivachenko A, et al. Genome-scale DNA methylation maps of pluripotent and differentiated cells. Nature 2008; 454:766–70.
[40] Bhutani N, Brady JJ, Damian M, Sacco A, Corbel SY, Blau HM. Reprogramming towards pluripotency requires AID-dependent DNA demethylation. Nature 2010; 463:1042–7.
[41] Wang P, Qu J, Wu M-Z, Zhang W, Liu GH, Izpisua Belmonte JC. "TET-on" pluripotency. Cell Res 2013;23:863–5.
[42] Simonsson S, Gurdon J. DNA demethylation is necessary for the epigenetic reprogramming of somatic cell nuclei. Nat Cell Biol 2004;6:984–90.
[43] Baba T, Convery PA, Matsumura N, Whitaker RS, Kondoh E, Perry T, et al. Epigenetic regulation of CD133 and tumorigenicity of $CD133^+$ ovarian cancer cells. Oncogene 2009;28:209–18.

[44] Tabu K, Sasai K, Kimura T, Wang L, Aoyanagi E, Kohsaka S, et al. Promoter hypomethylation regulates CD133 expression in human gliomas. Cell Res 2008; 18:1037–46.
[45] Yi JM, Tsai HC, Glückner SC, Lin S, Ohm JE, Easwaran H, et al. Abnormal DNA methylation of CD133 in colorectal and glioblastoma tumors. Cancer Res 2008; 68:8094–103.
[46] You H, Ding W, Rountree CB. Epigenetic regulation of cancer stem cell marker CD133 by transforming growth factor-beta. Hepatology 2010;51:1635–44.
[47] Chik F, Machnes Z, Szyf M. Synergistic anti-breast cancer effect of a combined treatment with the methyl donor S-adenosyl methionine and the DNA methylation inhibitor 5-aza-2′-deoxycytidine. Carcinogenesis 2014;35:138–44.
[48] Doi A, Park IH, Wen B, Murakami P, Aryee MJ, Irizarry R, et al. Differential methylation of tissue- and cancer-specific CpG island shores distinguishes human induced pluripotent stem cells, embryonic stem cells and fibroblasts. Nat Genet 2009;41:1350–3.
[49] van Vlerken LE, Hurt EM, Hollingsworth RE. The role of epigenetic regulation in stem cell and cancer biology. J Mol Med 2012;90:791–801.
[50] Stricker S, Pollard S. Reprogramming cancer cells to pluripotency: An experimental tool for exploring cancer epigenetics. Epigenetics 2014;9:798–802.
[51] Challen GA, Sun D, Jeong M, Luo M, Jelinek J, Berg JS, et al. Dnmt3a is essential for hematopoietic stem cell differentiation. Nat Genet 2012;44:23–31.
[52] Jin B, Ernst J, Tiedemann RL, Xu H, Sureshchandra S, Kellis M, et al. Linking DNA methyltransferases to epigenetic marks and nucleosome structure genome-wide in human tumor cells. Cell Rep 2012;2:1411–24.
[53] Liu CC, Lin JH, Hsu TW, Su K, Li AF, Hsu HS, et al. IL-6 enriched lung cancer stem-like cell population by inhibition of cell cycle regulators via DNMT1 upregulation. Int J Cancer 2015;136:547–59.
[54] Mahpatra S, Firpo MT, Bacanamwo M. Inhibition of DNA methyltransferases and histone deacetylases induces bone marrow-derived multipotent adult progenitor cells to differentiate into endothelial cells. Ethn Dis 2010;20: S1-60-64.
[55] Wang Y, Cardenas H, Fang F, Condello S, Taverna P, Segar M, et al. Epigenetic targeting of ovarian cancer stem cells. Cancer Res 2014;74:4922–36.
[56] Anderson OS, Sant KE, Dolinoy DC. Nutrition and epigenetics: an interplay of dietary methyl donors, one-carbon metabolism and DNA methylation. J Nutr Biochem 2012;23:853–9.
[57] Howland RH. Dietary supplement drug therapies for depression. J Psychosoc Nurs Ment Health Serv 2012;50:13–16.
[58] Papakostas GI, Cassiello CF, Iovieno N. Folates and S-adenosylmethionine for major depressive disorder. Can J Psychiatry 2012;57:406–13.
[59] Levkovitz Y, Alpert JE, Brintz CE, Mischoulon D, Papakostas GI. Effects of S-adenosylmethionine augmentation of serotonin-reuptake inhibitor antidepressants on cognitive symptoms of major depressive disorder. J Affect Disord 2012; 136:1174–8.
[60] Adwan L, Zawia NH. Epigenetics: a novel therapeutic approach for the treatment of Alzheimer's disease. Pharmacol Ther 2013;139:41–50.
[61] Fuso A, Nicolia V, Ricceri L, Cavallaro RA, Isopi E, Mangia F, et al. S-adenosylmethionine reduces the progress of the Alzheimer-like features induced by B-vitamin deficiency in mice. Neurobiol Aging 2012;33:e1481–1416.
[62] Anier K, Zharkovsky A, Kalda A. S-adenosylmethionine modifies cocaine-induced DNA methylation and increases locomotor sensitization in mice. Int J Neuropsychopharmacol 2013;16:2053–66.

[63] Hussain Z, Khan MI, Shahid M, Almajhdi FN. S-adenosylmethionine, a methyl donor, up regulates tissue inhibitor of metalloproteinase-2 in colorectal cancer. Genet Mol Res 2013;12:1106–18.
[64] Luo J, Li YN, Wang F, Zhang WM, Geng X. S-adenosylmethionine inhibits the growth of cancer cells by reversing the hypomethylation status of c-myc and H-ras in human gastric cancer and colon cancer. Int J Biol Sci 2010;6:784–95.
[65] Da MX, Zhang YB, Yao JB, Duan YX. DNA methylation regulates expression of VEGF-C, and S-adenosylmethionine is effective for VEGF-C methylation and for inhibiting cancer growth. Braz J Med Biol Res 2014;47:1021–8.
[66] Anstee QM, Day CP. S-adenosylmethionine (SAMe) therapy in liver disease: a review of current evidence and clinical utility. J Hepatol 2012;57:1097–109.
[67] Ringdahl E, Pandit S. Treatment of knee osteoarthritis. Am Fam Physician 2011;83:1287–92.
[68] Lopez HL. Nutritional interventions to prevent and treat osteoarthritis. Part II: focus on micronutrients and supportive nutraceuticals. PM R 2012;4:S155–68.
[69] Kim J, Lee EY, Koh EM, Cha HS, Yoo B, Lee CK, et al. Comparative clinical trial of S-adenosylmethionine versus nabumetone for the treatment of knee osteoarthritis: an 8-week, multicenter, randomized, double-blind, double-dummy, Phase IV study in Korean patients. Clin Ther 2009;31:2860–72.
[70] Choi LJ, Huang JS. A pilot study of S-adenosylmethionine in treatment of functional abdominal pain in children. Altern Ther Health Med 2013;19:61–4.
[71] Porter NS, Jason LA, Boulton A, Bothne N, Coleman B. Alternative medical interventions used in the treatment and management of myalgic encephalomyelitis/chronic fatigue syndrome and fibromyalgia. J Altern Complement Med 2010;16.
[72] Pfalzer AC, Choi SW, Tammen SA, Park LK, Bottiglieri T, Parnell LD, et al. S-adenosylmethionine mediates inhibition of inflammatory response and changes in DNA methylation in human macrophages. Physiol Genomics 2014;46:617–23.
[73] Liu GY, Wang W, Jia WD, Xu GL, Ma JL, Ge YS, et al. Protective effect of S-adenosylmethionine on hepatic ischemia-reperfusion injury during hepatectomy in HCC patients with chronic HBV infection. World J Surg Oncol 2014;12:27.
[74] Li TW, Yang H, Peng H, Xia M, Mato JM, Lu SC. Effects of S-adenosylmethionine and methylthioadenosine on inflammation-induced colon cancer in mice. Carcinogenesis 2012;33:427–35.
[75] Yang ML, Gee AJ, Gee RJ, Zurita-Lopez CI, Khare S, Clarke SG, et al. Lupus autoimmunity altered by cellular methylation metabolism. Autoimmunity 2013;46:21–31.
[76] Strickland FM, Hewagama A, Wu A, Sawalha AH, Delaney C, Hoeltzel MF, et al. Diet influences expression of autoimmune-associated genes and disease severity by epigenetic mechanisms in a transgenic mouse model of lupus. Arthritis Rheum 2013;65:1872–81.
[77] Neidhart M, Karouzakis E, Jungel A, Gay RE, Gay S. Inhibition of spermidine/spermine N1-acetyltransferase activity: a new therapeutic concept in rheumatoid arthritis. Arthritis Rheumatol 2014;66:1723–33.
[78] Brooks WH. X chromosome inactivation and autoimmunity. Clin Rev Allergy Immunol 2010;39:20–9.
[79] Brooks WH. Autoimmune diseases and polyamines. Clin Rev Allergy Immunol 2012;42:58–70.
[80] Brooks WH. Increased polyamines alter chromatin and stabilize autoantigens in autoimmune diseases. Front Immunol 2013;4:91.
[81] Karouzakis E, Gay RE, Gay S, Neidhart M. Increased recycling of polyamines is associated with global DNA hypomethylation in rheumatoid arthritis synovial fibroblasts. Arthritis Rheum 2012;64:1809–17.

[82] Pirnes-Karhu S, Jantunen E, Mäntymaa P, Mustjoki S, Alhonen L, Uimari A. Spermidine/spermine N(1)-acetyltransferase activity associates with white blood cell count in myeloid leukemias. Exp Hematol 2014;42:574–80.
[83] Brett-Morris A, Wright BM, Seo Y, Pasupuleti V, Zhang J, Lu J, et al. The polyamine catabolic enzyme SAT1 modulates tumorigenesis and radiation response in GBM. Cancer Res 2014;74:6925–34.
[84] Squassina A, Manchia M, Chillotti C, Deiana V, Congiu D, Paribello F, et al. Differential effect of lithium on spermidine/spermine N1-acetyltransferase expression in suicidal behaviour. Int J Neuropsychopharmacol 2013;16:2209–18.

CHAPTER

26

Demethylating Agents

OUTLINE

26.1 INTRODUCTION

Changes in heritable DNA methylation that alter phenotype are referred to as epimutations [1]. An important distinction between epigenetic and genetic alterations is intrinsic reversibility of the former, making cancer-associated changes in DNA methylation and histone modifications attractive targets for therapeutic intervention. As presented in Chapter 25, demethylating agents are compounds that can inhibit methylation, resulting in expression of the previously hypermethylated silenced genes. Cytidine analogs such as 5-azacytidine and 5-azadeoxycytidine (decitabine) are the most commonly used demethylating agents [2] (Figure 26.1). These compounds work by binding to the enzymes that catalyze the methylation reaction, DNA methyltransferases (DNMTs). 5-Aza-2′-deoxycytidine (decitabine) was first used as

M. Neidhart: DNA Methylation and Complex Human Disease.
DOI: http://dx.doi.org/10.1016/B978-0-12-420194-1.00026-9

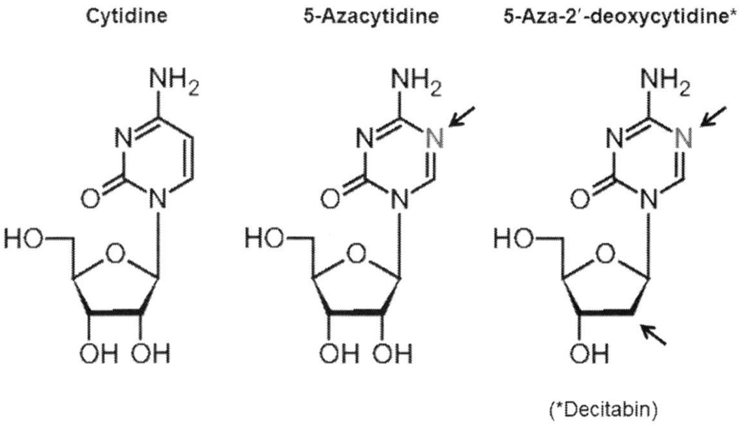

Cytidine 5-Azacytidine 5-Aza-2′-deoxycytidine*

(*Decitabin)

FIGURE 26.1 Structures of cytidine and cytidine analogs used as demethylating agents.

an antineoplastic agent in patients with leukemia [3] and multiple myeloma [4]. Demethylating agents are the standard of care for patients with higher risk myeloid malignancy, and the only agent known to improve the natural history of this disease [5,6]. Promoters silenced by DNA methylation, such as p16INK4, can be reactivated by treatment with such drugs. Combinations of demethylating agents and histone deacetylase inhibitors are employed in clinical trials to target multiple biological pathways, with the hope of synergistic pharmacodynamics [7].

26.2 MECHANISM OF ACTION

Following cellular uptake, azacitidine and decitabine are converted into their monophosphates, diphosphates, and triphosphates. Decitabine triphosphate is a deoxyribonucleotide that is incorporated only into DNA. Azacitidine is mainly converted to azacitidine triphosphate, which is incorporated into the RNA. A smaller portion of the administered azacitidine, about 10–20%, is converted to 5-aza-2′-deoxycytidine triphosphate via the enzyme ribonucleotide reductase, and is available for incorporation into the DNA. Incorporation into the DNA results in the formation of adducts between the DNA and DNMT1. Formation of a covalent DNMT–DNA complex leads to DNMT degradation and subsequent hypomethylation of CpG sites throughout the genome [8]. In human leukemic cell lines, DNMT1 levels are dramatically reduced with 100 nM decitabine or 1 μM 5-azacytidine. DNA hypomethylation is observed at even lower concentrations (30 nM decitabine or 300 nM 5-azacytidine) [9,10]. At high doses, the DNA is not able to recover and cell death occurs. However, at lower doses the formed adducts are degraded by the proteosome, after which the DNA is restored. DNA synthesis is then resumed in the absence of DNMT1. As a consequence, the aberrant DNA methylation pattern can no longer be reproduced toward the daughter strands [11]. In this way, a low dose of azacitidine or decitabine is able to induce re-expression of previously silenced genes. Reactivation of cell cycle-regulating genes that were initially silenced due to hypermethylation may induce cell differentiation, reduce proliferation, and/or increase apoptosis of the daughter cells [12].

26.3 NEW DRUG DEVELOPMENT

Despite the advantages of 5-azacytidine and decitabine, significant challenges stem from the instability of both compounds in aqueous solutions as well as *in vivo* deamination by cytidine deaminase. Efforts to circumvent problems of metabolic instability have yielded two different analogs of decitabine: NPEOC-decitabine and SGI-110. NPEOC-decitabine contains a moiety that protects the exocyclic amine of the drug. However, the new compound was no more potent than decitabine *in vivo* [13]. SGI-110 is a dinucleotide that includes a deoxyguanosine and is largely resistant to cytidine deaminase [14]. SGI-110 has exhibited DNA hypomethylation activity in primates [15].

In addition to improving the stability of the azanucleosides, efforts are under way to improve cellular delivery. CP-4200 is an elaidic acid ester analog of 5-azacytidine designed to render the drug less dependent on conventional nucleoside transport systems, and has shown superior efficacy in a leukemia mouse tumor model [16].

In addition, other cytidine analogs have been shown to inhibit DNMT activity once incorporated into DNA – for example, zebularine. This compound incorporated into DNA also inhibits DNMT1 through a covalent, though reversible, complex [17]. However, high concentrations of zebularine ($\sim 50\ \mu M$) are required to observe DNMT1 inhibition in cells due to inefficient incorporation into DNA [18].

The high toxicity of cytidine analogs prompted the search for non-nucleoside DNMT inhibitors. Some compounds assessed for their potential to induce hypomethylation in solid tumors include procaine, L-tryptophan derivative RG108, hydralazine, MG98, procainamide, and epigallocatechin-3-gallate, which is the main polyphenol compound in green tea [19]. We mentioned in Chapter 25 some natural compounds (Figure 26.2) that can be used as DNMT1 inhibitors, including black raspberry-derived anthocyanins [20,21], annurca apple polyphenol [22] (containing chlorogenic acid, catechin, and epicatechin as major components), acetyl-keto-β-boswellic acid [23], silibinin [24], genistein [25,26], wild thyme extracts [27], and kazinol Q [28].

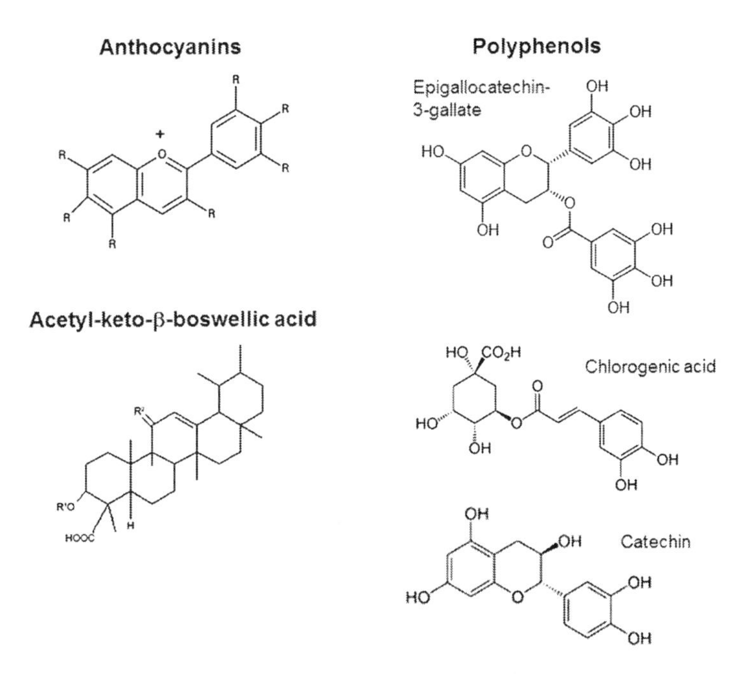

FIGURE 26.2 Some of the natural compounds that inhibit DNA methyltransferases.

26.4 COMBINATION THERAPIES

The potential anticancer activities of histone deacetylase (HDAC) inhibitors and DNMT inhibitors have been extensively studied for long time [29]. HDAC inhibitors suppress the activities of multiple HDACs, leading to an increase in histone acetylation. This histone acetylation induces enhancement of the expression of specific genes that elicit extensive cellular morphologic and metabolic changes, such as growth arrest, differentiation, and apoptosis. The combination of HDAC inhibitors with demethylating agents has become attractive since histones are connected to DNA by both physical and functional interactions. The combination of HDAC and DNMT inhibition is very effective (and synergistic) in inducing apoptosis, differentiation, and/or cell growth arrest in human lung, breast, thoracic, leukemia [7,30,31], myelodysplasia [7,31], and colon cancer cell lines. Nowadays, the majority of patients with cancer are treated with radiotherapy. To optimize the results obtained with this treatment modality, efforts are being put into strategies enhancing tumor response selectively in favor of normal tissue response. The combination of epigenetic drugs with radiotherapy is particularly valuable since a drug- and dose-dependent radiosensitizing potential of several classes of HDAC inhibitors has been proven *in vitro* and *in vivo*. Although promising so far, further research is needed before HDAC inhibitors administered alone or in combination with demethylating agents will be implemented in the clinic to act as radiosensitizers [32]. Combined epigenetic therapy may also "priming" patients to better respond to standard cytotoxic therapy or immunotherapy [33]. For example, if chemoresistance is the product of multiple non-genetic alterations that develop and accumulate over time in response to treatment, then the ability to epigenetically modify the tumor to reconfigure it back to its baseline non-resistant state holds tremendous promise for the treatment of advanced, metastatic cancer [34].

Drug combination approaches also have included epigenetic targeting together with either conventional chemotherapy or selective therapies such as Gleevec® [35]. The conclusions from trials using different combinations with demethylating agents (e.g., IL-2 [36], retinoic acid [37], thalidomide [38], etanercept [39], or cytarabine [40]) included both improved clinical benefit and hypomethylation of specific gene promoters, opening the possibility for evaluation of yet untested combinations.

26.5 STEM CELL DIFFERENTIATION

The demethylating agent 5-azacytidine was used as a tool to study trophoblast [41], stem cell [42], and cardiomyogenic [43] differentiation. Thus, 5-azacytidine can enhance the efficiency of iPS cell generation and the putative DNA demethylase protein activation-induced cytidine deaminase can erase DNA methylation at pluripotency gene promoters, thereby allowing cellular reprogramming [42]. In addition, demethylating agents have immunomodulatory effects that are under investigation in the allogeneic stem cell transplantation scenario [44]. Both drugs have been used in the perioperative period of allogeneic transplantations with varying degrees of success. It has been hypothesized that low-dose 5-azacytidine may increase the graft-versus-leukemia effect and have a role in the maintenance of remission after allogeneic transplantation for myeloid leukemias. It is also intriguing that this favorable effect might occur while mitigating graft-versus-host disease. SGI-110 also has shown activity in cancer xenograft models [45].

26.6 INDUCTION OF AUTOIMMUNITY

Autoimmune disorders have an estimated incidence of 10% among patients suffering from myodysplastic disorders and are causally related to increased morbidity and mortality, younger age at diagnosis, and more complex genetics. Conversely, systemic inflammatory disorders may be an early manifestation in this disease, show good response to immunosuppressive therapy, and frequently disappear during the course of specific hematologic therapy. Interestingly, 5-azacytidine is effective in controlling paraneoplastic inflammation [46]. Demethylating agents can also be used as tools to produce autoreactive T cells [47,48] and to mimic the activated phenotype of rheumatoid arthritis synovial fibroblasts [49]. Murine T cells treated with 5-azacytidine spontaneously lysed syngeneic macrophages and secreted IL-4, IL-6, and IFN-γ [47]. Adoptive transfer of such cells into unirradiated syngeneic recipients induced an immune complex glomerulonephritis and IgG anti-DNA and antihistone antibodies. These T cells expressed more LFA-1 (CD11a/CD18) [50] and TNFSF7 (CD70) [51]. This suggested that $CD4^{+}$ T-cell DNA hypomethylation may contribute to the development of drug induced and idiopathic human lupus [52]. These observations suggested that T-cell cytotoxicity against tumor cells can be enhanced by demethylating agents [53,54]. Indeed, decitabine can upregulate the expression of cancer–testis antigens, HLA molecules, and intracellular cell adhesion molecule-1 (CD154) on pediatric sarcoma cell lines, resulting in enhanced killing of tumor cells by specific cytotoxic T lymphocytes [55].

26.7 CONCLUSION

Initial successes demonstrate the potential effectiveness of epigenome-targeted therapies as monotherapy in hematologic malignancies, and should be followed by studies focused on combination therapies. Although much has been learned about the relationship between the epigenome and cancer, many questions remain unanswered at this time. The next step is to continue to translate emerging epigenetic knowledge into anticancer drug development. Combination therapies are promising and will provide novel information about not only the interplay between DNA methylation and histone modifications but also the link to other molecular mechanisms involved in health and disease. There have also been interesting developments in the understanding of the mechanism of action at molecular as well at cellular levels. First, demethylating agents were found to exert their beneficial effects in oncology by inducing the expression of silenced "tumor suppressor" genes. They mainly exert their action by favoring the degradation of DNMT1. Secondly, it has been recognized that demethylating agents also induce autoimmunity that might be beneficial in the case that cytotoxic T cells are able to more efficiently destroy cancer cells.

References

[1] Holliday R. Mutations and epimutations in mammalian cells. Mutat Res 1991;250:351–63.

[2] Derissen EJ, Beijnen JH, Schellens JH. Concise drug review: azacitidine and decitabine. Oncologist 2013;18:619–24.

[3] Momparler RL, Cote S, Eliopoulos N. Pharmacological approach for optimization of the dose schedule of 5-Aza-2′-deoxycytidine (Decitabine) for the therapy of leukemia. Leukemia 1997;11:175–80.

[4] Ng MH, Wong IH, Lo KW. DNA methylation changes and multiple myeloma. Leuk Lymphoma 1999;34:463–72.

[5] Garcia-Manero G. Demethylating agents in myeloid malignancies. Curr Opin Oncol 2008;20:705–10.

[6] Deneberg S. Epigenetics in myeloid malignancies. Methods Mol Biol 2012; 863:119–37.

[7] Bishton M, Kenealy M, Johnstone R, Rasheed W, Prince HM. Epigenetic targets in hematological malignancies: combination therapies with HDACis and demethylating agents. Expert Rev Anticancer Ther 2007;7:1439–49.

[8] Santi DV, Norment A, Garrett CE. Covalent bond formation between a DNA-cytosine methyltransferase and DNA containing 5-azacytosine. Proc Natl Acad Sci USA 1984;81:6993–7.

[9] Qin T, Youssef EM, Jelinek J, Chen R, Yang AS, Garcia-Manero G, et al. Effect of cytarabine and decitabine in combination in human leukemic cell lines. Clin Cancer Res 2007;13:4225–32.

[10] Hollenbach PW, Nguyen AN, Brady H, Williams M, Ning Y, Richard N, et al. A comparison of azacitidine and decitabine activities in acute myeloid leukemia cell lines. PloS One 2010;5:e9001.

[11] McCabe MT, Brandes JC, Vertino PM. Cancer DNA methylation: molecular mechanisms and clinical implications. Clin Cancer Res 2009;15:3927–37.

[12] Christman JK. 5-Azacytidine and 5-aza-2′-deoxycytidine as inhibitors of DNA methylation: mechanistic studies and their implications for cancer therapy. Oncogene 2002;21:5483–95.

[13] Byun HM, Choi SH, Laird PW, Trinh B, Siddiqui MA, Marquez VE, et al. 2′-Deoxy-N4-[2-(4-nitrophenyl)ethoxycarbonyl]-5-azacytidine: a novel inhibitor of DNA methyltransferase that requires activation by human carboxylesterase 1. Cancer Lett 2008;266:238–48.

[14] Foulks JM, Parnell KM, Nix RN, Chau S, Swierczek K, Saunders M, et al. Epigenetic drug discovery: targeting DNA methyltransferases. J Biomol Screen 2012;17:2–17.

[15] Lavelle D, Saunthararajah Y, Vaitkus K, Singh M, Banzon V, Phiasivongsva P, et al. S110, a novel decitabine dinucleotide, increases fetal hemoglobin levels in baboons (P. anubis). J Transl Med 2010;8:92.

[16] Brueckner B, Rius M, Markelova MR, Fichtner I, Hals PA, Sandvold ML, et al. Delivery of 5-azacytidine to human cancer cells by elaidic acid esterification increases therapeutic drug efficacy. Mol Cancer Ther 2010;9:1256–64.

[17] Champion C, Guianvarch D, Sénamaud-Beaufort C, Jurkowska RZ, Jeltsch A, Ponger L, et al. Mechanistic insights on the inhibition of c5 DNA methyltransferases by zebularine. PloS One 2010;5:e12388.

[18] Flotho C, Claus R, Batz C, Schneider M, Sandrock I, Ihde S, et al. The DNA methyltransferase inhibitors azacitidine, decitabine and zebularine exert differential effects on cancer gene expression in acute myeloid leukemia cells. Leukemia 2009;23.

[19] Ren J, Singh BN, Huang Q, Li Z, Gao Y, Mishra P, et al. DNA hypermethylation as a chemotherapy target. Cell Signal 2011;23:1082–93.

[20] Subramaniam D, Thombre R, Dhar A, Anant S. DNA methyltransferases: a novel target for prevention and therapy. Front Oncol 2014;4:80.

[21] Wang LS, Kuo CT, Cho SJ, Seguin C, Siddiqui J, Stoner K, et al. Black raspberry-derived anthocyanins demethylate tumor suppressor genes through the inhibition of DNMT1 and DNMT3B in colon cancer cells. Nutr Cancer 2013;65:118–25.

[22] Fini L, Selgrad M, Fogliano V, Graziani G, Romano M, Hotchkiss E, et al. Annurca apple polyphenols have potent demethylating activity and can reactivate silenced tumor suppressor genes in colorectal cancer cells. J Nutr 2007;137:2622–8.

[23] Shen Y, Takahashi M, Byun HM, Link A, Sharma N, Balaguer F, et al. Boswellic acid induces epigenetic alterations by modulating DNA methylation in colorectal cancer cells. Cancer Biol Ther 2012;13:542–52.

[24] Kauntz H, Bousserouel S, Gosse F, Raul F. Epigenetic effects of the natural flavonolignan silibinin on colon adenocarcinoma cells and their derived metastatic cells. Oncol Lett 2013;5:1273–7.

[25] Fang MZ, Jin Z, Wang Y, Liao J, Yang GY, Wang LD, et al. Promoter hypermethylation and inactivation of O(6)-methylguanine-DNA methyltransferase in esophageal squamous cell carcinomas and its reactivation in cell lines. Int J Oncol 2005; 26:615–22.

[26] Huang YW, Kuo CT, Stoner K, Huang TH, Wang LS. An overview of epigenetics and chemoprevention. FEBS Lett 2011;585:2129–36.

[27] Bozkurt E, et al. Effects of *Thymus serpyllum* extract on cell proliferation, apoptosis and epigenetic events in human breast cancer cells. Nutr Cancer 2012;64:1245–50.

[28] Weng JR, Lai IL, Yang HC, Lin CN, Bai LY. Identification of kazinol Q, a natural product from Formosan plants, as an inhibitor of DNA methyltransferase. Phytother Res 2014;28:49–54.

[29] Zhu WG, Otterson GA. The interaction of histone deacetylase inhibitors and DNA methyltransferase inhibitors in the treatment of human cancer cells. Curr Med Chem Anticancer Agents 2003;3:187–99.

[30] Leone G, Voso MT, Teofili L, Lubbert M. Inhibitors of DNA methylation in the treatment of hematological malignancies and MDS. Clin Immunol 2003;109:89–102.

[31] Platzbecker U, Germing U. Combination of azacitidine and lenalidomide in myelodysplastic syndromes or acute myeloid leukemia-a wise liaison?. Leukemia 2013;27:1813–19.

[32] De Schutter H, Nuyts S. Radiosensitizing potential of epigenetic anticancer drugs. Anticancer Agents Med Chem 2009;9:99–108.

[33] Juergens RA, Rudin CM. Aberrant epigenetic regulation: a central contributor to lung carcinogenesis and a new therapeutic target. Am Soc Clin Oncol 2013. Available from: http://dx.doi.org/10.1200/EdBook_AM.2013.33.e295.

[34] Oronsky B, Oronsky N, Knox S, Fanger G, Scicinski J. Episensitization: therapeutic tumor resensitization by epigenetic agents: a review and reassessment. Anticancer Agents Med Chem 2014;14:1121–7.

[35] Yang X, Lay F, Han H, Jones PA. Targeting DNA methylation for epigenetic therapy. Trends Pharmacol Sci 2010;31:536–46.

[36] Gollob JA, Sciambi CJ, Peterson BL, Richmond T, Thoreson M, Moran K, et al. Phase I trial of sequential low-dose 5-aza-2′-deoxycytidine plus high-dose intravenous bolus interleukin-2 in patients with melanoma or renal cell carcinoma. Clin Cancer Res 2006;12:4619–27.

[37] Soriano AO, Yang H, Faderl S, Estrov Z, Giles F, Ravandi F, et al. Safety and clinical activity of the combination of 5-azacytidine, valproic acid, and all-trans retinoic acid in acute myeloid leukemia and myelodysplastic syndrome. Blood 2007;110:2302–8.

[38] Raza A, Mehdi M, Mumtaz M, Ali F, Lascher S, Galili N. Combination of 5-azacytidine and thalidomide for the treatment of myelodysplastic syndromes and acute myeloid leukemia. Cancer 2008;113:1596–604.

[39] Scott BL, Ramakrishnan A, Storer B, Becker PS, Petersdorf S, Estey EH, et al. Prolonged responses in patients with MDS and CMML treated with azacitidine and etanercept. Br J Haematol 2010;148:944–7.

[40] Borthakur G, Huang X, Kantarjian H, Faderl S, Ravandi F, Ferrajoli A, et al. Report of a phase 1/2 study of a combination of azacitidine and cytarabine in acute myelogenous leukemia and high-risk myelodysplastic syndromes. Leuk Lymphoma 2010;51:73–8.

[41] Serman L, Dodig D. Impact of DNA methylation on trophoblast function. Clin Epigenetics 2011;3:7.

[42] De Carvalho DD, You JS, Jones PA. DNA methylation and cellular reprogramming. Trends Cell Biol 2010;20:609–17.

[43] Sreejit P, Verma RS. Natural ECM as biomaterial for scaffold based cardiac regeneration using adult bone marrow derived stem cells. Stem Cell Rev 2013;9:158–71.

[44] Bashir Q, William BM, Garcia-Manero G, de Lima M. Epigenetic therapy in allogeneic hematopoietic stem cell transplantation. Rev Bras Hematol Hemoter 2013;35:126–33.

[45] Chuang JC, Warner SL, Vollmer D, Vankayalapati H, Redkar S, Bearss DJ, et al. S110, a 5-aza-2′-deoxycytidine-containing dinucleotide, is an effective DNA methylation inhibitor *in vivo* and can reduce tumor growth. Mol Cancer Ther 2010;9:1443–50.

[46] Frietsch JJ, Dornaus S, Neumann T, Scholl S, Schmidt V, Kunert C, et al. Paraneoplastic inflammation in myelodysplastic syndrome or bone marrow failure: case series with focus on 5-azacytidine and literature review. Eur J Haematol 2014;93:247–59.

[47] Quddus J, Johnson KJ, Gavalchin J, Amento EP, Chrisp CE, Yung RL, et al. Treating activated $CD4^+$ T cells with either of two distinct DNA methyltransferase inhibitors, 5-azacytidine or procainamide, is sufficient to cause a lupus-like disease in syngeneic mice. J Clin Invest 1993;92:38–53.

[48] Yung R, Powers D, Johnson K, Amento E, Carr D, Laing T, et al. Mechanisms of drug-induced lupus. II. T cells overexpressing lymphocyte function-associated antigen 1 become autoreactive and cause a lupuslike disease in syngeneic mice. J Clin Invest 1996;97:2866–71.

[49] Karouzakis E, Gay RE, Michel BA, Gay S, Neidhart M. DNA hypomethylation in rheumatoid arthritis synovial fibroblasts. Arthritis Rheum 2009;60:3613–22.

[50] Richardson B, Powers D, Hooper F, Yung RL, O'Rourke K. Lymphocyte function-associated antigen 1 overexpression and T cell autoreactivity. Arthritis Rheum 1994; 37:1363–72.

[51] Zhou Y, Qiu X, Luo Y, Yuan J, Li Y, Zhong Q, et al. Histone modifications and methyl-CpG-binding domain protein levels at the TNFSF7 (CD70) promoter in SLE $CD4^+$ T cells. Lupus 2011;20:1365–71.

[52] Richardson B, Sawalha AH, Ray D, Yung R. Murine models of lupus induced by hypomethylated T cells. Methods Mol Biol 2012;900:169–80.

[53] Li B, Zhu X, Sun L, Yuan L, Zhang J, Li H, et al. Induction of a specific $CD8^+$ T-cell response to cancer/testis antigens by demethylating pre-treatment against osteosarcoma. Oncotarget 2014;5:10791–802.

[54] Serrano A, Tanzarella S, Lionello I, Mendez R, Traversari C, Ruiz-Cabello F, et al. Eexpression of HLA class I antigens and restoration of antigen-specific CTL response in melanoma cells following 5-aza-2′-deoxycytidine treatment. Int J Cancer 2001; 94:243–51.

[55] Krishnadas DK, Bao L, Bai F, Chencheri SC, Lucas K. Decitabine facilitates immune recognition of sarcoma cells by upregulating CT antigens, MHC molecules, and ICAM-1. Tumour Biol 2014;35:5753–62.

CHAPTER

27

Methyl Donors

OUTLINE

27.1 INTRODUCTION

Over the past few years, nutriepigenetics – the influence of dietary components on mechanisms influencing the epigenome – has emerged as an exciting new field in current epigenetic research [1,2]. The most studied pathway in this field regards the methionine/homocysteine pathway, in, for example, oncology [3], psychiatry [4], and cardiology [5]. Specifically, L-methionine [6,7], betaine [8], and S-adenosylmethionine (SAM) [5], used as nutrient supplements, are the principle methyl donors that belong to this pathway. They were largely studied *in vivo* and *ex vivo* for their effects on

M. Neidhart: DNA Methylation and Complex Human Disease.
DOI: http://dx.doi.org/10.1016/B978-0-12-420194-1.00027-0

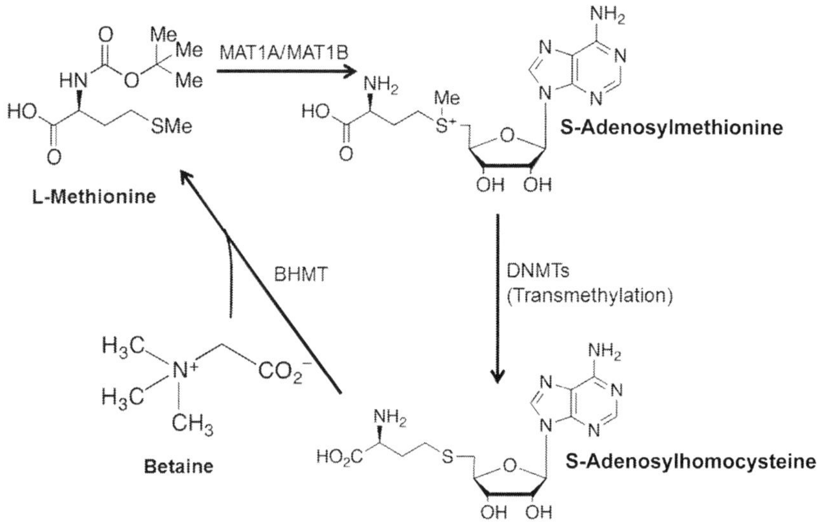

FIGURE 27.1 Methionine metabolism and recycling after transmethylation.

DNA methylation and resulting phenotypes, as shown, for example, for agouti gene expression in viable yellow mice [9]. Methyl donors are important in various cellular processes such as DNA methylation, transmethylation reactions, metabolism of amino acids lipids and polyamines, etc. After transmethylation catalyzed by DNA methyltransferases (DNMTs), the product S-adenosylhomocysteine (SAH) can be recycled using betaine (Figure 27.1). In this chapter, we present the situations in which supplementation with methyl donors is beneficial and focus on the differences between them.

27.2 L-METHIONINE

L-methionine is an essential, polar, alpha amino acid. Improper conversion of L-methionine into SAM has been linked to, for example, atherosclerosis [5]. The methionine metabolism is catalyzed by several enzymes in the presence of dietary micronutrients, including folate, choline, betaine, and other B vitamins [10]. For this reason, nutrition status, particularly micronutrient intake, has been a focal point when investigating epigenetic mechanisms. Remarkably, a subtle difference in methionine supplementation in the maternal diet was sufficient to cause significant changes in the transcriptome of the embryos [7]. The methyl groups transferred in mammalian DNA methylation reactions are ultimately derived from methionine. High dietary methionine intake might therefore be expected to increase DNA methylation. Because of the circular nature of the methionine cycle, however, methionine excess may actually impair DNA methylation by inhibiting remethylation of homocysteine [6]. Although little is known regarding the effect of dietary methionine supplementation on mammalian DNA methylation, the available data suggest that methionine supplementation can induce hypermethylation of DNA in specific genomic regions. Because locus-specific DNA hypomethylation is implicated in the etiology of various cancers and developmental syndromes, clinical trials of "promethylation" dietary supplements are already under way. However, aberrant hypermethylation of DNA could be deleterious. It is therefore important to determine whether dietary supplementation with methionine can effectively support therapeutic maintenance of DNA methylation without causing excessive and potentially adverse locus-specific hypermethylation. In the viable yellow agouti (A^{vy}) mouse, maternal diet affects the coat color distribution of offspring by perturbing the establishment of methylation at the A^{vy} metastable epiallele. On the other hand, methionine dependence can be observed in some cancers; this may be due to single or a combination of deletions, polymorphisms or alterations in the expression of genes in the methionine *de novo* and salvage pathways. Cancer cells with these defects are unable to regenerate methionine via these pathways. In such cases, vegan diets, which can be low in methionine, may prove to be a useful nutritional strategy in cancer growth control [11].

27.3 BETAINE

When methionine intake is high, an increase in SAM is expected. DNA methyltransferases convert SAM to SAH, which in turn increases the level of HCy. A high intracellular SAH and/or HCy concentration could inhibit the activity of DNA methyltransferases. This increase of SAH can be avoided by using betaine as a methyl donor. Betaine, N,N,N-trimethylglycine, was named after its discovery in sugar beet (*Beta vulgaris*) in the 19th century. It acts as a methyl donor for remethylation of homocysteine into L-methionine, thereby maintaining homocysteine homeostasis. Thus, betaine supplementation during lactation prevents the hyperhomocysteinemia induced by a high fat and sucrose intake [12]. In general, betaine supplement can be used to lower homocysteine levels [13]. Betaine is distributed widely in animals, plants, and microorganisms, and rich dietary sources include seafood, wheat germ, and spinach [8]. The principal physiologic role of betaine is as an osmolyte and methyl donor. As a methyl donor, betaine participates in the methionine cycle – primarily in the human liver and kidneys. Inadequate dietary intake of methyl groups leads to hypomethylation in many important pathways, including disturbed hepatic protein metabolism due to a deficiency in methionine, associated with elevated plasma homocysteine concentrations and decreased SAM levels. This alteration in liver metabolism may contribute to various diseases, including coronary, cerebral, hepatic, and vascular diseases. Betaine has been shown to protect internal organs, improve vascular risk factors, and enhance performance. In colon cancer, betaine-containing diets inhibit the incidence of tumors and hyperplasia, and downregulate inflammation [14].

27.4 S-ADENOSYLMETHIONINE

S-Adenosylmethionine is the final methyl donor for methylation of DNA [15], proteins [16], and phospholipids [17]. It interacts with SAM-dependent methyltransferases [18], such as DNA methyltransferases (DNMTs), and drives DNA methylation [15]. SAM is also involved in the synthesis of polyamines [19], and in the biosynthesis of several hormones and neurotransmitters that affect mood. Chronic alcohol consumption leads to significant reductions in SAM levels, thereby contributing to DNA hypomethylation [20]; SAM supplementation may prevent or restore specific promoter hypomethylation [21]. In gastric cancer, SAM can effectively induce vascular endothelial growth factor-C (VEGF-C) methylation and reduce the expression of VEGF-C, inhibiting neoangiogenesis and tumor growth [22]. Similarly, in colorectal cancer, SAM results in activation of the tissue inhibitor of metalloproteinase-2 (TIMP-2) and inhibition of the expression of genes such as matrix metalloproteinases (MMP-2, MT1-MMP), urokinase plasminogen activator, and vascular endothelial growth factors [23]. SAM has been studied in the treatment of osteoarthritis [24], and reduces the pain associated with this condition. Although an optimal dose has yet to be determined, SAM appears to be as effective as non-steroidal anti-inflammatory drugs. SAM also reduces the aggressiveness of rheumatoid arthritis synovial fibroblasts [25].

27.5 METHIONINE METABOLISM

The first step in (L)-methionine metabolism, the synthesis of SAM, is catalyzed by methionine adenosyltransferase I alpha (MAT1A) and methionine adenosyltransferase II alpha (MAT2A) [26]. SAM can transfer a methyl group to a variety of substrates (Figure 27.1). A mathematical model of transmethylation pathways provided information regarding product inhibition patterns [27]. MAT1A and MAT2A are known to be regulated via product inhibition by SAM, and are nearly saturated under normal methionine levels. In hepatocellular carcinoma, MAT1A/MAT2A switch is associated with global DNA hypomethylation, decreased in DNA repair, and genomic instability, signaling a rise in polyamine biosynthesis [28]. Following the transmethylation process and removal of the adenosine group, SAM produces S-adenosylhomocysteine (SAH) and, finally, homocysteine (HCy), which can be re-methylated to methionine or catabolized into cysteine. The remethylation of HCy into methionine is of particular interest, since it involves either the folic acid- or betaine-dependent mechanism.

27.5.1 Folic Acid-Dependent Mechanism

Folate metabolism is influenced by several processes, especially dietary intake and the polymorphisms of the associated genes involved. Aberrant folate metabolism therefore affects both methylation and the DNA synthesis processes [29]. The enzyme methionine synthase catalyzes the transfer of a methyl group from methylated folic acid (called methyltetrahydrofolate, MTHF) to HCy assisted by vitamin B12, which takes the methyl group from MTHF and adds it to the HCy molecule. The irreversible conversion of MTHF into 5-methyltetrahydrofolate (5MTHF) is catalyzed by methylenetetrahydrofolate reductase (MTHFR). This conversion is dependent on the availability of folic acid, in mammals an essential vitamin from the diet. Mutations of MTHFR lead to hyperhomocysteinemia. A common C677T C → T polymorphism in MTHFR has been associated with an increased risk for the development of cardiovascular disease [30], Alzheimer's disease [31], depression [32] in adults, and rheumatoid arthritis [33], as well as neural tube defects in the fetus [34,35]. However, the mutation also confers protection for certain types of cancers [35,36]. Finally, methionine synthase (MS), which is dependent on vitamin B12 availability, catalyzes transfer of the methyl group from 5MTHF to HCy, forming tetrahydrofolate (THF) and L-methionine. SAM allosterically inhibits MTHFR, while 5MTHF formed by MTHFR can inhibit glycine N-methyltransferase (GNMT) that catalyzes the synthesis of sarcosine from glycine using SAM as the methyl donor [37]. During deficiencies in methyl donors, SAM levels are reduced and no longer inhibit MTHFR, which in turn increases the levels of 5MTHF. 5MTHF is an inhibitor of GNMT, thereby lowering transmethylation and increasing intracellular SAM. When methyl group supply is high, elevated SAM inhibits MTHFR, which lowers the production of 5MTHF, thus increasing GNMT activity and the consumption of SAM. Thus, the reciprocal regulation of transmethylation and folate-dependent remethylation allows for finely tuned control of methyl group availability.

27.5.2 Betaine-Dependent Mechanism

In another pathway (mostly studied in the liver), HCy is reconverted into methionine using betaine. A methyl group from betaine is transferred to HCy in the presence of the enzyme betaine-homocysteine S-methyltransferase (BHMT), producing L-methionine and dimethylglycine (DMG). This so-called "folate-independent remethylation" is regulated by dietary intake of methyl groups, and is dependent of the availability of vitamin B6/B12, as well as the SAM/SAH ratio. In a negative feedback mechanism, elevated levels of SAM inhibit BHMT activity. On the other hand, BHMT mutation, SAM deficiency, and DNA hypomethylation appeared to be coupled in hepatocarcinoma [38]. Betaine supplementation in humans reduces concentration of total Hcy. In addition, plasma betaine is a strong predictor of plasma Hcy in individuals with low plasma concentrations of folate and other B vitamins in individuals with the MTHFR 677 C → T polymorphism [39].

27.5.3 S-Adenosylmethionine/S-Adenosylhomocysteine Ratio

Donation of the methyl group transforms SAM into SAH, which is a potent inhibitor of DNA methylation. In general, a low SAM/SAH ratio can be associated with inhibition of SAM-dependent reactions [40]. Glycine N-methyltransferase is an enzyme primarily expressed in liver, kidney, and pancreas that can help maintain the SAM/SAH ratio [37]. The SAM/SAH ratio is regarded as a gauge of the transmethylation potential. Ethanol inhibits methionine synthase, decreasing the SAM/SAH ratio in various organs [41]; this is associated with decreased nitric oxide synthase (NOS) gene promoter methylation and increased protein expression [42]. Nitric oxide synthases are a family of enzymes catalyzing the production of nitric oxide from L-arginine. Furthermore, in vascular smooth muscle cells, HCy decreased SAM and increased SAH; the consequent decrease in the ratio of SAM/SAH led to promoter hypomethylation of the platelet-derived growth factor (PDGF) gene [43]. These findings provided a novel insight into the molecular association between aberrant PDGF promoter hypomethylation and the proliferation of vascular smooth muscle cells in Hcy-associated atherosclerosis. In apolipoprotein E deficiency mice (a model for artherosclerosis), a high cholesterol and methionine diet for 15 weeks induced hyperlipidemia and hyperhomocysteinemia; this is associated with a decreased SAM/SAH ratio and global hypomethylation in aortic tissues [44]. However, the SAM/SAH ratio also decreased in septic animals [45] and patients [46]; inflammation led to considerable changes in the methylation metabolism without apparent functional consequences on HCy levels or global DNA methylation, as measured in plasma by liquid chromatography tandem mass spectrometry [46].

27.5.4 Homocysteine

Homocysteine (HCy) is formed from SAH following the removal of the adenosyl group by the enzyme SAH hydrolase. The two major methyltransferases that contribute to the formation of HCy are phospatidylethanolamine N-methyltransferase (PEMT), which synthesizes phosphatidylcholine, and guanidinoacetate N-methyltransferase (GAMT), which synthesizes creatine. Collectively, PEMT and GAMT account for roughly 85% of SAM-dependent transmethylation [47]. Interestingly, in bladder cancer PEMT polymorphisms are associated with the degree of global DNA hypomethylation [48]. HCy can be converted into cysteine by the vitamin B6-dependent enzymes cystathionine β-synthase (CBS) and cystathionine γ-lyase, or recycled by two separate remethylation pathways. Conversion to cysteine represents a catabolic process by which homocysteine is converted first into cystathionine via condensation with serine, then to cysteine or other metabolites, such as glutathione, pyruvate, and taurine. CBS activity is positively and allosterically regulated by SAM; therefore, increased production of HCy should correspond to increased catabolism. High plasma HCy has been shown to compromise the blood–brain barrier in mice [49]. HCy promotes atherosclerosis through fibrin deposition, oxidant stress, cytokine release, inflammation, and other mechanisms [50]. Atherosclerosis associated with high plasma HCy may also be partly due to homocysteine-induced stress to the endoplasmic reticulum of endothelial cells [51]. HCy damages endothelial cells [52]. Endothelial dysfunction has been shown to increase incrementally with incrementally higher doses of oral methionine (and subsequent incrementally higher plasma HCy) in normal human subjects [53]. Currently, HCy-lowering interventions in the form of supplements of folic acid and vitamins B6 or B12, given alone or in combination, are used for preventing cardiovascular events [54].

27.6 POLYAMINE METABOLISM

Pathways that compete for SAM may affect global DNA methylation; in particular, this has been hypothesized for the PMET-related pathway [48] (mentioned above) and the polyamine recycling pathway [55,56]. Polyamines are organic compounds with two or more primary amino groups. Polyamine metabolism is closely connected with transmethylation pathways via SAM. S-adenosyldecarboxylase (AMD) converts SAM into decarboxy SAM (dSAM). Ornithine decarboxylase (ODC) converts ornithine into putrescine. On the other hand, spermidine synthase (SdS) and spermine synthase (SmS) forms spermidine and spermine, respectively. SdS and SmS require dSAM produced by AMD in order to catalyze the biosynthesis of polyamines. In the case of increased ODC activity, for example after stimulation with cytokines or growth factors, putrescine levels are further increased and stabilize AMD, leading an increased consumption of SAM and production of dSAM. In addition, spermine and spermidine can be reconverted into putrescine by the sequential action of two enzymes, namely spermine/spermidine N1-acethyltransferase (SSAT1) and polyamine oxidase. Increased recycling of polyamines can affect the level of SAM and altered global DNA methylation [56]. The polyamine recycling pathway also is enhanced by cytokines and growth factors. Conversely, inhibition of this pathway increases the effectiveness of SAM supplementation.

27.7 CONCLUSION

Methyl donors can be given in cases of global DNA hypomethylation. There are important differences, however, between methionine and betaine, in particular regarding the beneficial effect of the latter on the levels of HCy. Even in cancer, in particular cases, methyl donors, alone or in combination with demethylating agents (Chapter 26), could help to restore normal gene expression. In humans, following the experimentation in animals and in cell cultures, many clinical trials are now required, particularly in psychiatric and rheumatic diseases.

References

[1] Gerhauser C. Cancer chemoprevention and nutriepigenetics: state of the art and future challenges. Top Curr Chem 2013;329:73–132.

[2] Remely M, Lovrecic L, de la Garza AL, Migliore L, Peterlin B, Milagro FI, et al. Therapeutic perspectives of epigenetically active nutrients. Br J Pharmacol 2015; 175:2756–68.

[3] de Vogel S, Dindore V, van Engeland M, Goldbohm RA, van den Brandt PA, Weijenberg MP. Dietary folate, methionine, riboflavin, and vitamin B-6 and risk of sporadic colorectal cancer. J Nutr 2008;138:2372–8.

[4] Dauncey MJ. Nutrition, the brain and cognitive decline: insights from epigenetics. Eur J Clin Nutr 2014;68:1179.

[5] Glier MB, Green TJ, Devlin AM. Methyl nutrients, DNA methylation, and cardiovascular disease. Mol Nutr Food Res 2014;58:172–82.

[6] Waterland RA. Assessing the effects of high methionine intake on DNA methylation. J Nutr 2006;136:1706S–10S.

[7] Penagaricano F, Souza AH, Carvalho PD, Driver AM, Gambra R, et al. Effect of maternal methionine supplementation on the transcriptome of bovine preimplantation embryos. PloS One 2013;8:e72302.
[8] Craig SA. Betaine in human nutrition. Am J Clin Nutr 2004;80:539–49.
[9] Cooney CA, Dave AA, Wolff GL. Maternal methyl supplements in mice affect epigenetic variation and DNA methylation of offspring. J Nutr 2002;132:2393S–400S.
[10] Anderson OS, Sant KE, Dolinoy DC. Nutrition and epigenetics: an interplay of dietary methyl donors, one-carbon metabolism and DNA methylation. J Nutr Biochem 2012;23:853–9.
[11] Cavuoto P, Fenech MF. A review of methionine dependency and the role of methionine restriction in cancer growth control and life-span extension. Cancer Treat Rev 2012;38:726–36.
[12] Cordero P, Milagro FI, Campion J, Martinez JA. Maternal methyl donors supplementation during lactation prevents the hyperhomocysteinemia induced by a high-fat-sucrose intake by dams. Int J Mol Sci 2013;14:24422–37.
[13] Obeid R. The metabolic burden of methyl donor deficiency with focus on the betaine homocysteine methyltransferase pathway. Nutrients 2013;5:3481–95.
[14] Kim DH, Sung B, Chung HY, Kim ND. Modulation of colitis-associated colon tumorigenesis by baicalein and betaine. J Cancer Prev 2014;19:153–60.
[15] Svedruzic ZM. Mammalian cytosine DNA methyltransferase Dnmt1: enzymatic mechanism, novel mechanism-based inhibitors, and RNA-directed DNA methylation. Curr Med Chem 2008;15:92–106.
[16] Lipson RS, Clarke SG. S-adenosylmethionine-dependent protein methylation in mammalian cytosol via tyrphostin modification by catechol-O-methyltransferase. J Biol Chem 2007;282:31094–102.
[17] Hirata F, Axelrod J. Phospholipid methylation and biological signal transmission. Science 1980;209:1082–90.
[18] Struck AW, Thompson ML, Wong LS, Micklefield J. S-adenosyl-methionine-dependent methyltransferases: highly versatile enzymes in biocatalysis, biosynthesis and other biotechnological applications. Chembiochem 2012;13:2642–55.
[19] Roje S. S-Adenosyl-L-methionine: beyond the universal methyl group donor. Phytochemistry 2006;67:1686–98.
[20] Zakhari S. Alcohol metabolism and epigenetics changes. Alcohol Res 2013;35:6–16.
[21] Khachatoorian R, Dawson D, Maloney EM, Wang J, French BA, French SW. SAMe treatment prevents the ethanol-induced epigenetic alterations of genes in the Toll-like receptor pathway. Exp Mol Pathol 2013;94:243–6.
[22] Da MX, Zhang YB, Yao JB, Duan YX. DNA methylation regulates expression of VEGF-C, and S-adenosylmethionine is effective for VEGF-C methylation and for inhibiting cancer growth. Braz J Med Biol Res 2014;47:1021–8.
[23] Hussain Z, Khan MI, Shahid M, Almajhdi FN. S-adenosylmethionine, a methyl donor, up regulates tissue inhibitor of metalloproteinase-2 in colorectal cancer. Genet Mol Res 2013;12:1106–18.
[24] Ringdahl E, Pandit S. Treatment of knee osteoarthritis. Am Fam Physician 2011; 83:1287–92.
[25] Neidhart M, Karouzakis E, Jungel A, Gay RE, Gay S. Inhibition of spermidine/spermine N1-acetyltransferase activity: a new therapeutic concept in rheumatoid arthritis. Arthritis Rheumatol 2014;66:1723–33.
[26] Kotb M, Geller AM. Methionine adenosyltransferase: structure and function. Pharmacol Ther 1993;59:125–43.
[27] Reed MC, Nijhout HF, Sparks R, Ulrich CM. A mathematical model of the methionine cycle. J Theor Biol 2004;226:33–43.

[28] Frau M, Feo F, Pascale RM. Pleiotropic effects of methionine adenosyltransferases deregulation as determinants of liver cancer progression and prognosis. J Hepatol 2013;59:830–41.

[29] Nazki FH, Sameer AS, Ganaie BA. Folate: metabolism, genes, polymorphisms and the associated diseases. Gene 2014;533:11–20.

[30] Husemoen LL, Skaaby T, Jørgensen T, Thuesen BH, Fenger M, Grarup N, et al. MTHFR C677T genotype and cardiovascular risk in a general population without mandatory folic acid fortification. Eur J Nutr 2014;53:1549–59.

[31] Zhang MY, Miao L, Li YS, Hu GY. Meta-analysis of the methylenetetrahydrofolate reductase C677T polymorphism and susceptibility to Alzheimer's disease. Neurosci Res 2010;68:142–50.

[32] Lok A, et al. Interaction between the MTHFR C677T polymorphism and traumatic childhood events predicts depression. Transl Psychiatry 2013;3:e288.

[33] Song GG, Bae SC, Lee YH. Association of the MTHFR C677T and A1298C polymorphisms with methotrexate toxicity in rheumatoid arthritis: a meta-analysis. Clin Rheumatol 2014;33:1715–24.

[34] Kondo A, Fukuda H, Matsuo T, Shinozaki K, Okai I. C677T mutation in methylenetetrahydrofolate reductase gene and neural tube defects: should Japanese women undergo gene screening before pregnancy? Congenit Anom (Kyoto) 2014;54:30–4.

[35] Trimmer EE. Methylenetetrahydrofolate reductase: biochemical characterization and medical significance. Curr Pharm Des 2013;19:2574–93.

[36] Hosseini-Asl SS, Pourfarzi F, Barzegar A, Mazani M, Farahmand N, Niasti E, et al. Decrease in gastric cancer susceptibility by MTHFR C677T polymorphism in Ardabil Province, Iran. Turk J Gastroenterol 2013;24:117–21.

[37] Wagner C, Briggs WT, Cook RJ. Inhibition of glycine N-methyltransferase activity by folate derivatives: implications for regulation of methyl group metabolism. Biochem Biophys Res Commun 1985;127:746–52.

[38] Pellanda H. Betaine homocysteine methyltransferase (BHMT)-dependent remethylation pathway in human healthy and tumoral liver. Clin Chem Lab Med 2013;51:617–21.

[39] Ueland PM. Choline and betaine in health and disease. J Inherit Metab Dis 2011;34:3–15.

[40] Williams KT, Schalinske KL. New insights into the regulation of methyl group and homocysteine metabolism. J Nutr 2007;137:311–14.

[41] Schalinske KL, Nieman KM. Disruption of methyl group metabolism by ethanol. Nutr Rev 2005;63:387–91.

[42] Medici V, Schroeder DI, Woods R, LaSalle JM, Geng Y, Shibata NM, et al. Methylation and gene expression responses to ethanol feeding and betaine supplementation in the cystathionine beta synthase-deficient mouse. Alcohol Clin Exp Res 2014;38:1540–9.

[43] Han XB, Zhang HP, Cao CJ, Wang YH, Tian J, Yang XL, et al. Aberrant DNA methylation of the PDGF gene in homocysteinemediated VSMC proliferation and its underlying mechanism. Mol Med Rep 2014;10:947–54.

[44] Jiang Y, Zhang H, Sun T, Wang J, Sun W, Gong H, et al. The comprehensive effects of hyperlipidemia and hyperhomocysteinemia on pathogenesis of atherosclerosis and DNA hypomethylation in ApoE −/− mice. Acta Biochim Biophys Sin (Shanghai) 2012;44:866–75.

[45] Semmler A, Smulders Y, Struys E, Smith D, Moskau S, Blom H, et al. Methionine metabolism in an animal model of sepsis. Clin Chem Lab Med 2008;46:1398–402.

[46] Semmler A, Prost JC, Smulders Y, Smith D, Blom H, Bigler L, et al. Methylation metabolism in sepsis and systemic inflammatory response syndrome. Scand J Clin Lab Invest 2013;73:368–72.

[47] Yeo EJ, Wagner C. Tissue distribution of glycine N-methyltransferase, a major folate-binding protein of liver. Proc Natl Acad Sci USA 1994;91:210–14.
[48] Tajuddin SM, Amaral AF, Fernández AF, Chanock S, Silverman DT, Tardón A, et al. LINE-1 methylation in leukocyte DNA, interaction with phosphatidylethanolamine N-methyltransferase variants and bladder cancer risk. Br J Cancer 2014;110:2123–30.
[49] Kamath AF, Chauhan AK, Kisucka J, Dole VS, Loscalzo J, Handy DE, et al. Elevated levels of homocysteine compromise blood–brain barrier integrity in mice. Blood 2006;107:591–3.
[50] McCully KS. Homocysteine, vitamins, and vascular disease prevention. Am J Clin Nutr 2007;86:1563S–8S.
[51] Werstuck GH, Lentz SR, Dayal S, Hossain GS, Sood SK, Shi YY, et al. Homocysteine-induced endoplasmic reticulum stress causes dysregulation of the cholesterol and triglyceride biosynthetic pathways. J Clin Invest 2001;107:1263–73.
[52] Xu D, Neville R, Finkel T. Homocysteine accelerates endothelial cell senescence. FEBS Lett 2000;470:20–4.
[53] Chambers JC, Obeid OA, Kooner JS. Physiological increments in plasma homocysteine induce vascular endothelial dysfunction in normal human subjects. Arterioscler Thromb Vasc Biol 1999;19:2922–7.
[54] Marti-Carvajal AJ, Sola I, Lathyris D, Karakitsiou DE, Simancas-Racines D. Homocysteine-lowering interventions for preventing cardiovascular events. Cochrane Database Syst Rev 2013;1:CD006612.
[55] Brooks WH. Autoimmune diseases and polyamines. Clin Rev Allergy Immunol 2012; 42:58–70.
[56] Karouzakis E, Gay RE, Gay S, Neidhart M. Increased recycling of polyamines is associated with global DNA hypomethylation in rheumatoid arthritis synovial fibroblasts. Arthritis Rheum 2012;64:1809–17.

CHAPTER

28

Methylome Analysis of Complex Diseases

OUTLINE

M. Neidhart: *DNA Methylation and Complex Human Disease.*
DOI: http://dx.doi.org/10.1016/B978-0-12-420194-1.00028-2

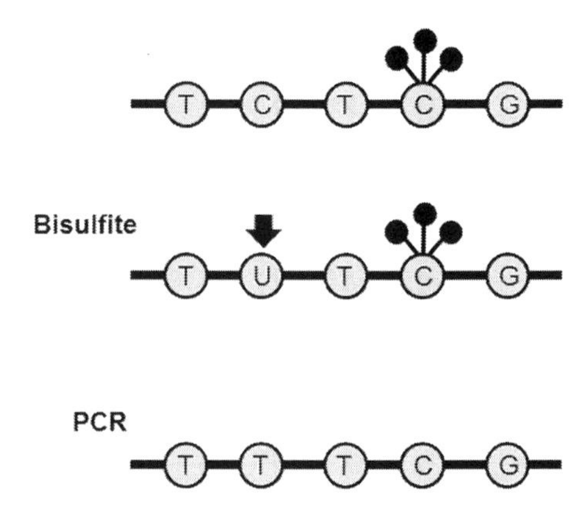

FIGURE 28.1 Bisulfite modification: methylated cytosines are protected, while unmethylated cytosine are modified to uracil. After PCR, uracil is amplified as thymine. Thus, a C→T modification shows unmethylated cytosine in the native DNA strand.

28.1 INTRODUCTION

A major advance in DNA methylation analysis was the development of a method for sodium bisulfite modification of DNA to convert unmethylated cytosines to uracil, leaving methylated cytosines unchanged [1] (Figure 28.1). This allows one to distinguish methylated from unmethylated DNA via polymerase chain reaction (PCR) amplification and analysis of the PCR products. During PCR amplification, unmethylated cytosines amplify as thymine, and methylated cytosines amplify as cytosine. Most methods for analyzing DNA methylation at specific loci are based on this approach. There has been a revolution in DNA methylation analysis technology over recent years [2]. Analyses that previously were restricted to specific loci can now be performed on a genome scale. Entire methylomes can be scanned at single-base-pair resolution. In this chapter, we present the different approaches.

28.2 GLOBAL DNA METHYLATION ANALYSIS

In mammals, 70–80% of all CpG dinucleotides are methylated and this methylation occurs predominantly in repetitive elements and regions in which CpG density is low. Conversely, CpG-rich regions ("CpG islands"), often found in gene promoters, are generally unmethylated. Interest in correlating the genomic 5-methylcytosine (5-meC) content with diet, lifestyle, and clinical outcomes, such as in cancer patients after treatment with hypomethylating agents, is widespread [3]. High-performance liquid chromatography is a classical method to quantify global DNA methylation, and is highly quantitative and reproducible [4]. This method requires large amounts of high-quality genomic DNA, however, and is not suitable for high-throughput analyses. To circumvent these problems, several bisulfite-based PCR methods have been developed to approximate global DNA methylation by assessing repetitive DNA elements such as LINE-1 and ALU elements [5,6]. These methods require very little DNA and can be applied to isolated primary cells and paraffin-embedded tissue. As global methylation analyses provide no information on the genomic positions at which methylation is altered, however, it is difficult to link such changes to functional outcomes.

Global DNA methylation can be assessed as the capacity of the DNA to incorporate [3H]methyl groups from [3H]-S-adenosylmethionine in the presence of sssl methylase [7]. A higher incorporation reflects a lower state of intrinsic methylation. DNA ploidy, S-phase fraction, and the proliferative index (propidium iodide = S + G_2M) of the cell cycle can be analyzed by flow cytometry.

Another possibility is to isolate the nuclei, stain them with anti-5meC antibodies and measure the binding by flow cytometry [8]. Cells growing in suspension are submitted to successive low-speed (300 g) pelleting and gentle resuspension steps in order to minimize clumping before analysis by flow cytometry. They are washed twice with pH 7.4 phosphate buffered saline (PBS) supplemented with 1% bovine serum albumin (BSA) and 0.1% Tween 20 (PBST-BSA) and then fixed with 0.25% paraformaldehyde in PBS for 10 minutes at 37°C. Specimens are then cooled and maintained at 4°C for 10 minutes before addition of 9 vol of methanol/PBS (88% methanol/12% PBS vol/vol) refrigerated at −20°C. After two washes with PBST-BSA at room temperature, cells are treated successively with 2-N HCl at 37°C for increasing periods of time, then with 0.1-M borate buffer pH 8.5 for 5 minutes. The HCl concentration has to be optimized for each cell type; for example, in the case of adherent synovial fibroblasts, 1-N HCl is sufficient [9]. The cells are then treated for 20 minutes at 37°C with a blocking solution made of PBST-BSA supplemented with 10% fetal calf serum, before incubation with murine anti-5-MeC antibodies. Specimens are then successively rinsed three times with PBS and incubated for 45 minutes at 37°C with goat anti-mouse immunoglobulins conjugated to fluorescein isothiocyanate diluted in PBST-BSA. Finally, samples are washed three times with PBS and are either mounted in phosphate-buffered glycerol for microscopy or stained with propidium iodide for 30 minutes before flow cytometry.

A protocol has also been developed for immunohistochemistry [9]. Formalin-fixed, paraffin-embedded sections of synovial tissue are deparaffinized and treated at 80°C for 30 minutes with citrate buffer (pH 3.4). The tissue slides were incubated with 2-N HCl for 2 hours at 37°C. After acid treatment, the slides are washed well with phosphate buffered saline with 0.05% Tween 20 (PBST). To determine the methylation in the tissue, murine anti-5meC antibodies are used. Mouse IgG serves as a negative control. Double staining with another specific marker for the cell population can be performed. The antibodies are incubated overnight at 4°C. Bound antibodies are detected after addition of biotinylated anti-mouse IgG for 1 hour. The slides are incubated with avidin-conjugated horseradish. The staining is visualized using, for example, 3,3-diaminobenzidine and HistoGreen. The images can even be quantified using specialized software. Similar protocols for flow cytometry and immunohistochemistry can be used with antibodies against methylation-binding proteins, such as MeCP2 [10,11].

28.3 GENE-SPECIFIC METHYLATION ANALYSIS

Gene-specific methylation analysis methods can be characterized as either candidate-gene or genome-wide approaches. Candidate-gene approaches can be further divided into "sensitive" and "quantitative" approaches [12]. In the sensitive methods, methylated or unmethylated alleles are detected by designing primers overlapping multiple CpG dinucleotides. In quantitative methods, primers are designed to amplify both methylated and unmethylated alleles with equal efficiency, and the methylation level is analyzed by various approaches.

28.3.1 Methylation-Sensitive Polymerase Chain Reaction

Methylation sensitive PCR (MSP) is a rapid and very sensitive technique to screen for methylation [13]. Following bisulfite modification, PCR is performed using two sets of primers designed to amplify either methylated or unmethylated alleles. MSP has the advantage of being highly sensitive (able to detect one methylated allele in a population of more than 1000 unmethylated alleles), and can be used on DNA samples of limited quantity and quality. MSP is not quantitative, however. Variations of MSP called MethyLight [6,14,15] or quantitative analysis of methylated alleles (QAMA) [16] were developed using real-time PCR for methylation detection. These assays were designed to detect completely methylated or unmethylated alleles, denying the reality of partial allelic methylation. The design of primers is essential for reliable results; ideally, the methylated and ummethylated primer sets should be designed for the same CpG sites and include multiple CpG sites at the 3′ ends. Simultaneous gene promoters can be analyzed at the same time using a special approach [17] in which digestion of genomic DNA with methylation-sensitive restriction enzyme (MSRE) is followed by multiplexed PCR with gene-specific primers (MSRE-PCR).

28.3.2 Denaturing Gradient Gel and Melting Differences

Denaturing gradient gel blots (DGG blots) were used in previous experiments to identify single base change mutations and polymorphisms in genomic fragments from several genes as sequence restriction fragment melting polymorphisms. In an earlier study [18], genomic DNA fragments from paired human sperm and leukocyte samples were compared using DGG blots. Blots are hybridized with native genomic DNA and PCR-amplified DNA of the same genes. Cell type- and allele-specific melting patterns can be detected between native DNA fragments of different cells from the same individuals, i.e. within the same genes. The absence of melting differences on DGG blots in PCR-amplified DNA fragments suggested that these melting variants are caused by differential cell type-specific DNA methylation. Cell type-specific native DNA fragment melting differences are common, and are often caused by differential methylation at sites not assayable by MSP analysis [19]. The newest method in this category is called methylation-sensitive high resolution melting (MS-HRM) [15].

28.3.3 Quantitative Methods

Bisulfite modification is the gold standard for mapping allele-specific methylation across CpG sites within postbisulfite PCR products [20]. In this method, a region of interest is amplified from bisulfite-modified DNA using PCR primers, not overlapping CpG sites, to amplify both methylated and unmethylated alleles. PCR products are ligated to a cloning vector and transfected to competent cells. Antibiotic resistant colonies are grown on agar plates, selected individually, and expanded by growing in LB medium. The plasmid DNA is isolated and sequenced. If a sufficient number of clones is sequenced, this method can be quantitative. As each clone represents a single allele, the data provide information on allele-specific methylation. Such information is particularly useful in the study of differentially methylated regions at genomic imprinted loci [21]. This technique is labor-intensive, however, and costly for large sets of samples.

An alternative approach employing direct radioactive sequencing of postbisulfite PCR products and quantitation by phosphor imaging was recently validated [22]. Instead of sampling a subset of alleles, as in the clone-and-sequence approach, direct sequencing averages across all alleles produced from the PCR step and is therefore more sensitive and quantitative. This method is cumbersome to perform on a large number of samples.

Bisulfite PCR followed by restriction analysis (COBRA) has been used as a quantitative technique for methylation detection [23]. After bisulfite modification and PCR amplification, the PCR product is digested with a restriction enzyme (whose recognition sequence is affected by bisulfite modification) and quantitated using gel electrophoresis and densitometry. This method provides data only for specific restriction enzyme cutting sites and is relatively time-consuming compared with MSP, due to the additional enzyme digestion step.

In all of these PCR-based methods, primer design is the key for successful amplification. Primers for quantitative bisulfite-PCR methods should ideally be free of CpGs. If this is not possible due to high CpG density, one CpG site can be included at the 5′ end of each primer. Such "degenerate" primers should be synthesized as Y (C/T) in the forward strand and R (G/A) in the reverse stand. Postbisulfite PCR primers should incorporate sufficient cytosines in the original sequence to prevent amplification of unmodified DNA. The main concern for bisulfite-PCR is PCR bias, which occurs because methylated and unmethylated DNA molecules sometimes amplify with greatly differing efficiencies [24]. To test for PCR bias, a mixing experiment using known quantities of methylated and unmethylated genomic DNA is highly recommended for each assay [25]. Selection of the annealing temperature is possibly the most critical variable for minimization of PCR bias.

28.4 METHYLOME-WIDE ANALYSIS

As gene expression microarrays accelerated and revolutionized the study of transcriptional regulation, rapidly improving technologies have increasingly enabled researchers to assess locus-specific DNA methylation on a genome-wide scale [2]. These can be divided into non-microarray and microarray analyses.

28.4.1 Non-Microarray-Based Genome-Wide Analysis

Several genome-wide DNA methylation analysis methods do not require microarrays. A classical method is Restriction Landmark Genome Scanning (RLGS) [26] – a two-dimensional DNA gel electrophoresis technique. In combination with methylation-sensitive restriction enzymes (NotI or AscI), this technique provides methylation profiles of thousands of loci at once. However, it has limited genome coverage and sensitivity. Nevertheless, this technique has been widely used to identify imprinted loci, as well as genes showing tissue-specific methylation in normal tissues [27] and genes aberrantly methylated in cancer [28]. It has been used to scan the genome of lung cancer cells [29] and leukemic cell lines [30], as well as to differentiate between tissues – for example, neural versus non-neural tissues [31].

Another non-array-based method, methylation-specific digital karyotyping (MSDK), has been developed [32]. MSDK is conceptually similar to serial analysis of gene expression (SAGE), and relies on the cleavage of genomic DNA with a methylation-sensitive enzyme (AscI). After NlaIII digestion, short sequence tags are sequenced and mapped to genomic locations. Although it requires a large number of sequencing reactions, MSDK requires no special device other than a DNA sequencer. Just like RLGS, however, it has limited genome coverage and requires a relatively large amount of DNA. MSDK has been used to show epigenetic dysregulation in, for example, breast cancer [33] and atherosclerosis [34].

An interesting method to annotate the genomic methylation landscape is to construct unmethylated and methylated domains using limiting digestion with McrBC or other restriction endonucleases. This method was an adaptation of differential methylation hybridization (DMH). DNA fragments are transfected into *Escherichia coli* and plasmid DNA from individual colonies analyzed by sequencing [35]. This method is relatively unbiased and high-resolution, but time consuming. Such massively parallel sequencing technology has been successfully applied in a genome-wide annotation of multiple histone modifications in a high-resolution and high-throughput manner [36]. The technique provides greater sensitivity to densely methylated regions than using a methylation-sensitive enzyme [2]. The new development in this area is called "comprehensive high-throughput arrays for relative methylation" (CHARM) [37,38]. This method has been used to detect substantial DNA methylation differences between tissues and across different brain regions in rats [39]. A further modification, known as MethylScope®, is to use McrBr to cut randomly sheared DNA [2].

28.4.2 Microarray-Based Genome-Wide Analysis

A first possible choice is methylated DNA immunoprecipitation (MeDIP), in which DNA is immunoprecipitated using an anti-5-meC antibody, then hybridized to a microarray [40]. This technique is independent of the specific methylation-sensitive restriction sites within the target sequence, but requires large amounts of genomic DNA and antibody. Ligation-mediated PCR (LM-PCR) has been used to perform MeDIP with limited quantities of DNA. LM-PCR is very inefficient, however, using blunt ends in the adaptor ligation, and potentially causes bias toward GC-poor regions. Thereafter, methylated CpG island recovery assay (MIRA) has been used for genome-wide methylation analysis in cancers [41]. This method is based on purification of methylated DNA by methyl CpG binding protein columns. Both MeDIP and

MIRA may lack sensitivity in genomic regions with a relatively low density of CpG sites. In addition, the cells or tissues analyzed have to be very homogenous; this is the reason why it is very rarely used outside cancer research. A new development, MeDIPonChip, uses the MeDIP approach in combination with a tiling array for the investigation of genome-wide DNA methylation patterns [42].

This method showed that the zinc finger protein 582 gene is frequently methylated in cervical cancer [43]. Bisulfite-PCR and specially designed oligonucleotide arrays have been used to quantify the bisulfite-induced C to T changes at defined genomic positions. Although this method requires gene-specific PCR, it can interrogate multiple CpG sites within hundreds of genes at once [44]. This approach does not represent the entire genome, and primer design is challenging due to the T richness of DNA sequences after bisulfite conversion. The combination of bisulfite-PCR and methylation CpG island arrays has been reported under the name "methylated CpG island amplification and microarray" (MCAM) [45,46]. This method was used to explore the methylome in, for example, human salivary gland adenoid cystic carcinoma [47].

Various array-based strategies have been developed using combinations of methylation-sensitive and methylation-insensitive restriction enzyme digestion, followed by ligation-mediated PCR to enrich for methylated or unmethylated fragments. Differential methylation hybridization (DMH) is one of the first described methods in this category [48]. Genomic DNA is digested with MseI (methylation-independent), ligated with linkers, then digested with BstUI or HpaII (both methylation-sensitive) to remove unmethylated fragments. The digested DNA is amplified by primers complementary to the linker sequence, and the products are labeled and hybridized to arrays. This method is relatively simple and requires little DNA. Only CpGs within the restriction enzyme sites are analyzed, however, and incomplete digestion could lead to false positive results. Nevertheless, the method is widely used; for example, it has recently been used to explore the methylome in breast [49] and prostate [50] cancer, as well as metabolic disorders [51], and to study the response of dermal fibroblasts to lipopolysaccharides [52].

Finally, a strategy named methylated CpG island amplification combined with microarray (MCA) addresses the problem of false positivity by using methylation-sensitive and insensitive isoschizomers (enzymes with the same recognition sequence) [53]. The principle is simple: DNA is first incubated with a methylation-sensitive restriction enzyme (SmaI) that digests unmethylated DNA but leaves methylated sites intact. The same DNA is then digested with a methylation-insensitive SmaI isoschizomer (XmaI). The key is that whereas SmaI leaves blunt ends, XmaI cleavage produces "sticky ends." Following ligation of adapters that anneal to the XmaI cut sites, adapter-specific

PCR results in amplification of methylated regions. This MCA product is then labeled and hybridized to microarrays [12]. This method has distinct advantages – genome amplification is limited to the target molecules (methylated DNA), and each SmaI fragment is amplified only if the two SmaI sites are both unaffected by the methylation-sensitive enzyme and digested by the methylation-insensitive enzyme (i.e., adapter ligation is truly methylation-sensitive). This dramatically reduces false positives. Coverage is still limited, however, to CpGs within SmaI sites. This method demonstrated that extensive hypermethylation in promoters of polycomb target genes is a characteristic of follicular lymphoma [54].

A further variation of DMH is to use a cocktail of methylation-sensitive restriction enzymes to digest one pool of DNA, and McrBC to digest the other pool. This is referred to a microarray-based methylation assessment of single samples (MMASS) [2]. Another method, known as HpaII tiny fragment enrichment by ligation-mediated PCR (HELP), uses ligation-mediated PCR for the amplification of HpaII or MspI genomic restriction fragments followed by array hybridization.

28.5 SEQUENCING APPROACHES

Restriction enzyme enrichment techniques are currently being adapted so that the readout can be obtained by next-generation sequencing techniques instead of array hybridization. Sequence-based analysis is more flexible and powerful as it allows for allele-specific DNA methylation analysis, does not require an appropriately designed microarray, can cover more of the genome with less input DNA, and avoids hybridization artifacts, although it is still subject to sequence library biases. Next-generation sequencing has been used to analyze the output of the HELP assay [55]. Sequencing-by-synthesis of libraries constructed from size-fractionated HpaII or MspI digests that are compared with randomly sheared fragments [56] is known as Methyl–seq; sequence-based analysis of HpaII digestion followed by the use of a flanking cut with a type-IIS restriction enzyme (MmeI) and adaptor ligation is known as methylation-sensitive cut counting (MSCC) [57]. The technique is used for analyzing functional DNA sequences [58].

28.5.1 Bisulfite Pyrosequencing

Like other methods described above, bisulfite pyrosequencing relies on bisulfite conversion and PCR amplification [59]. To facilitate the conversion of PCR products to single-stranded DNA for later pyrosequencing, the PCR reaction is performed either with one primer biotinylated

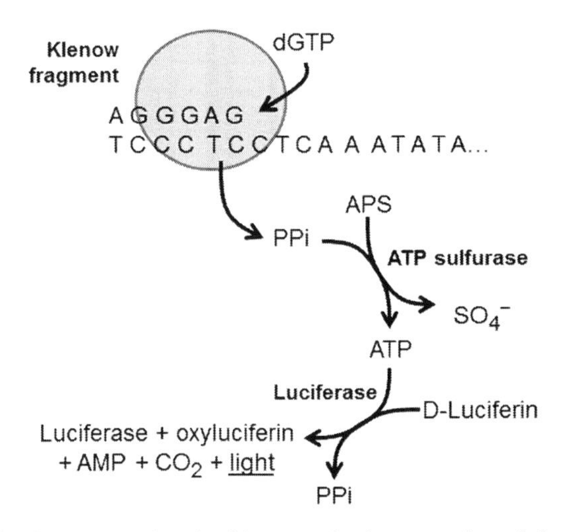

FIGURE 28.2 Pyrosequencing: in this example, incorporation of the complementary S-dGTP by the Klenow fragment of DNA polymerase I at the 3′-end of the pyrosequencing primer results in the release of PP_i, which is in turn used to convert adenosine phosphosulfate into ATP. The ATP provides the energy to form an unstable luciferase–luciferin–AMP complex, which in the presence of oxygen results in the release of light in an amount proportional to the available ATP and thus PP_i.

or using a tailed primer in combination with a biotin-labeled universal primer in the same reaction; this avoids biotin-labeling each primer for each assay. The sequencing primer is then annealed to single-stranded DNA and the samples are ready for pyrosequencing analysis. Pyrosequencing is a primer extension method for the analysis of short to medium-length DNA sequences. Incorporation of a nucleotide into the template strand leads to the release of pyrophosphate, which is quantified with a luciferase reaction (Figure 28.2). The signal produced is proportional to the amount of pyrophosphate released; therefore, the percentage of unconverted C and converted T nucleotides at each CpG site can be detected and quantified. This method has the advantages of introducing an internal control (DNA sequence including a control for unconverted cytosines) and allowing accurate quantitation of multiple CpG methylation sites in the same reaction. The only significant drawback is that only 25–30 base pairs can be sequenced in each reaction, limiting the number of CpG sites that can be assessed. Ultra-deep sequencing of a limited number of loci has been achieved by direct pyrosequencing of PCR products [60] and by sequencing of more than 100 PCR products in a single run at an average coverage of more than 1600 reads per locus on the Roche 454 platform [61]. The technique has been widely used recently in, for example, in breast cancer [62], acute myeloid leukemia [63], type 2 diabetes [64], and depression [65], as well as to study the adverse effects of maternal smoking [66], the leukocyte subpopulations in rheumatoid arthritis [67], and the mechanism of epithelial–mesenchymal transition [68].

28.5.2 Matrix-Assisted Laser Desorption Ionization Time-of-Flight Mass Spectrometry

Another high-throughput method utilizes base-specific cleavage of nucleic acids and analyzes samples by matrix-assisted laser desorption ionization time-of-flight mass spectrometry (MALDI-TOF MS) [69]. In this method, regions of interest are PCR amplified from bisulfite-modified DNA using primers with a T7 RNA polymerase tag. The PCR products are *in vitro* translated into a single-stranded RNA using T7 polymerase and subsequently cleaved by an endoribonuclease such as RNase A. Different cleavage patterns for the methylated and unmethylated CpG positions are then quantitated by mass spectrometry. Compared with pyrosequencing, this method offers the possibility of detecting more CpG sites in a single amplicon (a maximum of about 800 bp). This sophisticated method, however, is technically very challenging. The technique has been employed recently to show LINE1 hypomethylation in peripheral blood leukocytes from Chinese patients with juvenile systemic lupus erythematosus [70].

28.5.3 Reduced Representation Bisulfite Sequencing

The main challenge in sequencing bisulfite-converted DNA arises from its low sequence complexity. Reduced representation bisulfite sequencing (RRBS) was introduced to reduce sequence redundancy by selecting only some regions of the genome for sequencing by size-fractionation of DNA fragments after *Bg*/II digestion [71] or after MspI digestion [72]. These choices of restriction enzymes enrich for CpG-containing segments of the genome, but do not target specific regions of interest in the genome. Targeting has been accomplished by array capture before sequencing. Targeted capture on fixed arrays or by solution hybrid selection can enrich for sequences targeted by a library of DNA or RNA oligonucleotides and can be performed before or after bisulfite conversion. RRBS has been used recently to show the methylome of early human embryo [73] and how dynamic DNA methylation can be [74].

28.5.4 Whole-Genome Shotgun Bisulfite Sequencing

The ultimate comprehensive single base-pair resolution DNA methylation analysis technique is whole-genome bisulfite sequencing.

Whole-genome shotgun bisulfite sequencing (WGSBS) has been achieved on the Illumina Genome Analyzer platform [75]. Increased read lengths and paired-end sequencing strategies have aided the implementation of WGSBS [76–78], although approximately a tenth of the CpG dinucleotides in the mammalian genome remain refractory to alignment of bisulfite-converted reads. This technique has recently been used regarding atherosclerosis [79], stem cell differentiation [80], and rheumatoid arthritis synovial fibroblasts [81].

28.6 CONCLUSION

DNA methylation analysis has undergone a revolution over the past decade. From single sequence-based DNA methylation data, the trend is more and more toward whole methylome analysis. These data should be coordinated in international consortia and combined with transcriptome, microRNA, and proteinome data. Sequencing costs will drop; the difficulty will remain at the bioinformatic level.

References

[1] Clark SJ, Harrison J, Paul CL, Frommer M. High sensitivity mapping of methylated cytosines. Nucleic Acids Res 1994;22:2990–7.

[2] Laird PW. Principles and challenges of genomewide DNA methylation analysis. Nat Rev Genet 2010;11:191–203.

[3] Yang AS, Doshi KD, Choi SW, Mason JB, Mannari RK, Gharybian V, et al. DNA methylation changes after 5-aza-2′-deoxycytidine therapy in patients with leukemia. Cancer Res 2006;66:5495–503.

[4] Ehrlich M, Gama-Sosa MA, Huang LH, Midgett RM, Kuo KC, McCune RA, et al. Amount and distribution of 5-methylcytosine in human DNA from different types of tissues of cells. Nucleic Acids Res 1982;10:2709–21.

[5] Neidhart M, Rethage J, Kuchen S, Künzler P, Crowl RM, Billingham ME, et al. Retrotransposable L1 elements expressed in rheumatoid arthritis synovial tissue: association with genomic DNA hypomethylation and influence on gene expression. Arthritis Rheum 2000;43:2634–47.

[6] Weisenberger DJ, Campan M, Long TI, Kim M, Woods C, Fiala E, et al. Analysis of repetitive element DNA methylation by MethyLight. Nucleic Acids Res 2005; 33:6823–36.

[7] Gloria L, Cravo M, Pinto A, de Sousa LS, Chaves P, Leitão CN, et al. DNA hypomethylation and proliferative activity are increased in the rectal mucosa of patients with long-standing ulcerative colitis. Cancer 1996;78:2300–6.

[8] Habib M, Fares F, Bourgeois CA, Bella C, Bernardino J, Hernandez-Blazquez F, et al. DNA global hypomethylation in EBV-transformed interphase nuclei. Exp Cell Res 1999;249:46–53.

[9] Karouzakis E, Gay RE, Michel BA, Gay S, Neidhart M. DNA hypomethylation in rheumatoid arthritis synovial fibroblasts. Arthritis Rheum 2009;60:3613–22.

[10] Giraldo AM, Lynn JW, Purpera MN, Godke RA, Bondioli KR. DNA methylation and histone acetylation patterns in cultured bovine fibroblasts for nuclear transfer. Mol Reprod Dev 2007;74:1514–24.
[11] Tochiki KK, Cunningham J, Hunt SP, Geranton SM. The expression of spinal methyl-CpG-binding protein 2, DNA methyltransferases and histone deacetylases is modulated in persistent pain states. Mol Pain 2012;8:14.
[12] Shen L, Waterland RA. Methods of DNA methylation analysis. Curr Opin Clin Nutr Metab Care 2007;10:576–81.
[13] Herman JG, Graff JR, Myohanen S, Nelkin BD, Baylin SB. Methylation-specific PCR: a novel PCR assay for methylation status of CpG islands. Proc Natl Acad Sci USA 1996;93:9821–6.
[14] Eads CA, Danenberg KD, Kawakami K, Saltz LB, Blake C, Shibata D, et al. MethyLight: a high-throughput assay to measure DNA methylation. Nucleic Acids Res 2000;28:E32.
[15] Hernandez HG, Tse MY, Pang SC, Arboleda H, Forero DA. Optimizing methodologies for PCR-based DNA methylation analysis. Biotechniques 2013;55:181–97.
[16] Zeschnigk M, Böhringer S, Price EA, Onadim Z, Masshöfer L, Lohmann DR. A novel real-time PCR assay for quantitative analysis of methylated alleles (QAMA): analysis of the retinoblastoma locus. Nucleic Acids Res 2004;32:e125.
[17] Melnikov AA, Gartenhaus RB, Levenson AS, Motchoulskaia NA, Levenson Chernokhvostov VV. MSRE-PCR for analysis of gene-specific DNA methylation. Nucleic Acids Res 2005;33:e93.
[18] Reindollar RH, Fusaris KW, Gray MR. Methylation-dependent melting polymorphisms in genomic fragments of deoxyribonucleic acid. Am J Obstet Gynecol 2000;182:785–92.
[19] Laprise SL, Gray MR. Covalent genomic DNA modification patterns revealed by denaturing gradient gel blots. Gene 2007;391:45–52.
[20] Frommer M, McDonald LE, Millar DS, Collis CM, Watt F, Grigg GW, et al. A genomic sequencing protocol that yields a positive display of 5-methylcytosine residues in individual DNA strands. Proc Natl Acad Sci USA 1992;89:1827–31.
[21] Hammoud SS, Purwar J, Pflueger C, Cairns BR, Carrell DT. Alterations in sperm DNA methylation patterns at imprinted loci in two classes of infertility. Fertil Steril 2010;94:1728–33.
[22] Waterland RA, Lin JR, Smith CA, Jirtle RL. Post-weaning diet affects genomic imprinting at the insulin-like growth factor 2 (Igf2) locus. Hum Mol Genet 2006;15:705–16.
[23] Xiong Z, Laird PW. COBRA: a sensitive and quantitative DNA methylation assay. Nucleic Acids Res 1997;25:2532–4.
[24] Warnecke PM, Stirzaker C, Melki JR, Millar DS, Paul CL, Clark SJ. Detection and measurement of PCR bias in quantitative methylation analysis of bisulphite-treated DNA. Nucleic Acids Res 1997;25:4422–6.
[25] Shen L, Guo Y, Chen X, Ahmed S, Issa JP. Optimizing annealing temperature overcomes bias in bisulfite PCR methylation analysis. Biotechniques 2007;42:48–52.
[26] Costello JF, Plass C, Cavenee WK. Restriction landmark genome scanning. Methods Mol Biol 2002;200:53–70.
[27] Song F, Smith JF, Kimura MT, Morrow AD, Matsuyama T, Nagase H, et al. Association of tissue-specific differentially methylated regions (TDMs) with differential gene expression. Proc Natl Acad Sci U S A 2005;102:3336–41.
[28] Costello JF, Frühwald MC, Smiraglia DJ, Rush LJ, Robertson GP, Gao X, et al. Aberrant CpG-island methylation has non-random and tumour-type-specific patterns. Nat Genet 2000;24:132–8.
[29] Tessema M, Belinsky SA. Mining the epigenome for methylated genes in lung cancer. Proc Am Thorac Soc 2008;5:806–10.

[30] Dunwell TL, Dickinson RE, Stankovic T, Dallol A, Weston V, Austen B, et al. Frequent epigenetic inactivation of the SLIT2 gene in chronic and acute lymphocytic leukemia. Epigenetics 2009;4:265–9.
[31] Ghosh S, Yates AJ, Frühwald MC, Miecznikowski JC, Plass C, Smiraglia D. Tissue specific DNA methylation of CpG islands in normal human adult somatic tissues distinguishes neural from non-neural tissues. Epigenetics 2010;5:527–38.
[32] Hu M, Yao J, Cai L, Bachman KE, van den Brûle F, Velculescu V, et al. Distinct epigenetic changes in the stromal cells of breast cancers. Nat Genet 2005;37:899–905.
[33] Li J, Gao F, Li N, Li S, Yin G, Tian G, et al. An improved method for genome wide DNA methylation profiling correlated to transcription and genomic instability in two breast cancer cell lines. BMC genomics 2009;10:223.
[34] Zawada AM, Rogacev KS, Hummel B, Grün OS, Friedrich A, Rotter B, et al. SuperTAG methylation-specific digital karyotyping reveals uremia-induced epigenetic dysregulation of atherosclerosis-related genes. Circ Cardiovasc Genet 2012;5:611–20.
[35] Rollins RA, Haghighi F, Edwards JR, Das R, Zhang MQ, Ju J, et al. Large-scale structure of genomic methylation patterns. Genome Res 2006;16:157–63.
[36] Barski A, Cuddapah S, Cui K, Roh TY, Schones DE, Wang Z, et al. High-resolution profiling of histone methylations in the human genome. Cell 2007;129:823–37.
[37] Irizarry RA, Ladd-Acosta C, Carvalho B, Wu H, Brandenburg SA, Jeddeloh JA, et al. Comprehensive high-throughput arrays for relative methylation (CHARM). Genome Res 2008;18:780–90.
[38] Ladd-Acosta C, Aryee MJ, Ordway JM, Feinberg AP. Comprehensive high-throughput arrays for relative methylation (CHARM). Curr Protoc Hum Genet 2010; **Chapter 20**, Unit 20–19.
[39] Lee RS, Tamashiro KL, Aryee MJ, Murakami P, Seifuddin F, Herb B, et al. Adaptation of the CHARM DNA methylation platform for the rat genome reveals novel brain region-specific differences. Epigenetics 2011;6:1378–90.
[40] Weber M, Davies JJ, Wittig D, Oakeley EJ, Haase M, Lam WL, et al. Chromosome-wide and promoter-specific analyses identify sites of differential DNA methylation in normal and transformed human cells. Nat Genet 2005;37:853–86.
[41] Rauch T, Li H, Wu X, Pfeifer GP. MIRA-assisted microarray analysis, a new technology for the determination of DNA methylation patterns, identifies frequent methylation of homeodomain-containing genes in lung cancer cells. Cancer Res 2006;66:7939–47.
[42] Hsu YW, Huang RL, Lai HC. MeDIPonChip for Methylation Profiling. Methods Mol Biol 2015;1249:281–90.
[43] Huang RL, Chang CC, Su PH, Chen YC, Liao YP, Wang HC, et al. Methylomic analysis identifies frequent DNA methylation of zinc finger protein 582 (ZNF582) in cervical neoplasms. PloS One 2012;7:e41060.
[44] Shi H, Maier S, Nimmrich I, Yan PS, Caldwell CW, Olek A, et al. Oligonucleotide-based microarray for DNA methylation analysis: principles and applications. J Cell Biochem 2003;88:138–43.
[45] Estecio MR, Yan PS, Ibrahim AE, Tellez CS, Shen L, Huang TH, et al. High-throughput methylation profiling by MCA coupled to CpG island microarray. Genome Res 2007;17:1529–36.
[46] Colyer HA, Dellett M, Mills KI. Detecting DNA methylation using the methylated CpG island amplification and microarray technique. Methods Mol Biol 2012; 863:329–39.
[47] Bell A, Bell D, Weber RS, El-Naggar AK. CpG island methylation profiling in human salivary gland adenoid cystic carcinoma. Cancer 2011;117:2898–909.
[48] Yan PS, Chen CM, Shi H, Rahmatpanah F, Wei SH, Huang TH. Applications of CpG island microarrays for high-throughput analysis of DNA methylation. J Nutr 2002;132:2430S–4S.

[49] Halvorsen AR, Helland A, Fleischer T, Haug KM, Grenaker Alnaes GI, et al. Differential DNA methylation analysis of breast cancer reveals the impact of immune signaling in radiation therapy. Int J Cancer 2014;135:2085–95.
[50] White-Al Habeeb NM, Ho LT, Olkhov-Mitsel E, Kron K, Pethe V, Lehman M, et al. Integrated analysis of epigenomic and genomic changes by DNA methylation dependent mechanisms provides potential novel biomarkers for prostate cancer. Oncotarget 2014;5:7858–69.
[51] Fan C, Dong H, Yan K, Shen W, Wang C, Xia L, et al. Genome-wide screen of promoter methylation identifies novel markers in diet-induced obese mice. Nutr Hosp 2014;30:42–52.
[52] Green BB, Kerr DE. Epigenetic contribution to individual variation in response to lipopolysaccharide in bovine dermal fibroblasts. Vet Immunol Immunopathol 2014;157:49–58.
[53] Toyota M, Ho C, Ahuja N, Jair KW, Li Q, Ohe-Toyota M, et al. Identification of differentially methylated sequences in colorectal cancer by methylated CpG island amplification. Cancer Res 1999;59:2307–12.
[54] Bennett LB, Schnabel JL, Kelchen JM, Taylor KH, Guo J, Arthur GL, et al. DNA hypermethylation accompanied by transcriptional repression in follicular lymphoma. Genes Chromosomes Cancer 2009;48:828–41.
[55] Oda M, Glass JL, Thompson RF, Mo Y, Olivier EN, Figueroa ME, et al. High-resolution genome-wide cytosine methylation profiling with simultaneous copy number analysis and optimization for limited cell numbers. Nucleic Acids Res 2009;37:3829–39.
[56] Brunner AL, Johnson DS, Kim SW, Valouev A, Reddy TE, Neff NF, et al. Distinct DNA methylation patterns characterize differentiated human embryonic stem cells and developing human fetal liver. Genome Res 2009;19:1044–56.
[57] Ball MP, Li JB, Gao Y, Lee JH, LeProust EM, Park IH, et al. Targeted and genome-scale strategies reveal gene-body methylation signatures in human cells. Nat Biotechnol 2009;27:361–8.
[58] Colaneri AC, Jones AM. Genome-wide quantitative identification of DNA differentially methylated sites in *Arabidopsis* seedlings growing at different water potential. PLoS One 2013;8:e59878.
[59] Colella S, Shen L, Baggerly KA, Issa JP, Krahe R. Sensitive and quantitative universal Pyrosequencing methylation analysis of CpG sites. Biotechniques 2003;35:146–50.
[60] Korshunova Y, Maloney RK, Lakey N, Citek RW, Bacher B, Budiman A, et al. Massively parallel bisulphite pyrosequencing reveals the molecular complexity of breast cancer-associated cytosine-methylation patterns obtained from tissue and serum DNA. Genome Res 2008;18:19–29.
[61] Taylor KH, Kramer RS, Davis JW, Guo J, Duff DJ, Xu D, et al. Ultradeep bisulfite sequencing analysis of DNA methylation patterns in multiple gene promoters by 454 sequencing. Cancer Res 2007;67:8511–18.
[62] Barrio-Real L, Benedetti LG, Engel N, Tu Y, Cho S, Sukumar S, et al. Subtype-specific overexpression of the Rac-GEF P-REX1 in breast cancer is associated with promoter hypomethylation. Breast Cancer Res 2014;16:441.
[63] Hajkova H, Fritz M, Ha Kovec C, Schwarz JI, Álek C, Marková J, et al. CBFB-MYH11 hypomethylation signature and PBX3 differential methylation revealed by targeted bisulfite sequencing in patients with acute myeloid leukemia. J Hematol Oncol 2014;7:66.
[64] Cheng J, Tang L, Hong Q, Ye H, Xu X, Xu L, et al. Investigation into the promoter DNA methylation of three genes (CAMK1D, CRY2 and CALM2) in the peripheral blood of patients with type 2 diabetes. Exp Ther Med 2014;8:579–84.

[65] Oh G, Wang SC, Pal M, Chen ZF, Khare T, Tochigi M, et al. DNA modification study of major depressive disorder: beyond locus-by-locus comparisons. Biol Psychiatry 2015;77:246–55.
[66] Stroud LR, Papandonatos GD, Rodriguez D, McCallum M, Salisbury AL, et al. Maternal smoking during pregnancy and infant stress response: test of a prenatal programming hypothesis. Psychoneuroendocrinology 2014;48:29–40.
[67] Glossop JR, Emes RD, Nixon NB, Haworth KE, Packham JC, Dawes PT, et al. Genome-wide DNA methylation profiling in rheumatoid arthritis identifies disease-associated methylation changes that are distinct to individual T- and B-lymphocyte populations. Epigenetics 2014;9:1228–37.
[68] Tahara T, Shibata T, Okubo M, Ishizuka T, Nakamura M, Nagasaka M, et al. DNA methylation status of epithelial-mesenchymal transition (EMT) – related genes is associated with severe clinical phenotypes in ulcerative colitis (UC). PLoS One 2014;9:e107947.
[69] Ehrich M, Nelson MR, Stanssens P, Zabeau M, Liloglou T, et al. Quantitative high-throughput analysis of DNA methylation patterns by base-specific cleavage and mass spectrometry. Proc Natl Acad Sci U S A 2005;102:15785–90.
[70] Huang X, Su G, Wang Z, Shangguan S, Cui X, et al. Hypomethylation of long interspersed nucleotide element-1 in peripheral mononuclear cells of juvenile systemic lupus erythematosus patients in China. Int J Rheum Dis 2014;17:280–90.
[71] Meissner A, Gnirke A, Bell GW, Ramsahoye B, Lander ES, Jaenisch R. Reduced representation bisulfite sequencing for comparative high-resolution DNA methylation analysis. Nucleic Acids Res 2005;33:5868–77.
[72] Meissner A, Mikkelsen TS, Gu H, Wernig M, Hanna J, et al. Genome-scale DNA methylation maps of pluripotent and differentiated cells. Nature 2008;454:766–70.
[73] Guo H, Zhu P, Yan L, Li R, Hu B, Lian Y, et al. The DNA methylation landscape of human early embryos. Nature 2014;511:606–10.
[74] Ziller MJ, Gu H, Müller F, Donaghey J, Tsai LT, Kohlbacher O, et al. Charting a dynamic DNA methylation landscape of the human genome. Nature 2013;500:477–81.
[75] Lister R, Pelizzola M, Dowen RH, Hawkins RD, Hon G, Tonti-Filippini J, et al. Human DNA methylomes at base resolution show widespread epigenomic differences. Nature 2009;462:315–22.
[76] Dunn JJ, McCorkle SR, Everett L, Anderson CW. Paired-end genomic signature tags: a method for the functional analysis of genomes and epigenomes. Genet Eng (N Y) 2007;28:159–73.
[77] Korbel JO, Urban AE, Affourtit JP, Godwin B, Grubert F, Simons JF, et al. Paired-end mapping reveals extensive structural variation in the human genome. Science 2007;318:420–6.
[78] Johnson MD, Mueller M, Game L, Aitman TJ. Single nucleotide analysis of cytosine methylation by whole-genome shotgun bisulfite sequencing. Curr Protoc Mol Biol 2012; **Chapter 21**, Unit 21–23.
[79] Zaina S, Heyn H, Carmona FJ, Varol N, Sayols S, Condom E, et al. A DNA methylation map of human atherosclerosis. Circ Cardiovasc Genet 2014;7:692–700.
[80] Chen PY, Feng S, Joo JW, Jacobsen SE, Pellegrini M. A comparative analysis of DNA methylation across human embryonic stem cell lines. Genome Biol 2011;12:R62.
[81] Whitaker JW, Shoemaker R, Boyle DL, Hillman J, Anderson D, Wang W, et al. An imprinted rheumatoid arthritis methylome signature reflects pathogenic phenotype. Genome Med 2013;5:40.

CHAPTER

29

Methylome Analysis in Cancer

OUTLINE

29.1 INTRODUCTION

It is known that inactivation of certain tumor-suppressor genes occurs as a consequence of hypermethylation within the promoter regions, and numerous studies have demonstrated a broad range of genes silenced by DNA methylation in different cancer types [1]. On the other hand, global hypomethylation, inducing genomic instability, also contributes to cell transformation (see Chapters 2, 7, and 8). Many techniques have been developed to detect global DNA methylation in cancer cells compared to normal tissues. This knowledge helps us to better understand cancer progression, and also aids in the development of new biomarkers for early cancer detection. New prognostic tools for monitoring drug efficacy during cancer treatment can also be developed [2]. There are

M. Neidhart: DNA Methylation and Complex Human Disease.
DOI: http://dx.doi.org/10.1016/B978-0-12-420194-1.00029-4

over 28 million CpG sites in the human genome. Assessing the methylation status of each of these sites will be required to understand fully the role of DNA methylation in health and disease. Genome-wide analysis, using arrays and high-throughput sequencing, has enabled assessment of large fractions of the methylome, but each protocol comes with unique advantages and disadvantages [3]. In this chapter we will focus on genome-wide approaches in cancer. Advances in the robustness, throughput, and accuracy of epigenomic technologies over the past decade have not only co-occurred with a revolution in cancer research [4] and therapeutic strategies [5,6], but also provided important information in other related fields – for example, stem cell biology and regenerative medicine [7]. The first completed epigenomes documented the features of cultured, immortalized, or cancerous cell lines. While this enabled correlation of epigenomic features with genomic elements, the resulting epigenomic landscapes were a static view. Now that lineage-specific differentiation protocols produce populations of cells with increasing purity, we can analyze dynamic epigenomic changes during development, in cancer as well as in other fields [8,9]. These comprehensive studies are giving insight into how cell-state transitions are influenced by chromatin states at promoters and enhancers, DNA methylation dynamics at enhancers, and the expansion of repressive domains [10].

29.2 BISULFITE-BASED METHODS

It is obvious that the best way to determine the global methylation status of samples is to analyze the methylation status of a large number of single CpG sites. These types of analyses are currently possible thanks to the next-generation sequencing and DNA array-based technologies. A type of DNA array particularly useful in DNA methylation analyses is DNA methylation arrays. As mentioned in Chapter 28, methylation arrays often use bisulfite modified DNA and determine the methylation status of single CpG sites using fluorescent assays. The development of different types of microarray-based methylation analysis, such as Illumina [3], HumanMethylation27 DNA Analysis BeadChip™ in esophagal carcinoma [11], the Golden Gate® Methylation Cancer Panel in uretelial cancer [12], and VeraCode platforms for colorectal cancer [13] have enabled the definition of methylation profiles in many other tumors: chronic lymphocytic leukemia [14], various hematological neoplasms [15], B-cell lymphoma [16], and hepatocellular carcinoma [17]. Further microarray-based methylation studies comparing tumors will be required to evaluate the variation in hypomethylation patterns between individual tumor types in order for them to be exploited as prognostic markers. To date, these global DNA methylation screenings have helped to identify single markers with great clinical potential, such as predictors of recurrence in lung cancer [18], metastasis in colorectal cancer [19], or progression in virus-associated neoplasms [20].

29.3 NON-BISULFITE METHODS

Genome-wide DNA methylation can also be studied using non-bisulfite-based methods. Examples include methods based on the digestion of DNA with restriction enzymes, such as methylation CpG island amplification (MCA) in colorectal cancer [21], differential methylation hybridization (DMH) in breast cancer [22], restriction landmark genomic scanning (RLGS) in acute myeloid leukemia [23], amplification of inter-methylated sites (AIMS) in colorectal carcinomas [24], methylation-specific digital karyotyping (MSDK) in breast cancer [25,26], HpaII tiny fragment enrichment by ligation-mediated PCR (HELP) assay in acute leukemia [27,28], and CpGlobal in lung cancer [29]. For example, the application of CpGlobal to measure the changes in global DNA methylation in lung cancer showed substantial DNA methylation differences between healthy and tumor tissues.

29.4 METHYLATED DNA BINDING COLUMN

Other approaches include those based on the affinity of MBD proteins, such as methylated DNA binding column (MBD column) chromatography, in which methylated and unmethylated DNA fragments can be discriminated and separated using methyl-CpG-binding protein 2 (MeCP2) and then analyzed with appropriate PCR primers [30]. However, this only results in a rough determination of DNA methylation status, which must be compared with elution profiles of standard DNA [31]. MBD column chromatography could be used in clinical studies for the detection of DNA methylation biomarkers in colorectal cancer [32]. Another example derived from the affinity of MBD proteins is the methylated CpG island recovery assay (MIRA), which recognizes methylated CpG dinucleotides via the methyl-CpG-binding domain protein 2b (MBD2b) [33]. This method is quite simple to use and yields few false positives, making it useful in genome-wide analysis in various cancers [33,34], such as renal carcinoma [35], lung cancer [36–38], breast cancer [39], hematological and epithelial cancers [40], gastric cancer [41], Epstein-Barr virus-associated gastric carcinomas [42], and astrocytomas [43].

29.5 COMBINATION WITH MeDiP-seq OR RRBS

Finally, the currently widely used methods based on next-generation sequencing technology produce tremendous amounts of information on DNA methylation. Examples include the combination of this ultra-deep sequencing with methyl-DNA immunoprecipitation (MeDiP-seq) or its combination with digestion using methylation sensitive restriction enzymes (reduced representation bisulfite sequencing [RRBS]) [44]. Both are effective methods to analyze methylomes; RRBS has been used to analyze mammalian methylomes, and MeDiP-seq for human tumor [45] and sperm methylomes [46]. However, MeDiP-seq suffers from the same dependence on antibody quality as the MeDiP approach, and RRBS offers less genome-wide coverage [34,47,48]. RRBS protocol is shown in Figure 29.1.

RRBS allows the sequencing of methylated areas that are unable to be properly profiled using conventional bisulfite sequencing techniques. Current sequencing technologies are limited in regard to profiling areas of repeated sequences. This is unfortunate in regard to methylation studies, as these repeated sequences often contain methylated cytosines. This is especially limiting for studies involving profiling cancer genomes, as a loss of methylation in these repeated sequences is

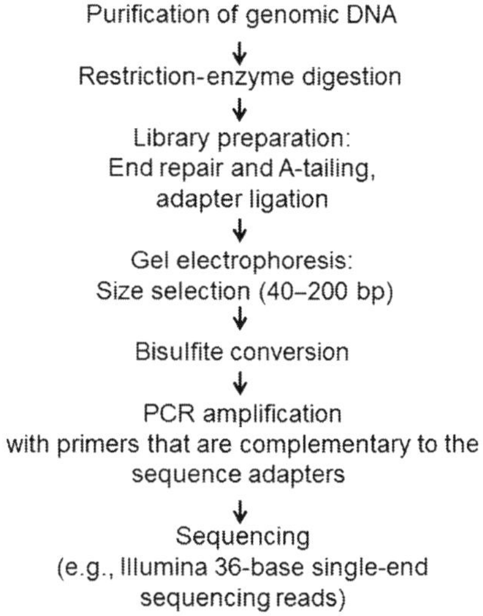

Purification of genomic DNA
↓
Restriction-enzyme digestion
↓
Library preparation:
End repair and A-tailing,
adapter ligation
↓
Gel electrophoresis:
Size selection (40–200 bp)
↓
Bisulfite conversion
↓
PCR amplification
with primers that are complementary to the
sequence adapters
↓
Sequencing
(e.g., Illumina 36-base single-end
sequencing reads)

FIGURE 29.1 RRBS uniquely uses a specific restriction enzymes to enrich for CpGs. MspI digestion, or any restriction enzyme that recognizes CpGs and cuts them, produces only fragments with CGs at the end [1]. This approach enriches CpG regions of the genome, so it can decrease the amount of sequencing required as well as decrease the cost. This technique is cost-effective, especially when focusing on common CpG regions.

observed in many cancer types [49]. RRBS eliminates the problems encountered due to these large areas of repeated sequences and thus allows these regions to be fully analyzed.

The four-nucleotide cutters, MspI and TaqαI, are restriction enzymes commonly used in RRBS that, when combined, achieve about 12% genomic coverage [50]. The increase in throughput of next-generation sequencers allows for novel combinations of restriction enzymes that provide higher CpG coverage. Recently, a study [51] performed a near-neighbor analysis of the four nucleotide sequences most frequently found within 50 nucleotides of all genomic CpGs. This resulted in the identification of seven methylation-insensitive restriction enzymes (AluI, BfaI, HaeIII, HpyCH4V, MluCI, MseI, and MspI) that share similar restriction conditions suitable for RRBS library preparation. They reported that the use of two or three enzyme combinations increases the theoretical epigenome coverage to almost half of the human genome. Free RRBS-Analyser, on the Internet, can be widely used by researchers [52,53].

29.6 DNA METHYLATION IN FORMALIN-FIXED PARAFFIN-EMBEDDED TISSUES

Recently, various platforms for genome-wide analysis of promoter DNA methylation have been developed, including the Illumina Infinium platform. A technique has been developed to identify DNA methylation signatures in complex tissues, using the HumanMethylation27 BeadChip assay [11,54,55] or laser capture microdissection-reduced representation bisulfite sequencing (LCM-RRBS) [56]. BeadChip was used to differentiate esophagal [54] or adrenocortical [55] premalignant and malignant tissues from normal tissues [54]. Regarding LCM-RRBS, in the mouse, changes in DNA methylation associated with gonadectomy-induced adrenocortical neoplasia were characterized [56]. Compared with adjacent normal tissue, the adrenocortical tumors showed reproducible gains and losses of DNA methylation at genes involved in cell differentiation and organ development. BeadChip and LCM-RRBS are rapid, cost-effective, and sensitive techniques for analyzing DNA methylation in heterogeneous tissues.

29.7 CONCLUSION

In general, genome-wide approaches are very useful tools to measure global DNA methylation and to identify biomarkers with clinical outcome. As the cost reduces, the amount of analysis performed with the new methylome-wide technologies will increase. The problem will be to manage the large amount of information and thus to increase the bioinformatics capacities, as well as to provide adequate professional help for researchers and clinicians.

References

[1] Kulis M, Esteller M. DNA methylation and cancer. Adv Genet 2010;70:27–56.
[2] Nai HS, Lau QC. Advent of the cancer methylome. Comb Chem High Throughput Screen 2012;15:216–20.
[3] Stirzaker C, Taberlay PC, Statham AL, Clark SJ. Mining cancer methylomes: prospects and challenges. Trends Genet 2014;30:75–84.
[4] Ma X, Wang YW, Zhang MQ, Gazdar AF. DNA methylation data analysis and its application to cancer research. Epigenomics 2013;5:301–16.
[5] Kim TM, Lee SH, Chung YJ. Clinical applications of next-generation sequencing in colorectal cancers. World J Gastroenterol 2013;19:6784–93.
[6] Papageorgiou EA, Koumbaris G, Kypri E, Hadjidaniel M, Patsalis PC. The Epigenome view: an effort towards non-invasive prenatal diagnosis. Genes 2014;5: 310–29.
[7] Jang H, Shin H. Current trends in the development and application of molecular technologies for cancer epigenetics. World J Gastroenterol 2013;19:1030–9.
[8] Piperi C, Papavassiliou AG. Strategies for DNA methylation analysis in developmental studies. Dev Growth Differ 2011;53:287–99.
[9] Bogdanovic O, Gomez-Skarmeta JL. Embryonic DNA methylation: insights from the genomics era. Brief Funct Genomics 2014;13:121–30.
[10] Rivera CM, Ren B. Mapping human epigenomes. Cell 2013;155:39–55.
[11] Thirlwell C, Eymard M, Feber A, Teschendorff A, Pearce K, Lechner M, et al. Genome-wide DNA methylation analysis of archival formalin-fixed paraffin-embedded tissue using the Illumina Infinium HumanMethylation27 BeadChip. Methods 2010;52:248–54.
[12] Chihara Y, Kanai Y, Fujimoto H, Sugano K, Kawashima K, Liang G, et al. Diagnostic markers of urothelial cancer based on DNA methylation analysis. BMC Cancer 2013; 13:275.
[13] Ostendorff HP, Awad A, Braunschweiger KI, Liu Z, Wan Z, Rothschild KJ, et al. Multiplexed VeraCode bead-based serological immunoassay for colorectal cancer. J Immunol Methods 2013;400-401:58–69.
[14] Kanduri M, Cahill N, Göransson H, Enström C, Ryan F, Isaksson A, et al. Differential genome-wide array-based methylation profiles in prognostic subsets of chronic lymphocytic leukemia. Blood 2010;115:296–305.
[15] Martin-Subero JI, Ammerpohl O, Bibikova M, Wickham-Garcia E, Agirre X, Alvarez S, et al. A comprehensive microarray-based DNA methylation study of 367 hematological neoplasms. PLoS One 2009;4:e6986.
[16] Martin-Subero JI, Kreuz M, Bibikova M, Bentink S, Ammerpohl O, Wickham-Garcia E, et al. New insights into the biology and origin of mature aggressive B-cell lymphomas by combined epigenomic, genomic, and transcriptional profiling. Blood 2009;113: 2488–97.
[17] Hernandez-Vargas H, Lambert MP, Le Calvez-Kelm F, Gouysse G, McKay-Chopin S, Tavtigian SV, et al. Hepatocellular carcinoma displays distinct DNA methylation signatures with potential as clinical predictors. PLoS One 2010;5:e9749.
[18] Brock MV, Hooker CM, Ota-Machida E, Han Y, Guo M, Ames S, et al. DNA methylation markers and early recurrence in stage I lung cancer. N Engl J Med 2008;358: 1118–28.

[19] Lujambio A, Calin GA, Villanueva A, Ropero S, Sánchez-Céspedes M, Blanco D, et al. A microRNA DNA methylation signature for human cancer metastasis. Proc Natl Acad Sci U S A 2008;105:13556–61.
[20] Fernandez AF, Rosales C, Lopez-Nieva P, Graña O, Ballestar E, Ropero S, et al. The dynamic DNA methylomes of double-stranded DNA viruses associated with human cancer. Genome Res 2009;19:438–51.
[21] Toyota M, Ho C, Ahuja N, Jair KW, Li Q, Ohe-Toyota M, et al. Identification of differentially methylated sequences in colorectal cancer by methylated CpG island amplification. Cancer Res 1999;59:2307–12.
[22] Huang TH, Perry MR, Laux DE. Methylation profiling of CpG islands in human breast cancer cells. Hum Mol Genet 1999;8:459–70.
[23] Rush LJ, Plass C. Restriction landmark genomic scanning for DNA methylation in cancer: past, present, and future applications. Anal Biochem 2002;307:191–201.
[24] Frigola J, Ribas M, Risques RA, Peinado MA. Methylome profiling of cancer cells by amplification of inter-methylated sites (AIMS). Nucleic Acids Res 2002;30.
[25] Hu M, Yao J, Cai L, Bachman KE, van den Brûle F, Velculescu V, et al. Distinct epigenetic changes in the stromal cells of breast cancers. Nat Genet 2005;37:899–905.
[26] Hu M, Yao J, Polyak K. Methylation-specific digital karyotyping. Nat Protoc 2006;1: 1621–36.
[27] Khulan B, Thompson RF, Ye K, Fazzari MJ, Suzuki M, Stasiek E, et al. Comparative isoschizomer profiling of cytosine methylation: the HELP assay. Genome Res 2006;16:1046–55.
[28] Figueroa ME, Melnick A, Greally JM. Genome-wide determination of DNA methylation by Hpa II tiny fragment enrichment by ligation-mediated PCR (HELP) for the study of acute leukemias. Methods Mol Biol 2009;538:395–407.
[29] Anisowicz A, Huang H, Braunschweiger KI, Liu Z, Giese H, Wang H, et al. A high-throughput and sensitive method to measure global DNA methylation: application in lung cancer. BMC Cancer 2008;8:222.
[30] Cross SH, Charlton JA, Nan X, Bird AP. Purification of CpG islands using a methylated DNA binding column. Nat Genet 1994;6:236–44.
[31] Shiraishi M, Sekiguchi A, Oates AJ, Terry MJ, Miyamoto Y, Tanaka K, et al. Variable estimation of genomic DNA methylation: a comparison of methyl–CpG binding domain column chromatography and bisulfite genomic sequencing. Anal Biochem 2002;308:182–5.
[32] Zou H, Harrington J, Rego RL, Ahlquist DA. A novel method to capture methylated human DNA from stool: implications for colorectal cancer screening. Clin Chem 2007;53:1646–51.
[33] Rauch T, Pfeifer GP. Methylated-CpG island recovery assay: a new technique for the rapid detection of methylated-CpG islands in cancer. Lab Invest 2005;85:1172–80.
[34] Mitchell N, Deangelis JT, Tollefsbol TO. Methylated-CpG island recovery assay. Methods Mol Biol 2011;791:125–33.
[35] Okuda H, Toyota M, Ishida W, Furihata M, Tsuchiya M, Kamada M, et al. Epigenetic inactivation of the candidate tumor suppressor gene HOXB13 in human renal cell carcinoma. Oncogene 2006;25:1733–42.
[36] Rauch T, Li H, Wu X, Pfeifer GP. MIRA-assisted microarray analysis, a new technology for the determination of DNA methylation patterns, identifies frequent methylation of homeodomain-containing genes in lung cancer cells. Cancer Res 2006;66:7939–47.
[37] Rauch TA, Wang Z, Wu X, Kernstine KH, Riggs AD, Pfeifer GP. DNA methylation biomarkers for lung cancer. Tumour Biol 2012;33:287–96.
[38] Kapitskaya KY, Azhikina TL, Ponomaryova AA, Cherdyntseva NV, Vlasov VV, Laktionov PP, et al. MIRA analysis of RARbeta2 gene methylation in DNA circulating in the blood in lung cancer. Bull Exp Biol Med 2014;157:516–19.

[39] Tommasi S, Karm DL, Wu X, Yen Y, Pfeifer GP. Methylation of homeobox genes is a frequent and early epigenetic event in breast cancer. Breast Cancer Res 2009;11:R14.
[40] Dunwell T, Hesson L, Rauch TA, Wang L, Clark RE, Dallol A, et al. A genome-wide screen identifies frequently methylated genes in haematological and epithelial cancers. Mol Cancer 2010;9:44.
[41] Park JH, Park J, Choi JK, Lyu J, Bae MG, Lee YG, et al. Identification of DNA methylation changes associated with human gastric cancer. BMC Med Genomics 2011;4:82.
[42] Okada T, Nakamura M, Nishikawa J, Sakai K, Zhang Y, Saito M, et al. Identification of genes specifically methylated in Epstein-Barr virus-associated gastric carcinomas. Cancer Sci 2013;104:1309–14.
[43] Wu X, Rauch TA, Zhong X, Bennett WP, Latif F, Krex D, et al. CpG island hypermethylation in human astrocytomas. Cancer Res 2010;70:2718–27.
[44] Nagarajan A, Roden C, Wajapeyee N. Reduced representation bisulfite sequencing to identify global alteration of DNA methylation. Methods Mol Biol 2014;1176:23–31.
[45] Feber A, Wilson GA, Zhang L, Presneau N, Idowu B, Down TA, et al. Comparative methylome analysis of benign and malignant peripheral nerve sheath tumors. Genome Res 2011;21:515–24.
[46] Smith ZD, Gu H, Bock C, Gnirke A, Meissner A. High-throughput bisulfite sequencing in mammalian genomes. Methods 2009;48:226–32.
[47] Harris RA, Wang T, Coarfa C, Nagarajan RP, Hong C, Downey SL, et al. Comparison of sequencing-based methods to profile DNA methylation and identification of monoallelic epigenetic modifications. Nat Biotechnol 2010;28:1097–105.
[48] Torano EG, Petrus S, Fernandez AF, Fraga MF. Global DNA hypomethylation in cancer: review of validated methods and clinical significance. Clin Chem Lab Med 2012;50:1733–42.
[49] Ehrlich M. DNA hypomethylation in cancer cells. Epigenomics 2009;1:239–59.
[50] Wang J, et al. Double restriction-enzyme digestion improves the coverage and accuracy of genome-wide CpG methylation profiling by reduced representation bisulfite sequencing. BMC genomics 2013;14:11.
[51] Martinez-Arguelles DB, Lee S, Papadopoulos V. In silico analysis identifies novel restriction enzyme combinations that expand reduced representation bisulfite sequencing CpG coverage. BMC Res Notes 2014;7:534.
[52] Wang T, Liu Q, Li X, Wang X, Li J, Zhu X, et al. RRBS-analyser: a comprehensive web server for reduced representation bisulfite sequencing data analysis. Hum Mutat 2013;34:1606–10.
[53] Liang F, Tang B, Wang Y, Wang J, Yu C, Chen X, et al. WBSA: web service for bisulfite sequencing data analysis. PLoS One 2014;9:e86707.
[54] Zhai R, Zhao Y, Su L, Cassidy L, Liu G, Christiani DC. Genome-wide DNA methylation profiling of cell-free serum DNA in esophageal adenocarcinoma and Barrett esophagus. Neoplasia 2012;14:29–33.
[55] Fonseca AL, Kugelberg J, Starker LF, Scholl U, Choi M, Hellman P, et al. Comprehensive DNA methylation analysis of benign and malignant adrenocortical tumors. Genes Chromosomes Cancer 2012;51:949–60.
[56] Schillebeeckx M, Schrade A, Löbs AK, Pihlajoki M, Wilson DB, Mitra RD. Laser capture microdissection-reduced representation bisulfite sequencing (LCM-RRBS) maps changes in DNA methylation associated with gonadectomy-induced adrenocortical neoplasia in the mouse. Nucleic Acids Res 2013;41:e116.

CHAPTER

30

Methylome Analysis in Non-Neoplastic Disease

OUTLINE

M. Neidhart: DNA Methylation and Complex Human Disease.
DOI: http://dx.doi.org/10.1016/B978-0-12-420194-1.00030-0

30.1 INTRODUCTION

The experimental approaches, techniques, and advances described in Chapter 29 highlight the critical role and significance of epigenetic modification in cellular processes that include cancer, development, homeostasis, and disease. These types of modifications, manifest as methylation of CpG dinucleotides in DNA and as histone tail modification, are apparent in all cell types, including those within specific organs. The rapid advances made in next-generation sequencing technologies make it possible to map DNA cytosine methylation at single-base resolution. In non-neoplastic diseases, the Illumina sequencing platform is currently the most widely used method [1–18] (Figure 30.1); the second is methylated DNA immunoprecipitation (MeDIP) followed by microarray hybridization [19–22], and the third the Comprehensive High-throughput Array for Relative Methylation (CHARM) platform [23,24]. In addition to the costs of generating the raw data, the generation and storage of many gigabytes to terabytes of sequence read data may require consideration of infrastructure or improvement of computational power. For these reasons, it is recommended that these experiments are best conducted in collaboration with bioinformaticians with experience in both data handling and methods. In this chapter, we review recent findings in the field of methylome-wide analysis of non-neoplastic diseases.

30.2 ENVIRONMENTAL FACTORS

Early pregnancy may be characterized by widespread hypomethylation compared with non-pregnant states; there is no apparent permanent methylation imprint after a normal-term gestation. Using an

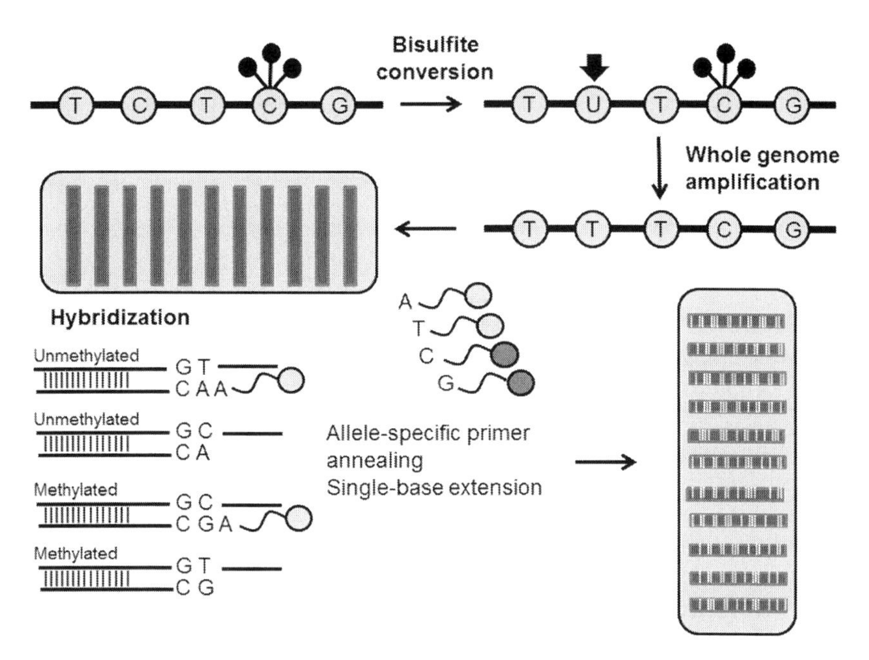

FIGURE 30.1 **Workflow of the Illumina Infinium HumanMethylation BeadChip™ assay.** Bisulfite conversion is followed by whole-genome amplification. On the chip, there are two bead types for each CpG site per locus. Each locus tested is differentiated by different bead types; there are over 200,000 bead types available. Each of the bead types is attached to a single-stranded 50-mer DNA oligonucleotide; these differ in sequence only at the free end. This type of probe is known as an allele-specific oligonucleotide. One of the bead types will correspond to the methylated cytosine locus and the other will correspond to the unmethylated cytosine locus, which has been converted into uracil during bisulfite treatment and later amplified as thymine during whole-genome amplification. The bisulfite converted amplified DNA products are denatured into single strands and hybridized to the chip via allele-specific annealing to either the methylation specific probe or the non-methylation probe. Hybridization is followed by single-base extension with hapten-labeled dideoxynucleotides. The chip is scanned to show the intensities of the unmethylated and methylated bead types.

Illumina BeadChip sequencing platform, a study identified nine potential candidate genes as differentially methylated in early pregnancy [7]; they may be related to the maternal adaptation to pregnancy. As mentioned in Chapter 13, prenatal development and early childhood are critical periods for establishing the tissue-specific epigenome, and may have a profound impact on health and disease in later life. Comparison between groups of monozygotic and dizygotic twins enables an estimation of the relative contribution of genetic and shared and non-shared environmental factors to phenotypic variability. Using DNA methylation profiling of about 20,000 CpG sites as a phenotype, a study examined discordance levels in three neonatal tissues from 22 monozygotic and 12 dizygotic twin pairs [25]. Regression analysis of methylation on birth weight revealed a general association between methylation of genes involved in metabolism and biosynthesis, providing further support for epigenetic change in the previously described link between low birth weight and increasing risk for cardiovascular, metabolic, and other complex diseases. Another genome-wide DNA methylation analysis was performed in 105 Black American children (59 males and 46 females) from birth through the first 2 years of life [1]. DNA isolated from cord blood at birth and venous blood samples thereafter were analyzed using Illumina Infinium HumanMethylation27 BeadChip™. A wide range of interindividual variations in genome-wide methylation was observed at each time point, including lower levels at CpG islands near transcription sites (called TSS200). Specifically, 159 CpG sites were identified in males and 149 CpG sites in females with significant longitudinal changes. These CpG sites appeared to be located within genes with important biological functions, including immunity and inflammation.

Gestational age at birth even strongly predicts neonatal, adolescent, and adult morbidity and mortality through mostly unknown mechanisms. A study performed a genome-wide analysis of DNA methylation, using microarrays (specifically CHARM 2.0), in 141 newborns [23]. The authors identified three differentially methylated regions at genome-wide significance levels adjacent to the nuclear factor 1 X-type (NFIX), Rap guanine nucleotide exchange factor-2 (RAPGEF2), and methionine sulfoxide reductase B3 (MSRB3) genes.

A disadvantaged socio-economic position in childhood is associated with increased adult mortality and morbidity. A study aimed to establish whether childhood socio-economic position is associated with differential methylation of adult DNA [19]. Forty adult males were selected from socio-economic extremes in both early childhood and mid-adulthood. The authors performed genome-wide methylation analysis on blood DNA taken at age 45 years using methylated DNA immunoprecipitation (MeDIP), and mapped using an array of the methylation state of promoters of approximately 20,000 genes and 400 microRNAs. Methylation levels for 1252 gene promoters were associated with childhood socio-economic position. Functionally, associations appear in promoters of genes enriched in key cell-signaling pathways.

Cigarette smoking is an environmental risk factor for many chronic diseases, and disease risk can often be managed by smoking control. Smoking can induce cellular and molecular changes, including epigenetic modification. Recent studies have identified smoking-related DNA methylation sites in Caucasians. To determine whether the same sites associate with smoking in African Americans, and to identify novel smoking-related sites, a recent study [26] conducted a methylome-wide association study of cigarette smoking using a discovery sample of 972 African Americans and a replication sample of 239 African Americans, with two array-based methods. The top two smoking-related DNA methylation sites in factor II receptor-like 3 (F2RL3) and G protein-coupled receptor 15 (GPR15) were confirmed in both Caucasians and African Americans. A similar work [27] using a 14,000 gene promoter array showed that the gene coding for a potential drug target of cardiovascular importance features altered methylation patterns in smokers, including the F2RL3 gene, coding for protease-activated receptor-4 (PAR4). Differences in F2RL3 methylation have been confirmed by a third genome-wide study [2] using the Illumina HumanMethylation27K BeadChip. F2RL3 methylation is strongly associated with mortality due to cardiovascular diseases [28]. Another study [9], using the same platform, investigated genome-wide methylation in cord blood of newborns in relation to maternal smoking during pregnancy. Differently methylated genes identified by using the Illumina platform include aryl-hydrocarbon receptor repressor (AHRR), cytochrome P450 family 1 subfamily A polypeptide-1 (CYP1A1), and growth factor independent 1 transcription repressor (GFI1). AHRR and CYP1A1 play a key role in the detoxification of the components of tobacco smoke. Thus, the methylation of several genes changed in children whose mothers smoked during pregnancy.

30.3 CARDIOVASCULAR DISEASES

30.3.1 Congenital Heart Defects

The majority of congenital heart defects are thought to result from the interaction between multiple genetic, epigenetic, environmental, and lifestyle factors. Epigenetic mechanisms are attractive targets in the study of complex diseases because they may be altered by environmental factors and dietary interventions. A study [6] conducted a population-based case–control study of genome-wide maternal DNA methylation to determine if alterations in gene-specific methylation were associated with this disease. Using the Illumina Infinium HumanMethylation27 BeadChip, they assessed maternal gene-specific methylation in over 27,000 CpG sites from DNA isolated from peripheral blood lymphocytes. The study sample included 180 mothers with non-syndromic congenital heart defect-affected pregnancies and 187 mothers with unaffected pregnancies. The analysis revealed that differently methylated genes are involved in multiple biological processes during fetal development.

30.3.2 Dilated Cardiomyopathy

Genome-wide cardiac DNA methylation was compared between patients with idiopathic dilated cardiomyopathy and controls [29]. Differently methylated genes included lymphocyte antigen 75 (LY75), tyrosine kinase-type cell surface receptor HER3 (ERBB3), homeobox B13 (HOXB13), and adenosine receptor A2A (ADORA2A).

30.3.3 Essential Hypertension

A genome-wide methylation analysis was conducted with peripheral blood leukocytes taken from eight hypertensive cases and eight normotensive age-matched controls aged 14–23 years [30]. A CpG site in the sulfactase-1 (SULF1) gene showed higher methylation levels in cases than in healthy controls.

30.3.4 Coronary Heart Disease

Epigenetic alteration is expected to make a significant contribution to the development of cardiovascular disease where environmental interactions play a key role in disease progression. A genome-wide analysis [31] identified 72 differentially methylated regions that were hypermethylated in patients with coronary heart disease in the background of varying homocysteine levels. Six CpG sites in three differentially methylated regions included the intronic region of complement C1q-Like protein-4 (C1QL4) gene and upstream region of coiled-coil domain containing-47 (CCD47) and transforming growth factor receptor-3 (TGFBR3) genes.

30.3.5 Atherosclerosis

A genome-wide analysis of DNA methylation in the atherosclerotic human aorta has recently been performed [16]. A total of 48 post-mortem human aortic intima specimens were examined. To avoid the effects of interindividual variation, we performed intraindividual paired comparisons. Bisulfite-modified genomic DNA was analyzed for DNA methylation with the Illumina HumanMethylation450 BeadChip. Three of the hypomethylated genes (Drosophila headcase [HECA], early B-cell factor 1 [EBF1], and nucleotide-binding oligomerization domain containing 2 [NOD2]) and three of the hypermethylated genes (human mitogen-activated protein kinase kinase kinase kinase 4 [MAP4K4], zinc finger E-box binding homeobox 1 [ZEB1], and proto-oncogene tyrosine-protein kinase FYN) had previously been implicated in atherosclerosis. The overexpression of HECA, EBF1, or NOD2 or the suppression of

MAP4K4, ZEB1, or FYN expression in cultured HEK293 cells resulted in significant changes in the expression of atherosclerosis-related genes, as determined with an expression microarray. These findings suggested that HECA, EBF1, and NOD2 were significantly hypomethylated, whereas MAP4K4, ZEB1, and FYN were hypermethylated, in atheromatous plaque lesions compared with plaque-free intima.

30.4 INFLAMMATORY SKIN DISEASES

30.4.1 Psoriasis

Psoriasis, a chronic inflammatory skin disorder, is characterized by aberrant keratinocyte proliferation and differentiation in the epidermis. Methylated DNA immunoprecipitation sequencing (MeDIP-Seq) was used to characterize whole-genome DNA methylation patterns in involved and uninvolved skin lesions from patients with psoriasis [20]. Gene ontology (GO) analysis of the data showed that the aberrantly methylated genes belonged to several different ontological domains, such as the immune system, cell cycle, and apoptosis, the most affected genes being programmed cell death protein-5 (PDCD5) and tissue metallopeptidase inhibitor-2 (TIMP2). GO is a major bioinformatics initiative to unify the representation of gene and gene product attributes across all species [32]. Moreover, monozygotic twins discordant for psoriasis were used to search for genome-wide differences in DNA methylation and gene expression in $CD4^+$ and $CD8^+$ T cells using Illumina's HumanMethylation27 [5]. As expected, several of the top-ranked genes according to significance of the correlation in $CD4^+$ T cells are known to be associated with psoriasis. Further, GO analysis revealed enrichment of biological processes associated with the immune response and clustering of genes in a biological pathway comprising cytokines and chemokines. By using a ChIP-seq method, it has been shown that the promoter regions of 121 genes on the X chromosome have dramatically elevated methylation levels in psoriasis patient T cells, compared to those from healthy controls [33]. These findings imply that methylation changes may affect $CD4^+$ T-cell polarization, especially in the pathogenesis of psoriasis. A recent methylome-wide analysis [34] reported that phosphatidic acid phosphatase type 2 domain containing 3 (PPAPDC3) is one of the most affected genes by promoter hypermethylation in psoriatic $CD4^+$ T cells. Furthermore, mesenchymal stem cells are likely involved in pathological processes of immune-related diseases, including psoriasis, because of their immunoregulatory and pro-angiogenic effects. Dermal mesenchymal stem cells from psoriatic patients and normal controls were isolated and expanded, then submitted to a genome-wide DNA methylation and gene ontology analyses [35]. The genome-wide promoter methylation profile of cells from psoriatic derma was markedly different from the normal derma derived cells. Genes involved in cell communication, the surface receptor signaling pathway, cellular response to stimulus, and cell migration were differently methylated. Several aberrantly methylated genes related to epidermal proliferation, angiogenesis, and inflammation were also found to be differently expressed in patients.

30.4.2 Systemic Sclerosis

The etiology of systemic sclerosis is not clear, but there is emerging evidence of gene-specific epigenetic dysregulation in its pathogenesis. A genome-wide DNA methylation study [36] was performed in dermal fibroblasts from 6 diffuse cutaneous systemic sclerosis patients and 6 limited cutaneous systemic sclerosis patients, and compared with 12 age-matched, sex-matched, and ethnicity-matched healthy controls. Cytosine methylation was quantified in more than 485,000 methylation sites across the genome. The authors identified 2710 and 1021 differentially methylated CpG sites in diffuse and limited systemic sclerosis, respectively. There were only 203 CpG sites differentially methylated in both diffuse and limited systemic sclerosis, representing 118 hypomethylated and 6 hypermethylated genes. Common hypomethylated genes include ITGA9, encoding $\alpha 9$ integrin. Other relevant genes such as disintegrin and metalloproteinase domain-containing protein-12 (ADAM12), collagens (COL23A1 and COL4A2), and myosin-1e (MYO1E), as well as transcription factor genes runt-related transcription factors 1, 2, and 3 (RUNX1, RUNX2, and RUNX3) were also hypomethylated in both forms of systemic sclerosis. Pathway analysis of differentially methylated genes in both forms revealed enrichment of genes involved in extracellular matrix–receptor interaction and focal adhesion.

30.4.3 Systemic Lupus Erythematosus

Systemic lupus erythematosus is an autoimmune disease characterized by multisystem involvement and autoantibody production. We mentioned in Chapter 19 that abnormal T-cell DNA methylation plays an important role in its pathogenesis. A genome-wide DNA methylation study [37] has been performed recently in two independent sets of lupus patients and matched healthy controls to characterize the DNA methylome of these $CD4^+$ T cells in the disease. DNA methylation was quantified for over

485,000 methylation sites across the genome. In naïve $CD4^+$ T cells from lupus patients hypomethylation in interferon-regulated genes was found, including interferon-induced protein with tetratricopeptide repeats-1 and -3 (IFIT1, IFIT3), interferon-induced GTP-binding protein (MX1), signal transducers and activators of transcription-1 (STAT1), interferon-induced protein 44-like (IFI44L), ubiquitin specific peptidase-18 (USP18), tripartite motif-containing 22 (TRIM22), and tetherin (BST2), suggesting epigenetic transcriptional accessibility in these genetic loci. These results suggest epigenetic "poising" of interferon-regulated genes. A more recent methylome-wide study [13] showed that the interferon hypersensitivity was apparent also in memory, naïve, and regulatory T cells.

30.5 ASTHMA

Asthma is a common respiratory disease that is characterized by bronchial hyper-responsiveness and airway obstruction due to chronic airway inflammation. The genome-wide DNA methylation levels in the bronchial mucosa tissues of 10 atopic asthmatics, 7 non-atopic asthmatics, and 7 normal controls were examined using microarrays [38]. A study [39] evaluated the methylation status of whole genome in blood and polyp tissues with and without aspirin hypersensitivity. Genome-wide DNA methylation levels in nasal polyps and peripheral blood cells were examined by microarray analysis. In the nasal polyps of the patients with aspirin-tolerant asthma, hypermethylation was detected at 332 loci in 296 genes, while hypomethylation was detected at 158 loci in 141 genes. In the arachidonate pathway, prostaglandin-D synthase (PGDS), arachidonate 5-lipoxygenase-activating protein (ALOX5AP), and leukotriene B4 receptor (LTB4R) were hypomethylated, whereas prostaglandin E synthase (PTGES) was hypermethylated. Furthermore, CpG DNA methylation differences associated with wheezing phenotypes were screened in children using the Illumina® GoldenGate Panel I [4].This screening identified lower DNA methylation at a CpG site in the arachidonate 12-lipoxygenase (ALOX12).

30.6 INFLAMMATORY BOWEL DISEASE

The impact of differences in methylation patterns in the intestine with regard to inflammatory bowel disease susceptibility and activity has been investigated by genome-wide methylation profiling using the HumanMethylation27 BeadChip microarray [40,41]. In biopsies [18] multiple genes showed significant evidence of differential methylation, several appearing in both ulcerative colitis and Crohn's disease comparisons, including thyroid hormone receptor associated protein-2 (THRAP2), Fanconi anemia (FANCC), globoside α-1,3-N-acetylgalactosaminyltransferase-1 (GBGT1), docking protein-2 (DOK2), tumor necrosis factor ligand superfamily member 4 (TNFSF4/CD252), tumor necrosis factor ligand superfamily member 12 (TNFSF12), and α-(1,3)-fucosyltransferase (FUT7). Many more than expected by chance overlapped with genes previously implicated as playing a role in inflammatory bowel disease susceptibility in genome-wide association scans, including caspase recruitment domain-containing protein-9 (CARD9), intercellular adhesion molecule-3 (ICAM3/CD50), and interleukin 8 receptor β (IL8RB). In peripheral blood leukocytes [19] the microarray identified differentially methylated genes related to testis-, prostate-, and placenta-expressed proteins. The significance of such findings is as yet unclear.

30.7 RHEUMATIC DISEASES

30.7.1 Rheumatoid Arthritis

A genome-wide evaluation of DNA methylation loci in fibroblast-like synoviocytes isolated from the site of disease in rheumatoid arthritis was performed using the Illumina HumanMethylation450 BeadChip [8]. Genomic DNA was isolated from six rheumatoid arthritis and five osteoarthritis synovial fibroblast cultures, which served as "healthy" controls. Hypomethylated loci were identified in genes, for example chitinase-3-like protein-1 (CHI3L1), caspase-1 (CASP1), signal transducer and activator of transcription 3 (STAT3), mitogen-activated protein kinase kinase kinase-5 (MAP3K5), pyrin (MEFV), and WNT1-inducible-signaling pathway protein-3 (WISP3). Hypomethylation was increased in multiple pathways related to cell migration, including focal adhesion, cell adhesion, transendothelial migration, and extracellular matrix interactions. Hypermethylated genes were also observed, for example transforming growth factor β receptor-2 (TGFBR2) and forkhead box protein O1 (FOXO1). Recently, a methylation array also identified transcription factor T-box transcription factor 5 (TBX5) as less methylated in rheumatoid arthritis synovium and synovial fibroblasts than in osteoarthritis samples [42]. Demethylation of the TBX5 promoter is accompanied by higher TBX5 expression. Taken together, these observations may explain the intrinsically activated phenotype of rheumatoid arthritis synovial fibroblasts.

30.7.2 Osteoarthritis

Genome-wide methylation profiles of bone from patients with hip osteoarthritis and those with osteoporotic hip fractures have been compared using Illumina methylation arrays [10]. Trabecular bone pieces were obtained from the central part of the femoral head of 27 patients with hip fractures and 26 patients with hip osteoarthritis. The differentially methylated genes were enriched for association with bone traits in the genome-wide association study catalog. Pathway analysis and text-mining analysis with Gene Relationships Across Implicated Loci (GRAIL) software [43] revealed enrichment in genes participating in glycoprotein metabolism or cell differentiation, and particularly in the homeobox superfamily of transcription factors. This suggests the existence of a developmental component in the predisposition to these disorders. Another study [11] compared the genome-wide DNA methylation profiles of human articular chondrocytes from osteoarthritic cartilage. DNA methylation profiling was performed using Illumina Infinium HumanMethylation27 in 25 patients with osteoarthritis and 20 healthy controls; 91 differentially methylated loci were identified, mostly associated with inflammatory processes. Finally, a very recent genome-wide DNA methylation profiling of $>$ 485,000 methylation sites identified 550 differentially methylated sites in osteoarthritis chondrocytes [44], including runt-related transcription factor 1 (RUNX1), serine protease HTRA1, fibroblast growth factor receptor-2 (FGFR2), and collagen XI (COL11A2), that are also overexpressed at the protein levels.

30.8 METABOLIC DISEASES

30.8.1 Obesity and Type 2 Diabetes

Cells of the endocrine organs, which include the pituitary, thyroid, thymus, pancreas, ovaries, and testes, have been shown to be susceptible to epigenetic alteration, leading to both local and systemic changes often resulting in life-threatening metabolic disease [45]. Obesity is a major health problem that is determined by interactions between lifestyle and environmental and genetic factors. The whole-blood DNA from 479 individuals of European origin was analyzed with the Infinium HumanMethylation450 array [17]. The authors also examined whether methylation levels at identified sites also showed an association with body mass index in DNA from 635 adipose tissue and 395 skin samples obtained from White female individuals participating in a German study. They found that increased body mass index is associated with increased methylation at the hypoxia-inducible factor 3α (HIF3A) locus in blood cells and in adipose tissue. Furthermore, a recent study [46] analyzed genome-wide methylation profiles of over 470,000 CpGs in peripheral blood samples from 48 obese and 48 lean Black American youths aged 14–20 years. Differentially variable CpG sites generally exhibited an outlier structure and were more variable in cases than in controls. The genes involved showed relationships with hypertension, dyslipidemia, and type 2 diabetes, supporting their roles in the etiology and pathogenesis of obesity. Furthermore, another work [47] hypothesized that DNA methylation changes are involved in obesity-induced immune dysfunction. A genome-wide methylation analysis was conducted on seven obese cases and seven lean controls. In the obese patients the UBASH3A gene showed higher methylation, while the TRIM3 gene was hypomethylated. Further studies should investigate whether such methylation changes can lead to immune dysfunction. A recent study [22] investigated the potential role of DNA methylation in mediating the increased risk of developing type 2 diabetes in offspring of mothers who had diabetes during pregnancy. Pima Indians showed increased risk of developing obesity and type 2 diabetes. In this investigation, peripheral blood leukocytes were collected from non-diabetic Pima Indians who were either offspring of diabetic mothers ($n = 14$) or offspring of nondiabetic mothers ($n = 14$). Differentially methylated regions were determined using a MeDIP-chip assay on an Affymetrix Human Tiling 2.0R Array. Pathway analysis of genes with differentially methylated promoters identified the top three enriched pathways as maturity onset diabetes of the young (MODY), type 2 diabetes, and Notch signaling. Several genes in these pathways are known to affect pancreatic development and insulin secretion. These findings support the hypothesis that epigenetic changes may increase the risk of type 2 diabetes via an effect on β-cell function in the offspring of mothers with diabetes during pregnancy.

30.8.2 Chronic Kidney Disease

A study [48] evaluated a total of 485,577 unique features in 255 individuals with chronic kidney disease and 152 individuals without evidence of renal disease. Following stringent quality control, raw data were quantile normalized and β values calculated to reflect the methylation status at each site. Twenty-three genes demonstrated significant methylation changes associated with this disease, including Cut-like homeobox-1 (CUX1), engulfment and cell motility protein 1 (ELMO1), FK506 binding protein 5 (FKBP5), inhibin βA anti-sense RNA (INHBA-AS1), receptor-type tyrosine-protein phosphatase N2 (PTPRN2), and 5′-AMP-activated protein kinase subunit γ2 (PRKAG2) genes.

30.9 NEURODEGENERATIVE DISEASES

30.9.1 Alzheimer's Disease

A study [3] investigated human post-mortem frontal cortex genome-wide DNA methylation profiles of 12 late-onset Alzheimer's disease and 12 cognitively normal age- and gender-matched subjects. Quantitative DNA methylation was determined at 27,578 CpG sites spanning 14,475 genes via the Illumina Infinium HumanMethylation27 BeadArray. The transmembrane protein-59 (TMEM59) promoter was found to be hypomethylated. TMEM59 is implicated in amyloid-β protein precursor post-translational processing.

30.9.2 Amyotrophic Lateral Sclerosis

In amyotrophic lateral sclerosis, loci-specific differentially methylated and expressed genes in spinal cord were identified by genome-wide 5-methyl-cytosine and expression profiling using high-throughput microarrays [49]. DNA hyper- or hypomethylations with parallel changes in gene expression were observed in 112 genes highly associated with biological functions related to immune and inflammation responses.

30.9.3 Parkinson's Disease

A study [50] investigated genome-wide DNA methylation in brain and blood samples from Parkinson's disease patients and observed a distinctive pattern of methylation involving many genes previously associated with the disease. The authors claimed to found concordant methylation alterations in brain and blood, suggesting that blood might hold promise as a surrogate for brain tissue to detect DNA methylation.

30.9.4 Multiple Sclerosis

Multiple sclerosis is thought to be caused by T-cell mediated auto-immune dysfunction. Risk of developing MS is influenced by environmental and genetic factors. Applying an Illumina array, a recent [15] study reported hypomethylation of the HLA-DRB1 locus in multiple sclerosis T cells. Using a similar array and a stringent statistical analysis with age and gender correction, a recent study [14] reported genome wide differences in DNA methylation between pathology-free regions derived from human multiple sclerosis-affected and control brains. Genes regulating oligodendrocyte survival, such as Bcl-2-like protein-2 (BCL2L2) and N-Myc downstream regulated-1 (NDRG1), are hypermethylated and expressed at lower levels in multiple sclerosis-affected brains than in controls, while genes related to proteolytic processing (for example, legumain [LGMN] and cathepsin Z [CTSZ]) are hypomethylated and expressed at higher levels. Applying an Illumina array, a recent [15] study reported hypomethylation of the HLA-DRB1 locus in multiple sclerosis T cells.

30.10 PSYCHIATRIC DISORDERS

30.10.1 Schizophrenia

Analyses were conducted using DNA from peripheral blood leukocytes of patients with schizophrenia, and controls. Global methylation results revealed highly significant hypomethylation in the disease [51]. A genome-wide DNA methylation revealed, in accordance with the dopamine hypothesis of psychosis, that catechol-O-methyltransferase (S-COMT) was hypermethylated in patients. The enzyme introduces a methyl group to the catecholamine, which is donated by S-adenosylmethionine (SAM). Another study [52] performed a genome-wide analysis of DNA methylation on peripheral blood DNA samples obtained from a unique sample of monozygotic twin pairs discordant for major psychosis. A genome-wide analysis of DNA methylation on peripheral blood DNA samples was performed. The top psychosis-associated, differentially methylated region, significantly hypomethylated in affected twins, was located in the promoter of α-N-acetylgalactosaminide α-2,6-sialyltransferase-1 (ST6GALNAC1), overlapping a previously reported rare genomic duplication observed in schizophrenia. A more recent study [53] used methyl-CpG-binding domain protein-enriched genome sequencing of the methylated genomic fraction, followed by next-generation DNA sequencing, to identify schizophrenia DNA methylation biomarkers in blood. The sample consisted of 759 schizophrenia cases and 738 controls collected in Sweden. Many top methylome-wide association study results could be linked to hypoxia and, to a lesser extent, infection, suggesting that a record of pathogenic events may be preserved in the methylome. The findings also confirmed a site in the reelin (RELN) gene, one of the most frequently studied candidates in methylation studies of schizophrenia. Another study [54] in a genome-wide meta-analysis replicated former findings in blood samples of patients with schizophrenia regarding the methylation of serologically defined colon cancer antigen 8 (SDCCAG8), cAMP responsive element binding protein 1 (CREB1), and ataxin-7 (ATXN7) genes.

30.10.2 Depressive Disorders

A recent study [12] investigated whether specific methylation profiles in white blood cells could be associated with major depressive disorders. The participants included 12 monozygotic twin pairs discordant for major depressive disorders, and 12 monozygotic twin pairs concordant for no disorders and low neuroticism. Bisulfite treatment and genome-wide interrogation of differentially methylated CpG sites using the Illumina HumanMethylation 450 BeadChip were performed in leukocyte-derived DNA. No overall difference in mean global methylation between cases and their unaffected co-twins was found; however, the differences in females were significant. Furthermore, using the CHARM platform, 39 post-mortem frontal cortexes of patients with major depression disorders were compared to 26 controls [24]. This array covers 3.5 million CpGs and identified 224 candidate regions with differences in DNA methylation. These regions are highly enriched for neuronal growth and development genes, such as proline-rich membrane anchor-1 (PRIMA1). Because PRIMA1 anchors acetylcholinesterase in neuronal membranes, decreased expression could result in decreased enzyme function and increased cholinergic transmission. A very recent study [18] determined whether epigenetic markers predict dimensional ratings of depression in maltreated children. A genome-wide methylation study was completed using the Illumina 450K BeadChip array and saliva-derived DNA of 94 maltreated and 96 healthy non-traumatized children. Methylation in three genes emerged as genome-wide significant predictors of depression: DNA-binding protein inhibitor ID-3 (ID3), glutamate receptor ionotropic N-methyl-D-aspartate-1 (GRIN1), and tubulin polymerization promoting protein (TPPP). These genes are all biologically relevant, with ID3 involved in the stress response, GRIN1 involved in neural plasticity, and TPPP involved in neural circuitry development.

30.10.3 Suicide

A genome-wide approach was used to investigate the extent of DNA methylation alterations in the brains of suicide completers [21]. Promoter DNA methylation was profiled using methylated DNA immunoprecipitation (MeDIP) followed by microarray hybridization in hippocampal tissue from 62 men (46 suicide completers and 16 comparison subjects). The authors identified 366 promoters that were differentially methylated in suicide completers relative to comparison subjects (273 hypermethylated and 93 hypomethylated). Functional annotation analyses revealed an enrichment of differential methylation in the promoters of genes involved, among other functions, in cognitive processes. These results suggest broad reprogramming of promoter DNA methylation patterns in the hippocampus of suicide completers.

30.11 NEUROLOGICAL DISORDERS

Prolonged seizures (status epilepticus) produce pathophysiological changes in the hippocampus that are associated with large-scale, wide-ranging changes in gene expression. Epileptic tolerance is an endogenous program of cell protection that can be activated in the brain by previous exposure to a non-harmful seizure episode before status epilepticus. A major transcriptional feature of tolerance is gene downregulation. A study in mice [55] performed a methylation analysis of 34,143 discrete loci representing all annotated CpG islands and promoter regions in the genome. The authors reported genome-wide DNA methylation changes in the hippocampus after status epilepticus and epileptic tolerance. A total of 321 genes showed altered DNA methylation after status epilepticus alone or status epilepticus that followed seizure preconditioning, with $> 90\%$ of the promoters of these genes undergoing hypomethylation. These profiles included genes not previously associated with epilepsy, such as the polycomb gene PHC2.

30.12 CONCLUSION

The Illumina HumanMethylation450 BeadChip is the most widely used array because of its simplicity. The technology relies on hybridization of genomic fragments to probes on the chip. However, certain genomic factors may compromise the ability to measure methylation using the array, such as single nucleotide polymorphisms, small insertions and deletions, repetitive DNA, and regions with reduced genomic complexity. Currently, there is no clear method or pipeline for determining which of the probes on this bead array should be retained for subsequent analysis in light of these issues [56]. Great amounts of information are generated by such technologies; the problem is to recognize the biological and pathophysiological relevance of the findings. It is essential to validate them with, for example, pyrosequencing and present more functional evidences.

References

[1] Wang D, Liu X, Zhou Y, Xie H, Hong X, Tsai HJ, et al. Individual variation and longitudinal pattern of genome-wide DNA methylation from birth to the first two years of life. Epigenetics 2012;7:594–605.

[2] Wan ES, Qiu W, Baccarelli A, Carey VJ, Bacherman H, Rennard SI, et al. Cigarette smoking behaviors and time since quitting are associated with differential DNA methylation across the human genome. Hum Mol Genet 2012;21:3073–82.

[3] Bakulski KM, Dolinoy DC, Sartor MA, Paulson HL, Konen JR, Lieberman AP, et al. Genome-wide DNA methylation differences between late-onset Alzheimer's disease and cognitively normal controls in human frontal cortex. J Alzheimers Dis 2012;29:571–88.

[4] Morales E, Bustamante M, Vilahur N, Escaramis G, Montfort M, de Cid R, et al. DNA hypomethylation at ALOX12 is associated with persistent wheezing in childhood. Am J Respir Crit Care Med 2012;185:937–43.

[5] Gervin K, Vigeland MD, Mattingsdal M, Hammerø M, Nygård H, Olsen AO, et al. DNA methylation and gene expression changes in monozygotic twins discordant for psoriasis: identification of epigenetically dysregulated genes. PLoS Genet 2012;8:e1002454.

[6] Chowdhury S, Erickson SW, MacLeod SL, Cleves MA, Hu P, Karim MA, et al. Maternal genome-wide DNA methylation patterns and congenital heart defects. PLoS One 2011;6:e16506.

[7] White WM, Brost BC, Sun Z, Rose C, Craici I, Wagner SJ, et al. Normal early pregnancy: a transient state of epigenetic change favoring hypomethylation. Epigenetics 2012;7:729–34.

[8] Nakano K, Whitaker JW, Boyle DL, Wang W, Firestein GS. DNA methylome signature in rheumatoid arthritis. Ann Rheum Dis 2013;72:110–17.

[9] Joubert BR, Håberg SE, Nilsen RM, Wang X, Vollset SE, Murphy SK, et al. 450K epigenome-wide scan identifies differential DNA methylation in newborns related to maternal smoking during pregnancy. Environ Health perspect 2012;120: 1425–31.

[10] Delgado-Calle J, Fernández AF, Sainz J, Zarrabeitia MT, Sañudo C, García-Renedo R, et al. Genome-wide profiling of bone reveals differentially methylated regions in osteoporosis and osteoarthritis. Arthritis Rheum 2013;65:197–205.

[11] Fernandez-Tajes J, Soto-Hermida A, Vázquez-Mosquera ME, Cortés-Pereira E, Mosquera A, Fernández-Moreno M, et al. Genome-wide DNA methylation analysis of articular chondrocytes reveals a cluster of osteoarthritic patients. Ann Rheum Dis 2014;73:668–77.

[12] Byrne EM, Carrillo-Roa T, Henders AK, Bowdler L, McRae AF, Heath AC, et al. Monozygotic twins affected with major depressive disorder have greater variance in methylation than their unaffected co-twin. Transl Psychiatry 2013;3:e269.

[13] Absher DM, Li X, Waite LL, Gibson A, Roberts K, Edberg J, et al. Genome-wide DNA methylation analysis of systemic lupus erythematosus reveals persistent hypomethylation of interferon genes and compositional changes to $CD4^+$ T-cell populations. PLoS Genet 2013;9:e1003678.

[14] Huynh JL, Garg P, Thin TH, Yoo S, Dutta R, Trapp BD, et al. Epigenome-wide differences in pathology-free regions of multiple sclerosis-affected brains. Nat Neurosci 2014;17:121–30.

[15] Graves M, Benton M, Lea R, Boyle M, Tajouri L, Macartney-Coxson D, et al. Methylation differences at the HLA-DRB1 locus in $CD4^+$ T-cells are associated with multiple sclerosis. Mult Scler 2013;20:1033–41.

[16] Yamada Y, Nishida T, Horibe H, Oguri M, Kato K, Sawabe M. Identification of hypo- and hypermethylated genes related to atherosclerosis by a genome-wide analysis of DNA methylation. Int J Mol Med 2014;33:1355–63.
[17] Dick KJ, Nelson CP, Tsaprouni L, Sandling JK, Aïssi D, Wahl S, et al. DNA methylation and body-mass index: a genome-wide analysis. Lancet 2014;383:1990–8.
[18] Weder N, Zhang H, Jensen K, Yang BZ, Simen A, Jackowski A, et al. Child abuse, depression, and methylation in genes involved with stress, neural plasticity, and brain circuitry. J Am Acad Child Adolesc Psychiatry 2014;53:417–24.
[19] Borghol N, Suderman M, McArdle W, Racine A, Hallett M, Pembrey M, et al. Associations with early-life socio-economic position in adult DNA methylation. Int J Epidemiol 2012;41:62–74.
[20] Zhang P, Zhao M, Liang G, Yin G, Huang D, Su F, et al. Whole-genome DNA methylation in skin lesions from patients with psoriasis vulgaris. J Autoimmun 2013;41:17–24.
[21] Labonte B, Suderman M, Maussion G, Lopez JP, Navarro-Sánchez L, Yerko V, et al. Genome-wide methylation changes in the brains of suicide completers. Am J Psychiatry 2013;170:511–20.
[22] del Rosario MC, Ossowski V, Knowler WC, Bogardus C, Baier LJ, Hanson RL. Potential epigenetic dysregulation of genes associated with MODY and type 2 diabetes in humans exposed to a diabetic intrauterine environment: an analysis of genome-wide DNA methylation. Metabolism 2014;63:654–60.
[23] Lee H, Jaffe AE, Feinberg JI, Tryggvadottir R, Brown S, Montano C, et al. DNA methylation shows genome-wide association of NFIX, RAPGEF2 and MSRB3 with gestational age at birth. Int J Epidemiol 2012;41:188–99.
[24] Sabunciyan S, Aryee MJ, Irizarry RA, Rongione M, Webster MJ, Kaufman WE, et al. Genome-wide DNA methylation scan in major depressive disorder. PLoS One 2012;7:e34451.
[25] Gordon L, Joo JE, Powell JE, Ollikainen M, Novakovic B, Li X, et al. Neonatal DNA methylation profile in human twins is specified by a complex interplay between intrauterine environmental and genetic factors, subject to tissue-specific influence. Genome Res 2012;22:1395–406.
[26] Sun YV, Smith AK, Conneely KN, Chang Q, Li W, Lazarus A, et al. Epigenomic association analysis identifies smoking-related DNA methylation sites in African Americans. Hum Genet 2013;132:1027–37.
[27] Breitling LP, Yang R, Korn B, Burwinkel B, Brenner H. Tobacco-smoking-related differential DNA methylation: 27K discovery and replication. Am J Hum Genet 2011;88:450–7.
[28] Breitling LP, Salzmann K, Rothenbacher D, Burwinkel B, Brenner H. Smoking, F2RL3 methylation, and prognosis in stable coronary heart disease. Eur Heart J 2012;33:2841–8.
[29] Haas J, Frese KS, Park YJ, Keller A, Vogel B, Lindroth AM, et al. Alterations in cardiac DNA methylation in human dilated cardiomyopathy. EMBO Mol Med 2013;5:413–29.
[30] Wang X, Falkner B, Zhu H, Shi H, Su S, Xu X, et al. A genome-wide methylation study on essential hypertension in young African American males. PLoS One 2013;8:e53938.
[31] Sharma P, et al. Genome wide DNA methylation profiling for epigenetic alteration in coronary artery disease patients. Gene 2014;541:31–40.
[32] Werner T. Bioinformatics applications for pathway analysis of microarray data. Curr Opin Biotechnol 2008;19:50–4.
[33] Han J, Park SG, Bae JB, Choi J, Lyu JM, Park SH, et al. The characteristics of genome-wide DNA methylation in naive $CD4^+$ T cells of patients with psoriasis or atopic dermatitis. Biochem Biophys Res Commun 2012;422:157–63.

[34] Park GT, Han J, Park SG, Kim S, Kim TY. DNA methylation analysis of CD4+ T cells in patients with psoriasis. Arch Dermatol Res 2014;306:259–68.
[35] Hou R, Yin G, An P, Wang C, Liu R, Yang Y, et al. DNA methylation of dermal MSCs in psoriasis: identification of epigenetically dysregulated genes. J Dermatol Sci 2013;72:103–9.
[36] Altorok N, Tsou PS, Coit P, Khanna D, Sawalha AH. Genome-wide DNA methylation analysis in dermal fibroblasts from patients with diffuse and limited systemic sclerosis reveals common and subset-specific DNA methylation aberrancies. Ann Rheum Dis 2014. Available from: http://dx.doi.org/10.1136/annrheumdis-2014-205303.
[37] Coit P, Jeffries M, Altorok N, Dozmorov MG, Koelsch KA, Wren JD, et al. Genome-wide DNA methylation study suggests epigenetic accessibility and transcriptional poising of interferon-regulated genes in naive CD4+ T cells from lupus patients. J Autoimmun 2013;43:78–84.
[38] Kim YJ, Park SW, Kim TH, Park JS, Cheong HS, Shin HD, et al. Genome-wide methylation profiling of the bronchial mucosa of asthmatics: relationship to atopy. BMC Med Genet 2013;14:39. Available from: http://dx.doi.org/10.1186/1471-2350-14-39.
[39] Cheong HS, Park SM, Kim MO, Park JS, Lee JY, Byun JY, et al. Genome-wide methylation profile of nasal polyps: relation to aspirin hypersensitivity in asthmatics. Allergy 2011;66:637–44.
[40] Cooke J, Zhang H, Greger L, Silva AL, Massey D, Dawson C, et al. Mucosal genome-wide methylation changes in inflammatory bowel disease. Inflamm Bowel Dis 2012;18:2128–37.
[41] Harris RA, Nagy-Szakal D, Pedersen N, Opekun A, Bronsky J, Munkholm P, et al. Genome-wide peripheral blood leukocyte DNA methylation microarrays identified a single association with inflammatory bowel diseases. Inflamm Bowel Dis 2012;18:2334–41.
[42] Karouzakis E, Trenkmann M, Gay RE, Michel BA, Gay S, Neidhart M. Epigenome analysis reveals TBX5 as a novel transcription factor involved in the activation of rheumatoid arthritis synovial fibroblasts. J Immunol 2014;193:4945–51.
[43] Raychaudhuri S, Plenge RM, Rossin EJ, Ng AC, International Schizophrenia Consortium, et al. Identifying relationships among genomic disease regions: predicting genes at pathogenic SNP associations and rare deletions. PLoS Genet 2009;5:e1000534.
[44] Jeffries MA, Donica M, Baker LW, Stevenson ME, Annan AC, Humphrey MB, et al. Genome-wide DNA methylation study identifies significant epigenomic changes in osteoarthritic cartilage. Arthritis Rheumatol 2014;66:2804–15.
[45] Emes RD, Farrell WE. Make way for the 'next generation': application and prospects for genome-wide, epigenome-specific technologies in endocrine research. J Mol Endocrinol 2012;49:R19–27.
[46] Xu X, Su S, Barnes VA, De Miguel C, Pollock J, Ownby D, et al. A genome-wide methylation study on obesity: differential variability and differential methylation. Epigenetics 2013;8:522–33.
[47] Wang X, Zhu H, Snieder H, Su S, Munn D, Harshfield G, et al. Obesity related methylation changes in DNA of peripheral blood leukocytes. BMC Med 2010;8:87.
[48] Smyth LJ, McKay GJ, Maxwell AP, McKnight AJ. DNA hypermethylation and DNA hypomethylation is present at different loci in chronic kidney disease. Epigenetics 2014;9:366–76.
[49] Figueroa-Romero C, Hur J, Bender DE, Delaney CE, Cataldo MD, Smith AL, et al. Identification of epigenetically altered genes in sporadic amyotrophic lateral sclerosis. PLoS One 2012;7:e52672.
[50] Masliah E, Dumaop W, Galasko D, Desplats P. Distinctive patterns of DNA methylation associated with Parkinson disease: identification of concordant epigenetic changes in brain and peripheral blood leukocytes. Epigenetics 2013;8:1030–8.

[51] Melas PA, Rogdaki M, Ösby U, Schalling M, Lavebratt C, Ekström TJ. Epigenetic aberrations in leukocytes of patients with schizophrenia: association of global DNA methylation with antipsychotic drug treatment and disease onset. FASEB J 2012;26:2712–18.

[52] Dempster EL, Pidsley R, Schalkwyk LC, Owens S, Georgiades A, Kane F, et al. Disease-associated epigenetic changes in monozygotic twins discordant for schizophrenia and bipolar disorder. Hum Mol Genet 2011;20:4786–96.

[53] Aberg KA, McClay JL, Nerella S, Clark S, Kumar G, Chen W, et al. Methylome-wide association study of schizophrenia: identifying blood biomarker signatures of environmental insults. JAMA Psychiatry 2014;71:255–64.

[54] Kumar G, Clark SL, McClay JL, Shabalin AA, Adkins DE, Xie L, et al. Refinement of schizophrenia GWAS loci using methylome-wide association data. Hum Genet 2015;134:77–87.

[55] Miller-Delaney SF, Das S, Sano T, Jimenez-Mateos EM, Bryan K, et al. Differential DNA methylation patterns define status epilepticus and epileptic tolerance. J Neurosci 2012;32:1577–88.

[56] Naeem H, Wong NC, Chatterton Z, Hong MK, Pedersen JS, Corcoran NM, et al. Reducing the risk of false discovery enabling identification of biologically significant genome-wide methylation status using the HumanMethylation450 array. BMC Genomics 2014;15:51.

CHAPTER

31

Outlook

OUTLINE

CONCLUDING THOUGHTS

The importance of DNA methylation as an epigenetic regulator in human health and disease is now well established. Until recently, protocols for reliable characterization of DNA methylation have lagged behind developments in genomic and transcriptomic fields. Considerable progress has been made in the past few years, and methylome characterization is likely to become a routine and biologically valid application of many laboratories worldwide. The search for methylation biomarkers continues, especially for cancer [1–4] and environmental risks [5,6] (Chapters 2–4), while many biomarkers suggested recently have yet to be validated [2,4].

Chapter 2 underlines the need for markers that can indicate good or poor outcomes. Prospective studies are necessary to evaluate their usefulness. Molecular biomarkers should help to improve personalized medicine, and in the near future could be used to characterize patients for specific therapeutic options.

In Chapter 3, we mention that although there is a vast potential for DNA methylation based biomarkers in non-neoplastic diseases, their use in clinical settings is still very limited. Again, prognostic and prophylactic screening is of great importance [7]. Diagnoses of complex diseases such as cancer, diabetes mellitus, or rheumatoid arthritis entail invasive biopsies or tissues from late or established disease; however, it is essential to cover the period before an individual becomes

M. Neidhart: DNA Methylation and Complex Human Disease.
DOI: http://dx.doi.org/10.1016/B978-0-12-420194-1.00031-2

symptomatic – i.e., the early phases of the diseases [8]. Biomarkers from peripheral blood are useful in a limited number of diseases, particularly autoimmune diseases such as type 1 diabetes and systemic lupus erythematosus. Unfortunately, this is rarely clinically relevant for infiltrative diseases such as rheumatoid arthritis, where a biopsy is essential. In the case of psychiatric diseases, for example, attempts have been made to do this using a surrogate system such as blood leukocytes or salivary gland epithelial cells. It is important to remember that blood plasma could be an attractive surrogate, too, because all cells shed DNA into the bloodstream. In the early disease phase, global DNA hypomethylation could be an interesting biomarker [9,10]. Additional unique methylation signatures could be discovered. The associations with diseased tissues, however, have to be validated. It is even possible that analysis of blood plasma DNA could allow localization of aberrant methylation patterns to a specific tissue or from a routine blood collection before an individual becomes symptomatic.

In Chapter 4, we conclude that the numbers of environmental factors capable of inducing epigenetic modifications are probably underestimated, and far greater efforts must be made in this field. Chapter 5, on epidemiology, presents two situations, namely endocrine disrupting chemicals and breast cancer, as well as smoking and various diseases, in which epigenetics provides a link between environment, lifestyle, diet, and disease development. In Chapter 6, we present evidence that viral oncoproteins target the elements of cellular epigenetic machinery, changing their expression and/or activity and thus leading to alterations in the epigenetic state of the host cell. This may be an underestimated pathway in many common human diseases, such as rheumatoid arthritis [11].

Chapters 7 and 8 present the main findings regarding DNA methylation in cancer, especially breast cancer; this analysis has generated new hypotheses regarding the molecular defect underpinning the epigenetic reprogramming, but many questions remain to be answered.

Chapter 9 addresses the question of acquired resistance to drugs, which is a major problem in the treatment of many diseases, including cancer. In cases where epigenetics is involved, it will be important to prevent or abolish drug resistance by reverting the methylome changes of non-responsive cells.

An often neglected regulator of the epigenome is the neuroendocrine system. Chapter 10 presents evidence from endocrinology that DNA methylation can be a dynamic process. This is also important in the context of actual discussion of the role of cytosine hydroxymethylation [12]. Chapter 11 shows that substantial evidence exists of links between epigenetic events and regulation of metabolic pathways leading to obesity and type 2 diabetes. Since several epigenetic marks are modifiable, not only by changing the exposure *in utero* but also by lifestyle changes in adult life, this implies that there is the potential for interventions to be introduced in postnatal life to modify or rescue unfavorable epigenomic profiles. Chapter 12 concludes that the links between epigenetics and endocrinology are rather incomplete; this is particularly obvious regarding the hypothalamus and pituitary gland, in spite of this system controlling a large part of the body's functions.

Chapter 13 illustrates the changes in DNA methylation during embryogenesis; the dynamic system of novel DNA demethylation/hydroxymethylation pathways provides important insights into the molecular processes that prepare the mammalian embryo for normal development. Chapter 14 discusses the great efforts that have been made to understand imprinted disorders that lead to growth retardation. The challenge will be to find out how *trans*-acting factors are perturbed in imprinting-related diseases in comparison with other causal mechanisms.

Chapter 15 presents the growing body of evidence showing that changes in the population's exposure to diet- and lifestyle-related factors induce epigenetic changes that in turn contribute to vascular complications. Future investigations are required to determine molecular mechanisms for environment–gene interactions, with the ultimate goal of explaining how environment-related risk factors for cardiovascular diseases act on the cell's transcriptional program.

The problem of cell heterogeneity in DNA methylation studies is addressed in Chapter 16, regarding neurology; thus, one particular difficulty arises from the complexity of neuronal subtypes, even within the same brain region. Surprisingly, there are relatively few studies that focus on the role of DNA methylation and drug addiction. Again, fast DNA demethylation/hydroxymethylation pathways could be involved. Chapter 17 illustrates the recent findings in severe psychosis and neurodegenerative diseases; however, the causal relationship between DNA methylation modifications and psychiatric diseases is uncertain, and the triggers are as yet unknown. It will be crucial to know more about the environmental factors that favor the pathogenesis, in particular during the perinatal period. A new trend will be to try epigenetic therapies in various diseases [13]; one short-term goal will be to reduce the invasiveness of cancer cells or synovial fibroblasts, for example, as well as to refine disease-forecasting by combining knowledge of all genetic and epigenetic marks. Chapter 18 describes studies in animal models of neurodegenerative diseases, highlighting the potential role of epigenetic drugs, including methyl donors, in ameliorating the cognitive symptoms and preventing or delaying the motor symptoms of the disease.

Chapter 19 states that the term "autoimmunity" is too broadly used; in an autoimmune disease, clear immune dysregulations have to be documented that involved antigen presentation, and T and B lymphocytes. Aberrant DNA methylation could change the transcription activity of genes involved in the development of autoimmune diseases, which contribute to skewed differentiation of T cells and proliferation and activation of other cells, such as endothelial cells and fibroblasts. Chapter 20 presents how epigenetics regulates important inherited events in the T-cell compartment. These range from some of the earliest events in developing thymocytes to possibly some of the latest events in T cells. Chapter 21 addresses the role of epigenetics in stem cell regulation. It has become clear that epigenetic modifications are not permanent; this raises the prospect that re-writing of the epigenetic marks of aging could be used to alter the potential of somatic stem cells from the elderly.

Chapters 22 and 23 present examples of changes in DNA methylation occurring in synovial fibroblasts during the disease process. There is increasing evidence implicating the epigenome with the development of inflammatory and age-related rheumatic diseases [14]. Initial attempts have concentrated on candidate genes in specific cell types that are known to be implicated in diseases, such as synovial fibroblasts in rheumatoid arthritis or dermal fibroblasts in systemic sclerosis, $CD4^{+}$ T cells in systemic lupus erythematosus, and chondrocytes in osteoarthritis. It is important to expand these investigations to other rheumatic diseases, such as dermatomyositis/polymyositis and hypermobility syndromes. Great efforts have been made to understand the intrinsically activated phenotype of rheumatoid arthritis synovial fibroblasts that leads to the aggressive behavior and joint destruction. A pioneering work showed that the effectiveness of methyl donors to restore DNA methylation is limited by at least two factors: an upregulation of the polyamine recycling pathway in some patients, and a counter-regulatory mechanism involving microRNAs that target DNA methyltransferases. Thus, the future will be trying combinations of methyl donors with specific inhibitors of the polyamine recycling pathway and/or antagomirs against specific microRNAs. Chapter 24 focuses on osteoporosis and the role of DNA methylation in osteoblast differentiation. An important question for the future regards individual variability in the response to vitamin D.

Chapters 25–27 address in detail the so-called "epigenetic therapies." New demethylating drugs are in development, many natural compounds can be tested, and a combination of demethylating drugs and methyl donors appears promising in the treatment of epimutations in cancer. Often, the differences between different methyl donors (e.g., S-adenosine methionine, L-methionine, and betaine) are neglected. Initial successes demonstrating the potential effectiveness of epigenome-targeted therapies as monotherapy in hematologic malignancies, mostly by using a demethylating agent, should be followed by studies focusing on combination therapies with methyl donors.

Finally, as presented in Chapters 28–30, high-throughput methylome scans are now possible. As changes will probably frequently be found outside CpG islands, the large interindividual variation in DNA methylation in these regions needs to be taken into account [15] and requires adequately powered epigenetic association studies as well as high-resolution DNA methylation technologies [16]. We can predict that studies gathering genomic, methylomic, transcriptomic (including microRNA and non-coding long RNAs [17]), and proteonomic data on the same samples will emerge in the next few years [18]. It is likely that the colossal effort required will be best suited to collaborative efforts, similar to those that have paved the way to success in human genome sequencing [19]. Large consortia are needed [20]. Epigenetic epidemiology may represent a logical evolution of genome/methylome-wide association studies (Chapter 5). Until these infrastructures materialize, methylomic data can form an additional layer to existing genotypic and transcriptomic data for well-defined phenotypes, in cancer as well as in non-neoplastic diseases. A very interesting approach will be to integrate polymorphisms and differentially methylated regions in the genome and to correlate these changes with particular diseases [21]. Many observations have suggested that some polymorphisms alter the epigenetic control of the transcription of specific genes. This can occur, for example, through the mutation of cytosine in the case that they are methylated in normal cells. The absence of methylated cytosines could induce a permissive histone modification and allow the gene to be expressed. This type of mechanism can change the phenotype and behavior of the cell [22]. Thus, by consolidating multiple layers of genome-wide profiles, we stand a good chance of accelerating progress towards improved personalized diagnostics, targeted protection, and, ultimately, prevention of common complex diseases.

References

[1] Ashbury JE, Taylor SA, Tse MY, Pang SC, Louw JA, Vanner SJ, et al. Biomarkers measured in buccal and blood leukocyte DNA as proxies for colon tissue global methylation. Int J Mol Epidemiol Genet 2014;5:120–4.

[2] Cesaroni M, Powell J, Sapienza C. Validation of methylation biomarkers that distinguish normal colon mucosa of cancer patients from normal colon mucosa of patients without cancer. Cancer Prev Res 2014;7:717–26.

[3] Mikeska T, Craig JM. DNA methylation biomarkers: cancer and beyond. Genes 2014;5:821–64.

[4] Chatterton Z, Burke D, Emslie KR, Craig JM, Ng J, Ashley DM, et al. Validation of DNA methylation biomarkers for diagnosis of acute lymphoblastic leukemia. Clin Chem 2014;60:995–1003.

[5] Laine JE, Bailey KA, Rubio-Andrade M, Olshan AF, Smeester L, Drobná Z, et al. Maternal arsenic exposure, arsenic methylation efficiency, and birth outcomes in the Biomarkers of Exposure to ARsenic (BEAR) pregnancy cohort in Mexico. Environ Health Perspect 2015;123:186–92.

[6] Philibert RA, Beach SR, Brody GH. The DNA methylation signature of smoking: an archetype for the identification of biomarkers for behavioral illness. Nebr Symp Motiv 2014;61:109–27.

[7] Sharma P, Stecklein SR, Kimler BF, Sethi G, Petroff BK, Phillips TA, et al. The prognostic value of promoter methylation in early stage triple negative breast cancer. J Cancer Ther Res 2014;3:1–11.
[8] Kuo IY, Chang JM, Jiang SS, Chen CH, Chang IS, Sheu BS, et al. Prognostic CpG methylation biomarkers identified by methylation array in esophageal squamous cell carcinoma patients. Int J Med Sci 2014;11:779–87.
[9] Henrique R, Jeronimo C. DNA hypomethylation in plasma as a cancer biomarker: when less is more? Expert Rev Mol Diagn 2014;14:419–22.
[10] Maghbooli Z, Hossein-Nezhad A, Larijani B, Amini M, Keshtkar A. Global DNA methylation as a possible biomarker for diabetic retinopathy. Diabetes Metab Res Rev 2015;31:183–9.
[11] Neidhart M, Karouzakis E, Schumann GG, Gay RE, Gay S. Trex-1 deficiency in rheumatoid arthritis synovial fibroblasts. Arthritis Rheum 2010;62:2673–9.
[12] Bakhtari A, Ross PJ. DPPA3 prevents cytosine hydroxymethylation of the maternal pronucleus and is required for normal development in bovine embryos. Epigenetics 2014;9:1271–9.
[13] Bae S, Ulrich CM, Bailey LB, Malysheva O, Brown EC, Maneval DR, et al. Impact of folic acid fortification on global DNA methylation and one-carbon biomarkers in the Women's Health Initiative Observational Study cohort. Epigenetics 2014;9:396–403.
[14] Gay S, Wilson AG. The emerging role of epigenetics in rheumatic diseases. Rheumatology 2014;53:406–14.
[15] Bock C, Walter J, Paulsen M, Lengauer T. Inter-individual variation of DNA methylation and its implications for large-scale epigenome mapping. Nucleic Acids Res 2008;36:e55.
[16] Rakyan VK, Down TA, Balding DJ, Beck S. Epigenome-wide association studies for common human diseases. Nat Rev Genet 2011;12:529–41.
[17] Anwar SL, Lehmann U. DNA methylation, microRNAs, and their crosstalk as potential biomarkers in hepatocellular carcinoma. World J Gastroenterol 2014;20:7894–913.
[18] White-Al Habeeb NM, Ho LT, Olkhov-Mitsel E, Kron K, Pethe V, Lehman M, et al. Integrated analysis of epigenomic and genomic changes by DNA methylation dependent mechanisms provides potential novel biomarkers for prostate cancer. Oncotarget 2014;5:7858–69.
[19] Butcher LM, Beck S. Future impact of integrated high-throughput methylome analyses on human health and disease. J Genet Genomics 2008;35:391–401.
[20] Bae JB. Perspectives of international human epigenome consortium. Genomics Inform 2013;11:7–14.
[21] Zabek T, Semik E, Wnuk M, Fornal A, Gurgul A, Bugno-Poniewierska M. Epigenetic structure and the role of polymorphism in the shaping of DNA methylation patterns of equine OAS1 locus. J Appl Genet 2015;56:231–8.
[22] Neidhart M, Karouzakis E. Genetics: a new interpretation of genetic studies in RA. Nat Rev Rheumatol 2014;10:199–200.

Glossary

5′-aza-2′-deoxycytidine (decitabine) a drug for the treatment of myelodysplastic syndromes, a class of conditions where certain blood cells are dysfunctional, and for acute myeloid leukemia.

5′-azacytidine thought to induce antineoplastic activity via two mechanisms' inhibition of DNA methyltransferase at low doses, causing hypomethylation of DNA, and direct cytotoxicity in abnormal hematopoietic cells in the bone marrow through its incorporation into DNA and RNA at high doses, resulting in cell death. 5′-azacytidine at low doses can be used to study the effect of DNA hypomethylation on a given cell type.

5-hydroxymethylcytosine formed from the DNA base cytosine by adding a methyl group and then a hydroxyl group. It is important in epigenetics, because the hydroxymethyl group on the cytosine can possibly switch a gene on and off. 5-methylcytosine dioxygenases (i.e., ten–eleven translocation, TET) proteins can catalyze 5-methylcytosine oxidation and generate derivatives, including 5-hydroxymethylcytosine. This may represent a mechanism of active demethylation.

5-methylcytosine dioxygenases (TETs) TETs catalyze the conversion of the modified DNA base 5-methylcytosine to 5-hydroxymethylcytosine.

5-methylcytosine a methylated form of the DNA base cytosine that is involved in the regulation of gene transcription. When cytosine is methylated, the DNA maintains the same sequence but the expression of methylated genes can be altered. It is an epigenetic modification formed by the action of DNA methyltransferases.

Acetylation process of introducing an acetyl group into a molecule, namely the substitution of an acetyl group for an active hydrogen atom. Histone acetylation and deacetylation, for example, are the processes by which the lysine residues within the N-terminal tail protruding from the histone core of the nucleosome are acetylated and deacetylated.

Acquired drug resistance rapid acquisition of resistance likely represents a physiologic mechanism of adaptation operative at the multicellular level rather than a stable genetic change and may be one of the reasons for the rapid development of drug resistance acquired by tumors *in vivo*.

Agouti mice The viable yellow agouti (A^{vy}) mouse model, in which coat color variation is correlated to epigenetic marks established early in development, has been used to investigate the impacts of nutritional and environmental influences on the fetal epigenome.

Allele number of alternative forms of the same gene or same genetic locus. Sometimes, different alleles can result in different observable phenotypic traits, such as different pigmentation.

Apoptosis resistance programmed cell death or apoptosis is the cell's intrinsic death program that is involved in the regulation of many physiological and pathological processes and is evolutionarily highly conserved. Evasion of apoptosis may contribute to carcinogenesis, tumor progression, and also treatment resistance.

Betaine (N,N,N-trimethylglycine) named after its discovery in sugar beet (*Beta vulgaris*). As a methyl donor, betaine participates in the methionine cycle, primarily in the human liver and kidneys.

Betaine-homocysteine S-methyltransferase (BHMT) a zinc metallo-enzyme that catalyzes the transfer of a methyl group from betaine to homocysteine to produce dimethylglycine and methionine.

Biomarker a measurable indicator of some biological state or condition. Epigenetic biomarkers either provide accurate measurements of alterations in a given patient sample or predict the severity, prognosis, etc., of a given disease.

Bisulfite modification the use of bisulfite treatment of DNA to determine its pattern of methylation. Treatment of DNA with bisulfite converts cytosine residues to uracil, but leaves 5-methylcytosine residues unaffected.

CCCTC-binding factor (CTCF) regulates the 3D structure of chromatin; binding of CTCF to DNA is inhibited by methylation on CpG dinucleotides.

Chromatin a complex of macromolecules found in cells, consisting of DNA, RNA, and protein. The lightly and tightly packed forms are called euchromatin and heterochromatin. Methyl-CpG-binding domain proteins recruit additional proteins to methylated loci, such as histone deacetylases that can modify histones, thereby forming compact, inactive heterochromatin.

CpG dinucleotides (CpG sites) regions of DNA where a cytosine occurs next to a guanine in the linear sequence of bases along its length. CpG is shorthand for "–C–phosphate–G–".

CpG island methylator phenotype (CIMP) widespread CpG island promoter methylation in various cancer cells, associated with loss-of-function mutations or epimutations (DNA hypermethylation) of the ten–eleven translocation (TET)-methylcytosine dioxygenase-2 (TET2) gene.

CpG islands regions with a high frequency of CpG sites. CpG islands typically occur at or near the transcription start site of genes, particularly housekeeping genes, in vertebrates. They are normally not methylated. However, methylation of CpG sites within the promoters of genes can lead to their silencing, a feature found in a number of human cancers.

CpG shores and shelves compared to CpG islands, methylation is more dynamic along the CpG shores (< 2 kb flanking CpG Islands) and CpG shelves (< 2 kb flanking outwards from a CpG shore).

Developmental Origin of Health and Disease Theory (DOHaD) evolved from epidemiological studies of infant and adult mortality. Extensive covalent modifications to DNA and related proteins occur from the earliest stages of mammalian development. These determine lineage-specific patterns of gene expression and so represent the most plausible mechanisms by which environmental factors can influence development during the life course.

Differentially methylated regions (DMRs) stretches of DNA that have different DNA methylation patterns compared to other samples. These samples can be different cells or tissues within the same individual, the same cell at different times, or cells from different individuals.

DNA demethylation can be achieved either passively, by the failure of maintenance methylation after DNA replication, or actively, by replication-independent processes.

DNA glycosylases family of enzymes involved in base excision repair. Thymine DNA glycosylase (TDG) is essential for active DNA demethylation and embryonic development.

DNA hydroxymethylation the conversion of 5-methylcytosine to 5-hydroxymethylcytosine is catalyzed by the ten–eleven translocation (TET) family of proteins or alternatively by activation-induced deaminase and thymine DNA glycosylase (TDG).

DNA methylation a biochemical process where a methyl group is added to the cytosine DNA nucleotides. In mammalian cells, DNA methylation occurs mainly at the C5 position of CpG dinucleotides and is carried out by two general classes of enzymatic activities – maintenance methylation and *de novo* methylation.

DNA methyltransferases (DNMTs) family of enzymes that catalyze the transfer of a methyl group to DNA. *De novo* methyltransferases (DNMT3A/3B) recognize something in the DNA that allows them to newly methylate cytosines. These are expressed mainly in early embryo development and they set up the pattern of methylation. The maintenance methyltransferase (DNMT1) adds methyl groups to cytosine when one DNA strand is already methylated. It works throughout the life of the organism to maintain the methylation pattern.

DNA mismatch repair gene belongs to a system for recognizing and repairing erroneous insertion, deletion, and mis-incorporation of bases that can arise during DNA replication and recombination, as well as repairing some forms of DNA damage.

Driver genes genes and pathways involved in disease development. In contrast, "passenger" genes are also modified, but tend to not contribute to the pathogenesis.

Drug efflux failure of cancer chemotherapy can occur through increased efflux of chemotherapeutic agents, leading to the reduction of intracellular drug levels and consequent drug insensitivity, often to multiple agents. A well-established cause of multidrug resistance involves the increased expression of members of the ATP binding cassette (ABC) transporter superfamily, many of which efflux various chemotherapeutic compounds from cells.

Endocrine disrupting chemicals (EDCs) chemicals that, at certain doses, can interfere with the endocrine system in mammals. These disruptions can cause cancerous tumors, birth defects, and other developmental disorders.

Endogenous retroviruses (ERVs) endogenous viral elements in the genome that closely resemble and can be derived from retroviruses. They are abundant in the genomes of vertebrates and occupy as much as 4.9% of the human genome.

Epiallele any of a group of otherwise identical genes that differ in the extent of methylation.

Epidemiology science that studies the patterns, causes, and effects of health and disease conditions in defined populations.

Epigenetics study of cellular and physiological traits that are *not* caused by changes in the DNA sequence; it describes the study of stable, long-term alterations in the transcriptional potential of a cell.

Epigenotypes inherited potential to express a particular type of differentiation. The predisposition of each cell to differentiate in a particular direction is a heritable characteristic is superimposed on the basic genetic program of that cell.

Epimutations heritable change in gene expression that does not affect the actual base-pair sequence of DNA. Primary epimutations often occur after fertilization and lead to somatic mosaicism. It has been estimated that the rate of primary epimutations is one or two orders of magnitude greater than for somatic DNA mutations.

Episome segment of DNA that can exist and replicate either autonomously in the cytoplasm or as part of a chromosome, mainly found in bacteria (plasmids). These viral DNA sequences can be methylated in the host.

Epithelial–mesenchymal transition (EMT) process by which epithelial cells lose their cell polarity and cell–cell adhesion, and gain migratory and invasive properties to become mesenchymal stem cells.

Euchromatic marks each chromatin state can be identified by defined marks, i.e., modifications affecting histones and DNA methylation. For example, in the euchromatin, active genes are hyperacetylated for histone 3 and histone 4, and hypermethylated at Lys 4 and Lys 79 of histone 3.

Exonic methylation methylation surrounding the transcription starting site (TSS) is tightly linked to transcriptional silencing, methylation of more downstream regions is unassociated with the magnitude of gene expression. DNA methylation downstream of the TSS, in the region of the first exon, is much more tightly linked to transcriptional silencing than is methylation in the upstream promoter region.

Folate (folic acid, vitamin B9) essential to synthesize, repair, and methylate DNA, as well as to act as a co-factor in numerous biological reactions. It is needed to carry one-carbon groups for methylation reactions and nucleic acid synthesis.

Haploinsufficiency occurs when an organism has only a single functional copy of a gene (with the other copy inactivated by mutation or epimutation) and the single functional copy does not produce enough of a gene product to bring about a wild-type condition, leading to an abnormal or diseased state.

Hemimethylated DNA only one of two complementary DNA strands is methylated. Any hemimethylated sites must have arisen during the last replication round, either because of a failure to faithfully propagate a parental methylation signal (by DNMT1) or because of a *de novo* methylation event (by DNMT3A/3B).

Histone acetyltransferases (HATs) enzymes that acetylate lysine on histone tails by transferring an acetyl group from acetyl CoA to form ε-N-acetyl lysine. Histone acetylation renders DNA more accessible to transcription factors.

Histone deacetylases (HDACs) enzymes that remove acetyl groups ($O=C-CH_3$) from an ε-N-acetyl lysine on histone tails, allowing the histones to wrap the DNA more tightly. Histone deacetylation is associated with gene silencing.

Histones highly alkaline proteins found in eukaryotic cell nuclei that package and order the DNA into structural units (nucleosomes). Histones undergo posttranslational modifications that alter their interaction with DNA and nuclear proteins. The transcription of genetic information is in part regulated by chemical modifications to histone proteins, e.g., acetylation and methylation of lysine located in the histone tails.

Homocysteine (Hcy) Hcy is biosynthesized from methionine via a multi-step process. First, methionine receives an adenosine group from ATP, a reaction catalyzed by S-adenosyl-methionine synthetase, to give S-adenosylmethionine (SAM). SAM then transfers the methyl group to an acceptor molecule (i.e., DNA methyltransferases). The adenosine is then hydrolyzed to yield L-homocysteine.

Hyperhomocysteinemia A high level of homocysteine in the blood makes a person more prone to endothelial cell injury, which leads to inflammation in the blood vessels, which in turn may lead to atherogenesis, which can result in ischemic injury. Hyperhomocysteinemia is therefore a possible risk factor for coronary artery disease.

Imprinted genes genes that are expressed in a parent-of-origin-specific manner. If the allele inherited from the father is imprinted and thereby silenced, then only (or primarily) the allele from the mother is expressed (e.g., in the case of H19). If the allele from the mother is imprinted, then only the allele from the father is expressed (e.g., in the case of IGF2).

Imprinting control region (ICR) chromosomal region that acts, in a methylation-sensitive way, to determine whether imprinted genes are expressed or not according to the parent from which the gene derived. The region is a regulated transcriptional insulator that binds CTCF.

Induced pluripotent stem cells (iPSCs) type of pluripotent stem cell that can be generated directly from adult cells. iPSCs are typically derived by introducing a specific set of pluripotency-associated genes into a given cell type. The original set of reprogramming factors are the genes OCT4 (POU5F1), SOX2, cMYC, and KIF4.

Intergenic germline differentially methylated region (IGDMR) a germ-line specific methylation mark in between transcripted genes.

LINE-1 Long INterspersed Elements belong to a group of genetic elements that are found in large numbers in eukaryotic genomes, composing 17% of the human genome (99.9% of which is no longer capable of transposition). LINE-1 gene hypomethylation is linked to the onset and poor prognosis of various diseases.

Locus specific location of a gene, DNA sequence, or position on a chromosome.

Methionine adenosyltransferases (MAT1A/2A) enzymes that catalyze the formation of S-adenosylmethionine by joining methionine and ATP.

Methionine cycle methionine is converted to S-adenosylmethionine, which is a methyl donor for numerous reactions. In losing its methyl group, SAM becomes S-adenosylhomocysteine, which is then converted to homocysteine. Homocysteine is either converted back to methionine, or enters the transsulfuration pathway to form other sulfur-containing amino acids.

Methionine synthase reductase (MTRR) catalyzes the synthesis of methionine from homocysteine; this enzyme requires vitamin B12 as a co-factor. Several polymorphisms in the MTRR gene have been identified.

Methionine one of the essential amino acids (building blocks of protein), meaning that it cannot be produced by the body and must be provided by the diet. It supplies sulfur and other compounds required by the body for normal metabolism and growth. The methionine-derivative S-adenosyl methionine is a co-factor that serves mainly as a methyl donor.

Methyl CpG binding proteins (MBPs) MBPs interpret the methylation of DNA and its components. The MBP family is divided into three branches: (1) methyl-binding domain containing proteins (MeCP2, MBDs, SETDBs), (2) methyl-CpG binding zinc fingers (Kaiso, ZBTs), and (3) the SRA domain containing proteins (UHRFs).

Methylated CpG island recovery assay (MIRA) MIRA exploits the intrinsic specificity and the high affinity of a methylated-CpG-binding protein complex to methylated CpG dinucleotides in genomic DNA. The MIRA approach works on double-stranded DNA and does not depend on the application of methylation-sensitive restriction enzymes. It can be performed on a few hundred nanograms of genomic DNA.

Methylated DNA immunoprecipitation (MeDIP) a large-scale genome-wide purification technique that is used to enrich for methylated DNA sequences. It consists of isolating methylated DNA fragments via an antibody raised against 5-methylcytosine.

Methylation sensitive polymerase chain reaction (MSP) a method used to analyze the methylation status of genes. It requires two pairs of primer sets, one each for amplification of unmethylated and methylated regions, respectively.

Methylation denotes the addition of a methyl group ($-CH_3$) to a substrate, or the substitution of an atom or group by a methyl group.

Methyl-binding domain (MBD) the part of methyl CpG binding proteins that interacts with DNA containing one or more symmetrically methylated CpG dinucleotides.

Methylenetetrahydrofolate reductase (MTHFR) MTHFR catalyzes the conversion of 5,10-methylenetetrahydrofolate to 5-methyltetrahydrofolate, a co-substrate for homocysteine remethylation to methionine. There are DNA polymorphisms associated with this gene.

Methylome the set of nucleic acid methylation (mostly cytosine) modifications in an organism's genome or in a particular cell.

Methyltetrahydrofolate (MTHF) the primary biologically active form of folic acid used at the cellular level for DNA replication, the cysteine cycle, and the regulation of homocysteine. It is used in the methylation of homocysteine to form methionine and tetrahydrofolate.

MicroRNA a small non-coding RNA molecule (containing about 21–22 nucleotides) that functions in RNA silencing and post-transcriptional regulation of gene expression. MicroRNAs can be regulation by DNA methylation and can interfere with the transcripts of, for example, DNA methyltransferases.

Microsatellite instability (MSI) the condition of genetic hypermutability that results from impaired DNA mismatch repair. Microsatellites are repeated sequences of DNA. Occasionally the mutation events leading to MSI are derived from the promoter hypermethylation of the DNA mismatch repair enzyme MLH1.

Monozygotic twins identical twins occur when a single egg is fertilized to form one zygote (hence "monozygotic"), which then divides into two separate embryos. They are genetically nearly identical and they are always the same sex unless there has been a mutation during development. A cause of difference between monozygotic twins is epigenetic modification, caused by differing environmental influences throughout their lives.

Nucleosomes basic units of DNA packaging, consisting of a segment of DNA wound in sequence around histone protein cores.

Nutriepigenomics the study of food nutrients and their effects on human health through epigenetic modifications. There is now considerable evidence that nutritional imbalances during gestation and lactation are linked to non-communicable diseases, such as obesity, cardiovascular disease, type 2 diabetes, hypertension, and cancer. If metabolic disturbances occur during critical time windows of development, the resulting epigenetic alterations can lead to permanent changes in tissue and organ structure or function and predispose individuals to disease.

Oncogene a gene that has the potential to cause cancer. In tumor cells, oncogenes are often mutated or expressed at high levels.

One-carbon metabolism comprises the cytosolic and mitochondrial folate cycles and the methionine cycle.

Pluripotency refers to a stem cell that has the potential to differentiate into any of the three germ layers: endoderm (interior stomach lining, gastrointestinal tract, the lungs), mesoderm (muscle, bone, blood, urogenital), or ectoderm (epidermal tissues and nervous system).

Polyamine recycling pathway The polyamine-acetylating enzyme, spermidine/spermine N1-acetyltransferase (SSAT1), participates in polyamine homeostasis by regulating polyamine export and catabolism. It is involved in the reconversion of spermidine and spermine into putrescine. The activation of this pathway results increased consumption of S-adenosylmethionine – the cell's methyl donor – that is used for the biosynthesis of polyamines. This enters into competition with the transmethylation pathway controlled by DNA methylatransferases and can favor DNA hypomethylation.

Polycomb repressive complex refers to a family of proteins first discovered in fruit flies that can remodel chromatin such that epigenetic silencing of genes takes place. Polycomb-group proteins are well known for silencing Hox genes through modulation of chromatin structure during embryonic development in fruit flies. Similarly, in mammals, polycomb group gene expression is important in many aspects of development. The polycomb repressive complex 2 has histone methyltransferase activity and primarily trimethylates histone 3 on lysine 27. It is required for long-term epigenetic silencing of chromatin and has an important role in stem cell differentiation.

Promoter region of DNA that initiates transcription of a particular gene. Promoters are located near the transcription start sites of genes, on the same strand and upstream on the DNA.

Proto-oncogenes genes that code for proteins responsible for proliferation. Mutations or epimutations in proto-oncogenes can lead to an increase in protein expression, hyperactivity (i.e., gain-of-function) and/or loss of regulation. This mutated form is called an oncogene.

Pyrosequencing method based on the sequencing by synthesis principle. It relies on the detection of pyrophosphate release on nucleotide incorporation. The desired DNA sequence is able to be determined by light emitted upon incorporation of the next complementary nucleotide by the fact that only one out of four of the possible A/T/C/G nucleotides are added and available at a time, so that only one letter can be incorporated on the single-stranded template. The intensity of the light determines if there are more than one of these letters in a row. The previous nucleotide letter is degraded before the next nucleotide letter is added for synthesis, allowing for the possible revealing of the next nucleotide through the resulting intensity of light. This process is repeated with each of the four letters until the DNA sequence of the single-stranded template is determined. Analysis of DNA methylation patterns by pyrosequencing combines this reaction protocol with measures of the degree of methylation at several CpGs in close proximity with high quantitative resolution. After bisulfite treatment and polymerase chain reaction, the degree of each methylation at each CpG position in a sequence is determined from the ratio of T and C.

Reduced representation bisulfite sequencing (RRBS) high-throughput technique used to analyze the genome-wide methylation profiles on a single nucleotide level. This technique combines restriction enzymes and bisulfite sequencing in order to enrich the areas of the genome that have a high CpG content.

Reprogramming refers to erasure and remodeling of epigenetic marks, such as DNA methylation, during mammalian development. After fertilization, some cells of the newly formed embryo migrate to the germinal ridge and will eventually become the germ cells. Due to the phenomenon of genomic imprinting, maternal and paternal genomes are differentially marked and must be properly reprogrammed every time they pass through the germline. Therefore, during the process of gametogenesis the primordial germ cells must have their original biparental DNA methylation patterns erased and re-established based on the sex of the transmitting parent. Reprogramming can also be induced artificially through the introduction of exogenous factors, usually transcription factors. In this context, it often refers to the creation of induced pluripotent stem cells from mature cells.

Restriction landmark genome scanning (RLGS) method that allows for rapid simultaneous visualization of thousands of landmarks, or restriction sites. Using a combination of restriction enzymes, some of which are specific to DNA modifications, the technique can be used to visualize differences in methylation levels across the genome.

Retroelements endogenous components of eukaryotic genomes that are able to amplify to new locations in the genome through a RNA intermediate.

Retrotransposons genetic elements that can amplify themselves in a genome and are ubiquitous components of the DNA of many eukaryotic organisms. They consist of two subtypes: the long terminal repeat (LTR) and the non-LTR retrotransposons. Endogenous retroviruses (ERVs) are the most important LTR retrotransposons in mammals, including humans, where ERVs make up 8% of the genome. Non-LTR retrotransposons consist of two subtypes: long interspersed elements (LINEs) and short interspersed elements (SINEs). They are widespread in eukaryotic genomes. Retrotransposons are often silenced by DNA methylation. Therefore, their expression is a sign of global DNA hypomethylation.

S-adenosyl-homocysteine (SAH) formed by the demethylation of S-adenosylmethionine in the transmethylation pathway.

S-adenosylmethionine (AdoMet, SAM) a common cosubstrate involved in methyl group transfers. Transmethylation, transsulfuration, and aminopropylation are the metabolic pathways that use it, i.e., by DNA methyltransferases. Another major role of SAM is in polyamine biosynthesis. SAM is decarboxylated by adenosylmethionine decarboxylase to form decarboxySAM. This compound then donates its n-propylamine group in the biosynthesis of polyamines from putrescine.

Somatic mosaicism occurs when the somatic cells are of more than one genotype. In the more common mosaics, different genotypes arise from a single fertilized egg cell, due to mitotic errors at first or later cleavages. It is common in embryogenesis due to retrotranspositin of LINE-1 and ALU transposable elements. In early development, DNA from undifferentiated cell types may be more susceptible to mobile element invasion due to long, unmethylated regions in the genome.

Spermidine/spermine N1-acetyltransferase (SSAT1) a rate-limiting enzyme in the catabolic pathway of polyamine metabolism. It catalyzes the acetylation of spermidine and spermine, and is involved in the regulation of the intracellular concentration of polyamines and their transport out of cells. It controls the polyamine recycling pathway.

Stem cells undifferentiated cells that can differentiate into specialized cells and can divide to produce more stem cells. In mammals, there are two broad types of stem cells: embryonic stem cells, which are isolated from the inner cell mass of blastocysts, and adult stem cells, which are found in various tissues. In adult organisms, stem cells and progenitor cells act as a repair system for the body, replenishing adult tissues. In a developing embryo, stem cells can differentiate into all the specialized cells – ectoderm, endoderm, and mesoderm – but also maintain the normal turnover of regenerative organs, such as blood, skin, or intestinal tissues. As the embryonic stem cells undergo differentiation, the level of DNA methylation increases.

Telomere region of repetitive nucleotide sequences at each end of a chromatid, which protects the end of the chromosome from deterioration or from fusion with neighboring chromosomes. As telomeres shorten with aging, the methylation levels of many gene promoters in subtelomeric regions may change, which in turn could cause changes in gene expression that increase the risk of age-related disease.

TET enzymes see 5-methylcytosine dioxygenases.

Tetrahydrofolate (THF) a co-factor in many reactions, especially in the metabolism of amino acids and nucleic acids. It acts as a donor of a group with one carbon atom. It gets this carbon atom by sequestering formaldehyde produced in other processes. It is produced from dihydrofolic acid by dihydrofolate reductase and is converted into 5,10-methylenetetrahydrofolate by serine hydroxymethyltransferase.

Tiling array a subtype of microarray chips. Like traditional microarrays, they function by hybridizing labeled DNA to probes fixed onto a solid surface. They differ from traditional microarrays in the nature of the probes. Instead of probing for sequences of known or predicted genes that may be dispersed throughout the genome, methyl-DNA immunoprecipitation followed by tiling array allows DNA methylation mapping and measurement across the genome.

Totipotency characteristic of early stem cells that have the potential to develop into any cell found in the body. In mammalian development, the egg cell in a female and the sperm cell from a male fuse together to form a single cell called the zygote. The zygote divides numerous times and forms cells that are precursors to the trillions of cells that will eventually constitute the human body.

Transcription factor (TF) a protein that binds to specific DNA sequences, thereby controlling the rate of transcription of genetic information from DNA to messenger RNA.

Transcription starting site (TSS) the location where transcription starts at the 5′-end of a gene sequence.

Transmethylation the chemical reaction in which a methyl group is transferred from one compound to another. An example of transmethylation is the recovery of methionine from homocysteine, or the DNA methylation by S-adenosylmethionine and DNA methyltransferases.

Trichostatin-A can be used to alter gene expression by interfering with the removal of acetyl groups from histones and therefore altering the ability of transcription factors to access the DNA molecules inside chromatin.

Tumor suppressor gene a gene that protects a cell from one step on the path to cancer. When this gene mutates to cause a loss or reduction in its function, or is silenced by DNA methylation, the cell can progress to cancer, usually in combination with other genetic and epigenetic changes.

Uniparental disomy occurs when a person receives two copies of a chromosome, or of part of a chromosome, from one parent and no copies from the other parent. Uniparental inheritance of imprinted genes can also result in phenotypical anomalies. Though few imprinted genes have been identified, uniparental inheritance of an imprinted gene can result in the loss of gene function which can lead to delayed development and mental retardation.

Vitamin B6 pyridoxal 5′-phosphate acts as a coenzyme in all transamination reactions. It is also an essential homocysteine re-methylation co-factor, and deficiency is associated with an increase in blood homocysteine levels.

Vitamin B9 see folic acid.

Vitamin B12 the structure of cobalamine is based on a corrin ring, which is similar to the porphyrin ring found in heme, chlorophyll, and cytochrome. The central metal ion is cobalt. Methionine synthase is a methyltransferase enzyme that uses the methylated cobalamine molecule to transfer a methyl group from 5-methyltetrahydrofolate to homocysteine, thereby generating tetrahydrofolate and methionine.

Whole-genome bisulfite sequencing (WGSBS) a sequencing-based methylation analysis application. The libraries are prepared for subsequent cluster generation starting from sample DNA through adaptor ligation, library purification, and quantification.

X-chromosome inactivation a process by which one of the two copies of the X chromosome present in female mammals is inactivated. The inactivated X chromosome has high levels of DNA methylation, low levels of histone acetylation, low levels of histone 3 lysine-4 methylation, and high levels of histone 3 lysine-9 methylation, all of which are associated with gene silencing.

Index

Note: Page numbers followed by "*f*" and "*t*" refer to figures and tables, respectively.

B

C

D

E

H

M

N

O

P

Q

R

U

V